To Brom, Chris and Sarah

Introduction
to Fungi *SECOND EDITION*

JOHN WEBSTER
Professor of Biological Sciences, University of Exeter

CAMBRIDGE UNIVERSITY PRESS

CAMBRIDGE

LONDON NEW YORK NEW ROCHELLE

MELBOURNE SYDNEY

Published by the Press Syndicate of the University of Cambridge
The Pitt Building, Trumpington Street, Cambridge CB2 1RP
32 East 57th Street, New York, N.Y. 10022, U.S.A.
296 Beaconsfield Parade, Middle Park, Melbourne 3206, Australia

First published 1970
Reprinted 1974, 1975
First paperback edition 1977
Second edition 1980
Reprinted 1983

Printed in Great Britain
at the Alden Press, Oxford

Library of Congress Cataloguing in Publication Data
Webster, John.
Introduction to fungi.
Bibliography: p.
Includes index.
1. Fungi. 2. Fungi—Classification. I. Title. QK603.W4 1980 589′.2 79-52856

ISBN 0 521 22888 3 hard covers
ISBN 0 521 29699 4 paperback

(First edition ISBN 0 521 07640 4 hard covers
ISBN 0 521 29232 8 paperback)

Contents

Preface to the first edition

There are several available good text-books of mycology, and some justification is needed for publishing another. I have long been convinced that the best way to teach mycology, and indeed all biology, is to make use, wherever possible, of living material. Fortunately with fungi, provided one chooses the right time of the year, a wealth of material is readily available. Also by use of cultures and by infecting material of plant pathogens in the glasshouse or by maintaining pathological plots in the garden, it is possible to produce material at almost any time. I have therefore tried to write an introduction to fungi which are easily available in the living state, and have tried to give some indication of where they can be obtained. In this way I hope to encourage students to go into the field and look for fungi themselves. The best way to begin is to go with an expert, or to attend a Fungus Foray such as those organised in the Spring and Autumn by mycological and biological societies. I owe much of my own mycological education to such friendly gatherings. A second aim has been to produce original illustrations of the kind that a student could make for himself from simple preparations of living material, and to illustrate things which he can verify for himself. For this reason I have chosen not to use electron micrographs, but to make drawings based on them.

The problem of what to include has been decided on the criterion of ready availability. Where an uncommon fungus has been included this is because it has been used to establish some important fact or principle. A criticism which I must accept is that no attempt has been made to deal with Fungi Imperfecti as a group. This is not because they are not common or important, but that to have included them would have made the book much longer. To mitigate this shortcoming I have described the conidial states of some Ascomycotina rather fully, to include reference to some of the form-genera which have been linked with them. A more difficult problem has been to know which system of classification to adopt. I have finally chosen the 'General Purpose Classification' proposed by Ainsworth, which is adequate for the purpose of providing a framework of reference. I recognise that some might wish to classify fungi differently, but see no great merit in burdening the student with the arguments in favour of this or that system.

Because the evidence for the evolutionary origins of fungi is so meagre I have made only scant reference to the speculations which have been made on this topic. There are so many observations which can be verified, and for this reason I have preferred to leave aside those which never will.

The literature on fungi is enormous, and expanding rapidly. Many undergraduates do not have much time to check original publications. However, since the book is intended as an introduction I have tried to give references to some of the more recent literature, and at the same time to quote the origins of some of the statements made.

Exeter, 27 April 1970   J.W.

Preface to the second edition

In revising the first edition, which was first published about ten years ago, I have taken the opportunity to give a more complete account of the Myxomycota, and to give a more general introduction to the Eumycota. An account has also been given of some conidial fungi, as exemplified by aquatic Fungi Imperfecti, nematophagous fungi and seed-borne fungi. The taxonomic framework has been based on Volumes IVA and IVB of Ainsworth, Sparrow & Sussman's *The Fungi: An Advanced Treatise* (Academic Press, 1973).

Exeter, January 1979   J.W.

Acknowledgements

I owe a debt of gratitude to many people for help in preparing this book. Glyn Woods and Mike Alexander helped with photography, and Tony Davey gave excellent technical assistance. I have acknowledged, in the text, figures which have been generously provided, or which I have traced. I am particularly grateful to Dr G. C. Ainsworth, as General Editor of *The Fungi: An Advanced Treatise*, and Academic Press for permission to make use of various keys to fungal taxa. The copyright of these keys is held by Academic Press. I am also grateful to Mrs H. Eva of the Exeter University Library who helped me to obtain publications on loan. Dr M. W. Dick and Dr M. J. Carlile kindly read the chapters on Oomycetes and on Myxomycetes, and made valuable constructive criticisms. Dr D. J. Alderman gave me useful advice on flagellar action. Above all, I am indebted to Joan Vaughan for her unfailing cheerfulness and skill in transforming a difficult manuscript into a readable typescript.

Introduction

It is difficult to give a precise definition of a fungus, largely because organisms which are regarded as fungi are very variable in form, behaviour and life cycle. Ainsworth (1973) has listed their main characteristics.

> *Nutrition:* heterotrophic (photosynthesis lacking) and absorptive (ingestion rare).
> *Thallus:* on or in the substratum and plasmodial amoeboid or pseudoplasmodial; or in the substratum and unicellular or filamentous (mycelial), the last, septate or nonseptate; typically nonmotile (with protoplasmic flow through the mycelium) but motile states (e.g. zoospores) may occur.
> *Cell wall:* well-defined, typically chitinised (cellulose in Oomycetes).
> *Nuclear status:* eukaryotic, multinucleate, the mycelium being homo- or heterokaryotic, haploid, dikaryotic, or diploid, the last being usually of limited duration.
> *Life cycle:* simple to complex.
> *Sexuality:* asexual or sexual and homo- or heterothallic.
> *Sporocarps:* microscopic or macroscopic and showing limited tissue differentiation.
> *Habitat:* ubiquitous as saprobes, symbionts, parasites, or hyperparasites.
> *Distribution:* cosmopolitan.

WHY STUDY FUNGI?

The absence of photosynthetic pigments enforces upon fungi a saprophytic or a parasitic existence. As saprophytes they share with bacteria and animals the rôle of decay of complex plant and animal remains in the soil, breaking them down into simpler forms which can be absorbed by further generations of plants. Without this essential process of decay, the growth of plants, upon which life is dependent, would eventually cease for lack of raw materials. Soil fertility is thus in part bound up with fungal activity. The roots of most green plants are infected with fungi and absorption of minerals may be enhanced following infection. Such infected root systems are termed mycorrhiza, and they are an example of a symbiotic relationship between green plants and fungi. In infertile natural soils the success of the higher plant may depend on infection (Harley, 1969). Harmful effects of saprophytic fungi on human economy are seen when food, timber and textiles are rotted. Fungi are also of importance in industrial fermentations as in brewing, production of anti-biotics, or citric acid fermentation. Food processing such as baking, cheese making, or wine fermentation is also dependent on fungi. Increasing use is made of fungi in carrying out chemical transformations in the pharmaceutical industry. These activities of fungi have been ably reviewed by Christensen (1965), Gray (1959) and Emerson (1973).

As parasites, fungi cause disease in plants and animals. Although fungal pests of crop plants have been known since human records began, it was the impact of potato blight on the population of Ireland in the mid-nineteenth century which gave the impetus to the scientific study of plant pathology (Large, 1958). As agents of disease in animals and man, fungi are commonly less severe than bacteria and viruses, although some are lethal. As the control of other diseases improves, the importance of fungal disease is being recognised (Ainsworth, 1952).

Apart from these applied aspects of the study of fungi, they have a claim to interest in their own right, and as tools for the physiologist, microbiologist, biochemist and geneticist, who often find them ideally suited for investigations of all kinds. Our general understanding of genetics owes much to investigations with *Neurospora,* and our understanding of respiration to studies on yeast. Investigations into the bakanae disease of rice caused by *Gibberella fujikuroi* led to the discovery of the group of plant growth hormones called gibberellins. These aspects of fungal biology will not be stressed in this book. They have been well described by Cochrane (1958), Fincham & Day (1971), Esser & Kuehnen (1967) and Burnett (1975).

CLASSIFICATION

Organisms do not classify themselves. They are classified by man for convenience of reference. Ideally a scheme of classification should reflect natural relationships, but in considering relationships mycologists may not attach the same weight to the criteria available. It should therefore not be surprising that different authorities do not use the same scheme of classification. I have chosen to adopt the scheme proposed by Ainsworth (1973). Fungi are divided into two Divisions, distinguished by the presence or absence of a plasmodium or pseudoplasmodium. A plasmodium is a mass of naked, multinucleate protoplasm, moving by amoeboid movement and usually feeding by ingesting particulate matter. Nuclear division in a plasmodium is usually simultaneous. A pseudoplasmodium is an aggregation of separate amoeboid cells. Fungi with plasmodia or pseudoplasmodia are classified in the Division Myxomycota, whilst the majority of fungi, which are usually filamentous, are classified in the Eumycota.

Key to divisions of fungi

Plasmodium or pseudoplasmodium present **Myxomycota (p. 5)**

Plasmodium or pseudoplasmodium absent, assimilative phase typically filamentous **Eumycota (p. 55)**

PART ONE

Myxomycota

Slime moulds and
similar organisms

Whether the Myxomycota are closely related to the Eumycota is doubtful. Possibly they are more closely related to protozoa. De Bary (1887) used the term Mycetozoa, indicating a relationship with animals. This view is shared by Olive (1970) who has reviewed the classification of the group. The classification below (Ainsworth, 1973), however, places them with the fungi. It is by no means certain that the organisms classified together in the Myxomycota are closely related to each other, and indeed some authorities would classify certain of the groups elsewhere.

Key to classes of Myxomycota

1 Assimilative phase free-living amoebae which unite as a pseudoplasmodium before reproduction   **Acrasiomycetes (p. 7)**
Assimilative phase a plasmodium   2

2 Plasmodium forming a network ('net plasmodium')
                                    Hydromyxomycetes (p. 19)
Plasmodium not forming a network   3

3 Plasmodium saprobic, free-living   **Myxomycetes (p. 22)**
Plasmodium parasitic within cells of the host plant
                              Plasmodiophoromycetes (p. 42)

1: ACRASIOMYCETES

The characteristic feature of this group is that they have a trophic (i.e. feeding) stage consisting of amoeboid cells, or myxamoebae, (or in a few cases minute plasmodia), which ingest bacteria or other prey. Raper (1973) has given a valuable outline. He has pointed out that even the organisms classified in the Acrasiomycetes may not represent a natural group. It is also uncertain whether the group is best considered as related to fungi or to protozoa. The main justification for considering them here is that they have been traditionally studied by mycologists.

The fructifications of Acrasiomycetes consist of **sporocarps** (delicate tubular stalks bearing one or a few spores) or **sorocarps** (multicellular fruiting structures in which a sorus of spores is borne at the tip of a unicellular or multicellular stalk).

The key below is from Raper (1973).

Key to subclasses of Acrasiomycetes

1 Sporulation not preceded by aggregation of myxamoebae; sporocarps one- to few-spored; trophic stage consisting of myxamoebae or minute plasmodia, both with filose pseudopodia and nuclei with single centrally positioned nucleoli; flagellate stage present in some genera, lacking in others   **Protostelidae (p. 8)**

Sporulation preceded by aggregation of myxamoebae to form pseudoplasmodia; sorocarps multispored; trophic stage consisting of uninucleate myxamoebae; pseudopodia filose in some genera and lobose in others; flagellate cells absent   2

2 Aggregating myxamoebae do not form streams in developing pseudoplasmodia; fructifications may or may not show definite sori and sorophores; myxamoebae with lobose pseudopodia and nuclei with single centrally positioned nucleoli   **Acrasidae (p. 10)**

Aggregating myxamoebae form convergent streams in developing pseudoplasmodia; sorocarps with well-defined sori and sorophores; myxamoebae with filose pseudopodia and nuclei with two or more peripheral nucleoli   **Dictyostelidae (p. 11)**

Protostelidae

PROSTELIALES

This group of organisms has only recently been discovered, probably because they are inconspicuous. They are, however, ubiquitous on decaying plant parts in soil, dung and also in fresh water. Olive (1967) has provided a monograph of the group. They can be cultivated on weakly nutrient agars such as lactose yeast extract agar (0·1% lactose, 0·05% yeast extract, 2% agar) or hay infusion agar (2·5 g hay/litre, 0·2% $K_2HPO_4.3H_2O$, 2% agar) in conjunction with other organisms ingested as food, such as the bacteria *Escherichia coli, Klebsiella aerogenes,* and the yeast *Rhodotorula mucilaginosa.*

Protostelium is a typical member of the group (Fig. 1). *Protostelium mycophaga* has been found on still-attached dead parts of plants. It grows and fruits well in culture, feeding at the expense of fungal cells such as yeasts, but not, apparently, bacteria. The fruiting body or sporocarp consists of a long, slender, tubular stalk about 75 μm long, bearing a single spherical spore about 4–10 μm in diameter. The spore is deciduous and readily detached. On germination, a single, uninucleate, amoeboid stage with thin pseudopodia

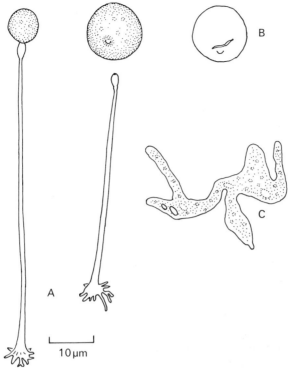

Fig. 1. *Protostelium* sp. A. Two sporocarps, one immature, the other with a detached spore. Note the apophysis beneath the spore. B. Empty spore case, after germination. C. Amoeboid phase.

emerges. The amoeboid stage feeds voraciously on yeast cells, and may also feed cannibalistically on amoebae of the same species. The general pattern of development in protostelids is illustrated in Fig. 2, and *P. mycophaga* conforms to this pattern. The generalised account of spore development below is based on Olive (1967). When feeding stops, the amoeba rounds off and heaps its protoplasm in the centre to form the so-called 'hat-shaped' stage (Fig. 2B). A membranous, pliable, impermeable sheath develops over the surface of the cell. The protoplast contracts into the central 'hump', leaving the sheath at the thin margin free to collapse on the substrate where it forms a disk-like base to the stalk of the sporocarp. Within the protoplast a granular basal core differentiates. This structure, which is destined to give rise to the stalk, is termed the **steliogen**. The steliogen begins to mould a narrow hollow tube (Fig. 2D, E). As the tube, which is added to at its upper end, grows in length, the protoplast migrates upwards, seated on the upper end of the tube. Meanwhile, the sheath becomes applied to the outside of the tube to form a second layer of the stalk (Fig. 2E). Finally, the steliogen is left at the tip of the stalk, while the protoplast withdraws from around it. It secretes a wall around itself and

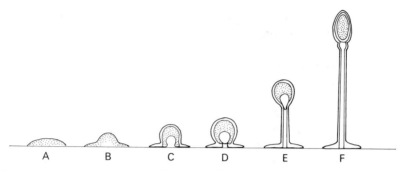

Fig. 2. Sporogenesis in a protostelid (after Olive, 1967). A. Early pre-spore stage. B. Hat-shaped stage. C. Appearance of steliogen. D. Beginning of stalk formation. E. Later stage in stalk development, with steliogen extending into upper part of stalk tube. F. Mature sporocarp, showing terminal spore, with subtending apophysis, outer sheath, and inner stalk tube.

becomes the spore. The remnant of the steliogen, which subtends the spore, is now devoid of cytoplasm, and persists as the **apophysis**. The whole process of sporulation may take less than half an hour in *P. mycophaga*. The fine structure of protostelid sporocarps has been studied by Furtado *et al.* (1971).

Although the development of protostelids follows this broad pattern, variations occur. In some species, the protoplasts are capable of anastomosis, and in *Schizoplasmodium cavostelioides* which forms plasmodia, protoplasts often coalesce to form larger plasmodia. In *Cavostelium,* the amoeboid stage may develop into a cell with a single anterior flagellum, or sometimes with two to four flagella. In this genus, sporocarps may be one to two spored. *Schizoplasmodium cavostelioides* is also of interest because its spores are violently discharged. Olive (1964) and Olive & Stoianovitch (1966) have interpreted the mechanism of spore discharge as due to the explosion of a bubble of gas between the spore wall and the outer sheath, and claim that a similar mechanism operates in the violent discharge of basidiospores of Basidiomycetes (see p. 400). Cysts, formed from amoeboid or plasmodial stages, are common in all protostelids, and may be uninucleate or multinucleate, varying in form from spherical to irregular. Sexual reproduction has, so far, not been reported.

Acrasidae

ACRASIALES

Because members of this group are not commonly encountered, they will not be described in detail. *Acrasis rosea,* which occurs on dead plant parts, such as florets of *Phragmites,* forms myxamoebae which feed independently. Eventually the myxamoebae aggregate to form a pseudoplasmodium. The fructification which develops from the pseudoplasmodium forms branched chains of spores (Olive & Stoianovitch, 1960; Raper, 1973).

Dictyostelidae

DICTYOSTELIALES

The best-known example of this group is *Dictyostelium*, which takes its name from the fact that the stalk of its multicellular fruit-body appears as a network, made up from the walls of the vacuolated myxamoebae from which it is formed. Species of *Dictyostelium* are common in soil, and can be isolated by smearing non-nutrient agar with cells of a suitable bacterial food such as *Klebsiella aerogenes* or *Escherichia coli*, and adding a small crumb of moistened soil to the centre of the bacterial smear (Raper, 1951). Myxamoebae of *Dictyostelium* will colonise the agar surface, feeding phagocytically on the bacterial food. When the bacteria, which are unable to multiply on the non-nutrient agar, have been consumed, the myxamoebae aggregate to form stellate, multicellular pseudoplasmodia. These pseudoplasmodia differentiate to form the fruit-bodies or sorocarps within about a week of adding the soil, and from the sorocarps it is a simple matter to streak out spores onto a fresh plate containing bacterial food and obtain a clean culture of the *Dictyostelium*. *Dictyostelium mucoroides* is the most commonly encountered, and it may also be found growing on dung incubated in moist conditions in the laboratory. However, the best-known and most widely studied species is *D. discoideum*, which has been the subject of extensive research on morphogenesis. Accounts of the group, which are sometimes referred to as cellular slime moulds, have been given by Bonner (1967, 1971) and Raper (1973), whilst morphogenesis in *Dictyostelium* has been reviewed by Gregg (1966), Newell (1971, 1975), Garrod & Ashworth (1973) and Loomis (1975). It is now possible to grow *Dictyostelium* in axenic (i.e. bacteria-free) culture (Sussman & Sussman, 1967; Ashworth & Watts, 1970; Schwalb & Roth, 1970; Watts & Ashworth, 1970).

The sorocarp of *D. discoideum* consists of three types of cell (Fig. 3), those forming the basal disk, the multicellular stalk and the spores, aggregated into a pear-shaped sorus at the tip of the stalk. The spores, which are cylindrical (Fig. 3), germinate within a few hours when placed on bacterial food. A split forms down one side of the spore and a myxamoeba emerges (Fig. 3c). The myxamoeba feeds by extending pseudopodia around bacteria and, after increasing in size, divides. The generation time for myxamoebae is 2–3 hours. The number of myxamoebae increases at the expense of the bacteria. During the feeding stage, the myxamoebae are more or less isodiametric. They are capable of amoeboid movement towards bacteria. When the bacterial food has been exhausted, the myxamoebae flow towards a common centre, a process termed **aggregation** and the resulting aggregates form stellate pseudoplasmodia (Fig. 4A, B), containing from a few hundreds to several thousand

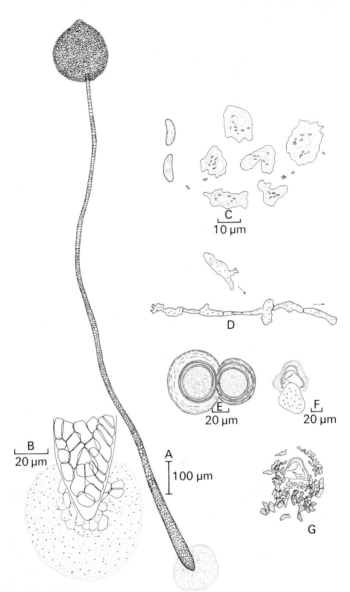

Fig. 3. *Dictyostelium discoideum*. A. Ripe sorocarp showing the sorus borne at the tip of a multicellular stalk supported on a basal disk. B. Magnified view of disk and base of stalk. C. Two spores, and six myxamoebae feeding amongst bacteria. The shape of these spores suggests that the strain is diploid. D. A few elongate myxamoebae making up part of the radial arm of an aggregating pseudoplasmodium. One myxamoeba is shown moving towards the arm. Arrows show direction of movement. Compare the shapes of the vegetative and aggregating myxamoebae in C and D. E. Two macrocysts. F. Germinating macrocyst showing liberation of contents still enclosed in the inner wall of the cyst. G. The same macrocyst 4 hours later, showing the escape of myxamoebae after rupture of the inner wall. C and D to same scale, F and G to same scale. E, F and G based on Nickerson & Raper (1973b).

cells. The term **grex** is sometimes used for the aggregation of myxamoebae. Individual myxamoebae move towards the centre of the aggregate, or towards one of the radiating arms. During the vegetative feeding phase, although the myxamoebae may make contact with each other, they do not adhere to each other, but during aggregation the myxamoebae change shape, becoming elongate (Figs. 3D, 4D). Their surface properties also change, so that they now adhere to each other. There is evidence that the stimulus to aggregate is a chemical one, and this substance, termed acrasin, has been identified as 3′-5′

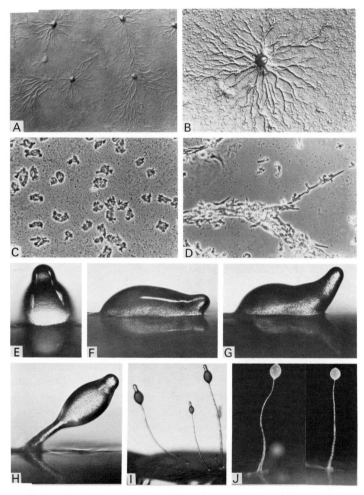

Fig. 4. *Dictyostelium discoideum* development. A. Aggregation of myxamoebae. B. Aggregation, enlarged. C. Myxamoebae feeding on bacteria: note isodiametric shape. D. Myxamoebae aggregating: note elongate shape. E. Late aggregation stage. F–G. Migration stage. H. Culmination: the spore mass is rising around the stalk. I. Spore mass almost at the apex of the stalk. J. Mature sorocarps.

cyclic adenosine monophosphate (cyclic AMP) by Konijn *et al.* (1967) (see Bonner, 1969). The aggregated myxamoebae pile up to form a bullet-shaped slug which in *D. discoideum* flops over onto the substratum and undergoes a period of **migration** towards the light. This phase of migration is not found in all Acrasiomycetes and does not occur, for example, in *D. mucoroides*. The distance of migration is often several millimetres at rates of up to 2 mm/hour. As the migrating pseudoplasmodium moves along, it leaves behind it a slime trail. At the early stage of migration, there is little visible difference between the cells making up the pseudoplasmodium, but later two different cell types can be distinguished; an anterior group of larger cells, and a posterior group of smaller cells with a sharp line of demarcation between them. There are also differences in staining properties. The anterior cells take up haematoxylin less readily than the posterior cells. They also have significantly larger nuclei. Grafting experiments have been done in which myxamoebae fed on the red-pigmented bacterium *Serratia marcescens* have been used as markers. By such means it has been possible to attach a red anterior to a white posterior, and to determine the fate of each group of cells. The larger anterior cells are destined to become stalk cells, whilst the smaller posterior cells are destined to become spores. The metabolic activities of these two cell types are different: the anterior cells are spenders of energy, whilst the posterior cells conserve it. The two types of cell are termed pre-spore cells and pre-stalk cells respectively.

The end of the migration phase is marked by the rounding off and erection of the pseudoplasmodium to form a flat-based, somewhat conical structure, which now undergoes further development by differentiating into a multicellular stalk and the sorus, which rises up the outside of the stalk. This final stage of development is termed **culmination**.

The cellular rearrangements which are associated with the differentiation of the sorocarp are illustrated in Fig. 5 (Bonner, 1944). A group of cells near the tip of the conical pseudoplasmodium amongst the pre-stalk cells becomes vacuolate, and forms a wedge-shaped mass which forces itself downwards through the pre-spore cells until it finally reaches the base and forms a complete stalk (Fig. 5ɪ, ᴊ). At its upper end, the stalk forms an open tube, and further cells migrate into the tube to add to the length of the stalk. The outer sheath of the stalk is made up of cellulose. During the final stages of differentiation, the cells making up the stalk vacuolate, their nuclei degenerate, and at maturity these cells are dead.

Studies on the fine structure of vegetative myxamoebae and developing sorocarps in *D. discoideum* have been made by George *et al.* (1972).

Two other reproductive structures which have been reported in some members of the Dictyostelidae, including *D. discoideum*, are **microcysts** and **macrocysts** (Blaskovics & Raper, 1957; Erdos *et al.* 1972, 1973a; Nickerson & Raper, 1973a, b). Microcysts are formed by encystment of individual myxamoebae. Macrocysts are, however, multicellular in origin. Such structures were thought to occur in only a low proportion of isolates, but Filosa & Chan

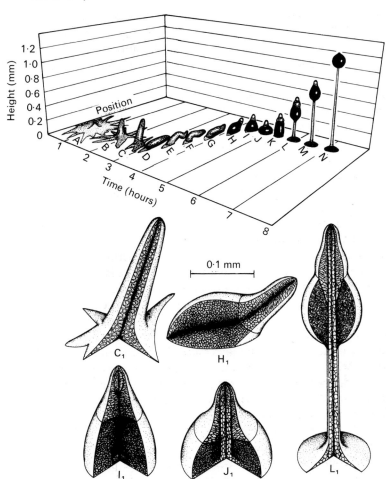

Fig. 5. *Dictyostelium discoideum*: development of sorocarp (after Bonner, 1944). A–C. Aggregation. D–H. Migration. I–N. Culmination. C_1. End of aggregation. H_1. End of migration. I_1. Beginning of culmination and stalk formation. J_1. Flattened stage of culmination. L_1. A later stage of culmination.

(1972) have claimed that, when appropriate techniques are used, about 50% of isolates of *D. mucoroides* develop macrocysts. Macrocysts are globose or ellipsoid, and arise from aggregated myxamoebae. They occur singly or in clusters, and are surrounded by a three-layered wall, the innermost of which is composed of cellulose. When mature, the macrocysts germinate to release myxamoebae (Fig. 3F, G) and, under starvation conditions, i.e. without bacteria present, these myxamoebae may aggregate to form small sorocarps (MacInnes & Francis, 1974). Macrocyst formation is enhanced by cultivation of myxamoebae under moist, warm, dark conditions. The cytology of macrocyst development has been studied by Dengler *et al.* (1970) who have shown that, after myxamoebae have aggregated, each aggregate subdivides into several spherical masses. In the centre of each mass, a large cell appears and, as it enlarges, this cell, termed the giant cell, ingests surrounding myxamoebae, enclosing each within a vacuole. These ingested cells had previously been recognised by other workers, who called them endocytes. The nucleus of the giant cell enlarges considerably and eventually all the peripheral myxamoebae are ingested. Subsequently, in place of the single large nucleus of the giant cell, numerous nuclei appear, so that the mature cyst is multinucleate.

Evidence of several kinds indicates that the macrocysts are involved in a true sexual process, i.e. one involving nuclear fusion and meiosis.

1 In *Polysphondylium violaceum* (see p. 19), developing macrocysts are binucleate and enlarge to become giant cells which engulf surrounding myxamoebae. Later, they contain only a single nucleus with several nucleoli. Later, this nucleus divides and, associated with the dividing nucleus, are single and paired axial elements, interpreted as **synaptonemal complexes** characteristic of cells undergoing meiosis (Erdos *et al.*, 1972).

2 In *Dictyostelium mucoroides*, MacInnes & Francis (1974) have provided evidence of genetical recombination in marked strains. The strain which they used (DM-7) is homothallic, but fusion between myxamoebae of different genotype to form a heterozygotic macrocyst may occur. On germination, the macrocysts may form up to 200 myxamoebae which can be induced to aggregate directly into sorocarps. When spores were plated out and the segregation of the markers among the progeny was analysed, the results were consistent with the occurrence of meiosis preceding macrocyst germination. Similar evidence using the heterothallic *D. giganteum* was obtained by Erdos *et al.* (1975). In this case, it was found that all the cells derived from a germinating macrocyst were identical in genotype, suggesting that only one haploid product of meiosis survives. It has been suggested that a single gene controls mating type, but that there are multiple alleles at the mating-type locus (see p. 40). Four alleles have, so far, been identified. *D. discoideum* has also been shown to be ambivalent in its sexual reactions. Some strains are homothallic, forming macrocysts in cultures derived from a single spore. In other strains, macrocysts only develop when certain cultures are paired together. Here, too, it is likely that mating competence is determined at a single locus with multiple alleles (Erdos *et al.*, 1973b).

It is also possible that sexual reproduction may occur which does not involve macrocyst formation. Skupienski (1918) described the pairing and fusion of myxamoebae of *D. mucoroides*, following which the zygotes underwent aggregation. C. M. Wilson (1952b, 1953) and Wilson & Ross (1957) claim to have confirmed this observation. According to these authors, in young aggregates of *D. discoideum*, one haploid myxamoeba may engulf another, and engulfment is followed by syngamy. Meiosis is believed to take place shortly afterwards; that is, syngamy, karyogamy and meiosis are all believed to occur during the aggregation phase. The haploid number of seven chromosomes is reported for *D. discoideum*. Later work by Ross (1960b) has shown that haploid and diploid strains of *D. discoideum* exist. Sussman & Sussman (1962) have confirmed the existence of stable haploid and diploid strains. The size of diploid myxamoebae is larger than that of haploids, and the size of diploid spores is significantly greater than that of haploids. In addition to the stable haploids and stable diploids, certain strains exist in which there is a varying proportion of haploid and diploid nuclei. Such isolates are termed metastable strains. It seems likely that the original isolate of *D. discoideum* by Raper (1935) was a stable haploid. However, a high proportion of laboratory stocks which have been maintained for some time in culture are diploid, suggesting that diploid strains may be selected by the conditions of cultivation. In some of the haploid strains, Ross has claimed that meiosis occurs at the *end* of the aggregation phase (the pre-migration phase), whilst in the diploid strains meiosis is rare. Other workers have interpreted the cytology of *D. discoideum* differently. Olive (1963b) has argued that the demonstration of diploid cells does not confirm sexual reproduction, and has pointed to the need for supporting genetical evidence.

Evidence for genetical recombination by an alternative mechanism to that described above has been obtained, and it seems likely that this is the result of a phenomenon termed **parasexual recombination**. Parasexual recombination was first discovered in the filamentous fungi *Aspergillus niger* and *A. nidulans*, and has since been demonstrated in a number of other fungi (Pontecorvo 1956; Roper, 1966), see p. 252. In filamentous fungi, parasexual recombination is brought about by a series of steps:

(*a*) Anastomosis between hyphae of two genetically distinct haploid strains resulting in the formation of a mycelium containing genetically dissimilar nuclei. Such a mycelium is termed a **heterokaryon.**

(*b*) Fusion of the genetically different nuclei to form heterozygous diploid nuclei which multiply along with the parental haploids.

(*c*) Separation of diploid colonies. This may arise by the formation of diploid asexual spores (i.e. diploid conidia) or possibly by the more rapid growth of diploid hyphal tips.

(*d*) During multiplication of the diploid nuclei a rare event, mitotic crossing over, occurs. This results in genetical exchange between homologous chromosomes.

(*e*) Haploidisation, a non-meiotic process in which, possibly by successive loss of single chromosomes during nuclear division, the diploid chromosome complement is reduced to the haploid. For example, if a diploid number is $2n$, it might be reduced to $2n-1$, $2n-2$, and so on.

Such nuclei with their abnormal chromosome complements are termed **aneuploid**.

(*f*) The establishment of new haploid colonies, some of which combine the genetical characteristics of the parental haploid strains.

The result of parasexual recombination is that, even in fungi which lack conventional sexual reproduction, i.e. a life cycle involving nuclear fusion and meiosis, genetical recombinants may arise which differ from the parental strains and may combine in different ways their genetical traits.

Evidence that a process very similar to parasexual recombination may occur in *Dictyostelium* has been presented by several workers (for example, Sinha & Ashworth, 1969; Fukui & Takeuchi, 1971). The essential steps in such a cycle would be:

1 *Cell fusion:* temporary cell anastomoses, and also complete cell fusions, have been reported (e.g. Huffmann & Olive, 1964; Kirk *et al.*, 1971).

2 *Nuclear fusion:* nuclear fusions have been reported (C. M. Wilson, 1952b, 1953). The variable size of nuclei is probably also correlated with chromosome content. There is also evidence for the formation of heterozygous diploid colonies (see below), implying nuclear fusion.

3 *Formation of diploid colonies:* as pointed out above, there is abundant evidence that diploid strains of *Dictyostelium* exist.

4 *Mitotic crossing over* has been demonstrated in heterozygous diploid strains of *D. discoideum* (Gingold & Ashworth, 1974).

5 *Haploidisation:* diploid strains are unstable during prolonged culturing, and tend to revert to haploid strains. Cytological investigations of an unstable heterozygous diploid strain have shown that it may contain a mixture of cells containing a range of chromosome complements, viz. diploid, i.e. 14 chromosomes, 30%; haploid, i.e. 7 chromosomes, 60%; aneuploid, i.e. 8–13 chromosomes, 10% (Sinha & Ashworth, 1969). Sinha & Ashworth have therefore concluded that haploidisation proceeds through transient aneuploidy, although the possibility that it could occur through other mechanisms has also been raised by Fukui & Takeuchi (1971).

The proportion of recombinant cells may be quite low. Sinha & Ashworth (1969) mixed myxamoebae of two genetically distinct haploid strains marked at five different loci. The myxamoebae were allowed to aggregate together and form fruit-bodies. When spores from these fruit-bodies were cultured, five colonies from amongst approximately 15,000 were identified as heterozygous diploids, with uninucleate spores larger than the parent strain. On further subculturing, these five colonies gave rise to a range of recombinant phenotypes. After five subcultures, ten different phenotypes were recovered, in a population of 173 strains, of which 47 were of parental phenotype and 126 of recombinant phenotype.

The story of fruit-body development in *Dictyostelium*, a comparatively simple organism, presents a challenge to developmental biologists, and raises many intriguing problems. One interesting question is the nature of the initiator of the aggregation process. The centre of an aggregate secretes cyclic AMP and starving myxamoebae respond chemotactically. It has been postulated by Ennis & Sussman (1958) that a single initiator cell, the *I* cell, is the founder of the aggregate, and that the proportion of *I* cells in a population is approximately constant for a given strain. Ashworth & Sackin (1969) have argued that in a population of myxamoebae there will be a proportion of aneuploid cells. If the genetic apparatus for cyclic AMP production is localised on a single chromosome, in a certain proportion of aneuploid cells there should be a double set of this genetical apparatus, and therefore twice the normal production of cyclic AMP. Ashworth & Sackin have calculated the frequency of such cells within the range of possible aneuploids, and their results are in good agreement with earlier estimates of the frequency of *I* cells.

Other members of the Dictyostelidae include *Polysphondylium* and *Acytostelium*. In *Polysphondylium,* as well as the terminal sorus, lateral whorls of sori occur on short stalks. *Acytostelium* is of interest because the tubular stalk on which the sorus is borne is acellular, in contrast with the multicellular stalk of *Dictyostelium* and *Polysphondylium* (Raper & Quinlan, 1958). It has been suggested that forms such as *Acytostelium* provide a link between the Dictyostelidae and the Protostelidae, and the observation that on pure water agar *Acytostelium* may form one-spored fruit-bodies has been interpreted as supporting evidence (Olive, 1967).

2: HYDROMYXOMYCETES

This is yet another group of uncertain affinity. The thallus is made up of a network of branched tubes, the so-called **net-plasmodium** or **filoplasmodium** within which amoeboid cells crawl.

LABYRINTHULALES

Labyrinthulales are mostly marine and estuarine organisms, considered by some to be related to slime moulds, by others to protozoa, and by others to the Mastigomycotina (zoosporic fungi) in the Eumycota or to Chrysophyceae (Pokorny, 1967; Perkins, 1974a). Members of the group can be isolated readily from marine Angiosperms such as *Zostera* and *Spartina* or from seaweeds, by floating a small portion of one of these substrata in sterile sea water poured over the surface of serum agar in the presence of a mixture of the antibiotics penicillin and streptomycin which aid in the elimination of bacterial contaminants. Two membered cultures of *Labyrinthula* have been established in the presence of yeasts or bacteria, and pure cultures have also been established (for references see Carlile, 1971).

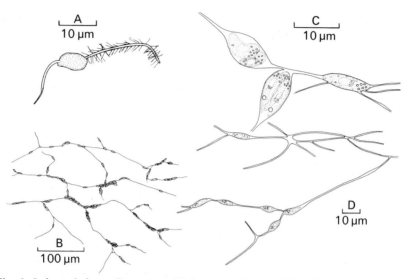

Fig. 6. *Labyrinthula*. A. Zoospore with long anterior tinsel flagellum and short posterior whiplash flagellum (after Amon & Perkins, 1968). B, C, D. Portions of colonies at different magnifications. In C, spindle cells are seen in swellings in the slime tracks.

Within a few days of inoculating serum seawater cultures, a fine network of strands can be seen extending over the agar surface (Fig. 6). Examination of these thalli shows that the network consists of branched tubes within which spindle-shaped cells move backwards and forwards. Such thalli are typical of *Labyrinthula* and *Labyrinthuloides*. The spindle-shaped cells are generally uninucleate, but in *Labyrinthuloides minuta* multinucleate, plasmodium-like structures have been reported (Watson & Raper, 1957). The tubular tracks within which the cells move are called slime tracks or ectoplasmic nets, and are made up of extracellular polysaccharide secreted by the cells. The material is somewhat elastic and can expand and contract as a cell moves through it. Rates of movement of up to 150 μm/minute have been reported. The mechanism of movement is not fully understood. Although the spindle-shaped cells are usually distributed singly within the tubes, occasionally clumps of two to three or even more may occur at the same point. In some cases, the occurrence of cells in groups is the result of mitosis. The fine structure of members of the Labyrinthulales has been studied by Hohl (1966), Stey (1968), Porter (1969), Perkins (1970, 1972, 1973a, b, 1974b) and Schwab-Stey & Schwab (1973). The cell structure conforms to the typical eukaryote plan. One unusual feature is a structure termed by Porter a bothrosome, and by Perkins a sagenetosome. These organelles are probably involved in the secretion of the slime tracks. The fine structure of the slime track shows that it contains a number of membranous and vesicular inclusions, but is devoid of structures such as ribosomes, endoplasmic reticulum or other cell organelles.

The outer layer of the slime track seems to be lined by a thin membrane, which is continuous with the plasmalemma of the cell which secretes it.

FRUITING

At least three different kinds of reproduction have been reported in *Labyrinthula:*

1. *Congregation:* the spindle cells may shorten and group together in a clump termed a sorus. In *Labyrinthula macrocystis,* the sorus may be made up of up to a hundred cells, around which a tough pellicle is secreted. When the pellicle is ruptured, a number of small, rounded cells are released. Pokorny (1967) has suggested that the term aggregation should not be used in *Labyrinthula* so as to avoid any imputation of relationship with cellular slime moulds.

2. *Cyst formation: L. macrocystis* forms cysts by the conversion of a single shortened spindle cell into four spherical daughter cells. From the cyst, ovoid cells are liberated.

3. *Zoospore formation:* in *Labyrinthuloides yorkensis*, Amon & Perkins (1968), Perkins & Amon (1969) and Perkins (1972) have described the release of biflagellate zoospores from a proportion of isolates. Zoospore formation is preceded by the congregation of spindle cells, which then round up and enlarge, forming presporangia. Within the large mucoid tubes which envelop the spindle cells, the cells become partitioned to form a sorus of sporangia, each sporangium of which releases eight zoospores through tears in the sporangial membranes. The zoospores are pear-shaped, narrow at the front and with two laterally attached flagella. The flagella are unequal in length, the anterior being about twice as long as the posterior. The two flagella also differ in structure. The anterior flagellum is of the 'tinsel' type, with two rows of fine lateral appendages or **mastigonemes,** whilst the posterior flagellum is of the 'whiplash' type, having a smooth shaft without lateral appendages (Fig. 6A). Zoospores with different flagella of this type are termed **heterokont.** In transverse section, both flagella conform to the classical $9 + 2$ subfibrillar pattern of construction typical of flagellate cells, with two central and nine peripheral strands (Gibbons & Grimstone, 1960). According to Perkins & Amon, the nuclear divisions which precede the formation of the eight zoospores probably involve meiosis followed by mitosis. Evidence for meiosis includes the formation of **synaptonemal complexes**. These are characteristic structures which probably mediate the pairing of homologous chromosomes during the zygotene stage of first meiotic prophase (Moses, 1968; John & Lewis, 1973). If this interpretation be correct, the congregated spindle cells are diploid, and the zoospores are haploid. At what stage in the life cycle the diplophase is restored is not yet known. The zoospores are capable of swimming for about 24 hours, propelled by the anterior flagellum. Upon settling, the zoospores lose their flagella and differentiate into motile spindle cells. No copulation of zoospores has been observed.

Labyrinthula macrocystis has been implicated as a possible agent of a 'wasting disease' of eelgrass, *Zostera marina* (Petersen, 1936; Young, 1943), but although the organism is commonly found associated with diseased plants, it cannot be regarded as conclusively proven that it is the sole cause of the disease (Pokorny, 1967). Perkins (1973a) has shown that the ectoplasmic nets of several labyrinthuloid organisms can bring about lysis of animal and plant cells, and that some can even degrade the chemically resistant sporopollenin which forms part of the wall of pollen grains.

The relationship between the Labyrinthulales and other organisms must remain, for the present, uncertain. There is, as yet, no general agreement. It is doubtful if 'net-plasmodia' are sufficiently similar to the true plasmodia of Myxomycetes to suggest close relationship, although multicellular, plasmodium-like structures have been reported. The structure of the zoospore with heterokont flagella is not characteristic of Myxomycetes which have zoospores with flagella of the whiplash type (p. 2). It would be of interest to have reports of zoospores in other species of *Labyrinthula*. Heterokont zoospores are found in the Oomycetes (p. 82), and Amon & Perkins (1968) have speculated that *Labyrinthula* might be related to the Thraustochytriaceae, a family of the Saprolegniales (p. 141). The Chrysophyceae is an algal group with which relationship may be possible.

3: MYXOMYCETES

These are the familiar slime moulds which are so common on moist, decaying wood and other organic substrata. The vegetative phase is a free-living plasmodium, a multinucleate naked mass of protoplasm varying in form from an inconspicuous microscopic structure (**aphanoplasmodium**) to large sheets or networks (**phaneroplasmodia**), within which the protoplasm shows rhythmic reversible streaming. The plasmodium feeds by engulfing bacteria, fungal cells, etc., and under favourable conditions gives rise to sporophores which are of diverse shapes. Beneath the developing sporophores, the plasmodium deposits a specialised layer, termed the hypothallus, which is very variable in form: disk-like, membranous, horny, spongy or calcareous. The spores are dispersed by wind and germinate to form myxamoebae or zoospores with one to two anterior whiplash flagella. The myxamoebae can reproduce asexually by fission, and the plasmodium is capable of fragmentation. Sexual reproduction occurs by fusion of myxamoebae or zoospores to form zygotes from which plasmodia develop.

An invaluable general account of the group has been given by Alexopoulos (1973) and Gray & Alexopoulos (1968), and a taxonomic account by Martin & Alexopoulos (1969). Discussions on the systematic position of the Myxomycetes have been provided by Martin (1960) and Olive (1970). The key below is taken from Alexopoulos (1973).

Key to subclasses of Myxomycetes

1 Spores borne singly at the tips of hair-like stalks, on columnar, dendroid, or morchelloid sporophores
                                    Ceratiomyxomycetidae (p. 23)
Spores borne in masses, within various types of sporophores, peridium persistent or early evanescent   2

2 Sporophore development subhypothallic, the plasmodial protoplast rising internally through the developing stalk in stipitate forms; peridium continuous with stalk and hypothallus; spores pallid, bright-coloured, ferruginous, purple-brown, or black; assimilative stage of various types, but never a true aphanoplasmodium
                                    Myxogastromycetidae (p. 26)
Sporophore development epihypothallic; stalk, when present, secreted internally, hollow, or partially filled with strands; spores violet-brown, lilac, ferruginous, or pallid by transmitted light; lime, if present, never on the capillitium; assimilative stage an aphanoplasmodium   **Stemonitomycetidae (p. 26)**

Ceratiomyxomycetidae

This subclass contains one order, the Ceratiomyxales, with a single genus *Ceratiomyxa,* which differs from the remaining Myxomycetes in that its spores are borne *externally* on the surface of column-like sporophores, whereas in other Myxomycetes the spores are borne *endogenously*, within the sporophores, usually enclosed until a late stage by an outer covering or

Fig. 7. Plasmodia and fructifications of *Ceratiomyxa fruticulosa.*

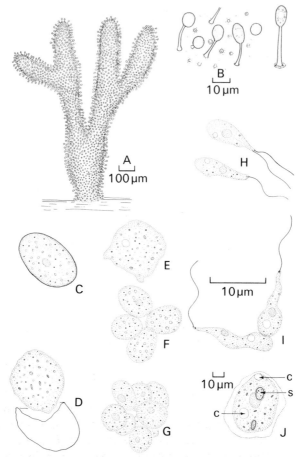

Fig. 8. *Ceratiomyxa fruticulosa*. A. Fruiting pillar bearing stalked spores. B. Portion of the surface of the fruit body showing spores and their attachments. C. Spore. D. Naked protoplast emerging from the spore at germination. E. Naked protoplast before cleavage. F. Cleavage of protoplast to form a tetrad of protoplasts. G. Octette stage: a clump of eight protoplasts. H. Uniflagellate and biflagellate swarmer released from the octette protoplasts. I. Copulation of swarmers by their posterior ends. J. Young plasmodium: c, contractile vacuole; s, ingested spore within a food vacuole. C, D, E, F, G, H, I to same magnification.

peridium. *Ceratiomyxa fruticulosa* forms thin, watery, whitish plasmodia in rotting wood. The plasmodium emerges and fruits by forming erect single or branched white pillars about 1–10 mm high (Figs. 7 and 8). The pillars consist of a central vacuolated matrix with a thin surface of protoplasm. All over the surface of these pillars, unicellular globose spores develop on short stalks. Scheetz (1972) has described the ultrastructure of developing spores. The spores are at first uninucleate, but by successive nuclear divisions they become

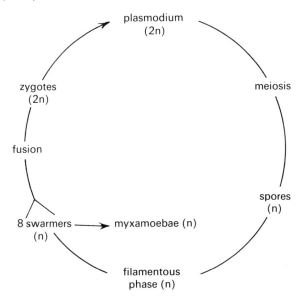

plasmodium
(2n)

zygotes
(2n)

meiosis

fusion

spores
(n)

8 swarmers ⟶ myxamoebae (n)
(n)

filamentous
phase (n)

Fig. 9. Possible life cycle of *Ceratiomyxa*.

quadrinucleate. Claims that the nuclear divisions are meiotic have been confirmed by the demonstration of synaptonemal complexes in developing spores (Furtado & Olive, 1971). Following dispersal by air currents, the spores germinate to release a single, naked, quadrinucleate protoplast which may elongate to form a short filament, then rounds up (Nelson & Scheetz, 1976). The filamentous stage is not always formed, and the naked protoplast which escapes from the spore on germination may remain approximately isodiametric in form, occasionally putting out pseudopodia (Fig. 8D, E). The four nuclei undergo mitosis, and the protoplast then cleaves into a tetrad of globose, uninucleate segments. Each cell of the tetrad undergoes a further mitotic division, after which the cells divide so that an eight-celled clump results (Fig. 8F, G). Each of the cells releases a swarmer which Gilbert claims to be uniflagellate, but which, according to McManus (1958) is unequally biflagellate. Swarmers, some of which are uniflagellate and some biflagellate, have been observed (Fig. 8H). Despite this variation, two kinetosomes are present at the anterior of the zoospore. The flagella are both of the whiplash type (Nelson & Scheetz, 1975). The swarmers may lose their flagella and become amoeboid. According to Gilbert (1935), the flagellate swarm cells fuse in pairs to initiate the diploid phase in the form of a plasmodium (Fig. 8I, J). Fusion of myxamoebae has not been reported. The life cycle is outlined in Fig. 9.

The relationships of *Ceratiomyxa* are uncertain. Martin & Alexopoulos (1969) classify it in a subclass of the Myxomycetes, whilst Olive (1970) has classified it in the Protostelidae, pointing to similarities in spore development.

The plasmodium of *Ceratiomyxa,* according to Olive, does not show reversal of cytoplasmic flow, a characteristic feature of the plasmodia of Myxomycetes, but this claim contradicts the observations of Gilbert (1935). A possible link between *Ceratiomyxa* and the Protostelidae is *Ceratiomyxella tahitiensis* (Olive & Stoianovitch, 1971). The sporocarps of this organism form globose unicellular spores, which, on germination, release a 'zoocyst' which may contain two, four or eight cells. Each cell of the zoocyst releases a swarmer with one or sometimes two anterior flagella, and such swarmers have been observed to fuse with each other. Non-flagellate amoeboid cells may also develop from the zoocyst, and fusions have been reported between amoebae and swarmers. Small reticulate plasmodia develop, resembling the aphanoplasmodia of Myxomycetes (p. 30), but these do not show rhythmic streaming. Sporulation occurs at the periphery of the plasmodium, by a process similar to that already described for the Protostelidae (p. 9). The cytological details of the life cycle remain to be worked out. However, the general similarity to *Ceratiomyxa* is obvious. It is unfortunate that it has not yet been possible to establish *Ceratiomyxa* in culture.

Myxogastromycetidae and Stemonitomycetidae

These organisms are commonly known as Myxomycetes, true slime moulds, or plasmodial slime moulds, to distinguish them from the cellular slime moulds. In contrast with *Ceratiomyxa*, they produce spores (endospores) internally, within a fructification which is at first enclosed in a **peridium,** a thin membrane.

Over 400 species of Myxomycetes are known (Martin & Alexopoulos, 1969), but of these only about 15% have been grown in culture, mostly members of a single order, the Physarales (Gray & Alexopoulos, 1968; Alexopoulos, 1969). Most grow on rotting wood or other vegetation, on dung, soil, sawdust heaps, etc. About 250 species have been found in Britain (Ing, 1968). A generalised life cycle is outlined in Fig. 10.

SPORES

The spores of Myxomycetes are unicellular, usually globose, with smooth, spiny or reticulately thickened walls, variously pigmented, from colourless or yellow to brown or purple. According to Schuster (1964), the wall of *Didymium nigripes* is two-layered. There is an outer, darker-pigmented layer which is spiny and may contain chitin, mucopolysaccharides, melanins or lipofuscins, and an inner, thicker layer which contains some cellulose. In *Physarum gyrosum,* however, Koevenig (1964) reports a three-layered wall. Most Myxomycete spores are uninucleate, although there are reports of two to eight nuclei in some (Gray & Alexopoulos, 1968). The spores are dispersed passively by air currents. In *Trichia* (Fig. 14), the spores are released by the hygroscopic twisting of spirally thickened **'elaters'**, one end of which is fixed and the other

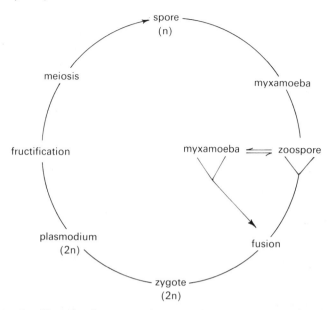

Fig. 10. Outline life cycle of a myxomycete.

rotates in response to changes in humidity (Ingold, 1971). In many other Myxomycetes, the spores are interspersed within the fructification by a network of tubular, branched, anastomosing threads which hold them in place and allow them to be sifted out slowly by air currents (Fig. 15). The network of threads is called the **capillitium**. Spores may remain viable for many years. Elliott (1949) germinated spores of *Physarum flavicomum* 60 years after collection.

Spore germination is by the cracking of the spore wall or by the formation of a pore. One or more naked protoplasts escape which may acquire one or two flagella (Fig. 11). The flagella are unequal in length (**anisokont**) and the shorter posterior flagellum may be difficult to see. The typical condition is biflagellate, but there are numerous reports of uniflagellate cells. Possibly these are due to the fact that the flagella may not develop simultaneously. Kerr (1960) has reported that newly flagellated cells of *Didymium nigripes* are uniflagellate, but a second flagellum develops later. Both flagella are of the whiplash type. Their fine structure has been studied by Gottsberger (1967) and Aldrich (1968). They show the usual $9+2$ arrangement of subfibrils. In some cases, the protoplast which emerges from the spore does not develop flagella. Conversely, flagellate cells may revert to an amoeboid stage in some species, and the two stages are readily interconvertible. The addition of water to a layer of myxamoebae will encourage development of flagella in *Physarum polycephalum*, and in *Didymium nigripes* (Kerr, 1960).

Both swarm cells and myxamoebae are capable of feeding phagocytically

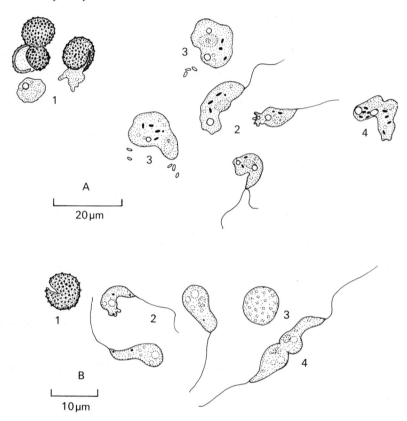

Fig. 11. Spore germination and swarmers in *Physarum* and *Reticularia*. A. *Physarum polycephalum*: 1, spores germinating to release myxamoebae; 2, uniflagellate and biflagellate swarmers, note the pseudopodia at the rear end of one swarmer; 3, myxamoebae; 4, fusion between two myxamoebae. B. *Reticularia lycoperdon*: 1, spore showing cracked wall; 2, swarmers, one with pseudopodia; 3, encystment stage; 4, fusion between two swarmers.

on bacteria, and dual-membered cultures of various Myxomycetes have been established on weakly nutrient agar such as half-strength corn-meal agar along with bacteria such as *Escherichia coli* or *Klebsiella aerogenes*. The myxamoebae multiply following mitotic nuclear division. The flagellate stage is apparently incapable of division: the flagella are withdrawn before or during prophase. Axenic culture of myxamoebae has been achieved by the use of dead or disrupted bacterial cells as food (Kerr, 1963), and Ross (1964) has succeeded in cultivating myxamoebae in the absence of dead bacterial cells in a complex medium containing extracts of chick embryo and corn meal, whilst Goodman (1972) has maintained the myxamoebae of *P. polycephalum* axenically on a medium containing mineral salts, glucose, peptone, liver infusion and bovine serum albumin.

The duration of the flagellate stage varies. Ross (1957b), who studied spore germination in nineteen Myxomycetes, has classified them into:

(a) *Briefly flagellate:* forms in which the flagellate stage is of brief duration, probably less than 24 hours. Protoplasts emerging from spores are at first amoeboid, and acquire flagella within 1–5 hours. After the relatively brief flagellate stage, Ross claims that the cells become irreversibly amoeboid, and quotes as examples of this type *Didymium squamulosum* and *Physarum oblatum*. However, Carlile & Odell (unpublished) have shown that amoebae of *D. squamulosum* can be converted to flagellate cells by the addition of water, and return to the amoeboid state when water is removed.

(b) *Flagellate:* germinating spores give rise to flagellate cells directly or to myxamoebae which quickly acquire flagella. The flagellate stage persists for several days. *Fuligo septica* and *Physarum polycephalum* behave in this way (Fig. 11A), but in the latter the occurrence of flagellate cells or amoebae can be controlled by the presence or absence of free water.

(c) *Completely flagellate:* spores germinate to produce flagellate cells, whilst myxamoebae are not observed. Flagellate cells of *Reticularia lycoperdon* and *Stemonitis nigrescens* have been maintained in this condition for over three weeks (Fig. 11B).

Under certain conditions, the amoebae of slime moulds may round off to form cysts.

SYNGAMY

Both the flagellate stage and the myxamoebae can function as gametes. The fusion of haploid cells leading to the formation of a diploid zygote is termed **syngamy**. Cell fusion preceding nuclear fusion is termed **plasmogamy**, whilst the nuclear fusion is termed **karyogamy**. In completely flagellate forms such as *Reticularia lycoperdon*, plasmogamy can only occur between swarmers (Fig. 11B), but in other forms fusion may be either between swarmers or between myxamoebae (Fig. 11A). Where fusion is between swarmers, the cells fuse at their posterior ends, and the flagella persist so that the zygote may bear two sets of flagella. Karyogamy occurs, following plasmogamy between swarmers or myxamoebae, often within minutes. The diploid zygote initiates the formation of the plasmodium. Evidence that the plasmodium contains diploid nuclei has been obtained by Therrien (1966). The mating systems of Myxomycetes will be described later.

THE PLASMODIUM

The diploid nucleus in the zygote divides repeatedly so that a multinucleate mass of protoplasm is formed. The nuclear divisions in the plasmodium are approximately synchronous. The plasmodium continues to feed phagocytically on bacteria, fungal cells, etc., and may also engulf myxamoebae. The plasmodium is capable of movement, and creeps over the substratum, often forming large sheets several centimetres in extent. It is of historical interest that the plasmodia of *Fuligo* were used in some of the earliest chemical analyses of cytoplasm. The plasmodium frequently organises itself into a fan-shaped network of veins or channels, within which the cytoplasm shows rhythmic flow, forwards and backwards. However, plasmodia are variable in their characters, and Alexopoulos (1960b, 1969) has classified them into three types:

1. *Protoplasmodia:* plasmodia of this type are found in *Echinostelium minutum*, a Myxomycete which forms tiny, stalked fructifications on bark and on leaves, usually less than 0·5 mm high, and bearing sporangia 40–50 µm in diameter. To quote Alexopoulos' (1960a) description:

The plasmodium of this species, even in an advanced stage of development, appears to be more or less homogeneous in structure. The ground substance is granular. Vacuoles are often conspicuously present, and a thin slime sheath may be evident at the margin, but there is no reticulation at any time, no veins, no advancing fan, no rhythmically reversible streaming. The streaming of the protoplasm is so slow as to be almost imperceptible except under high magnification and, as far as could be detected, is irregular.

As implied by the prefix 'proto', it is believed that this type of plasmodium may be primitive. Protoplasmodia have been found, in addition to *Echinostelium*, in *Clastoderma debaryanum* and in *Licea* spp. Probably most protoplasmodia give rise to only a single sporangium.

2. *Aphanoplasmodia:* the prefix 'aphano' means inconspicuous. The plasmodia of species of *Stemonitis* and *Comatricha*, both members of the Stemonitomycetidae, are of this type, and it appears likely that they are characteristic of the group. Apart from their inconspicuous appearance, aphanoplasmodia differ from the more obvious and better known phaneroplasmodia (see below) in having a thin, open network of plasmodial strands, and in being flattened and transparent. The plasmodial strands are only about 5–10 µm wide, resembling fungal hyphae, whilst the meshes which they compose may be 100–200 µm across (Fig. 15). Aphanoplasmodia remain colourless to a late stage, becoming yellowish in colour shortly before fruiting, when the strands become thicker and coralloid.

3. *Phaneroplasmodia:* the prefix 'phanero' means obvious or conspicuous, and applies to the large, conspicuous plasmodia characteristic of the order Physarales, many of which have been grown in culture. They include species from the genera *Fuligo, Badhamia, Physarum* and *Didymium*. The best-known

Fig. 12. *Physarum polycephalum* plasmodia. A. Margin of healthy plasmodium. B. Fusion between compatible plasmodia. Note complete fusion of veins. C. Lethal reaction following fusion between incompatible plasmodia. Photographs kindly supplied by Dr M. J. Carlile. A from Carlile (1971) by permission of Academic Press; B,C from Carlile & Dee (1967) by permission of Macmillan's Journals.

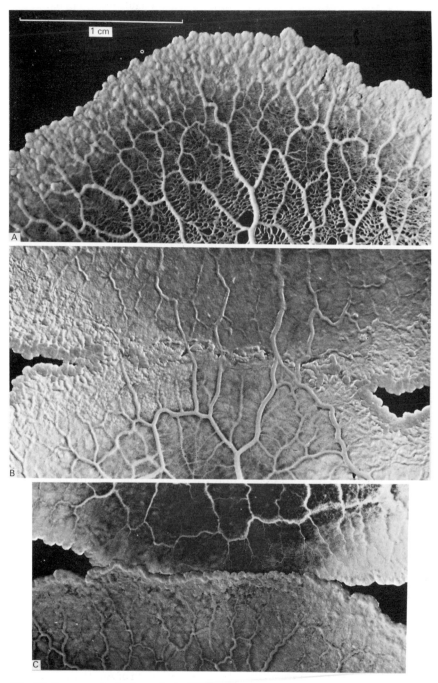

Fig. 12. *See opposite for legend.*

of these is *Physarum polycephalum* which grows in nature on dead wood and decaying, fleshy fungi in the U.S.A. and some other countries. It has been possible to free the plasmodium from other organisms and to grow it on semi-defined media either in liquid or in solid form (Fig. 12) (Gray & Alexopoulos, 1968; Carlile, 1971). The essential ingredients of the semi-defined media are glucose (or starch), peptone, mineral salts, vitamins and haematin. It is also possible to cultivate the plasmodium on living bacteria or yeasts or on corn-meal agar or oat-flake agar. Because the plasmodium can be readily cultivated and can be induced to sporulate, it has been widely used in a number of studies.

Phaneroplasmodia are typically organised into fan-shaped reticulate masses, yellow in colour (fig. 12A). The veins show reversible cytoplasmic flow (shuttle streaming), in which the cytoplasm moves in one direction for a period, and then reverses. The duration of flow in each direction varies, but is often of the order of 70–85 seconds, although flow for as long as 30 minutes in one direction has been reported. The observed rates of flow are of the order of 1 mm/second, which is considerably higher than the rates of cytoplasmic flow recorded for fungi or for plant cells. The mechanism of cytoplasmic flow, and especially the reversal of flow, is not fully understood, although it has been the subject of much research. It is possible that contractile protein fibrils in the plasmodium, capable of contraction when activated by ATP, are involved. For a fuller discussion, see Gray & Alexopoulos (1968), Korohoda *et al.* (1970), Alléra & Wohlfarth-Bottermann (1972).

It has already been mentioned that plasmodia may engulf swarm cells and myxamoebae. Curiously, a mature plasmodium may also give rise to swarmers or myxamoebae (Indira, 1964; Ross & Cummings, 1967). These cells should, presumably, be diploid, but this needs confirmation. Indira has suggested that they represent a means of asexual reproduction of the plasmodium, but Ross & Cummings have raised the possibility that swarmer formation may be a reaction of a plasmodium to unfavourable conditions.

When conditions are unsuitable for the plasmodium to sporulate, it may give rise to dark, horny sclerotia. These are thick-walled resting structures within which the cytoplasm is divided up into a large number of multinucleate cells, sometimes termed **spherules** (Jump, 1954). The sclerotia are apparently not viable for long and, in contrast with the spores which may retain viability for many years, their viability is often to be measured in months. On the return of suitable conditions, the sclerotia may quickly give rise to fresh plasmodia. Although, in some species, sclerotia may form in response to drying of the plasmodia, in others, e.g. *P. polycephalum*, sclerotia can be induced to form in liquid culture.

SPORULATION

After a period of feeding, given suitable conditions, the plasmodium will become converted to fruit-bodies. Usually, the whole of the plasmodium is involved. Before sporulation occurs, the plasmodium may move considerable distances over the substratum, often towards the light. It is possible that exhaustion of available food is a stimulus to sporulation, because it has been shown for several Myxomycetes, including *P. polycephalum,* that so long as adequate food is available, the plasmodium will persist. Light appears to stimulate fruiting in a number of cases, especially those with pigmented plasmodia, whilst for non-pigmented forms, sporulation may occur equally well in the absence of light (Gray & Alexopoulos, 1968). In nature, it seems that the actual process of sporulation occurs most readily at night, and may be completed within a few hours.

The form and colour of the mature fructifications are very varied, but in general they conform to three types:

(a) *Sporangia:* one to several thousand sporangia may develop from a single plasmodium. Sporangia may be stalked or sessile, globose, cylindrical or cup-shaped, and they are seated in most cases on a membranous layer, the **hypothallus**, which persists at the base. In the young sporangium, the spores are encased in a peridium, an outer membranous layer, and this may disappear at maturity, or may persist, opening by irregular cracks or special lines of dehiscence. Where the sporangium is stalked, the stalk may be extended into the body of the sporangium, as the **columella**. Branched capillitial threads may traverse the cavity of the sporangium, separating and supporting the spores, but in many species capillitia are absent. Sporangia are found in *Physarum* (Fig. 13D), *Trichia* (Fig. 14), *Arcyria* (Figs. 13C, 16), *Stemonitis, Comatricha* (Fig. 15) and numerous other genera.

(b) *Plasmodiocarps:* A plasmodiocarp resembles a sporangium, but forms a network over the substratum, often following the veins of the plasmodium from which it developed. This type of sessile fructification is found in *Hemitrichia.*

(c) *Aethalia:* aethalia are large pulvinate masses with a surface crust. Possibly they represent a mass of fused sporangia. The aethalia of *Reticularia lycoperdon* (Fig. 13B) are about the size of a hen's egg on rotting tree trunks. They are at first covered by a silvery peridium, and when this cracks open, the brown spores are exposed. *Fuligo septica* (flowers of tan) forms flat yellowish crusts several centimetres in extent over soil, sawdust or rotting wood (Fig. 13A). When the crust is broken, the blackish purple spores are freed. *Lycogala epidendrum* forms pink hemispherical aethalia about 1 cm in diameter on rotting wood.

Intermediate types of fructification are also known.

Fig. 13. Some different kinds of Myxomycete fructification. A. *Fuligo septica* on a sawdust heap overgrown by brambles. The fruit-bodies are bright yellow. B. *Reticularia lycoperdon* on a dead tree trunk. The whitish peridium bursts to release chocolate brown spores. C. *Arcyria denudata* on rotting wood. The fructifications are dull red in colour. They have released most of their spores and the capillitium network is exposed. D. *Physarum polycephalum* in Petri-dish culture on agar. Note the multi-headed sporangia.

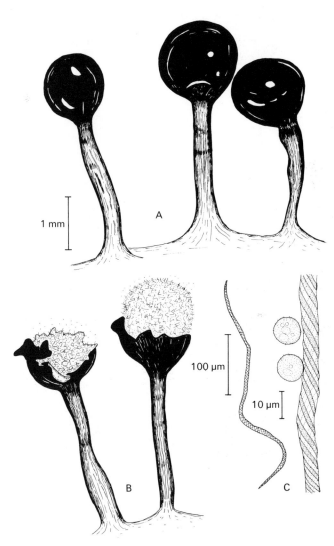

Fig. 14. *Trichia floriforme*. A. Undehisced sporangia. Note that the sporangial stalks are continuous with the hypothallus. B. Dehisced sporangia releasing spores by twisting of elaters. C. Elaters and spores.

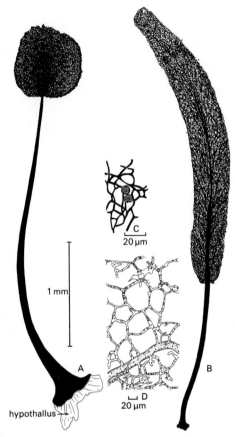

Fig. 15. Fructifications of Stemonitomycetidae. A. Sporangium of *Comatricha nigra*. Note the hypothallus. B. Sporangium of *Stemonitis fusca*. C. Portion of the capillitium of *Stemonitis fusca* and three spores. D. Portion of the plasmodium (aphanoplasmodium) of *Stemonitis fusca*. A and B to same scale.

DEVELOPMENT

The development of Myxomycete plasmodia follows a number of different patterns, and these will be illustrated by reference to two distinctive types, *Physarum* and *Stemonitis*.

In *Physarum polycephalum*, the plasmodial strands become knotted into irregular nodules, each of which becomes a sporangial primordium (Guttes *et al.*, 1961). The biochemical changes associated with differentiation of the fruit-body have been detailed by Sauer (1973). Sporangial development occurs by erection of pillars of protoplasm with slender stalks and terminal lobed sacs. The stalk is continuous below, with a residual sheet of plasmodial material termed the hypothallus, which persists as a deposit at the base of the

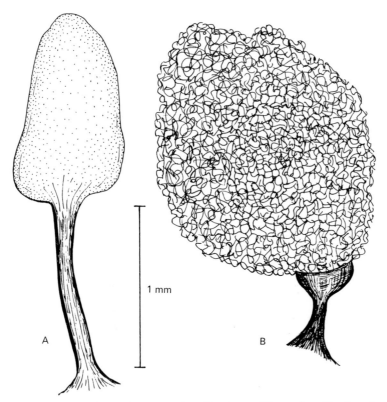

1 mm

A   B

Fig. 16. *Arcyria denudata*. A. Immature fructification. B. Mature fructification showing the fluffy capillitium.

fructification (Fig. 17A). The interior of the stalk is stuffed with amorphous granular material, which provides skeletal support. Above, the stalk is continuous with the peridium. This is formed from the outer layer of the protoplasm making up the sporangium. It is a non-cellular membrane, which in *Physarum* contains lime ($CaCO_3$) in amorphous state. Because the hypothallus, the outer covering of the stalk and the peridium which surrounds the sporangium are all continuous, it is clear that sporangial development takes place *beneath* the hypothallus. Such development, which is characteristic of the Myxogastromycetidae, is termed **subhypothallic** and distinguishes them from the Stemonitomycetidae (see below).

The bulk of the protoplasm flows into the sac where it differentiates into capillitium and spores. The capillitium is an anastomosing network of threads. Where the threads join, there is an accumulation of lime, at the so-called lime knots or lime nodes. There is need for a study of the fine structure of capillitial development in the Physarales. In the related genus *Didymium*, the capillitium is formed by the deposition of a tubular layered system which encloses portions of the cytoplasm. In *Arcyria*, a member of the

Trichiales, Mims (1969) has shown that the capillitium differentiates from a system of elongated vacuoles, which anastomose. The vacuolar structures are separated from the surrounding protoplasm by a membrane, which does not enclose recognisable organelles, but may contain bacterial remains. He has argued that this is evidence for the incorporation of food vacuoles in the vacuolar system which gives rise to the capillitium. A layer of fibrillar material to the inside of the membrane forms the wall of the capillitial thread. It is believed that the capillitium may form a conducting system through which calcium carbonate and other substances are transported towards the peridium where they are deposited. Fine structural studies (Ellis *et al.,* 1973) show that some capillitia are solid, whilst others are hollow.

The development of the sporangium in *Stemonitis* takes an entirely different course, described as **epihypothallic.** Whereas, in the Myxogastromycetidae, the hypothallus is formed on the surface of the plasmodium and only comes into contact with the substratum after the plasmodial contents have moved up into the sporangia, in *Stemonitis* and the related genera *Comatricha* (Fig. 15) and *Lamproderma,* the hypothallus is laid down on the surface of the sub-stratum, and the sporangia develop *above* it (Ross, 1957a, 1960a, 1973). A thin, continuous layer of protoplasm connects developing sporangia, and as the protoplasm from the plasmodium flows into the sporangia, the hypo-thallus formed beneath the developing sporangia is left behind in contact with the substratum. It dries out and remains as a thin membranous sheet connect-ing the bases of the sporangia, and continues upwards into each sporangial primordium as the stalk initial. The stalk is made up of a system of parallel interlaced tubules continuous with the hypothallus at the base. The stalk, which is a tube open above, elongates by addition to the length of the tubules at the upper end. It continues upwards into the sporangium as the columella. The peridium, surrounding the sporangium, differentiates relatively late. The capillitium develops as a system of tubes extending from the columella and bending outwards and branching. In *Stemonitis,* capillitium threads also originate in the protoplasm surrounding the columella, and the two sets of threads anastomose.

Differences between the two kinds of development are illustrated diagram-matically in Fig. 17.

Spore formation occurs a few hours after the formation of sporangial primordia, following one or two synchronous nuclear divisions. Although it is now generally accepted that the plasmodial nuclei are diploid and those of the spore haploid, the exact position of meiosis is not the same in all Myxomy-cetes, and may even be variable for a given species. Meiosis may precede spore cleavage or occur after cleavage, i.e. take place inside the spore. In *Didymium iridis,* there have been claims that meiosis precedes spore cleavage (Therrien, 1966; Carroll & Dykstra, 1966), but later work suggests that the cytological evidence for this is based on a misinterpretation, and that in this species meiosis takes place some 12–24 hours after spore cleavage (Aldrich & Carroll,

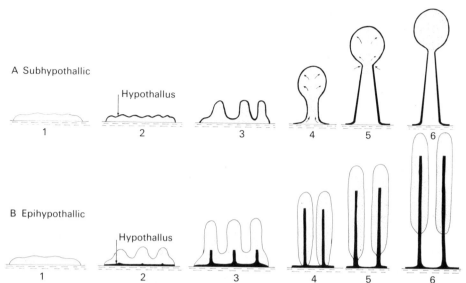

A Subhypothallic

Hypothallus

1   2   3   4   5   6

B Epihypothallic

Hypothallus

1   2   3   4   5   6

Fig. 17. Sporangial development in Myxomycetes (based on Ross, 1973). A. Subhypothallic development characteristic of Myxogastromycetidae: 1. Plasmodium. 2. Formation of supraplasmodial hypothallus. 3. Differentiation of primordia; protoplasmic flow into primordia. 4. Enlargement of primordia as protoplasmic flow continues; apex expands. 5. Continued expansion of apex as wall of stalk constricts. 6. Full-size sporangium, prior to capillitium formation. Note that the peridium, stalk and hypothallus are continuous. B. Epihypothallic development characteristic of Stemonitomycetidae: 1. Plasmodium. 2. Formation of subplasmodial hypothallus. 3. Differentiation into primordia; stalk initial secreted internally. 4. Protoplasm segregates into discrete primordia. 5. Continuous deposition of material at stalk apex causes stalk elongation; pre-spore mass of protoplasm is elevated. 6. Full-size sporangium prior to capillitium formation. Note that the interior stalk is continuous with the subplasmodial hypothallus.

1972). In three species of *Physarum, P. flavicomum, P. polycephalum* and *P. globuliferum,* Aldrich (1967) has shown that meiosis occurs in the spores after cleavage. Ross (1961), however, claimed that meiosis occurred prior to spore cleavage, and this may be a case where the position of meiosis is variable. In those cases where meiosis occurs in the spore, it is usual for three of the haploid products to abort, so that there is only one functional haploid nucleus in the ripe spore.

MATING SYSTEMS

The life cycles of most Myxomycetes (Fig. 10) are sexual, including karyogamy and meiosis. However, it is possible that in some the life cycle is completed without sexual fusion. Such a life cycle is described as **apogamous**. It is difficult to obtain conclusive proof of apogamy, but von Stosch *et al.* (1964) have claimed apogamy in certain strains of *Physarum polycephalum* and *Didymium nigripes*, based on chromosome counts. N. S. Kerr (1967) and S. Kerr (1968) have also claimed apogamous development of plasmodia in a strain of *D. nigripes*, on the grounds that the ploidy of amoebae and plasmodia were the same, and as a result of a time-lapse photographic record of plasmodial development.

Sexually reproducing Myxomycetes may behave in two ways. They may be:
(*a*) *Homothallic:* sexual reproduction occurs between swarmers or myxamoebae derived from a single spore. An example of this kind of behaviour is *Didymium squamulosum.*

(*b*) *Heterothallic:* sexual reproduction does not occur between swarm cells or myxamoebae derived from a single spore, but only when swarmers or myxamoebae derived from genetically different spores are brought together. Examples of heterothallic Myxomycetes are *Physarum polycephalum* (Dee, 1960, 1966) and *Didymium iridis* (Collins, 1963). Working with a strain of *P. polycephalum*, Dee (1960) produced evidence that, amongst a population of spores derived from a single plasmodium, two mating types were represented, which she termed $(+)$ and $(-)$. When $(+)$ myxamoebae were mixed with other $(+)$ myxamoebae, no plasmodia developed. Likewise, when $(-)$ myxamoebae were mixed together, plasmodial development did not take place. But when myxamoebae of the $(+)$ mating type were mixed with myxamoebae of the $(-)$ mating type, plasmodial formation was initiated. The usual explanation of such behaviour is that there is a single gene locus controlling mating type, with two alleles, $(+)$ and $(-)$. Later (1966), working with a different strain, she identified two further alleles at the same locus. This necessitated a change in the nomenclature of the mating types. Those of the original strain were designated mt_1 and mt_2, whilst those of the second strain were designated mt_3 and mt_4. As with the first strain, only two mating types were represented amongst a population of spores of the second strain, in approximately equal proportion. When myxamoebae of mt_3 were mated with mt_1 or mt_2, they formed plasmodia with both strains. Similarly, myxamoebae of mt_4 could mate with both mt_1 and mt_2. The explanation of this situation is that, for successful mating, two *different* alleles must be present in the myxamoebae, and this would be the case in all crosses between these strains. Because a given strain of *P. polycephalum* is diploid, only two alleles will be represented at the locus which controls mating type, but in the world population there is a multiple series of alleles at this locus. Dee discovered that matings took place more readily, i.e. plasmodia developed more quickly, in crosses between myxamoebae that were distantly related (i.e. inter-strain crosses) than in those between closely related strains (i.e. intra-strain crosses). She also discovered that occasionally plasmodia developed in a population of myxamoebae derived from a single spore. In view of the cytological observations that meiosis may occur in the spore followed by abortion of some of the haploid products, one possible explanation is that at least two nuclei might have survived, representing two mating types.

A similar situation has been discovered in *Didymium iridis*. Collins (1963), working with an isolate from Honduras, showed that it had two alleles at a single locus. Later, Collins & Ling (1964), working with an isolate from Costa Rica, showed that it had two further alleles at the same locus. Myxamoebae from either Costa Rican strain would develop plasmodia with *both* of the Honduran mating types. A third isolate from Panama contained yet another pair of alleles at the same locus. Since a single locus is involved, Collins designated the six alleles as A^1, A^2, A^3, A^4, A^5, A^6. It was found that, although most single-spore isolates failed to form plasmodia (i.e. behaved as heterothallic isolates), occasionally single-spore cultures gave rise to plasmodia (i.e. behaved as homothallic isolates). Collins (1965) explained one such case which arose from a single spore derived from a plasmodium containing the A^3 and A^4 alleles as due to mutation amongst the myxamoebae derived from an A^4 spore. He designated this mutant as A^{4m-1}. Alternative explanations in terms of contamination, or recombination within the A locus, although theoretically possible, were thought unlikely.

Multiple allelomorph incompatibility is not confined to Myxomycetes. It has been much more extensively studied in Basidiomycotina, notably amongst the Hymenomycetes (see p. 410). It can be interpreted as an outbreeding mechanism where 50% of inbreeding is successful, and, provided that the number of alleles at the locus for incompatibility is sufficiently large, outbreeding can be almost 100% successful.

PLASMODIAL INCOMPATIBILITY

When separate plasmodia of a given species of Myxomycete meet, two reactions are possible:

(*a*) *Compatible reaction:* the two plasmodia fuse, and their veins coalesce to form a common mass of protoplasm (Fig. 12B).

(*b*) *Incompatible reaction:* the plasmodia fail to fuse and may move apart from each other. Alternatively, fusion may begin, but is followed by destruction and death of parts of the fusing plasmodia (Fig. 12C), the *lethal reaction* (Carlile, 1972).

Genetical studies with *Physarum polycephalum* (Carlile & Dee, 1967; Poulter & Dee, 1968; Collins & Haskins, 1970; Wheals, 1970) and with *Didymium iridis* (Collins, 1966, 1963; Collins & Clark, 1968; Collins & Ling, 1972; Clark & Collins, 1973) show that for fusion to occur, there must be close genetic similarity between the fusing plasmodia. Fusion is under genetical control. In *P. polycephalum*, it has been claimed that there are two unlinked loci controlling fusion, with at least four alleles at one locus, and two at the other. Several loci appear to be involved in *D. iridis*, but in this case there is no suggestion of multiple alleles.

The type of incompatibility brought about by the interaction of genetically distinct plasmodia is an example of a widespread phenomenon termed *hetero-*

genic somatic incompatibility, found not only in slime moulds, but in fungi, vertebrates and many organisms. In humans, this phenomenon is well known in relation to blood-grouping, or to the failure of tissue transplantation. Slime moulds may prove to be ideal material for studying some of the biochemical reactions involved.

The differing behaviour of myxamoebae and plasmodia of heterothallic slime moulds presents a paradox (Carlile, 1973). In sexual reproduction, fusion between genetically identical myxamoebae is discouraged and that of unlike myxamoebae promoted by the occurrence of mating types. Vegetative fusion between genetically distinct plasmodia is discouraged by somatic heterogenic incompatibility. The advantages of encouraging sexual fusion between genetically unlike cells are easily explained in terms of enhanced outbreeding and genetical recombination. It is less obvious why vegetative fusion of genetically dissimilar cells is forbidden. Possible reasons are:

(*a*) The presence of genetically distinct nuclei in a common cytoplasm has harmful consequences.

(*b*) It may be desirable to discourage relatively inefficient systems of parasexual recombination in organisms where a sexual system occurs.

(*c*) The rejection of cytoplasm which is genetically dissimilar may be important as a defence mechanism against pathogenic organisms.

CLASSIFICATION

The detailed classification of Myxomycetes is beyond the scope of this book. Alexopoulos (1973) divides the Myxogastromycetidae into four orders: Physarales, Liceales, Echinosteliales and Trichiales. The Stemonitomycetidae contain a single order, the Stemonitales.

4: PLASMODIOPHOROMYCETES

PLASMODIOPHORALES

The Plasmodiophorales are obligate (i.e. biotrophic) parasites. The best-known examples attack higher plants, causing economically important diseases such as club-root of brassicas *(Plasmodiophora brassicae)*, powdery scab of potato *(Spongospora subterranea)* and crook-root disease of watercress *(S. subterranea* f. sp. *nasturtii)*. Others attack roots and shoots of non-cultivated plants, especially aquatic plants. Algae and fungi are also attacked. About nine genera are recognised, and they are separated from each other largely on the way in which the resting spores are arranged in the host cell. Accounts of the group have been given by Karling (1968), Sparrow (1960) and Waterhouse (1973a).

The zoospore in the Plasmodiophorales is biflagellate. The flagella are of unequal length and there is clear evidence from electron micrographs that the

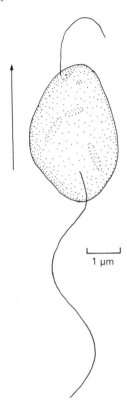

Fig. 18. Zoospore of *Plasmodiophora brassicae* (based on Aist & Williams, 1971). The zoospore has a short anterior flagellum and a long posterior flagellum. Both flagella are of the whiplash type. The arrow shows the direction of movement.

two flagella are both of the whip type (Fig. 18) (Kole & Gielink, 1961; Keskin, 1964). Zoospores of this type are said to be **anisokont**.

The wall of the resting spore in *P. brassicae* is believed to contain chitin and, although cellulose has occasionally been reported in some species, it has been stated not to occur in *Woronina polycystis* and *Octomyxa brevilegniae*, both parasitic on Saprolegniaceae (Pendergrass, 1950; Goldie-Smith, 1954).

The details of the life cycle of many members of the Plasmodiophorales are still uncertain. In most genera, there are two distinct plasmodial phases. The first, usually resulting from infection by a zoospore derived from a resting spore, gives rise to thin-walled zoosporangia. The second, which in its early development may be indistinguishable from the zoosporangial plasmodium, gives rise to resting spores. Although there have been suggestions that the development of the resting spore plasmodium may be preceded by a sexual fusion in some species, this cannot be regarded as proven nor may it be true of all species. In some cases, the plasmodium giving rise to resting spores appears

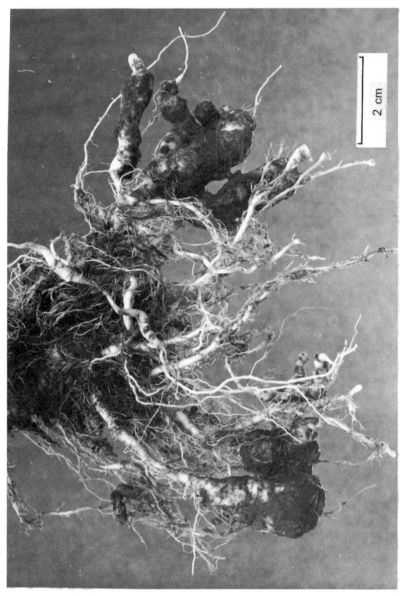

Fig. 19. Club-root of cabbage caused by *Plasmodiophora brassicae*.

to arise merely at a later stage of infection of the host than the zoosporangial phase.

Plasmodiophora

P. brassicae is the causal organism of club-root or finger-and-toe disease of brassicas (Fig. 19). The disease is common in gardens and allotments where cabbages are frequently grown, especially if the soil is acid and poorly drained. A wide range of cruciferous hosts is attacked, and root-hair infection of some non-cruciferous hosts can also occur. The disease is widely distributed throughout the world (Colhoun, 1958).

Infected crucifers usually have much-swollen roots. Both tap roots and lateral roots may be affected. Occasionally infection results in the formation of adventitious root buds which give rise to swollen stunted shoots. Above ground, however, infected plants may be difficult to distinguish from healthy ones. The first symptom is wilting of the leaves in warm weather, although often such wilted leaves recover at night. Later the rate of growth of infected plants is retarded so that they appear yellow and stunted. When plants are attacked in the seedling stage, they can be killed, but if infection is delayed, the effect is much less severe and well-developed heads of cabbage, cauliflower, etc. can form on plants with quite extensive root hypertrophy. Commonly, infected root hairs are hypertrophied, expanding at their tips to form club-shaped swellings which are sometimes lobed and branched (Fig. 20). Macfarlane & Last (1959) followed the growth of cabbage seedlings either uninfected or infected with *Plasmodiophora* at various times after germination. Within 35 days of infection of seedlings, significant retardation of the weight increase of the tops occurs as compared with healthy controls. In infected plants, the root/shoot ratio is appreciably higher, suggesting a diversion of materials to the clubbed roots. Swollen roots contain large numbers of spherical small resting spores, and when these roots decay, the spores are released into the soil. Electron micrographs show that the resting spores have spiny walls (Williams & McNabola, 1967). The resting spore germinates to produce a single zoospore with two flagella of unequal length, both of the whiplash type and with the usual $9+2$ arrangement of fibrils (see Fig. 19; Kole & Gielink, 1962; Aist & Williams, 1971). Germination is stimulated by substances diffusing from the cabbage roots (Macfarlane, 1952, 1959, 1970).

The primary zoospores (i.e. the first motile stage released from the resting spore) swim by means of their flagella, the long flagellum trailing and the short flagellum pointing forwards. The process of root-hair infection has been followed in an elegant study by Aist & Williams (1971). Their study parallels a similar one by Keskin & Fuchs (1969) on penetration by *Polymyxa betae,* a parasite of sugar beet. In this study, the authors used German terms to describe the organelles involved in the penetration process, but in the account

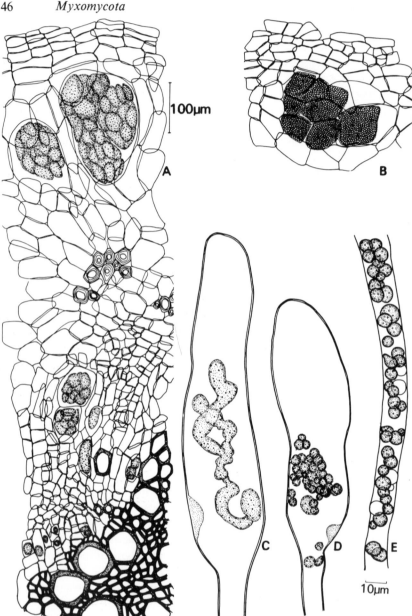

Fig. 20. *Plasmodiophora brassicae*. A. Transverse section through young infected cabbage root showing plasmodia in the cortex. Note the hypertrophy of some of the host cells containing plasmodia, and the presence of young plasmodia in cells immediately outside the xylem. B. Transverse section of a cabbage root at a later stage of infection showing the formation of resting spores. C. Zoosporangial plasmodium in cabbage root hair 4 days after planting in a heavily contaminated soil. D. Young zoosporangia in root hair. Note the club-shaped swelling of the infected root hair. E. Mature and discharged zoosporangia. A and B to same scale. C, D and E to same scale.

which follows, the English equivalents of these terms are used with the German terms in parentheses.

Primary zoospores of *P. brassicae* are released some 26–30 hours after placing a suspension of resting spores close to seedling roots of cabbage. The zoospores may collide several times with a root hair before becoming attached, and appear to be attached at a point opposite to the origin of the flagella.

The flagella coil around the zoospore body which becomes flattened against the host wall, and pseudopodium-like extensions of the zoospore develop, being continuously extended and withdrawn. The flagella are then withdrawn, and the zoospore encysts, attached to the root hair (Fig. 21D). A vacuole and lipid bodies develop within the zoospore cyst (Fig. 22). The most conspicuous unusual feature is a long **tube** (Rohr), with its outer end orientated toward the host wall (i.e. the root-hair wall). This end of the tube is occluded by a plug. Within the tube is a bullet-shaped **stylet** (Stachel), the outer part of which is made up of parallel fibrils. Behind the blunt end of the stylet, the tube narrows to form a **sac** (Schlauch).

Penetration of the root-hair wall takes place about 3 hours after encystment, and at about this time empty vacuolated cysts are observed. Penetration takes place rapidly. An interpretation of the penetration process is shown in Fig. 22. Firm attachment of the tube to the root hair is believed to be brought about by the **adhesorium**, which may develop by evagination (i.e. turning inside out) of the tube (Fig. 22B). The enlargement of the vacuole is presumably the driving force which brings about complete evagination of the tube,

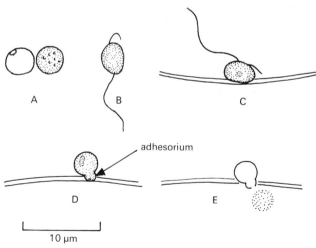

Fig. 21. *Plasmodiophora brassicae*. A. Resting spores, one full, one empty (showing a pore in the wall). B. Zoospore. C. Attachment of zoospore to root hair. D. Zoospore cyst with adhesorium following withdrawal of flagellar axonemes. E. Entry of amoeba into root hair. Based on Aist & Williams (1971).

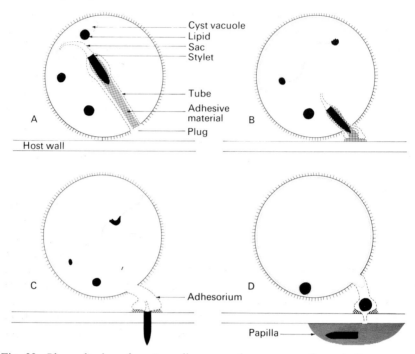

Fig. 22. *Plasmodiophora brassicae*: diagrammatic summary of penetration process (after Aist & Williams, 1971). The diagrams show a zoospore cyst attached to the wall of a root hair. A. Cyst vacuole not yet enlarged. B. Cyst vacuole enlarges and small adhesorium appears. C. Stylet punctures host cell wall. D. Penetration has occurred and the host protoplast has deposited a papilla at the penetration site.

thrusting the stylet through the host wall. The parasite is injected into the host cell as a small spherical amoeba which becomes caught up by cytoplasmic streaming. After penetration (Fig. 22D), a papilla of callose develops beneath the adhesorium, around the penetration point, possibly as a wound-healing response.

Within the infected root hair, the amoeba divides to give rise to several uninucleate amoebae. Later the nuclei divide and small multinucleate plasmodia are formed. These plasmodia are termed sporangial plasmodia or primary plasmodia. The plasmodia divide up to form a variable number of roughly spherical thin-walled zoosporangia lying packed together in the host cell (Fig. 20) possibly by coalescence of separate protoplasts. The production of zoosporangia may take place within 4 days. Each zoosporangium finally contains four to eight uninucleate zoospores. The mature zoosporangium becomes attached to the host cell wall and a pore develops at this point through which the zoospores escape. Occasionally zoospores are released into the lumen of the host cell. The behaviour of the released zoospores is not completely known, but it is possible that they function as gametes and fuse in pairs.

Quadriflagellate, binucleate swarmers have indeed been observed (Kole & Gielink, 1961), but swarmers with six flagella have also been seen, and whether such swarmers result from fusion of separate biflagellate swarmers or from incomplete separation of zoospore initials has not been established.

There have been differing interpretations of the subsequent part of the life cycle. Tommerup & Ingram (1971) have studied the behaviour of the parasite in callus tissue culture in *Brassica napus* and have presented an interpretation of the life cycle as shown in Fig. 23. It is possible that the behaviour of the parasite in tissue culture does not agree exactly with what happens in roots in soil, but Tommerup & Ingram carried out parallel studies on clubbed roots in soil and were able to confirm their tissue culture observations.

The callus tissue cultures were initiated from surface-sterilised clubbed root tissue containing resting spores. When infected callus tissue was transferred to fresh medium, the resting spores (which would normally have been released into the soil) germinated *in situ* to release uninucleate, unequally biflagellate zoospores. Numerous parasite protoplasts were released into each infected host cell, and these coalesced to form multinucleate primary plasmodia. These plasmodia later cleaved to form zoosporangia which are presumed to be homologous with those normally developed in root hairs. Plasmogamy between zoospores was observed, but this was not followed immediately by karyogamy. Similar behaviour was noted by Kole & Gielink (1963) by zoo-

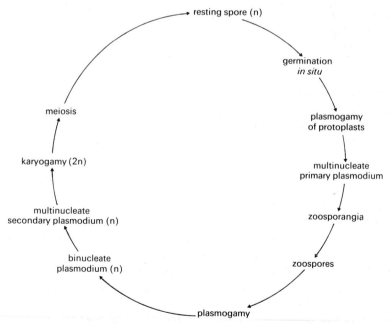

Fig. 23. Life cycle of *Plasmodiophora brassicae* in callus tissue culture (after Tommerup & Ingram, 1971).

spores released from cabbage root hairs. The binucleate zoospores are believed to re-infect the root and initiate the formation of binucleate plasmodia which penetrate the root cortex. Whether active penetration of cell walls occurs or whether the plasmodia are transferred passively from cell to cell during cell division is not certain. The plasmodium has no specialised feeding structures such as haustoria. It is immersed in the host cytoplasm surrounded by a thin plasmodial envelope. There is no evidence for phagocytic inclusion of host cell organelles within the plasmodium (Williams & Yukawa, 1967). The plasmodium enlarges, repeated nuclear divisions take place, and the cells containing them become hypertrophied (Fig. 20), although the host nucleus remains active. Hypertrophy of host cells is apparently brought about by blocking of the mechanism for cell division, and is accompanied by enhanced DNA synthesis. There is evidence that the presence of plasmodia in host cells is associated with increased ploidy of host nuclei, at least in callus cultures (Tommerup & Ingram, 1971). The auxin (IAA) content of clubbed cabbage roots is 50–100 times greater than that of uninfected roots (Raa, 1971), whilst the cytokinin content of clubbed turnip roots is 10–100 times greater than in healthy roots (Dekhuizen & Overeem, 1971). The tissues of healthy crucifers contain relatively large amounts of indole glucosinolates such as glucobrassicin and neoglucobrassicin. They also contain the enzyme glucosinolase which hydrolyses glucobrassicin to 3-indoleacetonitrile (IAN). A second enzyme, nitrilase, can convert IAN to 3-indoleacetic acid (IAA). Butcher *et al.* (1974) have presented evidence suggesting that the growth of *Plasmodiophora* interferes with the host's metabolism, leading to the release of relatively large amounts of IAN from indole glucosinolates present in the tissues, and that these increased levels are responsible for hypertrophy. Starch accumulates in infected cells (Williams, 1966). At first, only cortical cells of the young root are infected, but later small plasmodia can be found in the medullary ray cells and in the vascular cambium. Subsequently, tissues derived from the cambium are infected as they are formed. In large swollen roots, extensive wedge-shaped masses of hypertrophied medullary ray tissue may cause the xylem to be split. In this stage, the root tissue shows a distinctly mottled appearance. When the growth of the plasmodia is complete, they are transformed into masses of resting spores. According to Tommerup & Ingram (1971), the multinucleate mature plasmodia contain many haploid nuclei associated in pairs. Prior to resting spore formation, nuclear fusions occur to give diploid zygote nuclei, and meiosis is presumed to follow quickly, so that when resting spores are cleaved from the plasmodial cytoplasm they contain haploid nuclei. Only during late stages of resting spore development do the host nuclei begin to degenerate. The resting spores are at first naked, but later become surrounded by a thin cell wall. They are closely packed together inside the host cell and are released into the soil as the root tissues decay.

This account of the possible life cycle of *P. brassicae* differs from that suggested by Heim (1955). She has described young plasmodia in the cortical and medullary parenchyma, forming pseudopodia which actively penetrate from an infected cell to neighbouring cells. In such plasmodia most of the nuclei become associated in pairs and then fuse to form diploid nuclei around which the cytoplasm condenses. Following nuclear fusion the plasmodia lose their ability to invade cells. Their contours become regular and they no longer put out pseudopodia. The diploid nuclei undergo a meiosis followed by a mitosis; then the haploid nuclei and adjacent cytoplasm become surrounded by cell walls to form spores. According to Heim, the spores can germinate within the infected root. The resting spores lose their spherical shape, elongate and acquire the appearance of young plasmodia, putting out pseudopodia into uninfected neighbouring cells. Fusion of amoeboid parasite cells occurs within the host before the vegetative plasmodium develops. Heim did not study the process of infection following resting spore germination in the soil and she ascribes no role to the zoosporangia.

Spongospora

The life cycle of *S. subterranea,* the cause of powdery scab of potato, is possibly similar to that of *P. brassicae* (see Piard-Douchez, 1949; Kole, 1954; Kole & Gielink, 1963). Diseased tubers show powdery pustules at their surface, containing masses of resting spores clumped into hollow balls. The resting spores release unequally biflagellate zoospores which can infect root hairs of potatoes or tomatoes. In the root hairs plasmodia form which develop into zoosporangia. Zoospores from such zoosporangia are capable of infection resulting in a further crop of zoosporangia. Zoospores released from the

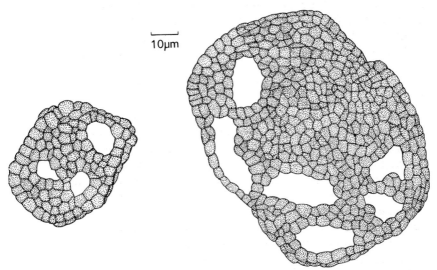

10μm

Fig. 24. *Spongospora subterranea* f. sp. *nasturtii*. Spore balls from watercress roots with crookroot disease.

zoosporangia have also been observed to fuse in pairs or occasionally in groups of three to form quadri- or hexaflagellate swarmers (Kole, 1954) but whether this represents a true sexual fusion is uncertain. *S. subterranea* f. sp. *nasturtii* causes a disease of watercress in which the most obvious symptom is a coiling or bending of the roots. Zoosporangia and spore balls are found in infected root cells (Fig. 24; Tomlinson, 1958b). Heim (1960) has described a life cycle for this organism which resembles that which she has proposed for *P. brassicae* involving the fusion of myxamoebae within the infected host cells. Nuclear fusion is believed to initiate a short diploid phase which is ended by meiosis and a mitosis preceding spore formation.

In addition to being the causal agent of powdery scab of potatoes, *S. subterranea* is also important as the vector of potato-mop-top virus disease, which can result in a decreased yield of tubers of over 20% in some potato varieties (Jones & Harrison, 1969, 1972). The virus is transmitted by the zoospores and can also persist for several years in the spore balls in soil. The fungus can infect and form zoosporangia in the root hairs of a wide range of host plants outside the family Solanaceae, and can transmit virus to them. Thus the fungus and numerous wild plants can provide a reservoir of infection even when potatoes have not been grown in a field for many years. Transmission of wheat mosaic virus has also been reported in another Plasmodiophoromycete, *Polymyxa graminis*.

CONTROL OF CLUB-ROOT, POWDERY SCAB AND CROOK-ROOT

The control of club-root disease is difficult. Because resting spores remain viable in soil for many years, short-term crop rotation may not eradicate the disease. The fact that *Plasmodiophora* can infect cruciferous weeds such as Shepherd's purse *(Capsella bursa-pastoris)* suggests that the disease may be carried over on such hosts and that weed control may be important. Moreover, it is known that root-hair infection can also occur on non-cruciferous hosts such as *Agrostis, Dactylis, Holcus, Lolium, Papaver* and *Rumex,* all of which are common garden or field weeds. Whether such infections play any part in maintaining the disease in the absence of a cruciferous host is not known. General measures aimed at mitigating the incidence of the disease include improved drainage and the application of lime. Lime probably inhibits the germination of the resting spores, but if the spore load in the soil is sufficiently high, lime will not prevent an outbreak of the disease. Since the effect of liming on the soil does not persist it is possible that it may delay the germination of the resting spores and thus prolong their persistence in the soil (Macfarlane, 1952). Early infection of seedlings can result in severe symptoms, so it is important to raise seedlings in non-infected or steam-sterilised soil. The young plants can then be transplanted to infested soil. Infection can be minimised or delayed by the application of mercurous chloride (calomel) or mercuric chloride (corrosive sublimate) to the soil at the time of transplantation. Various other fungicides such as benomyl have been used in control of club-root (Jacobsen & Williams, 1970). Systemic fungicides (i.e. substances taken up by the plant and translocated) have also been used successfully (van Assche & Vanachter, 1970). Since it is known that some resting spores survive animal digestion, manure from animals fed with diseased material should not be used for growing brassicas. Certain kinds of brassicas seem to have a natural resistance to the disease, e.g. some swede varieties, and plant breeders have also bred and selected resistant strains of other kinds of brassicas. The nature of the resistance is not known, but root-hair infection of resistant and susceptible varieties occurs with equal facility, so it is possible that it is during later stages of the cycle of infection that host resistance becomes effective. There are several physiological races of *P. brassicae* which vary in their ability to infect different kinds of brassicas, and this complicates the problem of breeding resistant varieties. Varieties bred for club-root resistance in one country may be completely susceptible to strains of the pathogen derived from other countries. It is clearly desirable to test selected brassica varieties against material of *P. brassicae* obtained from many different sources. It is probable that resistance to club-root is a polygenic character (Watson & Baker, 1969; Chiang & Crête, 1970; Johnston, 1970; Lammerink, 1970; Vriesinga & Honma, 1971). Buczacki *et al.* (1975) have shown how fifteen differential host varieties (five each of *Brassica campestris, B. napus* and *B. oleracea*) can be used to discriminate between different physiologic races of the pathogen, and, using these hosts, have identified thirty-four physiologic races in Europe.

Powdery scab of potatoes is normally of relatively slight economic importance, and amelioration of the incidence of the disease can be brought about by good drainage. Crook-root of watercress can be controlled by application of zinc to the water supply. The zinc can be applied by dripping zinc sulphate into the water supply to the watercress beds to give a final concentration of about 0·5 parts per million, or by the use of finely powdered glass containing zinc oxide (zinc frit) to the watercress beds. The slow release of zinc from the frit maintains a sufficiently high concentration to inhibit infection (Tomlinson, 1958a).

The affinities of the Plasmodiophorales are not clear. They resemble some other Myxomycota in having a plasmodium and zoospores with flagella of unequal length. They are sometimes classified with zoosporic fungi (Mastigomycotina), although not clearly related to any other group (Waterhouse, 1973a). Sparrow (1958) has proposed that they be regarded as a separate group, the Plasmodiophoromycetes. Possibly they are more closely related to Protozoa than to other fungi (Karling, 1968).

PART TWO

Eumycota

Introduction

Most fungi belong to this group. Their thallus organisation differs in that (with a very few exceptions) they do not possess plasmodia or pseudoplasmodia, but are unicellular or filamentous. It is probable that they have diverse evolutionary origins and relationships, i.e. they are polyphyletic. Nevertheless, they have certain common features of organisation, nutrition, physiology and reproduction.

THALLUS ORGANISATION

The thallus of a fungus is generally based on a uniseriate branched filament. In most fungi, the thallus is differentiated into a vegetative part which absorbs nutrients, and a reproductive part. Such thalli are described as **eucarpic**. However, in some, the thallus does not show this differentiation, and following a phase of vegetative growth, the entire thallus becomes converted into propagules. Thalli of this kind are termed **holocarpic**, and are found in certain parasitic fungi in which the thallus lives inside the host cell (e.g. *Synchytrium*, *Olpidium*), and in some unicellular free-floating organisms which can obtain nutrients from the surrounding medium. The unicellular type of thallus is typical of yeasts and yeast-like fungi (for example, *Saccharomyces* and *Sporobolomyces*). Some fungi, especially animal pathogens, can exist either in the filamentous or unicellular (i.e. yeast-like) phase, and this phenomenon is known as **dimorphism**. Change from the filamentous to the unicellular form can be brought about by manipulating the environmental conditions – for example, the composition of the medium and the CO_2 concentration (Romano, 1966). The vegetative filaments of fungi are termed **hyphae**, and the collective term for a thallus made up of hyphae is the **mycelium**.

An important distinguishing feature between groups of fungi is the presence or absence of cross-walls or **septa** in the hyphae. Certain groups of fungi, notably the Oomycetes and Zygomycetes, generally have non-septate hyphae, whilst Ascomycetes, Basidiomycetes and Deuteromycetes have septate hyphae. In non-septate forms, the mycelium contains numerous nuclei which are, of course, not separated from each other by cross walls, but lie in a common mass of cytoplasm. Such a condition is described as **coenocytic**. In septate forms, the hyphal segments may contain one, two or more nuclei. When the nuclei are genetically identical, as in a mycelium derived from a single uninucleate spore, the mycelium is said to be **homokaryotic**, but where a cell or mycelium contains nuclei of different genotype, possibly as a result of mutation, or anastomosis of hyphae, it is said to be **heterokaryotic**. A special condition is found in the mycelia of certain Basidiomycetes where each cell may contain two genetically distinct haploid nuclei. This condition is **dikaryotic**, to distinguish it from mycelia which are **monokaryotic**, with segments containing single, haploid, genetically identical nuclei.

WALL STRUCTURE

The filament owes its form to the turgor of the protoplast it contains and to the rigidity of its wall. The changing form of the fungal thallus as it undergoes development is caused by changes in shape and proportion of fungal cells, which are closely linked with changes in the wall structure. Thus a clear understanding of cell wall structure is important in understanding morphogenetic problems. There are also correlations between the chemical structure of fungal cell walls and taxonomy. Chemical analyses of cell walls which have been cleaned free of cytoplasm by physical and chemical techniques show that they contain 80–90% polysaccharides, with most of the remainder consisting of protein and lipid (Aronson, 1965; Bartnicki-Garcia, 1968, 1970). The 'skeletal' material of the wall consists, in most forms, of spirally orientated fibrils of chitin (a polymer of N-acetyl-glucosamine), cellulose (a polymer of D-glucose), or sometimes other glucans (see Fig. 25). Chitin is the most usual component, but cellulose is present in Oomycete cell walls along with glucan. Occasionally, chitin and cellulose are found together, for example in *Rhizidiomyces,* a member of the Hyphochytridiomycetes, and in *Ceratocystis,* an Ascomycete. Various other substances have been found associated together in

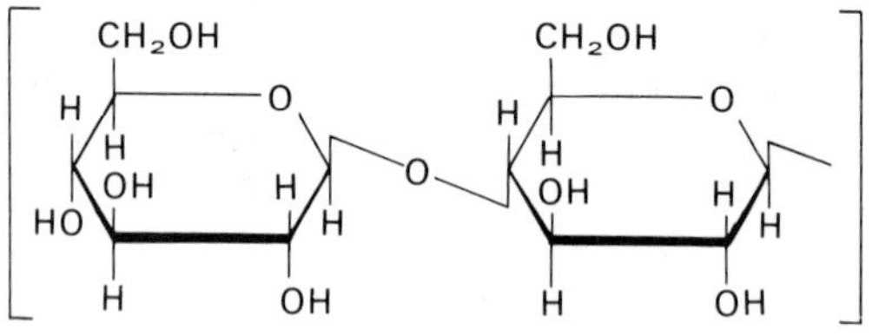

Cellulose

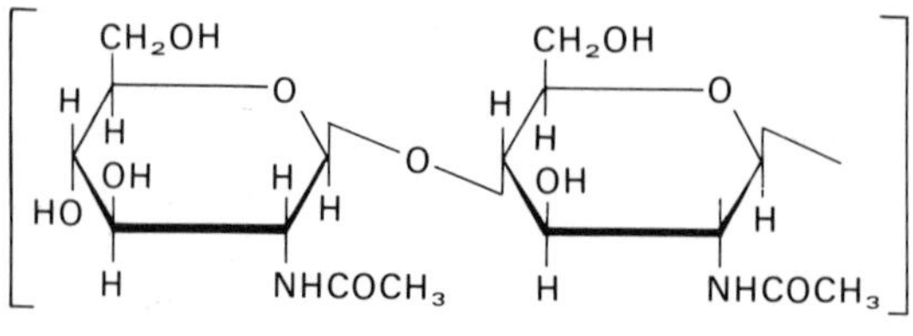

Chitin

Fig. 25. Structural formulae for the repeating units making up cellulose and chitin.

Table 1. *Cell wall taxonomy of fungi*

Chemical category	Taxonomic group	Distinctive features
I. Cellulose–glycogen	Acrasiales	pseudoplasmodia
II. Cellulose–glucan	Oomycetes	biflagellate zoospores
III. Cellulose–chitin	Hyphochytridiomycetes	anteriorly uniflagellate zoospores
IV. Chitosan–chitin	Zygomycetes	zygospores
V. Chitin–glucan	Chytridiomycetes	posteriorly uniflagellate zoospores
	Ascomycetes	septate hyphae, ascospores
	Basidiomycetes	septate hyphae, basidiospores
	Deuteromycetes	septate hyphae
VI. Mannan–glucan	Saccharomycetaceae	yeast cells, ascospores
	Cryptococcaceae	yeast cells
VII. Mannan–chitin	Sporobolomycetaceae	yeasts (carotenoid pigment) ballistospores
	Rhodotorulaceae	yeasts (carotenoid pigment)
VIII. Polygalactosamine–galactan	Trichomycetes	heterogeneous group, arthropod parasites

cell walls, and the association of cell-wall components appears to be correlated with their taxonomy, as shown in table 1 (simplified from Bartnicki-Garcia, 1968). A characteristic feature of Oomycete cell walls containing cellulose is the presence of the amino-acid hydroxyproline (Novaes-Ledieu *et al.*, 1967).

The microfibrillar components are embedded in a matrix of other substances. Protein is an important component. Some of this may represent enzymes closely bound to wall components. The presence of enzymes as an integral part of the wall explains why wall fragments are not inert, but demonstrate biochemical activity.

The walls of yeasts have distinctive chemical properties, being composed of a mannan–glucan complex (Phaff, 1963).

FINE STRUCTURE

The cells of Eumycota (Eumycetes) are eukaryotic and, apart from chloroplasts, contain many of the familiar organelles characteristic of eukaryotes (Moore, 1965; Bracker, 1967). The nucleus is surrounded by a double membrane which is continuous with the endoplasmic reticulum (ER). In multinucleate hyphae, the nuclei may be interconnected by ER. There are numerous pores in the nuclear envelope which may allow for interchange of materials between nucleus and cytoplasm. When mitotic nuclear division occurs, the nuclear membrane does not always break down as in most other organisms, but may constrict in the middle to separate two sister nuclei. This process has been termed by Moore (1965) **karyochorisis**. Mitochondria are of diverse shape, but are often elongate. They are sufficiently large in many fungi to be seen by light microscopy, and move about rapidly within the fungal protoplast. The endoplasmic reticulum may be smooth, or rough, i.e. with ribosomes, concerned with protein synthesis, attached to the surface. Cytoplasmic microtubules have been widely reported in fungal cells, and may be concerned with protoplasmic movement and the maintenance of cell shape. In many non-fungal eukaryotic cells, a Golgi apparatus, consisting of stacks of folded membranes or dictyosomes functioning in secretion, have been reported, but they are of comparatively rare occurrence in fungal cells. They have, however, been reported in Oomycetes such as *Pythium* (Grove *et al.,* 1970). Other characteristic cytoplasmic inclusions are lipid droplets, and glycogen, a typical fungal storage product. Lipid and glycogen are especially abundant in mature cells, food storage structures and spores. The surface of the fungal protoplast is the plasmalemma, a typical unit membrane consisting of lipoprotein. Where vacuoles are present, they are surrounded by a similar membrane, the tonoplast.

Septa in fungal cells are of three main kinds. Where septa delimit reproductive structures, they are complete, i.e. do not contain a pore. Septa of this type are rare in vegetative hyphae. In Ascomycetes and Deuteromycetes, the septum is a simple transverse plate lying at right angles to the axis of the hypha. The septum is usually perforated, and movement of cytoplasmic organelles such as mitochondria and nuclei can occur freely through the pores (Fig. 122). In some groups of Basidiomycetes (excluding the rust and smut fungi), the septum is more complex. Surrounding the central pore in the septum is a curved flange of wall material which is often thickened to form a barrel-shaped or cylindrical structure surrounding the pore. Septa of this type are termed **dolipore** septa. These septa are often overlaid by perforated endoplasmic reticulum (Fig. 225).

HYPHAL GROWTH

Growth of hyphae is, in most cases, apical (Burnett, 1976, Chapter 3; Bartnicki-Garcia, 1973), and as hyphae extend, there must obviously be rapid synthesis of new wall components, and membrane components to provide for the increase in the plasmalemma. The driving force which thrusts the growing apex forward is the turgor of the protoplast, possibly generated by increasing vacuolation of the hyphae distal to the apex. The fine structural changes which accompany growth of hyphae have been extensively studied, and there are minor variations in different groups of fungi. In *Pythium ultimum*, an Oomycete with cellulose instead of chitin as a principal wall component, Grove *et al.* (1970) have interpreted the deposition of wall and membrane material as shown in Figs. 26 and 27. The dictyosomes (D), which are possibly the equivalent of the Golgi apparatus of other organisms, are differentiated into a proximal pole, Dp, proximal to the nucleus or endoplasmic reticulum, and a

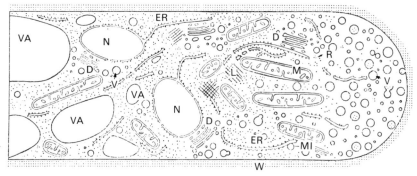

Fig. 26. Diagrammatic interpretation of hyphal apex in *Pythium ultimum* (from Grove *et al.*, 1970). D, dictyosome; ER, endoplasmic reticulum; L, lipid body; M, mitochondrion; MI, microbody; N, nucleus; R, ribosome; V, cytoplasmic vesicle; VA, vacuole; W, wall.

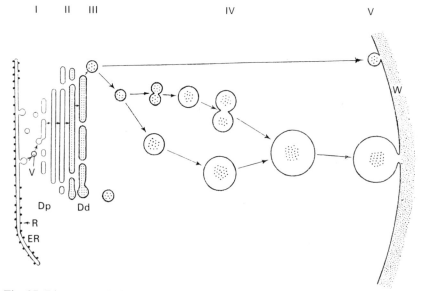

Fig. 27. Diagrammatic interpretation of hyphal extension in *Pythium ultimum* (from Grove *et al.*, 1970). I. Material is transferred from ER to dictyosome by blebbing of ER and refusion of vesicles to form a cisterna at the proximal pole of the dictyosome (Dp). II. Cisternal contents and membranes are transformed as the cisterna is displaced to the distal pole (Dd) by the continued formation of new cisternae. III. Cisternae vesiculate to form secretory vesicles as they approach and reach the distal pole. IV. Secretory vesicles migrate to the hyphal apex. Some may increase in size or fuse with other vesicles to form large secretory vesicles, while others are carried directly to the cell surface. V. Vesicles accumulate in the apex and fuse with the plasma membrane, liberating their contents into the wall region. C, centriole; D, dictyosome; Dd, distal pole of dictyosome; Dp, proximal pole of dictyosome; ER, endoplasmic reticulum; R, ribosome; V, cytoplasmic vesicle; W, wall.

E R

distal pole, Dd, which faces away from the nucleus towards the plasmalemma. The dictyosomes are in a state of dynamic equilibrium, receiving membrane material derived from blebs of the endoplasmic reticulum on the proximal face, and giving rise to secretory vesicles on the distal face. The secretory vesicles migrate towards the hyphal apex directly, or may coalesce to form larger vesicles. Ribosomes are scarce in the vicinity of the apex. At the hyphal apex, the vesicles fuse with the plasma membrane and release their contents into the wall region. It is presumed that the secretory vesicles contain polysaccharides needed in wall construction, and possibly material for the plasma membrane and enzymes (Grove & Bracker, 1970, 1978). Various workers have shown that the incorporation of new wall material is confined to the very tips of growing hyphae (Gooday, 1971; Gull & Trinci, 1974). Grove *et al.* have made calculations of the numbers of vesicles needed to provide for extension growth of a hypha 5 µm in diameter, growing at the rate of 1 mm/hour. They estimate that about 10,000 vesicles are required to support 1 minute of growth.

There are certain differences in the arrangement of vesicles in different fungi (Fig. 28). Dictyosomes have not been reported in all taxonomic groups of fungi. In Zygomycetes, large secretory vesicles form a crescent-shaped accumulation immediately behind the apex. Marchant *et al.* (1967) claim that in *Phycomyces,* a Zygomycete, the vesicles are compound (multivesicular bodies) and believe that this is a feature of fungi with chitinous walls. Cell disruption and fractionation has enabled the isolation of a specialised organelle, the **chitosome**, which, when incubated *in vitro* with the precursor of

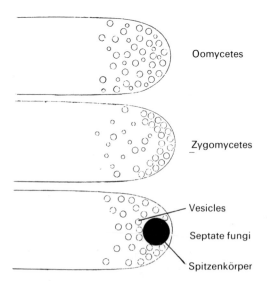

Oomycetes

Zygomycetes

Vesicles

Septate fungi

Spitzenkörper

Fig. 28. Diagrammatic comparisons of the principal forms of apical organisation in hyphae, based on representatives from major taxonomic groups (after Grove & Bracker, 1970).

chitin, uridine diphosphate *N*-acetyl-D-glucosamine, produced chitin microfibrils. It is believed that chitosomes are the cytoplasmic containers of chitin synthetase zymogen which convey it to its destination at the cell surface. In shape, chitosomes are spheroidal, and predominantly of two kinds – proctoid (i.e. resembling an anus) or cycloid. They measure 40–70 nm in diameter (Bracker *et al.*, 1976; Barnicki-Garcia *et al.*, 1978). In septate fungi, using the light microscope, a dark apical body (**Spitzenkörper**) has been reported. The interpretation of this structure is at present uncertain. Grove & Bracker (1970) have shown that it corresponds to a regional differentiation of the apical zone, intimately associated with apical vesicles and with hyphal tip growth. The Spitzenkörper disappears when growth ceases, and reappears when growth is resumed. The forces which cause the secretory vesicles to move forwards are at present unknown. Since vesicles lack the power of self-propulsion, the possibility of electrophoresis has been put forward. It has been claimed that there is a sufficient potential difference between the apex and subapical regions of *Neurospora* hyphae to generate an electric current sufficiently strong to account for electrophoretic movement of the vesicles (Slayman & Slayman, 1962; Jaffe, 1968). Bartnicki-Garcia (1973) has speculated that a possible function of the Spitzenkörper might be to generate electric potential, but it should be stressed that there is, at present, no evidence for this.

At the surface of the plasmalemma, in the region where fusion of the cytoplasmic secretory vesicles is taking place, specialised organelles have been reported, such as **plasmalemmasomes** and **lomasomes**. Heath & Greenwood (1970a) have defined a plasmalemmasome as 'the various membrane configurations which are external to the plasmalemma, often in a pocket projecting into the cytoplasm, and less obviously embedded in wall material', whilst a lomasome has been defined as 'membranous vesicular material embedded in the wall external to the line of the plasmalemma'. However, there is a gradation between these two kinds of structure. They are not peculiar to fungi, but occur in other kinds of walled cells (Marchant & Robards, 1968). Heath & Greenwood (1970a) have suggested that the plasmalemmasomes are produced when the balance between wall plasticity and turgor pressure is disturbed in such a way that more plasmalemma is produced than is needed to line the cell wall. When plasmalemmasomes are extruded (sequestered) in the developing wall, they become lomasomes.

DIFFERENTIATION BEHIND THE HYPHAL APEX

As we have seen, hyphal growth takes place at the apex. With increasing distance behind the apex, differentiation takes place. Visually, this is obvious by the formation of septa in septate fungi, and of vacuoles. Zalokar (1959) has shown, by means of histochemical staining in *Neurospora crassa* hyphae, that there is a distinct zonation behind the apex of enzymes and of other cell components. For example, although the older regions of hyphae are rich in glycogen, there is none demonstrable near the apex. He interprets this as due to the hydrolysis of glycogen near the apex to sugar which is used in building the cell wall. Ribonuclease is concentrated into the hyphal apex about 15 μm from the tip, whilst alkaline phosphatase is most abundant in a subapical zone some 20–60 μm behind the apex. The maximum growth rate of *N. crassa* hypha is about 100 μm/minute or 6 mm/hour, and Zalokar has estimated that, in order to supply the growing apex with its requirement, the synthetic activities of some 12 mm of hyphal length behind the apex would be needed to assure continued growth. The vigorous streaming of the cytoplasm and the formation of vacuoles, which first appear about 150 μm behind the apex, are important in pushing material forward to the growing apex. The importance of the perforations in the septa in allowing translocation of materials is also obvious.

Differentiation of hyphae also occurs to form a great variety of structures adapted to particular functions. For example, in many parasitic fungi, the tip of the hypha which penetrates the host may form a specialised adhesive structure. Within the host cell, specialised absorptive organs called **haustoria** are formed, surrounded by an invagination of the host plasmalemma. These structures are especially well developed in the so-called obligate (or biotrophic) parasites such as the downy mildews (Oomycetes, Peronosporales), powdery mildews (Ascomycetes, Erysiphales) and the rusts (Basidiomycetes, Uredinales). The fine structure of the haustoria of these groups is described in the account given for each group. Hyphae are considerably modified to form reproductive structures, and these modifications are discussed later.

BRANCHING

The structure of most fungal thalli, including vegetative and reproductive structures, is based on the differentiation of branched hyphae. In most cases, the growth of a fungal hypha is monopodial, with a single main axis of potentially unlimited growth. Branches arise at a considerable distance behind the apex, suggesting some form of apical dominance, i.e. the presence of the main apex inhibits the development of lateral branches close to it. Dichotomous branching is rare, but does occur in *Allomyces* (see p. 126, Fig. 55). In septate fungi, lateral branches arise immediately behind a septum. Branches usually arise singly in vegetative hyphae, although whorls of branches (i.e. branches arising near a common point) occur in reproductive structures. Branching is caused by de-differentiation of a hyphal wall, resulting from a thinning and softening of the wall structure. The softened wall balloons out as a result of the turgor of the protoplast, and extends as a new apex, with an organisation resembling that of the main axis. There is evidence that branching is under genetical (i.e. internal) and external control (Butler, 1961, 1966; Plunkett, 1966; Burnett, 1976). Chemotropic growth towards a source of diffusible nutrients, and inhibited growth caused by staling products produced by hyphae which have colonised a substratum, tend to result in an even spacing of hyphae. The circular growth of fungal colonies in Petri-dish cultures arises because certain lateral branches grow out and fill the space between the leading radial branches, keeping pace with their rate of growth.

Most macroscopic fungal structures are formed by hyphal aggregation. Aggregations of vegetative hyphae may give rise to **mycelial strands** or to **rhizomorphs**. Thick-walled resting bodies containing food reserves are termed **sclerotia**. The familiar sporophores of the larger Ascomycetes and Basidiomycetes are made up of pseudoparenchymatous arrangements of hyphae, often showing much tissue differentiation. The details of such structures will be described later in relation to the different groups of fungi.

(a) Mycelial strands

The formation of aggregates of parallel, relatively undifferentiated hyphae is quite common in Basidiomycetes and in some Ascomycetes and Deuteromycetes. Mycelial strands form the familiar 'spawn' of the cultivated mushroom *Agaricus bisporus*. These strands form most readily when the mycelium has developed on a food base and extends into nutrient-poor surroundings. Mathew (1961) has described their development. Robust leading hyphae extend from the food base and branch at fairly wide intervals to form finer laterals, most of which grow away from the parent hypha. A few branch hyphae, however, form at an acute angle to the parent hypha and tend to grow parallel to it. Occasionally, hyphae grow alongside another hypha which they chance to encounter. A later stage in strand development is characterised by the formation of numerous fine, non-septate 'tendril hyphae', as branches from the older regions of the main hyphae. The tendril hyphae, which may extend either forwards or backwards, become appressed to the main hypha, and branch frequently to form even finer tendrils which grow round the main hyphae and ensheath them. Anastomosis between the hyphae of the major strands helps to consolidate them, and they increase in thickness by the accretion of minor strands. Very similar development has been noted in the strands of *Serpula lacrimans*, the dry-rot fungus (Fig. 29), which are capable of extending for several metres across brick-work from a food base in decaying wood (Butler, 1957, 1958; Watkinson, 1971). Mycelial strands are capable of translocating materials in both directions. They are believed to afford a means by which a fungus can extend from an established food base and colonise a new substratum, by increasing the inoculum potential of the fungus at the point of colonisation (Garrett, 1954, 1956, 1960, 1970).

(b) Rhizomorphs

In contrast with mycelial strands which are relatively undifferentiated and have no well-defined apical meristem, certain fungi have highly differentiated

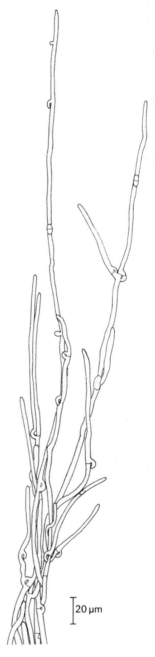

20 µm

Fig. 29. *Serpula lacrimans*: formation of hyphal strands. Note the formation of lateral branches which grow parallel to the direction of the main hyphae. The buckle-shaped structures at the septa are clamp connections.

aggregations of hyphae with a well-developed apical meristem, a central core of larger, thin-walled, elongated cells, and a rind of smaller, thicker-walled cells which are often darkly pigmented. These root-like aggregations of hyphae are well illustrated by *Armillariella mellea* (= *Armillaria mellea*), the honey fungus or honey agaric, which is a serious parasite of trees and shrubs (Fig. 30). The fungus may spread underground from one root system to another by means of the rhizomorphs. In nature, two kinds are found – a dark, cylindrical type, or a paler, flatter type. The latter type is particularly

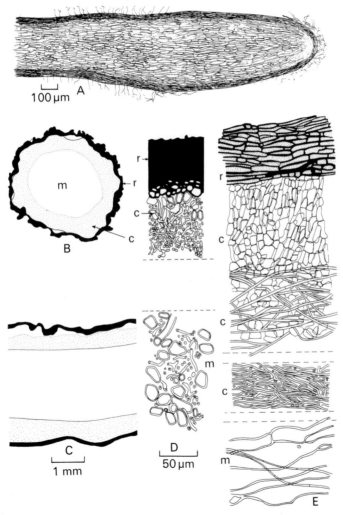

Fig. 30. Rhizomorph structure of *Armillariella mellea*. A. L.S. B. T.S. diagrammatic. C. L.S. diagrammatic. D. T.S. showing details of cells. E. L.S. showing details of cells. r, rind; c, cortex; m, medulla.

common beneath the bark of infected trees. Rhizomorphs on dead trees measure up to about 4 mm in diameter. It has been estimated that a rhizomorph only 1 mm in diameter must contain over 1,000 hyphae aggregated together. The development of rhizomorphs in agar culture has been described by Garrett (1953, 1970) and Snider (1959). Initiation of rhizomorphs can first be observed after about 7 days' mycelial growth, at the interface between the air and the agar, as a compact mass of hypertrophied cells aggregated together, and darkly pigmented over most of the surface by melanin. These pigmented structures have been termed **microsclerotia**. From white, non-pigmented points on their surface, the rhizomorphs develop. The growth of the rhizomorphs can be several times greater than that of unorganised hyphae (Rishbeth, 1968). The terminal half-millimetre of the rhizomorph contains a compact growing-point of small isodiametric cells, protected by a cap of intertwined hyphae in a slimy matrix. Mitotic activity has been reported in this meristematic zone (Motta, 1967). Behind is a zone of elongation. The middle of the rhizomorph may be hollow and filled with air, or may be solid. Surrounding the central lumen or making up the central medulla is a zone of enlarged, elongated cells four to five times wider than the vegetative hyphae (Fig. 30E). Possibly these cells serve in translocation. Towards the periphery of the rhizomorph, the cells become smaller, darker, and thicker-walled. Extending outwards between the outer layer of cells of the rhizomorphs, there may be a growth of vegetative hyphae somewhat resembling the root-hair zone in a higher plant. Rhizomorphs may develop on monokaryotic mycelia derived from single basidiospores, or on dikaryotic mycelia formed following fusion of compatible monokaryotic mycelia. Dikaryotic rhizomorphs of *Armillariella* do not possess clamp connections (Hintikka, 1973).

Mycelial strands and rhizomorphs represent extremes in a range of hyphal aggregation, and several intergrading forms can be recognised (Townsend, 1954; Garrett, 1970).

(c) Sclerotia

Sclerotia are pseudoparenchymatous aggregations of hyphae which serve the function of survival. They are found particularly amongst plant pathogens. Some common plant pathogens which possess sclerotia are *Thanatephorus cucumeris, Sclerotinia* spp. and *Claviceps purpurea. Thanatephorus cucumeris* is a Basidiomycete (usually referred to in the mycelial state as *Rhizoctonia solani*), which causes sharp eyespot disease of wheat, black scurf disease of potato, and damping-off of seedlings of many kinds. *Sclerotinia* spp. (see p. 358) are Ascomycetes which cause diseases of a wide range of plants. *Sclerotinia fuckeliana* is a sexual stage of *Botrytis cinerea,* the cause of grey mould of numerous host plants, whilst *S. sclerotiorum* causes disease of plants such as potato, broad beans, cauliflower, carrots and hops. *Claviceps purpurea* (see p. 349) is an Ascomycete which causes ergot of grasses and cereals.

The form of sclerotia is very variable (Butler, 1966). The sclerotium of the Australian *Polyporus mylittae* can reach the size of a man's head, and is known as native bread or blackfellow's bread. At the other extreme, they may be of microscopic dimensions consisting of a few cells only. Townsend & Willetts (1954) and Willetts & Wong (1971) have distinguished several kinds of development in sclerotia.

(i) *The loose type:* exemplified by *Rhizoctonia solani.* Sclerotia of this type can readily be seen as the thin brown scurfy scales so common on the surface of potato tubers. In pure culture, sclerotial initials arise by branching and septation of hyphae. The cells become barrel-shaped and considerably wider than vegetative hyphae. These cells become filled with dense contents and numerous vacuoles, and darken to reddish-brown. The mature sclerotium does not show well-defined zones of 'tissues'. It is made up of a central part, which is pseudoparenchymatous, but its hyphal nature can be seen. Towards the outside, the hyphae are more loosely arranged, but are not made up of thicker cells (Willetts, 1969).

(ii) *The terminal type:* this type of sclerotial development is characterised by a well-defined pattern of branching. It is exemplified by *Botrytis cinerea* and *Botrytis allii. Botrytis cinerea* is the cause of grey mould diseases of an enormous range of plants (Moore, 1959), whilst *Botrytis allii* causes neck rot disease of onion. Sclerotia of *B. cinerea* can be found on overwintering stems of herbaceous plants, especially umbellifers such as *Anthriscus, Heracleum* and *Angelica.* They can also be induced to form readily in culture, especially those with a high carbon:nitrogen ratio. When growing on host tissue, the sclerotia of *Botrytis* may include host cells, a feature shared also by the sclerotia of *Sclerotinia* spp. (see p. 358). As shown on p. 359, *Botrytis* is the name given to the conidial (i.e. asexual spore) stages of certain fungi. These include some species of *Sclerotinia.* In *B. allii,* the sclerotia arise by repeated dichotomous branching of hyphae, accompanied by cross-wall formation. The hyphae then coalesce to give the appearance of a solid tissue. A mature sclerotium may be about 10 mm long and 3–5 mm wide, and is usually flattened, measuring 1–3 mm in thickness. It is often parallel to the long axis of the host plant. It is differentiated into a rind composed of several layers of rounded, dark, thickened cells, and a narrow cortex of thin-walled pseudoparenchymatous cells with dense contents, and a medulla made up of loosely arranged filaments.

(iii) *The strand type:* this is illustrated by *Sclerotinia gladioli,* the causal agent of dry rot of corms of *Gladiolus, Crocus* and other plants. Sclerotial initials commence with the formation of numerous side branches which arise from one or more main hyphae. Where several hyphae are involved, they lie parallel. They are thicker than normal vegetative hyphae, and become divided by septa into chains of short cells. These cells may give rise to short branches, some of which lie parallel to the parent hypha, whilst others grow out at right angles and branch again before coalescing. The hyphae at the margin continue to branch, and the whole structure darkens. The mature sclerotium is about 0·1–0·3 mm in diameter, and differentiated into a rind of small thick-walled cells and a medulla of large thin-walled cells. More complex sclerotia are found in *Sclerotium rolfsii* (the sclerotial state of the Basidiomycete *Pellicularia rolfsii*). Here the mature sclerotium is differentiated into four zones: a fairly thick skin or cuticle, a rind made up of 2–4 layers of tangentially flattened cells, a cortex of thin-walled cells with densely staining contents, and a medulla of loose filamentous hyphae with dense contents. Ultrastructural studies by Chet *et al.* (1969) have shown that the skin or cuticle is made up of the remnants of cell walls attached to the outside of the empty, melanised, thick-walled rind cells. All the cells of the sclerotium have thicker walls than those of vegetative hyphae. Cells of the outer cortex contain large vesicles which leave little room for cytoplasm or other organelles. The inner cortex is densely packed with granules which may be of polysaccharide.

(iv) *Other types:* it is probable that other kinds of sclerotial development occur (Butler, 1966) The sclerotia of *Claviceps purpurea* (the 'ergots' of grasses and cereals, see p. 349) develop from a pre-existing mass of mycelium which fills the ovary, starting from the base and extending towards the apex. The outer layers form a violet-coloured rind, within which are colourless, thick-walled cells containing lipid. *Cordyceps militaris*, an insect parasite, forms a dense mass of mycelium in the buried insect's body (p. 355). This mass of mycelium, from which fructifications develop, is enclosed by the exoskeleton of the host, not by a fungal rind. Many wood-rotting fungi enclose colonised woody tissue with a black zone-line of dark, thick-walled cells, and the whole structure may be regarded as a kind of sclerotium.

The giant sclerotium of *Polyporus mylittae* is alveolate in structure, comprising white strata and translucent tissue. It has an outer, smooth, thin, black rind. Three distinct types of hyphae make up the tissues: thin-walled, thick-walled and 'layered' hyphae. Thin- and thick-walled hyphae are abundant in the white strata, and sparse in the translucent tissue, whilst the layered hyphae occur only in the translucent tissue. Detached sclerotia are capable of forming basidiocarps without wetting. It is believed that the translucent tissue functions as an extracellular nutrient and water store (Macfarlane *et al.*, 1978). The structure of the sclerotium in this fungus appears to be related to its ability to fruit in dry conditions such as occur in Western Australia.

Sclerotia may survive for long periods, sometimes for several years (Sussman, 1968; Coley-Smith & Cooke, 1971; Willetts, 1971). Germination may take place in three ways: by the development of mycelium, conidia, or by the formation of ascocarps or basidiocarps. Mycelial germination occurs in *Sclerotium cepivorum*, the cause of white rot of onion, and is stimulated by volatile exudates from onion roots. Conidial development occurs in *Botrytis cinerea* and can be readily demonstrated by placing overwintered sclerotia in moist, warm conditions. The development of ascocarps is seen in *Sclerotinia,* where stalked cups or **apothecia,** bearing asci, arise from sclerotia under suitable conditions (p. 360), and in *Claviceps purpurea,* where the overwintered sclerotia give rise to drumstick-like structures termed **perithecial stromata** which contain **perithecia,** flask-shaped cavities within which the asci are formed.

(d) Other vegetative mycelial aggregates

One special kind of aggregation of mycelium is the mantle or sheath which surrounds the roots of trees possessing sheathing mycorrhizas (to use the term proposed by Lewis, 1973). The root tips of many coniferous and deciduous trees growing in relatively infertile soils are covered with a continuous sheet, several cell layers thick, of fungal cells. The mycelium extends outwards into the litter layer of the soil, and inwards, between the cortical cells of the root, to form the so-called 'Hartig network'. This external layer of fungal tissue effectively replaces the root hairs as a system for the absorption of minerals from the soil (Clowes, 1951), and there is good evidence that, in most normal forest soils of low to moderate fertility, the performance and nutrient status of mycorrhizal trees is superior to that of uninfected trees (Harley, 1969). The fungi which cause mycorrhizal infection are usually toadstools belonging to the Agaricales (p. 429). Many of them have a fairly specific symbiotic relationship with particular kinds of tree. Within the soil, or in pure culture, the mycelium is not aggregated into the tissue-like structure of the sheath, and presumably this development is caused by the close association between the tree root and the mycelium. Read & Armstrong (1972) have produced experimental evidence that the formation of fungal pseudoparenchyma making up the sheath is in response to three stimuli, a surface, available oxygen and a supply of nutrients. In soil, root surfaces would provide this combination of factors.

(e) Aggregation to form reproductive structures

In the larger fungi, mostly Ascomycotina and Basidiomycotina but also in the Deuteromycotina, hyphae may become aggregated together to form fruiting structures, bearing spores of various kinds. In Ascomycotina, the sexually produced spores (i.e. spores formed as a result of nuclear fusion and meiosis) are termed **ascospores**, and they are contained in globose or cylindrical sacs or **asci** (singular: **ascus**). In most cases, the asci can discharge the ascospores explosively, The asci, although occasionally naked, are usually enclosed in an aggregation of hyphae termed an **ascocarp**. Ascocarps are very variable in form, and a number of distinct types have been distinguished (Booth, 1966b). These are described more fully later. Forms in which the asci are totally enclosed, and in which the ascocarp has no special opening, are termed **cleistocarps** or **cleistothecia**. They are found in the Eurotiales (p. 298) and in the powdery mildews (Erysiphales, p. 283). Cup-fungi (Pezizales and Helotiales, p. 358) possess saucer-shaped ascocarps termed **apothecia**, with a mass of non-fertile hyphae supporting a layer of asci lining the upper side of the fruit-body. The non-fertile elements of the apothecium often show considerable differentiation of structure. The asci are free to discharge their spores at the same time. In other Ascomycotina, the asci are surrounded by ascocarps with a very narrow opening or **ostiole**, through which the asci can discharge their spores one at a time. Ascocarps of this type are termed **perithecia** or **pseudothecia**. Perithecia are found in Sphaeriales and Hypocreales (p. 314), whilst pseudothecia occur in the Loculoascomycetes (p. 389). The two types of ascocarp develop in distinctive ways, but an important distinction is that in forms with pseudothecia the ascus wall is double, i.e. the ascus is **bitunicate**, whilst in forms with perithecia the ascus wall is single, i.e. **unitunicate**. In many of the Sphaeriales and the Hypocreales, the perithecia are borne on or embedded in a mass of fungal tissue termed the **perithecial stroma**, and these are well shown by such fungi as the Xylariaceae (p. 333) and by *Cordyceps* (p. 355) and *Claviceps* (p. 349). In some cases, in addition to the perithecial stroma, a fungus may develop a stromatic tissue on which or within which asexual spores or conidia develop. *Nectria cinnabarina* (p. 345), the coral-spot fungus, so common on freshly dead deciduous twigs, is an example of a fungus which forms pink conidial stromata, which later, under suitable conditions of humidity, give rise to perithecia, i.e. become converted into perithecial stromata (Fig. 180).

The fruit-bodies of mushrooms, toadstools, bracket-fungi, etc., are all examples of **basidiocarps** which bear the sexually produced spores or **basidiospores** on **basidia**. Most basidia, other than those of Gasteromycetes (puffballs, stinkhorns, bird's nest fungi, etc.), have a violent spore discharge mechanism and wind-dispersed spores, so that the basidia are arranged in such a way that the spores can be projected into the air. The basidiocarps are almost invariably constructed from dikaryotic hyphae. Within the pseudoparenchymatous tissues which make up the basidiocarp, there is considerable differentiation in structure and function of the hyphae (Smith, 1966). Differentiation of the hyphal elements making up the basidiocarp is seen at its most highly developed in the fruit-bodies of some Aphyllophorales, where a number of distinct hyphal elements have been recognised (p. 438).

In the Deuteromycotina, mycelial aggregations bearing conidia are seen in various genera. In some, there are tufts of parallel conidiophores termed **coremia** or **synnemata**, exemplified by fungi such as *Penicillium claviforme* (p. 311). In the Coelomycetes (p. 566), the conidia develop in flask-shaped cavities termed **pycnidia**. Various other kinds of mycelial fruiting aggregate are also known (Tubaki, 1966).

DIFFERENTIATION OF REPRODUCTIVE CELLS

The characteristic structures by which fungi reproduce and by which they may

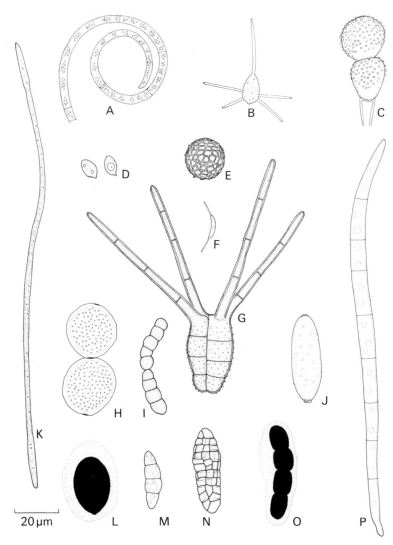

Fig. 31. Some different kinds of fungal spores: A, *Helicomyces scandens* conidium; B, *Nia vibrissa* basidiospore; C, *Tranzschelia discolor* teliospore; D, *Marasmius oreades* basidiospore; E, *Tilletia caries* chlamydospore; F, *Dinemasporium graminum* conidium; G, *Tetraploa aristata* conidium; H, *Puccinia poarum*, two aeciospores; I, *Torula herbarum*, two conidia; J, *Erysiphe graminis*, conidium; K, *Epichloe typhina*, ascospore; L, *Sordaria fimicola*, ascospore; M, *Leptosphaeria microscopica* ascospore; N, *Pleospora herbarum* ascospore; O, *Sporormia intermedia* ascospore; P, *Fusarium* sp. phialoconidium.

be dispersed are spores. Spores are very varied in structure. They are unicellular or multicellular, colourless or pigmented, thin- or thick-walled, self-propelled by means of flagella, or carried passively by wind, water or animal vectors. They vary greatly in form from globose, or elliptical, to filamentous or highly branched. They range in size from about 2 μm in diameter to over 150 μm. Some different kinds of fungus spore are illustrated in Fig. 31. The functions of spores are also varied. Often they are the means of dispersal. This is especially true of the various kinds of asexual reproductive unit found in fungi. They enable rapid dissemination of a pathogenic fungus from an initial focus of infection, or the rapid exploitation of a newly exposed substratum suitable for fungal growth. Sexually produced spores, i.e. spores produced as a result of nuclear fusion and meiosis, may also be involved in dispersal, as in the ascospores of many Ascomycetes or the basidiospores of many Basidiomycetes. In other cases, the sexually produced spores may be sedentary, and not readily dispersed. This is true of the oospores of many Oomycetes, or the zygospores of Zygomycetes, which are relatively large and remain dormant in the soil for long periods until suitable conditions, such as the proximity of an adequate substratum, provide the possibility for further active growth. Spores may also have a sexual rôle, i.e. they may function as gametes, and indeed some may be incapable of germination to form vegetative mycelium. This is the case with the pycniospores of rust fungi which can only germinate in close contact with receptive hyphae with which they make contact by short germ tubes (see p. 507). Following anastomosis with a receptive hypha, the pycniospore transfers to it a compatible nucleus which initiates further development.

In some cases, the unit of dispersal is not the spore, but the sporangium or other structure which surrounds it. Sporangial dispersal may occur in water moulds such as *Dictyuchus* (p. 150) in which the sporangia may be detached from the mycelium and be dispersed passively by water currents. Sporangial discharge may also occur in terrestrial fungi. In *Phytophthora infestans*, the cause of late blight of potato, the sporangia are borne on stalks which project from infected leaves (Fig. 72). The sporangia are dispersed to foliage of fresh hosts, and germinate by releasing zoospores which bring about infection. In the related genus *Peronospora*, a downy mildew, similar structures, resembling the sporangia of *Phytophthora*, develop. These are also dispersed by wind, but do not release zoospores. The dispersal unit, believed to be homologous to the sporangium of *Phytophthora*, is sometimes termed a **conidium** (see later p. 179). Violent sporangial discharge takes place in *Pilobolus*, so common on the dung of herbivorous animals (p. 214). The sporangia can be discharged violently for a distance of several metres, before becoming attached firmly to herbage, where they may later be ingested by animals. In the related genus *Pilaira* (p. 220), also found on herbivore dung, the sporangium is similar in all respects to that of *Pilobolus*, but is not discharged violently, merely slipping off its stalk to become attached to adjacent herbage. There are a few cases in which ascospores are dispersed whilst still within the ascus. An example is the powdery mildew *Podosphaera* (p. 294) in which the single ascus is projected by an elastic action of the expanded walls of the ascocarp. Truffles provide further examples in which the subterranean, fragrant fruit-bodies are excavated by rodents and ingested. The asci, with ascospores inside them, are dispersed by the rodents, and presumably the ascospores germinate after the process of digestion (p. 384). Some of the most striking examples of dispersion of spores within protective coverings are found in the Gasteromycetes. In *Sphaerobolus*, which forms globose orange fruit-bodies about 2 mm in diameter on decaying vegetation and on old herbivore dung, basidiospores and other reproductive cells termed gemmae are contained within a brown sticky projectile, termed the **glebal mass**, flung out for distances as great as 5–6 m. In the bird's nest fungus, *Cyathus* (p. 474), the basidiospores are contained in a lens-shaped **peridiole**, attached to the inner wall of a funnel-shaped fructification by a complex spirally coiled **funicular cord**. Rain splash triggers the unwinding of the funicular cord which is dispersed with the peridiole for 1 m or more, the sticky end of the funicular cord attaching the peridiole to vegetation.

SOME DIFFERENT KINDS OF FUNGAL SPORE

Some of the more commonly encountered types of fungal spore will now be described.

Zoospores

These are spores which are self-propelled by means of flagella. The fungal groups which possess flagella are mostly aquatic and are arbitrarily classified here in the Mastigomycotina. This does not necessarily imply that all fungi with zoospores are closely related to each other. Within Eumycota, zoospores are of three kinds:

(a) Posteriorly uniflagellate zoospores with flagella of the whiplash type, characteristic of the Chytridiomycetes.
(b) Anteriorly uniflagellate zoospores with flagella of the tinsel type, characteristic of the Hyphochytridiomycetes.
(c) Biflagellate zoospores with anteriorly or laterally attached flagella, one of which is of the whiplash type and the other of the tinsel type, characteristic of the Oomycetes.

(a) Zoospores with posterior whiplash flagella. Details of the fine structure of this kind of zoospore have been reviewed by Fuller (1966, 1976). The best-known are *Blastocladiella emersonii* (Cantino *et al.,* 1963; Reichle & Fuller, 1967; Lessie & Lovett, 1968) and *Allomyces macrogynus* (Hill, 1969; Fuller & Olson, 1971). The structure of the zoospore of *B. emersonii* is summarised diagrammatically in Fig. 32. The zoospore is tadpole-like with a

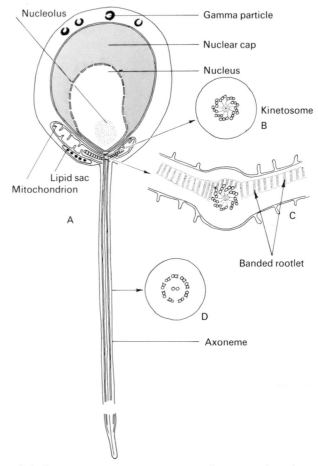

Fig. 32. *Blastocladiella emersonii*: zoospore structure, diagrammatic and not to scale.
A. L.S. zoospore along axis of flagellum. B. T.S. kinetosome showing nine triplets of
subfibrils. C. T.S. kinetosome at slightly lower level showing the origin of two of the
banded rootlets which extend into the mitochondrion. The cristae of the mitochondrion are shown close to the membrane which surrounds the banded rootlets. D. T.S.
axoneme showing the nine peripheral paired subfibrils and two central fibrils.

pear-shaped head about 7×9 μm and a single trailing flagellum about 20 μm
long. Under the light microscope, the most conspicuous internal structure is
the dense crescent-shaped nuclear cap which surrounds the more transparent
nucleus. The nuclear cap is rich in ribonucleic acid and protein, and is filled
with ribosomes. *Blastocladiella emersonii* is unusual in that its zoospore
contains only a single large mitochondrion, situated near the base of the
zoospore. The flagellum originates in a basal body or **kinetosome** (sometimes
also referred to as a **blepharoplast**). The kinetosome is itself derived from the
larger of two centrioles, formed at the time of the nuclear division which

precedes cleavage of the zoospores. The second, smaller centriole lies near to the kinetosome, adjacent to the nucleus. The kinetosome is roughly cylindrical in shape and made up of nine fibrils each consisting of a triplet of microtubules arranged in the manner of a cart-wheel (Fig. 32B). At its lower end, the kinetosome ends in a terminal plate. Below the terminal plate, the main shaft of the flagellum or **axoneme** continues as a ring of nine doublet microtubules surrounding a central pair of single microtubules. The whole group of eleven microtubules is surrounded by a flagellar sheath continuous with the zoospore membrane. This is the typical $9 + 2$ arrangement of flagellar microtubules in most motile and ciliated cells (Fig. 32D). The two central microtubules end at about the level of the transverse plate of the kinetosome. The upper part of the flagellum in the region of the kinetosome passes through a channel in the mitochondrion. A series of microtubules extends from the proximal end of the kinetosome towards the nucleus and nuclear cap, and probably provides structural rigidity. Extending into the mitochondrion, linking up the kinetosome with it, are three striated bodies variously referred to as **flagellar rootlets, striated rootlets** or **banded rootlets**. The banded rootlets are contained within separate channels, and each is surrounded by a unit membrane. Since the energy for propulsion is generated within the mitochondrion, it is possible that the banded rootlets are, in some way, responsible for transmitting energy to the base of the axoneme, but this is, at present, speculative. It is also possible that the banded rootlets serve to anchor the flagellum within the body of the zoospore. At its distal end, the axoneme narrows to a pointed or blunt-ended whiplash. The narrowing is due to a reduction in the number of microtubules extending into the tip of the axoneme.

There are two other obvious kinds of organelle within the body of the zoospore: the lipid sac attached to the mitochondrion, and a group of gamma-particles most frequently found at the anterior of the zoospore between the nuclear cap and the spore wall. The lipid sac contains several globules which stain with osmium tetroxide. The group of globules is bounded by a single membrane. It is not known whether the lipid forms the energy reserve used in swimming: indeed, it seems more probable that 'the important substrate for energy production is a sizeable pool of glycogen-like polysaccharide' (Cantino *et al.,* 1968). The gamma-particles are granules about 0.5 μm in diameter. Studies of their fine structure (Cantino & Mack, 1969; Myers & Cantino, 1974) suggest that a gamma-particle consists of an inner core, shaped like an elongated cup, bearing two unequal openings on opposite sides of the cup. The cup-shaped structure is enveloped in a unit membrane. The gamma-particles are rich in glycolipid. When isolated gamma-particles were incubated in the presence of the substrate for chitin synthetase, chitin was produced in the form of chitin fibrils (Mills & Cantino, 1978).

The zoospore is propelled forwards by rhythmic lashing of the flagellum. It is also capable of amoeboid changes of shape. It can swim for a period under anaerobic conditions. On coming to rest, the flagellum is retracted into the

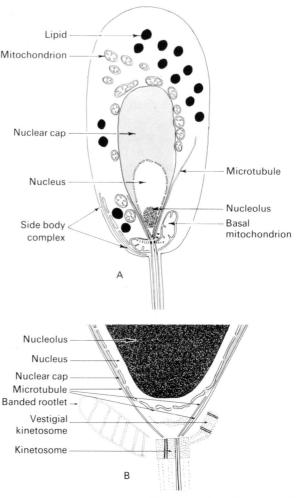

Fig. 33. *Allomyces macrogynus*: zoospore, diagrammatic and not to scale, based on electron micrographs by Fuller & Olson (1971). A. L.S. zoospore. B. Details of basal part of nucleus showing the two kinetosomes and a banded rootlet. Note the group of microtubules which extend upwards from the functional kinetosome.

body of the zoospore, and it has been suggested (Cantino *et al.*, 1968) that the retracted flagellum is coiled up inside the body of the zoospore, possibly as a result of a pushing action by the kinetosome on the nucleus which causes the latter to rotate in such a way that the flagellum is coiled around it.

The zoospore of *Allomyces* differs in several ways from that of *Blastocladiella*. Its structure is summarised in Fig. 33. Although there is a large basal mitochondrion, many smaller mitochondria are also present, generally located along the membrane of the nuclear cap in the anterior part of the cell.

There are no gamma-particles. A complex structure situated laterally at the base of the body of the zoospore, between the nucleus and the zoospore membrane, has been termed by Fuller & Olson the side body complex. It consists of two closely appressed membranes separated by an electron-opaque material. These membranes subtend numerous electron-opaque, membrane-bound bodies, lipid bodies and a portion of the basal mitochondrion. In addition, there are membrane-bound, non-lipid bodies, termed by Fuller & Olson Stüben bodies, whose function and composition are uncertain. The side body complex of the *Allomyces* zoospore does not therefore correspond exactly with the lipid sac of *Blastocladiella*.

The flagellum of the *Allomyces* zoospore has the usual arrangement of $9+2$ microtubules. The flagellum arises from a functional kinetosome, alongside which is a centriole, sometimes referred to as the non-functional or vestigial kinetosome. Both kinetosomes have the typical arrangement of nine sets of triplet microtubules. Extending from the proximal end of the functional kinetosome are twenty-seven (i.e. 9×3) microtubules which form a funnel-shaped structure, closely appressed to the posterior portion of the nucleus and the nuclear cap. They also extend into the body of the zoospore beyond the nuclear cap. Fuller & Olson have suggested that the conical shape of the nucleus and nuclear cap may be due to the constraints imposed by these microtubules. It is also possible that the microtubules confer structural rigidity to the body of the zoospore to allow the propulsive force of the flagellum to be transmitted.

Blastocladiella and *Allomyces* are both representatives of the Blastocladiales. The other two orders of the Chytridiomycetes are the Chytridiales (Koch, 1956, 1968) and Monoblepharidales (Fuller & Reichle, 1968). There are differences in the details of the structure of zoospores in these groups, but these will be described later (see also Fuller, 1966; Sparrow, 1973a).

(b) Zoospores with anterior tinsel-type flagella. The only group of fungi with this kind of zoospore is the Hyphochytridiomycetes, of which *Rhizidiomyces* is the best-known example (Fuller, 1966, 1976; Fuller & Reichle, 1965). The structure of a zoospore of *R. apophysatus* is illustrated in Fig. 34. The body of the zoospore is about 5 μm in diameter. The zoospore swims in a spiral path, being pulled along by the sine wave undulations of its anterior flagellum. The axoneme of the flagellum bears a series of fine **mastigonemes** along the whole of its length. The mastigonemes consist of a wider basal shaft for about two-thirds of their length, ending in a terminal part about one-fifth of the diameter of the basal part. The wider basal part shows transverse or spiral bands of alternating light and dark material. Sometimes there are two finer extensions of the apex of the mastigoneme. Sections of the zoospore (Fig. 34B) show that the nucleus is in the anterior. The flagellum arises from a kinetosome (**BB** for basal body in Fig. 34B) close to the nuclear membrane. An associated centriole C_2 lies close to the kinetosome. There is no

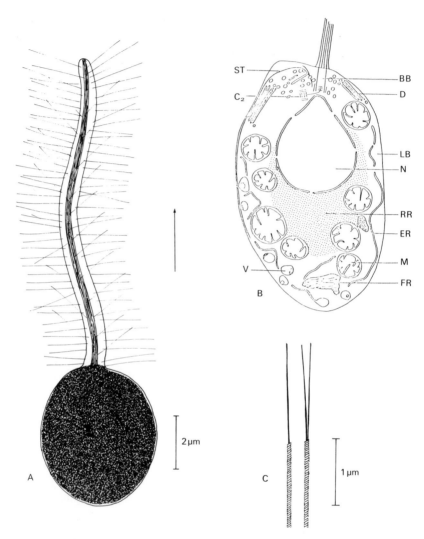

Fig. 34. *Rhizidiomyces apophysatus.* A. Zoospore as seen in a shadowed preparation. The main shaft of the flagellum or axoneme bears finer lateral mastigonemes. The arrow indicates the direction of zoospore movement. B. Diagrammatic representation of L.S. zoospore from Fuller & Reichle (1965). Abbreviations: ST, subsurface tubules; C_2, centriole associated with basal body; BB, basal body (=kinetosome); V, vacuole; D, dictyosome; LB, lipid body; N, nucleus; RR, ribosome-containing region; ER, endoplasmic reticulum; M, mitochondrion; FR, fibre-containing region. C. Appearance of mastigonemes at high magnification.

membrane-bound nuclear cap as in *Blastocladiella* or *Allomyces*, but there are numerous ribosomes (RR) near the posterior of the nucleus. Lipid bodies, vacuoles, microtubules, a dictyosome and endoplasmic reticulum are also recognisable. When the zoospore comes to rest, the axoneme is retracted and the mastigonemes are left attached to the membrane of the zoospore cyst.

In comparing the two kinds of uniflagellate zoospore, one is bound to ask

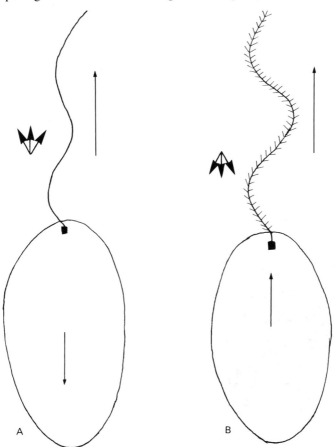

Fig. 35. Flagellar action. The thin arrows to the right of the flagella indicate the direction of movement of the sine wave along the flagellum. The thin arrows within the zoospore indicate the direction of movement of the zoospore. The thick arrows to the left of the flagellum indicate the direction of movement of the water as a result of flagellar activity. A. The sine wave moves away from the body of the zoospore causing water movement in the same direction. The zoospore body moves by reaction in the opposite direction, i.e. away from the flagellum. B. Plectonematic arrangement. The flagellum bears two rows of mastigonemes with a large surface area relative to the rest of the flagellum. The sine wave is directed away from the body of the zoospore, but the net water movement is towards the zoospore body, which therefore moves by reaction in the opposite direction, i.e. towards the flagellum.

why a posterior whiplash flagellum *drives* a spore forward, whilst an anterior tinsel flagellum *pulls*. The explanation is connected with the large surface area which mastigonemes present to the water as compared with the surface area of a whiplash flagellum. In Fig. 35, the two types of uniflagellate zoospore are compared diagrammatically. Fig. 35A represents a zoospore with a single flagellum of the whiplash type. A sinusoidal wave is generated along the flagellum from the kinetosome towards the tip of the flagellum. The movement of the flagellum creates a movement of water in the same direction as the sine wave, i.e. distally from the kinetosome. The spore is therefore propelled in the opposite direction to the water movement, in exactly the same way as a swimmer propels himself forward by creating backwardly directed currents of water. Fig. 35B represents a zoospore with a tinsel flagellum. The sine wave is generated from the kinetosome down the length of the flagellum. However, the large surface area presented by the mastigonemes results in the creation of water currents in the opposite direction to the movement of the sine wave, and the net result is that the zoospore is pulled in the direction of the flagellum.

(c) Biflagellate zoospores. The biflagellate type of zoospore is characteristic of the Oomycetes (Colhoun, 1966; Fuller, 1976; Bartnicki-Garcia & Hemmes, 1976). The zoospore is **heterokont,** i.e. it has one flagellum of the whiplash type and one of the tinsel type. The flagella may be inserted apically, as in the first motile stage of *Saprolegnia*, or laterally, arising from within a groove. The tinsel-type flagellum is directed forwards, whilst the whiplash flagellum is directed backwards. The detailed structure of this type of zoospore is best known in *Phytophthora* (Ho *et al.*, 1968a, b; Vujičić *et al.*, 1968; Reichle, 1969). The structure of a *Phytophthora* zoospore as seen by light microscopy is shown in Fig. 36. The zoospore is ovoid, bluntly pointed at the anterior end, with a longitudinal groove or furrow which is overarched by lip-like folds of the body of the zoospore (Fig. 36A). The depth of the groove varies along its length: it is shallow at the ends and deeper towards the middle of the zoospore. A large vacuole lies immediately to the inside of the groove. The two flagella arise from within the groove, seated on a small projection or ridge. The tinsel-type flagellum is directed forwards, i.e. in the direction of movement of the zoospore, whilst the whiplash flagellum is directed backwards. The posterior flagellum is longer than the anterior. Each arises from a kinetosome situated close to the nucleus. The structure of the flagella presents no unusual features: the two central microtubules end at the basal plate of the kinetosome, and the proximal part of the kinetosome consists of the usual array of nine triplet microtubules. The two kinetosomes are linked together by an interconnecting fibre, and each kinetosome also has two rootlets, composed of about eight microtubules, extending towards the cell surface. In the groove region is a large number of microtubules running parallel to the groove. These may perform a skeletal function, providing attachment to the flagella and transmission of their propulsive force.

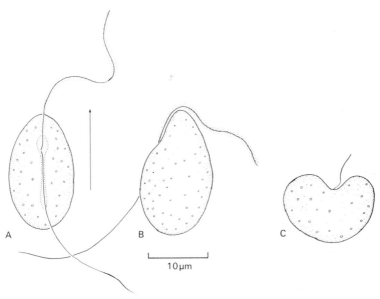

Fig. 36. *Phytophthora* zoospores. A. View looking into the longitudinal groove. The arrow indicates the direction of movement. B. Lateral view. C. End view showing the longitudinal groove. Note the reniform shape from this aspect. (Based on Ho *et al.*, 1968b)

The tinsel-type flagellum bears an array of mastigonemes or **flimmer hairs**, apparently attached to the *sheath* of the flagellum and not to the component microtubules. Studies at very high magnification of the tinsel flagellum of *Phytophthora* (Reichle, 1969) show that the basal two-thirds of the mastigoneme is tubular. This portion is about 17–19 nm in diameter, with a lumen 5–7 nm wide. The terminal one-third is thinner and appears to consist of two thin strands 2–5 nm in diameter which may be twisted around each other so that the terminal part appears as a single strand. In *Saprolegnia*, where the flimmer hairs occur in two rows, their attachment to the flagellar sheath has also been confirmed (Heath *et al.*, 1970). The flimmer hairs consist of a wider, basal portion, 1·5 μm long, and a finer terminal hair-point, 1·1 μm. In section, the basal part can be seen to be a tube about 15 nm in diameter attached to the flagellum sheath by a tapering structure about 0·3 μm long.

The whiplash flagellum of Oomycetes, although usually regarded as smooth, may in fact bear fine hair-like appendages visible only at very high magnification (about 20,000). In *P. erythroseptica*, Vujičić *et al.* (1968) describe the hairs on the whiplash flagellum as 0·6 μm long, whilst in *S. ferax*, Manton *et al.* (1951) have shown them to be 0·5 μm long. According to these authors, the whiplash flagellum of *S. ferax* may be flattened to form a fin-like extension along two sides. The tip of the whiplash flagellum is usually narrower than the rest of the flagellum and devoid of fine hairs.

Most fungus spores are not self-propelled. Dispersal may be *active* or *passive*. In active dispersal, the energy for separation of the spore from the parent fungus is generated by the fungus itself. Passive dispersal involves some other agent, e.g. wind currents, rain-splash or water currents, or animals.

Sporangiospores

In the Zygomycetes, and especially in the Mucorales (see p. 191), the asexual spores are contained in globose sporangia or cylindrical sacs termed **mero-sporangia**. Because they are non-motile, the spores are sometimes termed **aplanospores.** The spores may be uni- or multinucleate, and are unicellular, generally smooth-walled and globose or ellipsoid in shape. When mature, they may be surrounded by mucilage, in which case they are usually dispersed by rain-splash or insects, or they may be dry and dispersed by wind currents (Ingold & Zoberi, 1963). In some cases, entire sporangia are detached. The number of sporangiospores per sporangium may vary from several thousands to only one. The detachment and dispersal of intact sporangia containing a few sporangiospores or a single one is indicative of the way in which conidia may have evolved (see p. 224).

Ascospores

Ascospores are the characteristic spores of the large group of fungi known as Ascomycotina. They are formed as a result of nuclear fusion immediately followed by meiosis. The four haploid daughter nuclei then divide mitotically to give eight haploid nuclei around which the ascospores are cut out. Details of ascospore development are described more fully on p. 257. In most Asco-mycetes, the eight ascospores are contained within a cylindrical sac or ascus, from which they are forcibly ejected by a squirting process in which the ascus contents, consisting of ascospores and ascus sap, are ejected by explosive breakdown of the tip of the turgid ascus whose elastic walls contract. Asco-spores vary greatly in size, shape and colour. In size, the range is from about 4–5×1 μm in small-spored forms such as the minute cup-fungus *Dasyscyphus*, to 130×45 μm in the lichen *Pertusaria pertusa*, which is a symbiotic association between an Ascomycete and a green alga. Ascospore shape varies from globose to oval or elliptical, lemon-shaped, sausage-shaped, cylindrical, needle-shaped. Ascospores may be uninucleate or multinucleate, unicellular or multicellular, divided up by transverse or by transverse and longitudinal septa. The wall may be thin or thick, hyaline or coloured, smooth or rough, sometimes folded into reticulate folds, and may have a mucilaginous outer layer which may be extended to form simple or branched appendages. In many cases, ascospores are resting structures which survive adverse conditions. They may have extensive food reserves in the form of lipid and sugars such as trehalose.

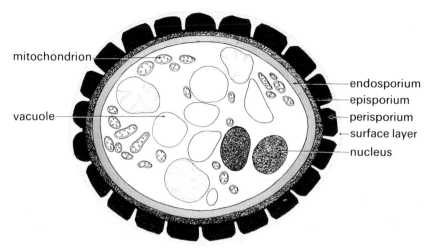

Fig. 37. *Neurospora tetrasperma*. T.S. ascospore. Simplified diagram based on an electron micrograph by Lowry in Sussman & Halvorson (1966).

It is obvious that there is no such thing as a typical ascospore. As an example of an ascospore whose structure has been extensively studied, *Neurospora tetrasperma* will serve (Lowry & Sussman, 1958, 1968; see also Austin *et al.*, 1974). *Neurospora tetrasperma* is somewhat unusual in that it has four-spored asci and the ascospores are binucleate. The spores are black, thick-walled and shaped rather like a rugby football, but with flattened ends. The name *Neurospora* refers to the ribbed spores, because the dark outer wall is made up of longitudinal raised ribs, separated by interrupted grooves. The structure of a spore in section is shown in Fig. 37. Within the cytoplasm of the spore are the two nuclei, fragments of endoplasmic reticulum (not illustrated), swollen mitochondria and two types of vacuole, bounded by single membranes, one kind appearing empty and the other containing coarsely precipitated material after permanganate fixation. The vacuoles presumably contain food reserves, which have been shown to be lipids and trehalose (Lingappa & Sussman, 1959). The wall surrounding the protoplast is composed of several layers. The innermost layer is the **endosporium**, outside which is the **episporium**. The ribbed layer is termed the **perisporium**. Between the ribs are lighter intercostal veins containing a material which is chemically distinct from the ribs. This material is continuous over the whole surface of the spore, giving it a relatively smooth surface. The spore germinates, after heat shock, by the extrusion of germ tubes from a pre-existing germ pore, or thin area in the episporium, at either end of the spore.

Basidiospores

In comparison with the morphological diversity of ascospores, basidiospores

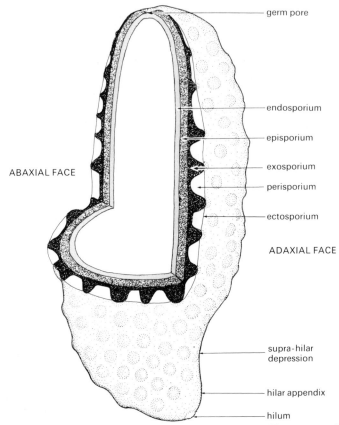

Fig. 38. Basidiospore structure. Diagram of cut-away view to illustrate terminology of the layers of the spore wall, based on Pegler & Young (1971). The contents of the spore are not shown.

are rather more uniform. Typically, they are unicellular, but transversely septate spores are found in certain groups, for example the Dacrymycetaceae (Reid, 1974). In shape, they vary from globose, sausage-shaped, fusoid, almond-shaped (i.e. flattened), and the wall may be smooth or ornamented with spines, ridges or folds. The colour of basidiospores is an important criterion of classification. They may be colourless, white, cream, yellowish, brown, pink, purple or black. The spore colour may be due to coloured substances in the cytoplasm of the spore or in the spore wall. This explains the change of colour of a mushroom gill from pink, when immature, due to cytoplasmic spore pigments, to purple, when mature, due to wall pigments. The generalised structure of a basidiospore is illustrated in Fig. 38. The spore is attached, usually as one of a tetrad, to the basidium at the tip of a **sterigma**, a curved horn-like prong projecting from the apex of the basidium (see Fig. 218). The spore is projected for a short distance from the basidium, by a mechanism which is still not fully understood. (For a discussion of possible mechanisms, see p. 400.) The point at which the spore was attached to the sterigma is the **hilum**, and it is usually found at the tip of a short conical projection, the **hilar appendix**. The term **ballistospore** is used to describe basidiospores which are violently projected from their sterigmata. Whilst most basidiospores are ballistospores, some are not. For example, in the Gasteromycetes, the basidiospores are not violently projected from their sterigmata, and a variety of alternative mechanisms have evolved to disperse the spores (see p. 469).

The cytoplasm of the spore usually contains a haploid nucleus, resulting from meiotic division in the basidium. Sometimes there may be two nuclei, especially in species with two-spored basidia, but even in four-spored species, there may be mitotic division of the original nucleus, so that the spore is binucleate. Other organelles within the spore include mitochondria.

The structure of the wall layers has been studied by transmission and scanning electron microscopy (Pegler & Young, 1971). The terminology of the wall layers corresponds to that proposed for *Neurospora*. Five layers have been distinguished in some kinds of spore. Working outwards from the surface of the spore protoplast, the layers are termed the **endosporium, episporium, exosporium, perisporium** and **ectosporium**. The wall layers develop progressively from the outside inwards, i.e. the endosporium is the last to develop. It is rather transparent to the electron beam. The episporium determines the size and form of the spore. It is electron-dense and shows lamellar organisation. Wall pigments are located in the episporium. The exosporium lies above the episporium. It is a colourless layer, chemically distinct from adjacent layers. In many spores, this is the outermost layer because the perisporium and ectosporium disintegrate. It may be lobed or folded to form the surface ornamentation of the spore, but in *Russula* and *Lactarius* the exosporium is penetrated by ornamentation developing from the episporium beneath. The perisporium envelops the other spore layers, including ornamentation derived from the exosporium. It may disappear as the spore develops, or may persist, filling the space between the ornaments. In a few cases, e.g. some species of *Coprinus*, the perisporium forms a loosely attached layer which surrounds the spore as a loose envelope, the **perisporial sac**. The outer membrane of the perisporium is termed the ectosporium.

Germination of many basidiospores is through a germ pore, usually at the apex of the spore. In the region of the germ pore, the wall layers with the exception of the endosporium are thin.

Zygospores

Zygospores are sexually produced spores formed as a result of plasmogamy between gametangia which are usually equal in size. They are resting structures. Zygospores are the typical sexually produced spores of Zygomycetes, e.g. Mucorales (p. 191) and Entomophthorales (p. 235). Zygospores are often large, thick-walled, warty structures with large food reserves and are unsuitable for long-distance dispersal. They usually remain in the position in which they are formed, awaiting suitable conditions for further development. The gametangia which fuse to form the zygospore may be uninucleate or multinucleate, and correspondingly the zygospore, within which nuclear fusion and meiosis eventually occur, may have one, two or many nuclei within it. Nuclear fusion may occur early, or may be delayed until shortly before zygospore germination. The structure of some different kinds of zygospore is illustrated in Figs. 92, 94 and 95.

Oospores

An oospore is a sexually produced spore which develops from unequal gametangial copulation or markedly unequal gametic fusion. It is the characteristic sexually produced spore of the Oomycetes. Oospores develop from fertilised **oospheres**, or sometimes parthenogenetically. One or more oospheres develop within **oogonia**, which are multinucleate, globose female gametangia. The number of oospores per oogonium may be important taxonomically. In the Saprolegniales, there are usually, but not invariably, several oospores per oogonium, whilst in the Peronosporales there is typically only one. The nuclear divisions which precede oosphere maturation in Oomycetes are meiotic and, following penetration of an oosphere by a fertilisation tube from an antheridium, a haploid male gamete is introduced. Nuclear fusion follows. The oospore or egg develops a thick outer wall, and lays down food reserves usually in the form of lipids. Some different kinds of oospore are illustrated in Figs. 64, 65, 71 and 73.

Chlamydospores

In most groups of fungi, terminal or intercalary segments of the mycelium may become packed with food reserves and develop thick walls. The walls may be colourless or pigmented with dark melanic pigments. Structures of this type have been termed chlamydospores. Generally there is no mechanism for detachment and dispersal of chlamydospores, but they become separated from each other by the disintegration of intervening hyphae. They thus remain *in situ* on the substratum which bore them. They form important organs of asexual survival, especially in soil fungi. For example, in *Absidia glauca*, a common soil fungus, large intercalary and terminal chlamydospores, packed with lipid reserves, develop on the mycelium following the formation of septa (Fig. 39B), and it is probable that when the substratum on which *A. glauca* grows is exhausted, the chlamydospores survive. Similar structures are found in old hyphae of the aquatic fungus *Saprolegnia* (see p. 153), either singly or in chains. In this case, the chlamydospores may break free from the mycelium

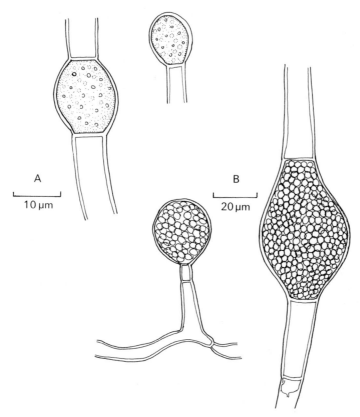

Fig. 39. Chlamydospores. A. *Trichoderma viride*: intercalary and terminal chlamydo-spores. B. *Absidia glauca*: intercalary and terminal chlamydospores.

and be dispersed in water currents. Chlamydospores which are dispersed in this way are termed **gemmae**.

The term chlamydospore is also used to describe the thick-walled dikaryotic spore characteristic of smut fungi, the Ustilaginales (p. 483).

Conidiospores

Conidiospores, commonly known as conidia, are asexual reproductive structures. They are found in many different groups of fungi, but especially in Ascomycotina, Basidiomycotina and Deuteromycotina. In the last-named group, they are the only means of reproduction: sexually reproduced spores are not known. The term conidium has, unfortunately, been used in a number of different ways, so that it no longer has any precise meaning. It has come to mean 'any asexual spore (but not a sporangiospore or intercalary chlamydo-spore)' (Ainsworth *et al.*, 1971). There is great variation in conidial ontogeny.

This topic will be dealt with more fully later when considering the conidial states of Ascomycotina and Deuteromycotina (p. 517), but at this stage it is sufficient to distinguish between the major types of conidial development, which may be either **thallic** or **blastic**. Cells which produce conidia are **conidiogenous cells**. The term thallic is used to describe development where there is no enlargement of the conidium initial (Fig. 40A), i.e. the conidium arises directly from a pre-existing segment of the fungal thallus without enlargement. An example of this kind is *Endomyces geotrichum*, in which the conidia are formed by dissolution of septa along a hypha (p. 269 and Fig. 133). In most conidia, development is blastic, i.e. there is enlargement of the conidium initial before it is delimited by a septum. Two main kinds of blastic development have been distinguished:

(*a*) **Holoblastic,** in which both the inner and outer wall layers of the conidiogenous cell contribute to conidium formation (Fig. 40B). Examples of

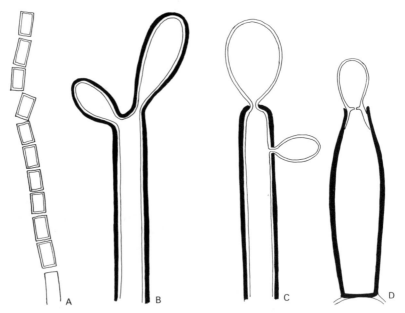

Fig. 40. Diagrams to illustrate different methods of conidial development. A. Thallic development: no enlargement of the conidium initial. B. Holoblastic development: all layers of the wall of the conidiogenous cell balloon out to form a conidium which is recognisably larger than the conidiogenous cell. C. Enteroblastic tretic development: only the inner wall layers of the conidiogenous cell are involved in conidium formation. The inner wall layers balloon outwards through a narrow channel in the outer wall. D. Enteroblastic phialidic development: the conidiogenous cell is a phialide. The wall of the phialide is not continuous with the wall surrounding the conidium. The conidial wall arises *de novo* from newly-synthesised material in the neck of the phialide. Diagrams based on Ellis (1971b).

this kind of development are shown by the conidia of *Pleospora herbarum* (Fig. 215) or *Cladosporium herbarum* (Fig. 290).

(*b*) **Enteroblastic,** in which only the inner wall, or a completely new wall layer, is involved in conidium formation. Where the inner wall layer balloons out through a narrow pore or channel in the outer wall layer, development is described as **tretic** (Fig. 40c). An example of enteroblastic tretic development is shown in *Helminthosporium velutinum* (Fig. 292). Another important method of enteroblastic development is termed **phialidic** development. Here the conidiogenous cell is a specialised cell termed the **phialide**. During the formation of the first-formed conidium, the tip of the phialide is ruptured. Further conidia develop by the extension of a mass of cytoplasm enclosed by a new wall layer distinct from the wall of the phialide, and laid down in the neck of the phialide. The protoplast of the conidium is pinched off by the formation of an inwardly growing flange of new wall material which closes to form a septum (Fig. 40d). New conidia develop beneath the earlier ones, so that a chain may develop with the oldest conidium at its apex and the youngest at its base. Details of phialidic development are discussed more fully in relation to *Aspergillus* (Fig. 153) and *Penicillium* (Fig. 154) which reproduce by means of chains of dry **phialoconidia** dispersed by wind. Sticky phialospores which accumulate in slimy droplets at the tips of the phialides are common in many genera, e.g. *Trichoderma* (Fig. 179), *Gliocladium* (Fig. 179), *Verticillium* (Fig. 184) and *Nectria* (Figs. 182–184). Sticky phialoconidia are usually dispersed by insects, rain-splash or other agencies.

As mentioned on p. 74, the term conidium is sometimes used for structures which are probably homologous to sporangia. A series can be erected in the Peronosporales in which there are forms with deciduous sporangia, which release zoospores when in contact with water (e.g. *Phytophthora*), and other forms which germinate directly, i.e. by the formation of a germ tube (e.g. *Peronospora*). A similar series can be erected in the Mucorales, where in some forms (e.g. *Thamnidium*, *Choanephora*) few-spored sporangia, termed **sporangiola**, may be dispersed, whilst in others there are unicellular 'conidia'. There is evidence to suggest that the conidia should be interpreted as one-spored sporangiola.

There are numerous other kinds of spore found in fungi, and these are described more fully later, in relation to the particular groups of fungi in which they are found.

LIFE CYCLES

Although there is enormous variation in the details of life cycles in fungi (Raper, 1966b), these can be reduced to a limited number of basic patterns.

1. Sexual cycle

The essential events in the life cycle of a sexually reproducing fungus are

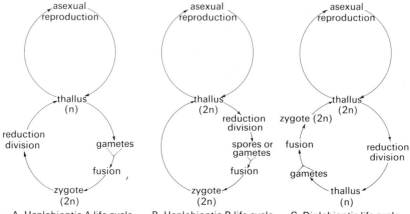

A. Haplobiontic A life cycle   B. Haplobiontic B life cycle   C. Diplobiontic life cycle

Fig. 41. Life cycles of sexually reproducing fungi.

nuclear fusion and meiosis (reduction division). These two events may occur at different times, and their relative position will determine the kind of life cycle. We may distinguish three kinds of sexual cycle.

(a) Haplobiontic A life cycle. The term haplobiontic means single generation. In this kind of life cycle (Fig. 41A), the nuclei follow the sequence

nuclear fusion→reduction→mitosis.

Because nuclear fusion is followed immediately by reduction division, the vegetative cells (or **soma**) are haploid. This condition is found in most fungi, except the Oomycetes and the Basidiomycotina, and is the typical condition of, for example, the Mucorales and the Ascomycotina.

(b) Haplobiontic B life cycle. In this kind of life cycle (Fig. 41B), the nuclei follow the sequence

nuclear fusion→mitosis→reduction.

Because nuclear fusion is followed by mitosis, the vegetative phase of the organism is diploid, i.e. it has a diploid soma. Life cycles of this type are found in many Oomycetes (p. 140), and in the yeast *Saccharomyces cerevisiae*.

In many Basidiomycotina, the haplobiontic B life cycle is modified. Reduction division or meiosis occurs in a specialised cell, the basidium, which then gives rise usually to four haploid basidiospores. In heterothallic forms, segregation of the genetic factors which control mating occurs during meiosis in the basidium, so that individual basidiospores differ in mating type. The basidiospore germinates to produce a mycelium with simple septa, dividing the mycelium into uninucleate segments. This mycelium is described as **monokaryotic** (i.e. it is a **monokaryon**). Since all the haploid nuclei in the mycelium are derived by mitosis from the original haploid nucleus of the basidiospore, they are genetically identical. The mycelium can therefore also be described as **homokaryotic** (or as a **homokaryon**). When compatible haploid monokaryotic mycelia make contact with each other, anastomosis of the hyphae (plasmogamy) occurs. However, nuclear fusion does not immediately occur. Nuclear division of the two genetically distinct nuclei occurs and the mycelium becomes segmented into cells each of which contains a pair of nuclei of compatible mating type. Such a mycelium is said to be **dikaryotic** (i.e. it is a **dikaryon**). A special process of nuclear rearrangement occurs at the hyphal apex. This is accompanied by the formation of a backwardly directed branch which grows around the septum and fuses with the penultimate cell, to allow nuclear transfer. This structure characteristic of a dikaryotic mycelium is termed a **clamp connection**. The formation of clamp connections is described more fully on p. 406. The basidia are formed on fruit-bodies which develop on the dikaryotic mycelium. Thus there may be a long delay between plasmogamy and karyogamy. The intervening period is occupied in the dikaryotic condition. Whilst the dikaryotic segments cannot be described as diploid in the sense that two homologous sets of chromosomes are enclosed in a single nuclear envelope, nevertheless two homologous sets of chromosomes are enclosed in a single cell. For this reason, the term **dikaryotic diploid** is sometimes used to describe this state. Thus the thallus or soma is a dikaryotic diploid, and conforms to the generalised pattern found in other organisms with the haplobiontic B life cycle.

(c) Diplobiontic life cycle. Here there are two generations in the life cycle, i.e. there is an alternation between a haploid and diploid generation. Life cycles of this type are comparatively rare in fungi, but have been described in certain members of the Blastocladiales, especially in the section *Eu-Allomyces* of the genus *Allomyces* (Fig. 57). The life cycle (Fig. 41c) may be written in the form

fusion→mitosis→reduction→mitosis.

In *Allomyces*, the alternation of generations is **isomorphic**, i.e. the haploid gametophytic generation is morphologically similar to the diploid sporophytic generation. The gametophytes bear male and female gametangia, which give rise to motile gametes of differing size. The zygote formed by fusion of the gametes develops directly into a sporophytic thallus which bears thin-walled sporangia and thick-walled sporangia. In the latter, meiosis occurs to give rise to haploid zoospores which develop into gametophytic thalli.

2. Non-sexual cycle

There are many fungi in which sexual reproduction has not been observed, and in which it probably does not occur. The large assemblage of fungi lassified as Deuteromycotina are all forms in which sexual reproduction is lacking. Because the sexually produced spore is termed the **perfect spore**, and the fruit-body which contains it is termed the **perfect state**, the Deuteromycotina are also known as the Fungi Imperfecti. Forms lacking true sexual reproduction have probably been derived from sexually reproducing ancestors, and in many of the major groups of sexual fungi, related forms are known in which sexual reproduction apparently does not occur. Examples of this kind include *Mucor ramannianus* (= *Mortierella ramanniana*), an extremely common fungus of acid soil for which no zygospore state has been described, and *Aspergillus niger,* a common fungus from warmer soils, for which no ascocarps are known. Both these fungi have been grown in culture on a wide range of media by numerous investigators, and it would be surprising if they were shown to reproduce sexually. Some fungi produce resting spores which resemble closely in their development the sexually produced spores of related fungi, but there is no evidence that nuclear fusion or meiosis are involved in their formation. Good examples of this are found in the Saprolegniales (Oomycetes), some species of which develop oogonia containing oospores, but without antheridia. Such a condition is described as **parthenogenetic, apogamous** or **amictic**.

3. Parasexual cycle

Despite the fact that Fungi Imperfecti apparently lack a true sexual cycle, they are amazingly abundant and successful in all kinds of habitats, terrestrial and aquatic, either as saprotrophs or as parasites. It is unlikely that such fungi could have survived in the absence of some mechanism for genetic recombination to allow for adaptation to environmental change both in the long and short term. The efforts of plant breeders to breed varieties of host plant resistant to Deuteromycete pathogens such as species of *Fusarium* and *Verticillium* have frequently been frustrated by the appearance of 'new' races of pathogens capable of causing severe disease symptoms in 'resistant' varieties. There is evidence that the new pathogenic forms may have arisen as a result of genetic recombination in a cycle termed the parasexual cycle. Reference has already been made to this sequence of events in relation to *Dictyostelium* (p. 17), and laboratory experiments point to its existence in many filamentous fungi including species of *Aspergillus, Penicillium, Fusarium, Verticillium* and in rust fungi (Uredinales). The essential events are anastomosis of genetically distinct hyphae to form a heterokaryon, rare nuclear fusion to form diploid nuclei, mitotic crossing over, and haploidisation (Pontecorvo, 1956; Roper, 1966). True meiosis does not occur. A result of the discovery of parasexual recombination is that it is possible to follow inheritance, i.e. study genetics of Fungi Imperfecti. In *Emericella nidulans*, which reproduces sexually by the formation of ascospores, and asexually by the formation of conidia, it is possible to manipulate culture conditions to prevent sexual reproduction. In this species, it is therefore possible to compare genetical recombination by means of the normal sexual process, involving the formation of ascospores following meiosis, with parasexual recombination involving conidia only and excluding ascospore formation and meiosis. Chromosome maps showing the distribution of marker genes along the different chromosomes show a remarkable degree of correspondence using the two different methods. The phenomenon of parasexual recombination is discussed more fully on p. 252.

PHYSIOLOGY AND NUTRITION

The Eumycota, like the Myxomycota, are **chemoheterotrophic**. They obtain carbon compounds from non-living organic material, as saprophytes, or from living tissues of animals and plants, as symbionts. The term symbiosis is used here in the sense of Lewis (1974) as 'associations of usually dissimilar organisms which exhibit permanent, or at least prolonged, intimate contact'. This definition returns to the original concept of the term by de Bary (1887). De Bary included, within the meaning of symbiosis, parasitic and mutualistic associations (i.e. associations in which both partners derive a mutual benefit). Lewis (1973, 1974) has classified fungi into five groups based on their ecological and nutritional behaviour as shown in table 2.

It is not possible in this book to discuss at length the nutrition and physiology of the Eumycota. The nutritional requirements of Group 1, the obligate saprotrophs, are the best understood. The ease with which they can be isolated and grown in pure culture has led to extensive studies of their nutrition (see Cochrane, 1958; Ainsworth & Sussman, 1965). Some of these fungi have been used in industrial fermentations to produce such diverse products as alcohol, organic acids, enzymes, antibiotics and other drugs (Gray, 1959; Hesseltine, 1965; Smith, 1969; Hawker & Linton, 1971). Whilst the majority of these fungi are aerobic, some are capable of anaerobic growth by fermentation.

Parasitic symbionts can be classified into two broad categories, necrotrophic and biotrophic parasites. Necrotrophic parasites are those which cause extensive tissue damage and usually bring about the death of their hosts. The facultative necrotrophs (Group 2 of table 2) appear capable of prolonged existence in the saprotrophic state, in soil. Given suitable conditions and available substrata (e.g. young tissues in poorly drained soil), they may cause diseases such as damping-off, foot rots and pre-emergence killing of seedlings. Species of *Pythium* (p. 163) and *Rhizoctonia* are excellent examples of facultative necrotrophs. For a fuller discussion of their biology, see Garrett (1970). Obligate necrotrophs (Group 3 of table 2) appear to have a limited capacity for saprotrophic growth, and may be largely confined in the vegetative phase to dead or dying colonised host plant tissue. They can be regarded as relatively specialised parasites (Garrett, 1970).

Biotrophic parasites are those which in ecological situations are dependent on living host tissues. The most important are the obligate biotrophs (Group 5).

Table 2. *Groups of fungi based on ecological and nutritional behaviour*

Group 1. *Ecologically obligately saprophytic saprotrophs (obligate saprotrophs)*
This group comprises the typical saprophytes with no capacity for parasitic or mutualistic symbiosis.

Group 2. *Ecologically facultative symbionts, whose nutrition is necrotrophic in the parasitic mode, but otherwise saprotrophic (facultative necrotrophs)*
This group embraces both facultative saprophytes and facultative parasites of de Bary (1887) and includes species classed as unspecialised parasites. The distinction between facultative saprophytes and facultative parasites has always been a quantitative one, ranging from those species which are normally saprophytes but can exist as parasites (e.g. *Rhizopus stolonifer*) to those, like the fungi responsible for damping-off (*Pythium, Rhizoctonia*, etc.), which are equally effective as saprophytes or parasites. The group also includes necrotrophic mycoparasites.

Group 3. *Ecologically obligately symbiotic necrotrophs (obligate necrotrophs)*
This group, essentially the extreme form of Group 2, includes those specialised parasites such as *Gaumannomyces graminis*, the cause of 'take-all' disease of wheat, vascular wilts (*Fusarium, Verticillium*, etc.) and wood-rotting fungi whose saprophytic life is largely restricted to mere survival in dead, infected host tissue. Their life thus alternates between an expanding parasitic phase and a declining saprophytic one.

Group 4. *Ecologically facultative symbionts, whose nutrition is biotrophic in the symbiotic mode, but otherwise saprotrophic (facultative biotrophs)*
This group consists of some facultative mycorrhizal fungi and some facultative lichens. The category is also available for any parasitic or mutualistic fungi in Group 5 which can be shown to have an independent mycelial existence *in the field*.

Group 5. *Ecologically obligately symbiotic biotrophs (obligate biotrophs)*
This group includes both mutualistic and parasitic members, embracing the fungi of vesicular–arbuscular and sheathing mycorrhizas, and lichens, smuts (Ustilaginales), leaf curls (Taphrinales), rusts (Uredinales), powdery mildews (Erysiphales), downy mildews and white blisters (of the Peronosporales), and the Plasmodiophorales.[a] It also includes both contact and haustorial mycoparasites and some members of the Chytridiales, Blastocladiales and Lagenidiales, as well as species such as *Phytophthora infestans, Venturia inaequalis, Epichloe typhina* and *Claviceps purpurea*.

After Lewis (1973).
[a] Classified here within the Myxomycota, see p. 42.

They include a number of serious groups of plant pathogens, notably the downy mildews and blister rusts (Peronosporales), powdery mildews (Erysiphales), smut fungi (Ustilaginales) and rust fungi (Uredinales). Although it has not yet proved possible to grow the downy or powdery mildews in the absence of living host tissue, limited growth of some smut fungi is possible. In recent times, a number of rust fungi have been cultivated away from their living hosts (Scott & Maclean, 1969), and this has blurred the distinction between 'obligate' biotrophic parasites and 'facultative' biotrophic parasites.

It is for this reason that Lewis has stressed the term 'ecologically' in his classification. It is virtually certain that none of the fungi in Group 5 exists in the free-living mycelial state in nature in the absence of a living host, although they may, of course, survive in various reproductive forms such as spores, sporangia, oospores, or sclerotia.

Biotrophic parasites have much in common nutritionally with various mutualistic symbionts, notably with certain kinds of mycorrhizal fungi (e.g. the vesicular–arbuscular type of mycorrhiza, caused by members of the Endogonaceae (p. 233, and Mosse, 1973) and with the sheathing mycorrhizas (often termed ectotrophic mycorrhizas) of forest trees (Harley, 1969; Smith *et al.,* 1969). For example, it has been shown that, although the products of photosynthesis of their green plant hosts normally occur in the form of sugars such as glucose, fructose or sucrose, transfer of carbohydrate to the fungus, both in the mutualistic and parasitic biotrophs, is accompanied by conversion to characteristic 'fungal' carbohydrates such as trehalose and glycogen, or polycyclic alcohols such as mannitol.

Lewis (1974) has argued that necrotrophic parasites evolved from saprotrophs. The biotrophic condition may have evolved directly from the saprotrophic state in some cases, but in others it is more likely that biotrophic forms arose from necrotrophic ancestors.

CLASSIFICATION OF THE EUMYCOTA

The Eumycota are divided into five major groups, distinguished from each other as shown in the Key below (Ainsworth, 1973). The Deuteromycotina is an artificial assemblage, probably consisting of forms related to Ascomycotina and Basidiomycotina which have lost the ability to reproduce sexually (i.e. by forming 'perfect' spores as a result of nuclear fusion and meiosis), or forms whose perfect stage has not yet been discovered or connected to the imperfect (conidial) state.

Key to subdivisions of Eumycota

1 Motile cells (zoospores) present; perfect-state
  spores typically oospores   **Mastigomycotina (p. 98)**
  Motile cells absent   2

2 Perfect state present   3
  Perfect state absent   **Deuteromycotina (p. 516)**

3 Perfect-state spores zygospores   **Zygomycotina (p. 191)**
  Zygospores absent   4

4 Perfect-state spores ascospores   **Ascomycotina (p. 248)**
  Perfect-state spores basidiospores   **Basidiomycotina (p. 395)**

1

Mastigomycotina
(Zoosporic Fungi)

The Mastigomycotina are zoosporic fungi. Three distinct kinds of zoospore occur; with a single posterior flagellum of the whiplash type, with a single anterior flagellum of the tinsel type, and biflagellate zoospores with apically or laterally attached flagella, one of the tinsel type and one of the whiplash type. Details of the structure of these three kinds of zoospore have been described on p. 75–83. Zoospore structure is an important feature in the classification of Mastigomycotina, and three classes have been separated, each characterised by their distinctive type of zoospore (Ainsworth, 1973).

Key to classes of Mastigomycotina

1 Zoospores posteriorly uniflagellate
  (flagella whiplash-type)   **Chytridiomycetes (p. 98)**
  Zoospores not posteriorly uniflagellate   2

2 Zoospores anteriorly uniflagellate
  (flagella tinsel-type)   **Hyphochytridiomycetes**
  Zoospores biflagellate (posterior flagellum whiplash-type;
  anterior tinsel-type); cell wall cellulosic   **Oomycetes (p. 140)**

The different structure of the zoospore suggests that the Mastigomycotina may be polyphyletic; that is their evolutionary origins are to be found in unrelated groups of organisms. This means that the group may not be a natural one, but is merely an assemblage of organisms which resemble each other only to the extent of reproduction by means of zoospores. An introduction to the literature on the taxonomy of the group, with particular reference to zoospore structure, has been given by Sparrow (1973a).

1: CHYTRIDIOMYCETES

The Chytridiomycetes have one feature in common, the posteriorly uniflagellate zoospore of the whiplash type, and it has been suggested (Sparrow, 1958) that the ancestry of these organisms lies in the posteriorly uniflagellate Monads. The characters by which the group is separated into three orders are shown below (Sparrow, 1960).

(*a*) Thallus either lacking a vegetative system and converted as a whole into reproductive structures (holocarpic) or with a specialised rhizoidal vegetative system (eucarpic) and one (monocentric) or more (polycentric) reproductive structures; zoospore usually bearing a single conspicuous oil globule, germination monopolar

Chytridiales

(*b*) Thallus nearly always differentiated into a well-developed vegetative system, often hypha-like, on which are borne numerous reproductive organs: zoospore without a conspicuous globule, germination bipolar.

(i) Thallus usually having a well-defined basal cell anchored in the substratum by a system of tapering rhizoids; resting structure an asexually formed, thick-walled, often punctate, resting spore; sexual reproduction by means of isogamous or anisogamous planogametes; alternation of generations present in some species

Blastocladiales

(ii) Thallus without a well-defined basal cell; composed of delicate much-branched hyphae; resting structure an oospore; sexual reproduction oogamous, the male gamete always free-swimming, the female devoid of a flagellum   **Monoblepharidales**

More recently, a fourth order, the Harpochytriales, has been added (Emerson & Whisler, 1968). For Keys to the families and important genera of the Chytridiomycetes, see Sparrow (1973b).

CHYTRIDIALES

Members of this group are mostly aquatic, growing saprophytically on plant and animal remains in water or parasitically in the cells of algae and small aquatic animals. Some are terrestrial, apparently growing saprophytically on various plant and animal substrata in soil, whilst some attack the underground parts and also the aerial shoots of higher plants, occasionally causing diseases which are of economic significance, as in the case of *Synchytrium endobioticum* which causes black wart disease in potato. The chytrids which parasitise algae may cause severe depletion in the population level of the host alga. Some of the soil- and mud-inhabiting chytrids can decompose cellulose, chitin and keratin, and 'baits' composed largely of these materials such as cellophane, shrimp exoskeleton, snake skin and hair, if floated in water to which soil has been added frequently become colonised by chytrid zoospores which give rise to mature thalli. Details of such methods for isolating these organisms have been given by Sparrow (1957, 1960), Willoughby (1956, 1958) and Emerson (1958). Some of the saprophytic forms have been grown in pure culture and information is accumulating about their nutrition and physiology. The nutritional requirements of most species which have been investigated are simple, and media containing mineral salts and carbohydrate in the form of sugar, starch or cellulose will support growth. As a group the chytrids have the capacity to reduce sulphate and to utilise both nitrate and ammonium salts for growth, and it has been argued that these are primitive nutritional characteristics (Cantino & Turian, 1959). A further 'primitive' feature is the ability of some chytrids to synthesise essential vitamins for growth, so that growth occurs on unsupplemented media, but it is also known that some members of the group are heterotrophic for vitamins, and their growth is stimulated by the addition of thiamine and other vitamins (Goldstein, 1960a,b, 1961).

The cell walls of some chytrid thalli have been examined microchemically and by X-ray diffraction, and chitin has been detected (Bartnicki-Garcia, 1968). The composition of the wall is of interest because chitin, a polymer of *N*-acetylglucosamine is also present in the walls of the Blastocladiales, Monoblepharidales, Zygomycotina, Ascomycotina, Basidiomycotina and Deuteromycotina whilst the cell walls of members of the Oomycetes are composed of cellulose, a linear polymer of D-glucose. Cellulose and chitin have been detected by X-ray diffraction methods occurring together in the walls of a species of *Rhizidiomyces*, a member of the Hyphochytridiomycetes (Aronson, 1965).

The form of the thallus in the Chytridiales is very variable. In the morphologically simpler types, such as *Olpidium* and *Synchytrium*, the mature thallus is a spherical or cylindrical sac surrounded by a wall. There are no rhizoids and the entire cytoplasmic contents of the thallus become transformed into reproductive structures, zoospores or gametes. Thalli of this type in which the whole structure is reproductive in function are described as holocarpic. In many other chytrids the thallus is differentiated into a vegetative part, concerned primarily with the collection of nutrients from the substratum, and a reproductive part which gives rise to zoospores or gametes. This type of thallus construction, which is the most common type, is termed eucarpic. The relationship between the rhizoidal system, the sporangium and the substratum is also variable. In some chytrids such as *Rhizophydium* the rhizoidal system only penetrates the host cell (often an algal cell or a pollen grain) and the sporangium is superficial or **epibiotic**. In others, such as *Diplophlyctis*, the whole thallus, rhizoids and sporangium are formed within the host cell, and are described as **endobiotic**. In *Physoderma* both epibiotic and endobiotic sporangia are found. Whilst in many types the zoospore on germination gives rise to a rhizoidal system bearing a single sporangium or resting spore (e.g. in *Entophlyctis*, *Rhizophlyctis* and *Diplophlyctis*), in others such as *Cladochytrium* and *Nowakowskiella* a more extensive rhizoidal system, sometimes termed the **rhizomycelium**, is established, on which numerous sporangia develop. Such thalli are polycentric, that is they form several reproductive centres instead of the single one where the thallus is termed monocentric. These types of thallus structure are illustrated in Fig. 42. When a range of isolates of certain fungi are studied, it may be found that some are monocentric, whilst others are polycentric. Monocentric and polycentric forms of

Holocarpic

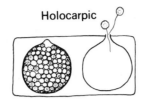

Eucarpic

Monocentric   Polycentric

Endobiotic

Epibiotic

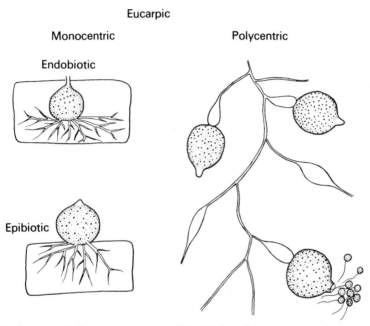

Fig. 42. Types of thallus structure in the Chytridiales, diagrammatic and not to scale.

Rhizophlyctis rosea have been described. Thus the terms monocentric and polycentric are useful in a descriptive sense, but the distinction between the two conditions may not always be clear-cut.

The zoosporangium is usually a spherical or pear-shaped sac bearing one or more discharge tubes or exit papillae. The method by which zoospore release is achieved is used in classification. In the **inoperculate** chytrids such as *Olpidium*, *Diplophlyctis* and *Cladochytrium* the sporangium forms a discharge tube which penetrates to the exterior of the host cell and its tip becomes gelatinous and dissolves away. In **operculate** chytrids such as *Chytridium* and *Nowakowskiella* the tip of the discharge tube breaks open at a special line of weakness and becomes detached as a special cap or operculum or in some forms may remain attached at one side of the discharge tube and fold back like a hinged lid to allow the zoospores to escape.

The numbers of zoospores formed inside zoosporangia varies with the size of the zoospore and the zoosporangium. Although the size of the zoospore is roughly constant for a given species the size of the sporangium may be very variable. In *Rhizophlyctis rosea* tiny sporangia containing only one or two zoospores have been reported from culture media deficient in carbohydrate but on cellulose-rich media large sporangia containing many thousands of spores are commonly formed. The release of zoospores is brought about by internal pressure which causes the exit papillae to burst open. In studies of the fine structure of mature sporangia of *R. rosea* and *Nowakowskiella profusa* (Chambers & Willoughby, 1964; Chambers *et al.*, 1967), it has been shown that the single flagellum is coiled round the zoospore like a watch-spring. The zoospores are separated by a matrix of spongy material which may absorb water and swell rapidly at the final stages of sporangial maturation. When the internal pressure has been relieved by the escape of some zoospores, those remaining inside the sporangium may escape by swimming or wriggling through the exit tube. In some species the spores are discharged in a mass which later separates into single zoospores, but in others the zoospores make their escape individually. The external form of the zoospore is similar in all chytrids, but in the details of internal structure there is considerable variation. There is a spherical body, which in some forms is capable of plastic changes in shape, and a long trailing whiplash type of flagellum. When swimming, the zoospores glide through the water, often making abrupt changes in direction and also showing characteristic jerky or 'hopping' movements. In some species amoeboid crawling of zoospores has been reported. The internal structure of the zoospore as revealed by the light microscope and the electron microscope is somewhat variable (see Koch, 1956, 1958, 1961). The tip of the flagellum is often tapered into a narrower portion which may end in a bulbous enlargement. An interesting feature of certain (but not all) chytrid zoospores is the presence of a second kinetosome attached or close to the one bearing the flagellum. Koch has termed the second kinetosome 'non-functional'. Its presence has aroused speculation that the ancestors of these chytrids may have been biflagellate, and that the 'non-functional' kinetosome is a relic of the second flagellum which has disappeared (see Olson & Fuller, 1968). However, it has been argued that the kinetosome represents a modified centriole, and since centrioles normally divide prior to mitotic nuclear division, and commonly persist in pairs, the presence of two kinetosomes cannot be used as evidence for an originally biflagellate condition. In some zoospores, the 'functional' kinetosome is connected to a fibrillar rhizoplast which may end in a flattened grid-like disk probably attached to a lipoid body close to the nucleus. Possibly the disk is a photoreceptor (Chambers *et al.*, 1967). The nucleus itself is single and is surrounded in many cases (but not all) by a 'nuclear cap' of uneven thickness. The nuclear cap is usually thinner near the point of attachment of the rhizoplast to the nucleus. Beyond the nuclear cap in chytrid zoospores there is a closely apposed group of mitochondria. The most conspicuous feature within the zoospore is a large refringent globule which is a lipoid body, and in some zoospores other smaller lipoid bodies may be found. It has been suggested that the lipoid body forms a concentrated food reserve to provide energy for zoospore movement, or that it may function as a lens, concentrating light on the grid-like plate.

The period of zoospore movement varies. Some flagellate zoospores seem to be incapable of active swimming, and amoeboid crawling may take place instead, or swimming may last for only a few minutes. In others motility may be prolonged for several hours. On germination the zoospore comes to rest and encysts. The flagellum may contract, it may be completely withdrawn or it may be cast off, but the precise details are often difficult to follow. The subsequent behaviour differs in different species. In holocarpic parasites the zoospore encysts on a host cell and the cyst wall and host-cell wall are dissolved and the cytoplasmic contents of the zoospore enter the host cell. In many monocentric chytrids rhizoids develop from one point on the zoospore cyst and the cyst itself enlarges to form the zoosporangium, but there are variants of this type of development in which the cyst enlarges into a prosporangium from which the zoosporangium later develops. In the polycentric types the zoospore on germination may form a limited rhizomycelium on which a swollen cell arises, giving off further branches of rhizomycelium. It has been claimed that germination from a single point on the wall of the zoospore cyst (monopolar germination) as distinct from two points, enabling growth to take place in two directions (bipolar germination), is an important character distinguishing the Chytridiales (monopolar) and the Blastocladiales (bipolar), but this distinction is not absolute. Indeed the distinction between the two groups is becoming less clear. The morphological resemblance between certain forms is certainly most striking (compare *Rhizophlyctis rosea* and *Blastocladiella emersonii*).

CLASSIFICATION

Sparrow (1960, 1973b) has provided keys separating the Chytridiales into families. The primary separation is based on the presence or absence of an operculum. Whiffen (1944) has objected to this method of classifying chytrids on the grounds that this character is of minor significance. She has emphasised instead the development of the thallus in relation to zoospore germination. For example, in *Rhizophydium* and *Rhizidium* the encysted zoospore enlarges into a zoosporangium, whilst in *Polyphagus* the encysted zoospore enlarges into a prosporangium from which the zoosporangium develops. In *Entophlyctis* the zoosporangium develops from an enlargement of the germ tube whilst in *Diplophlyctis* the germ tube enlarges to form a prosporangium from which the zoosporangium develops. On the basis of these and other differences Whiffen has proposed a division of the Chytridiales which she claims results in a more natural taxonomic grouping. However, the detailed classification need not concern us, and we shall study only selected examples.

Olpidiaceae

In this family the thallus is endobiotic and holocarpic and becomes entirely converted into a zoosporangium or resting sporangium. Sexual reproduction occurs by fusion of motile isogametes to form a biflagellate zygote which penetrates the host to form an endobiotic resting sporangium, which on germination gives rise to zoospores. The members of the family are mostly aquatic, but species of *Olpidium* are parasitic on the roots of higher plants.

Olpidium (Fig. 43)

Olpidium is a good example of the holocarpic type of thallus. About thirty species are known, most of them parasitic on aquatic algae, microscopic aquatic plants or the spores of various plants which fall into water or the soil. Other species are parasitic on moss protonemata and on leaves and roots of higher plants (Macfarlane, 1968; Johnson, 1969). *O. brassicae* is common on the roots of cabbages, especially when growing in wet soils, but is also found on a wide range of unrelated hosts. On lettuce a fungus morphologically similar to *O. brassicae* has been shown to be associated with symptoms of big-vein disease (yellow-vein banding sometimes associated with leaf puckering). However, not all lettuces showing big-vein symptoms contain *Olpidium*, and plants heavily infected with *Olpidium* may lack big-vein symptoms. Because of the characteristic syndrome and because the disease can be transmitted by grafting of lettuce shoots in the absence of *Olpidium* infection it has been concluded that a virus is involved. At first the virus was thought to be tobacco necrosis virus but some workers now regard it as a distinct big-vein virus. There is now much evidence to suggest that *Olpidium* can serve as a vector of several viruses. The virus is possibly carried either on or inside zoospores, and although there is no evidence that the virus can multiply inside the *Olpidium* thallus it can probably survive several months within the thick-walled resting sporangia and subsequently give rise to infection (Hewitt & Grogan, 1967).

The strain of *Olpidium* on lettuce is possibly not identical with that from cabbages since attempts to cross-inoculate the fungus from one host to the other have been unsuccessful. Sahtiyanci (1962) has described it as a distinct species. On cabbages infection seems to have little effect. When infected cabbage roots are washed in water the fungus can be seen under the low power of the microscope within the epidermal cells (and occasionally in the cortical cells) and root hairs (Fig. 43). The *Olpidium* thalli are spherical or cylindrical, and there may be one to several in a host cell. There are no rhizoids. Occasionally a large cylindrical thallus completely fills an epidermal cell. The cytoplasm of the thallus is granular and the entire contents divide into numerous posteriorly uniflagellate zoospores which escape through one or more discharge tubes from the thallus penetrating the outer wall of the host

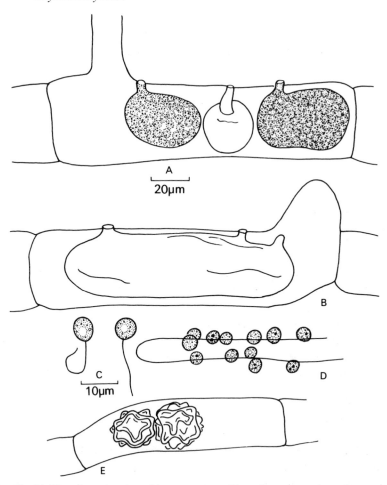

Fig. 43. *Olpidium brassicae* in cabbage roots. A. Two ripe sporangia and one empty sporangium in an epidermal cell. Each sporangium has a single exit tube. B. Empty sporangium showing three exit tubes. C. Zoospores. D. Zoospore cysts on a root hair. Note that some cysts are uninucleate and some are binucleate. E. Resting sporangia. A, B, D, E to same scale.

cell and opening to the exterior. The release of the zoospores takes place within a few minutes of washing the roots free from soil. The tip of the discharge tube breaks down and zoospores rush out and swim actively in the water. The zoospores are very small, tadpole-like, with a spherical head and a trailing flagellum. At high magnification under the light microscope the spherical body of the zoospore is seen to contain a single globose lipoid body but electron microscope studies show that there are several lipid globules closely associated together (see p. 107). The flagellum narrows at its tip to a whiplash point. The zoospores swim actively in water for about 20 minutes. If roots of cabbage seedlings are placed in the zoospore suspension the zoospores settle on the root hairs and lose their flagella. The root hair is penetrated and the cytoplasmic contents of the encysted zoospore are transferred to the inside of the root hair, whilst the empty zoospore cyst remains attached to the outside. The process of penetration can take place in less than 1 hour. Within 2 days of infection small spherical thalli can be seen in the root hairs and epidermal cells of the root often carried around the cell by cytoplasmic streaming. The thalli enlarge and become multinucleate and within 4–5 days discharge tubes are developed and the thalli are ready to discharge zoospores.

In some infected roots, in addition to the smooth zoosporangia with their discharge tubes, stellate bodies with thick folded walls, lacking discharge tubes, are also found. These are resting sporangia and it seems likely that they are formed following a sexual fusion. In *O. viciae* and in *O. trifolii* Kusano has shown that the zoospores may copulate outside the host plant and produce biflagellate zygotes which infect their respective host plants, producing thick-walled resting sporangia. In these species the resting sporangia break open after several months to release further zoospores. The resting sporangia are binucleate and nuclear fusion occurs shortly before germination. Meiosis probably takes place during the division of the fusion nucleus before zoospore formation. In *O. brassicae* there is similarly evidence of zoospore fusion; indeed compound zoospores with up to six flagella have been seen. However, it would be wrong to conclude that biflagellate swarmers are necessarily zygotes. Incomplete cleavage of zoospore initials may result in multiflagellate structures (Garrett & Tomlinson, 1967). A proportion of the zoospore cysts found on root hairs are binucleate (see Fig. 43), and the resting sporangia are also binucleate (see Sampson, 1939). Sahtiyanci (1962) has shown that if cultures on cabbage roots were started from a single zoosporangium, resting sporangia were not formed. When cultures were started from zoospores derived from several zoosporangia, resting spores were formed. By mixing zoospore suspensions from eight single zoosporangial lines in all possible combinations, it was shown that the fungus exists in two distinct strains, and resting sporangia occurred only when opposite strains were mixed. The resting sporangia were capable of germination 7–10 days after they were mature, and germinated by the formation of one of two exit papillae through which numerous zoospores were discharged. Figure 45 shows an outline of the probable life cycle of *Olpidium*.

The fine structure of zoospores, the course of infection, and the development of thalli within host cells, have been studied by Temmink & Campbell (1968, 1969a, b) and by Lesemann & Fuchs (1970a, b). The structure of the zoospore is summarised in Fig. 44. The body of the zoospore measures about 2×3 μm, and the whiplash flagellum is about 21 μm long. The flagellum is attached to a kinetosome. This differs from the kinetosome of other zoospores (see p. 75) in that the terminal plate on which the central fibrils of the axoneme normally terminate is inconspicuous. A centriole (not shown in Fig.

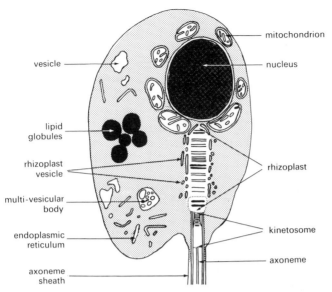

Fig. 44. *Olpidium brassicae*: diagrammatic representation of L.S. zoospore (after Temmink & Campbell, 1969a).

44) lies parallel to the kinetosome. The kinetosome is connected to the nucleus by a rhizoplast (banded rootlet) consisting of alternate bands of electron-transparent and electron-dense material. At the posterior end, the rhizoplast appears to be double, with one portion attached to the kinetosome and the other to the centriole. Near its attachment to the nucleus, at its anterior end, the rhizoplast is associated with a mitochondrion which is larger than the rest. Possibly this mitochondrion is tubular, and may form a collar around the rhizoplast. The remaining mitochondria surround the nucleus. There are several lipid globules grouped together in the cytoplasm. When the zoospore encysts on an epidermal cell of a suitable host, the flagellum is retracted, surrounded by its axonemal sheath, and sections of the cysts show the flagellum wound around inside the body of the zoospore. The cysts are attached by a mucilaginous secretion. Opposite the point of attachment, living host cells deposit a thickening of wall material. Penetration occurs through this thickened area by means of a narrow tube which breaks down at its tip to release a naked protoplast, which is probably expelled from the cyst by enlargement of a vacuole. The protoplast lies free within the host cytoplasm, surrounded only by a single membrane. Protoplasts destined to form sporangia form a wall outside the plasma-membrane, some 36–48 hours following penetration, and are fully grown by 60–72 hours. Cleavage vesicles develop which separate the zoospore protoplasts. Exit tubes develop as bulges on the thallus wall, and each bulge becomes filled with a 'spongy' material. The thallus wall ruptures and the plug of spongy material persists, blocking the

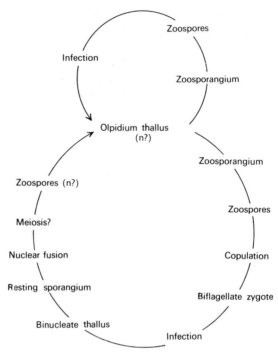

Fig. 45. The probable life history of *Olpidium brassicae*. Certain information is lacking for this fungus, and this is indicated by question marks. This kind of life history has, however, been described in *O. viciae* and *O. trifolii*.

exit tube. Presumably this plug of material dissolves or is pushed aside to allow zoospores to escape.

Interest has centred on the form in which virus particles are carried by zoospores. It has been shown that virus particles can be adsorbed onto zoospores *in vitro* (Campbell & Fry, 1966; Temmink *et al.*, 1970; Temmink, 1971). Since the greater portion of the zoospore cyst remains outside the host when the fungal protoplast is injected, it is thought likely that the virus particles adsorbed onto the axonemal sheath, which is retracted into the body of the cyst, are the particles which are transmitted to the host.

Synchytriaceae

In this family the thallus is endobiotic and holocarpic, and at reproduction it may become converted directly into a group (or sorus) of sporangia, or to a prosorus which later gives rise to a sorus of sporangia. Alternatively the thallus may become converted into a resting sporangium which can either function directly as a sporangium and give rise to zoospores, or can function as a prosorus, producing a vesicle whose contents cleave to form a sorus of sporangia. The zoospores are of the characteristic chytrid type. Sexual reproduction is by copulation of isogametes, resulting in the formation of thick-walled resting sporangia. Sparrow (1960) recognises three genera, *Synchytrium*, *Endodesmidium* and *Micromyces*. *Endodesmidium* and *Micromyces* are parasitic on green algae, but the largest genus is *Synchytrium*, with perhaps more than 100 species parasitic on flowering plants. Some species parasitise only a narrow range of hosts, e.g. *S. endobioticum* on Solanaceae, but others, e.g. *S. macrosporum*, may attack a wide range of hosts (Karling, 1964). Many species are not very destructive to the host plant, but result in the formation of galls on leaves, stems and fruits. The most serious parasite is *S. endobioticum*, the cause of black wart disease (or wart disease) of potato. *Synchytrium endobioticum* is a biotrophic parasite, which has so far not been successfully cultured outside living host cells. Attempts to maintain it in potato callus tissue culture have also been unsuccessful (Ingram, 1971).

Synchytrium (Figs. 46 and 47)

Wart disease of potato (see Fig. 46) is now distributed throughout the main potato-growing regions of the world, especially in mountainous areas and those with a cool, moist climate. Diseased potato tubers when lifted bear dark-brown, warty cauliflower-like excrescences. Galls may also be formed on

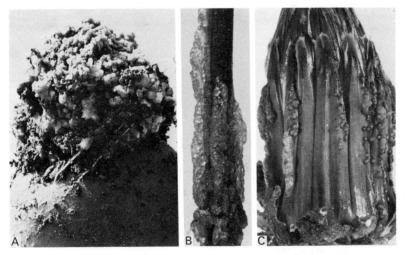

Fig. 46. A. *Synchytrium endobioticum*. A potato tuber (variety Arran Chieftain) artificially infected with wart disease. The cauliflower-like excrescences are hypertrophied masses of host tissue resulting from abnormal growth and repeated reinfection of a shoot. The diseased tissue when growing near the surface of the soil is bright green and contains both prosori and resting sporangia. B. *S. mercurialis*. Stem of *Mercurialis perennis* showing hypertrophy of the epidermal cells surrounding resting sporangia of the fungus. C. *S. taraxaci*. Involucral bracts of *Taraxacum officinale* showing blisters of hypertrophied cells surrounding sporangial sori of the fungus.

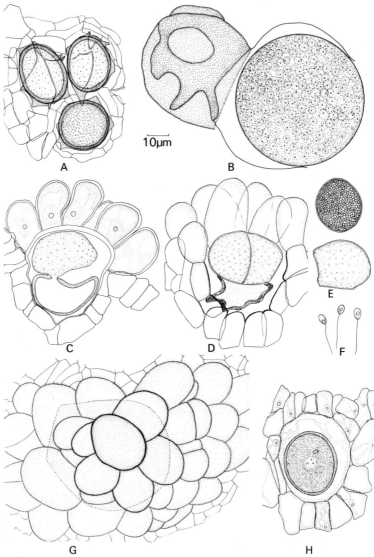

Fig. 47. *Synchytrium endobioticum.* A. Resting sporangia in section of wart. B. Germinating resting sporangium showing the formation of a vesicle containing a single globose sporangium (after Kole, 1965a). C. Section of infected host cell containing a prosorus. The prosorus is extruding a vesicle. Note the hypertrophy of the infected cell and adjacent uninfected cells. D. Cleavage of vesicle contents to form zoosporangia. E. Two extruded zoosporangia. F. Zoospores. G. Rosette of hypertrophied potato cells as seen from the surface. The outline of the infected host cell is shown dotted. H. Young resting sporangium resulting from infection by a zygote. Note that the infected cell lies beneath the epidermis due to division of the host cells.

the aerial shoots, and they are then green with convoluted leaf-like masses of tissue. Heavily infected tubers may have a considerable proportion of their tissues converted to warts. The yield of saleable potatoes from a heavily infected crop may be less than the actual weight of the seed potatoes planted. The disease is thus potentially a serious one, but fortunately varieties of potatoes are available which are immune from the disease, so that control is practicable. The life history of the causal fungus has been studied by Curtis (1921), Köhler (1923, 1931a,b) and by Heim (1956a,b). Heim's account differs substantially from those of Curtis and Köhler and has not been generally accepted (see Lingappa, 1958a,b). The dark warts on the tubers are galls in which the host cells have been stimulated by the presence of the fungus to divide. Many of the cells contain resting sporangia which are more or less spherical cells with thick dark-brown walls with folded, plate-like extensions (see Fig. 47A). The resting sporangia are released by the decay of the warts and they may remain alive in the soil for many years. The outer wall (epispore) bursts open by an irregular aperture and the endospore balloons out to form a vesicle within which a single sporangium differentiates (Kole, 1965; Sharma & Cammack, 1976). Thus the resting sporangium functions as a prosporangium on germination, and not directly as a sporangium as earlier workers had supposed.

The zoospores are capable of swimming for about 2 hours in the soil water. If they alight on the surface of a potato 'eye' or some other part of the potato shoot such as a stolon or a young tuber before its epidermis is suberised, they come to rest, and withdraw their flagella. Penetration of the epidermal cell of the host shoot occurs, and the contents of the zoospore cyst are transferred to the host cell, whilst the cyst membrane remains attached to the outside. When a dormant 'eye' is infected, dormancy may be broken and the tuber may begin to sprout at the infected 'eye'. If the potato variety is susceptible to the disease the small fungal thallus inside the host cell enlarges, and the host cell is stimulated to enlarge. Cells surrounding the infected cell also enlarge so that a rosette of hypertrophied cells surrounds a central infected cell (Fig. 47G). The walls of these cells adjacent to the infected cell are often thickened and assume a dark-brown colour. The infected cell remains alive for some time but eventually it dies. The parasite passes to the bottom of the host cell, enlarges, becomes spherical, and fills the lower part of the host cell. A double-layered chitinous wall which is golden-brown in colour is secreted around the thallus, and at this stage the thallus is termed a **prosorus** or summer spore. During its development until this stage the prosorus has remained uninucleate. Further development of the prosorus involves the protrusion of the inner wall through a pore in the outer wall, and the inner wall then expands as a vesicle which enlarges upwards and fills the upper half of the host cell (Fig. 47C). The cytoplasmic contents of the prosorus including the nucleus, are transferred to the vesicle. The process is quite rapid, and may be completed in 4 hours. During its passage into the vesicle the nucleus may divide, and division continues so that the vesicle contains about thirty-two nuclei, and at this stage the cytoplasmic contents of the vesicle become cleaved into a number of sporangia (Fig. 47D) forming a sorus. The number of sporangia varies from about four to nine. After the formation of the sporangial walls further nuclear divisions occur in each sporangium, and finally each nucleus with its surrounding mass of cytoplasm becomes differentiated to form a zoospore. As the sporangia ripen, they absorb water and swell, causing the host cell which contains them to burst open. Meanwhile, division of the host cells underlying the rosette has been taking place, and enlargement of these cells pushes the sporangia out on to the surface of the host tissue (Fig. 47E). The sporangia swell if water is available and burst open by means of a small slit through which the zoospores escape. There may be as many as 500–600 zoospores in large sporangia. The zoospores, which resemble those derived from resting sporangia, swim in the water film by a characteristic jerky, hopping movement and are capable of swimming for up to 20 hours. If suitable host tissues are available they encyst on the epidermis, and penetrate it within a few hours. Sometimes several zoospores succeed in penetrating a single cell so that it contains several fungal protoplasts. Within the host cell the thallus enlarges to form a prosorus, whilst the surrounding host cells enlarge to form the rosette. Eventually a further crop of sporangia is produced from which zoospores are released. This cycle of infection resulting in the formation of several generations of prosori can be continued throughout the spring and early summer.

According to Curtis, Köhler and a number of other workers the resting sporangia of *S. endobioticum* are formed following copulation. These workers have noticed that the zoospores released from the soral sporangia may fuse in pairs (or occasionally in groups of three or four) to form zygotes which retain their flagella and swim actively for a time. The zoospores which function as gametes do not differ in size and shape, so copulation can be described as isogamous. There are however indications that the gametes may differ physiologically. Curtis has suggested that fusion may not occur between zoospores derived from a single sporangium, but only between zoospores from separate sporangia. Köhler (1956) has claimed that the zoospores are at first sexually neutral. Later they mature and become capable of copulation. Maturation may occur either outside the sporangia or within, so that in over-ripe sporangia the zoospores are capable of copulation on release. At first the zoospores are 'male', and swim actively. Later the swarmers become quiescent ('female') and probably secrete a substance which attracts 'male' gametes to them chemotactically. After swimming by means of the two flagella the zygote encysts on the surface of the host epidermis and penetration of the host cell may then follow by a process essentially similar to zoospore penetration. Multiple infections by several zygotes penetrating a single host cell can also occur. Nuclear fusion occurs in the young zygote, before penetration. The results of zygote infections differ from infection by azygotes (zoospores). When infection by an azygote occurs, the host cell reacts by undergoing *hypertrophy*, i.e. increase in cell volume, and adjacent cells also enlarge to form the characteristic rosette which surrounds the resulting prosorus. When a zygote infects, the host cell undergoes *hyperplasia*, i.e. repeated cell division. The parasite lies towards the bottom of the host cell, and division occurs in such a way that the fungal protoplast is transferred to the innermost daughter cell. As a result of repeated divisions of the host cells the fungal protoplasts may be buried several cell-layers deep beneath the epidermis (see Fig. 47H). During these divisions of the host tissue the zygote thallus enlarges and becomes surrounded by a two-layered wall, a thick outer layer which eventually becomes dark-brown in colour and is thrown into folds or ridges which appear as spines in section, and a thin hyaline inner wall surrounding the granular cytoplasm. The host cell eventually dies and some of its contents may also be deposited on the outer wall of the resting sporangium. During its development the resting sporangium remains uninucleate. The resting sporangia are released into the soil and are capable of germination within about two months. Before germination the nucleus divides repeatedly to form the nuclei of the zoospores whose further development has already been described. Some cytological details of the life history are still in doubt. Presumably the zygote and the young resting sporangium are diploid, and it has been assumed that meiosis occurs during germination of the resting sporangia prior to the formation of zoospores, so that these zoospores, the prosori and the soral zoospores are also believed to be haploid. These assumptions seem plausible in the light of knowledge of the life history and cytology of *Synchytrium fulgens*, a parasite of *Oenothera*, described by Kusano (1930a,b). In *S. fulgens* Kusano has described the occurrence of meiosis during the germination of

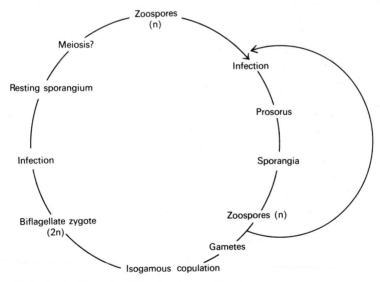

Fig. 48. Probable life cycle of *Synchytrium endobioticum*.

resting sporangia, and Lingappa has also suggested that meiosis occurs at this point (Lingappa, 1958b; Karling, 1958). An outline of the probable life history of *S. endobioticum* is shown in Fig. 48.

CONTROL OF WART DISEASE

Control is based largely on the breeding of resistant varieties of potato. It was discovered that certain varieties of potato such as Snowdrop were immune from the disease and could be planted on land heavily infected with *Synchytrium* without developing warts. Following this discovery plant breeders have developed a number of immune varieties such as Arran Pilot, Arran Banner, Arran Peak and Majestic. A number of potato varieties suceptible to the disease are still widely grown, however, including King Edward, Eclipse and Duke of York. In most countries where wart disease occurs legislation has been introduced requiring that only approved immune varieties are planted on land where wart disease has been known to occur, and prohibiting the movement and sale of diseased material (Anon, 1973). In Britain the growing of immune varieties on infested land has prevented the spread of the disease, and it is confined to a small number of foci in the west Midlands, north-west England and mid- and south Scotland. The majority of the outbreaks are found in allotments, gardens and small-holdings. As new varieties of potatoes are developed by plant breeders they are tested for susceptibility to wart disease by the Ministry of Agriculture Plant Pathology Laboratory and most of those that are made available commercially are immune.

The reaction of immune varieties to infection varies (Noble & Glynne, 1970). In some cases when 'immune' varieties are exposed to heavy infection in the laboratory they may become slightly infected, but infection is often confined to the superficial tissues which are soon sloughed off. In the field such slight infections would probably pass unnoticed. Occasionally infections of certain potato varieties may result in the formation of resting sporangia, but without the formation of noticeable galls. Penetration of the parasite seems to occur in all potato varieties, but when a cell of an immune variety is penetrated it may die within a few hours, and since the fungus is a biotrophic parasite, further development is checked. In other cases the parasite may persist in the host cell for up to 2–3 days, apparently showing normal development, but after this time the fungal thallus undergoes disorganisation and disappears from the host cell.

Other methods of control are less satisfactory. Attempts to kill the resting sporangia of the fungus in the soil have been made, but this is a costly and difficult process, and requires large applications of fungicides to the soil. In America applications of copper sulphate or ammonium thiocyanate at the rate of 1 ton/acre have been used, and local treatment with mercuric chloride or with formaldehyde and steam has been used to eradicate foci of infection. Control measures based on the use of resistant varieties seem more satisfactory. Unfortunately, it has been discovered that new physiological races (or biotypes) of the parasite have arisen which are capable of attacking varieties previously considered immune. About ten biotypes of *S. endobioticum* are now known. The implications of the discovery of these new races are obvious. It is vital to prevent their spread, and prompt action has been taken to do this. If the dispersal of the new races cannot be prevented, or if, as seems likely, new races arise independently elsewhere, much of the work of the potato breeder during the last 60 years may have to be started all over again. So far there are no reports of outbreaks of new races in Britain, but this is not to say that they do not exist, and intensive research in areas where wart disease has persisted may bring them to light.

OTHER SPECIES OF *Synchytrium*

Not all species of *Synchytrium* show the same kind of life history as that of *S. endobioticum*. *S. fulgens*, a parasite of *Oenothera*, resembles. *S. endobioticum*.

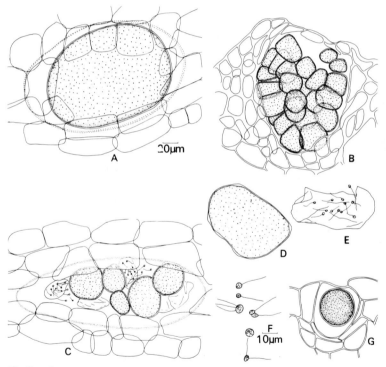

Fig. 49. *Synchytrium taraxaci*. A. Undivided thallus in epidermal cell of *Taraxacum* scape. Outline of host cell shown dotted. B. Section of *Taraxacum* scape showing thallus divided into a sorus of sporangia. C. A sorus of sporangia seen from above. Two sporangia are releasing zoospores. D. A ripe sporangium. E. Sporangium releasing zoospores. F. Zoospores and zygotes. The triflagellate zoospore probably arose by incomplete separation of zoospore initials. G. Section of host leaf showing a resting sporangium. A, B, C, D, E, G to same scale.

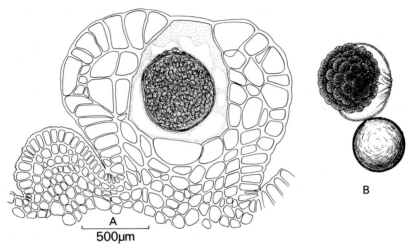

Fig. 50. *Synchytrium mercurialis.* A. Section of stem of *Mercurialis perennis* showing hypertrophied cells surrounding a resting sporangium. B. Germination of the resting sporangia to release a sorus of zoosporangia. Thus in *S. mercurialis* the resting sporangium functions as a prosorus (after Fischer, 1892).

Both the summer spore and the resting sporangia function as prosori, extruding sporangia (Lingappa, 1958a,b). But in this species it has also been shown that the zoospores from resting sporangia can function as gametes and give rise to zygote infections from which further resting sporangia arise (Lingappa, 1958b; Kusano, 1930a). Köhler (1931a) has suggested that the same phenomenon may occasionally occur in *S. endobioticum*. In *S. taraxaci*, parasitic on *Taraxacum* (Fig. 49) and in a number of other species the mature thallus does not function as a prosorus but cleaves directly to form a sorus of sporangia and the resting sporangium also gives rise to zoospores directly. In some species, e.g. *S. aecidioides*, resting sporangia are unknown whilst in others, e.g. *S. mercurialis*, a common parasite on leaves and stems of *Mercurialis perennis* (Fig. 50), only resting sporangia are known and summer sporangial sori do not occur. *Mercurialis* plants collected from March to June often show yellowish blisters on leaves and stems. The blisters are galls made up of one or two layers of hypertrophied cells, mostly lacking chlorophyll, which surround the *Synchytrium* thallus, which matures to form a resting sporangium. In this species the resting sporangium functions as a prosorus during the following spring. The undivided contents are extruded into a spherical sac which becomes cleaved into a sorus containing as many as 120 sporangia, from which zoospores arise. The variations in the life histories of the various species of *Synchytrium* form a useful basis for classifying the genus (Karling, 1964).

Rhizidiaceae

In this family the thallus is typically monocentric, with a single, usually globose inoperculate sporangium or resting sporangium arising from an extensive richly branched rhizoidal system. The resting sporangia may be formed asexually, or sexually by fusion of isogamous or anisogamous **aplano-gametes** (i.e. non-motile gametes), and on germination the resting sporangia function either as sporangia or as prosporangia. The family includes forms parasitic on fresh-water algae and a large number saprophytic on the cast exoskeleton of insects and on plant debris in water and in soil.

Rhizophlyctis (Fig. 51)

Rhizophlyctis rosea grows on cellulose-rich substrata in a variety of habitats such as soil and lake mud, and it undoubtedly plays an active role in cellulose decay. It can readily be isolated and grown in culture. This has made possible studies on its nutrition and physiology (Haskins, 1939; Stanier, 1942; Quantz, 1943; Haskins & Weston, 1950; Cantino & Hyatt, 1953; Davies, 1961). Its nutritional requirements are simple and it shows vigorous growth on cellulose as the sole carbon source, although it can utilise a range of other carbo-hydrates such as glucose, cellobiose and starch. If boiled grass leaves or cellophane are floated in water to which soil or mud containing *R. rosea* has been added, the bright pink globose sporangia often up to 200 μm in diameter can be found after about 7 days, and are easily visible under a dissecting microscope. The pink colour is due to the presence of carotenoid pigments such as γ-carotene, lycopene, and a xanthophyll. The sporangia are attached to coarse rhizoids which commonly arise at several points on the sporangial wall, and extend throughout the substratum, tapering to fine points. Although usually monocentric there are also records of some polycentric isolates. When ripe the sporangia have pink granular contents which differentiate into numerous uninucleate posteriorly uniflagellate zoospores (Fig. 51A). One to several discharge tubes are formed, and the tip of each tube contains a clear mucilaginous plug which prior to discharge is exuded in a mass from the tip of the tube (Fig. 51C). Whilst the plug of mucilage is being discharged the zoospores within the sporangium show active movement and then escape by swimming through the tube. In some specimens of *R. rosea* it has been found that a membrane may form over the cytoplasm at the base of the discharge tubes. If the sporangia do not discharge their spores immediately the mem-brane may thicken. When spore discharge occurs these thickened membranes can be seen often floating free within the sporangia, and the term endo-oper-culum has been applied to them. The genus *Karlingia* was erected for forms possessing such endo-opercula, including *R. rosea*, which is therefore some-times referred to as *Karlingia rosea*, but the validity of this separation is questionable because the presence or absence of endo-opercula is a variable character. The zoospores are capable of swimming for several hours. The head of the zoospore is often globose, but can become pear-shaped or show amoeboid changes in shape. It contains a prominent lipoid body, several

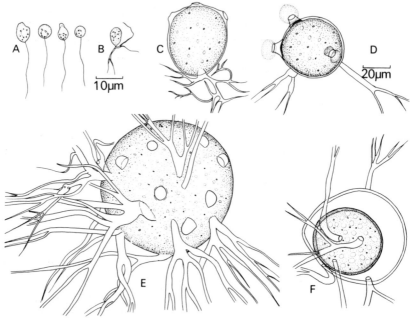

Fig. 51. *Rhizophlyctis rosea*. A. Zoospores. B. Young thallus formed on germination of zoospore. The zoospore cyst has enlarged and will form the sporangium. C. An older plant showing three dehiscence papillae. D. A plant showing mucilage plugs at the tips of the dehiscence papillae and thickenings of the cell membrane at the bases of the papillae. Such thickenings are sometimes termed endo-opercula. E. A mature plant with a globose sporangium and seven visible papillae. F. A resting spore formed inside an empty zoosporangium. A, B to same scale; C, D, E, F, G to same scale.

bright refringent globules, and bears a single trailing flagellum. On coming to rest on a suitable substratum the flagellum is withdrawn and the body of the zoospore usually enlarges to form the rudiment of the sporangium, whilst rhizoids appear at various points on its surface. The fine structure of *R. rosea* sporangia has been studied by Chambers & Willoughby (1964). Within the sporangium, the flagella are tightly wrapped around the zoospores. It seems likely that it is in this condition that the zoospores emerge.

Resting sporangia are also found. They are brown, globose or angular and have a thickened wall. Whether they are formed sexually in *R. rosea* is not known. Couch (1939) has, however, put forward evidence that the fungus is heterothallic (i.e. will only reproduce sexually following interaction of two differing thalli). Single isolates grown in culture failed to produce resting sporangia, but when certain cultures were paired resting sporangia were formed. Stanier (1942) has reported the occurrence of biflagellate zoospores, but whether these represented zygotes seemed doubtful. On germination the resting sporangium functions as a prosporangium. In the homothallic chitino-philic fungus *Rhizophlyctis oceanis*, Karling (1969) has described frequent fusions between zoospores. These fusions are possibly sexual, but unfortunately Karling was unable to cultivate the thalli which resulted from fusion to the stage of resting spore development. In *Karlingia dubia*, Willoughby (1957) was able to follow the development of fused zoospores (gametes) into a thallus within which a thick-walled, smooth resting spore was formed.

Cladochytriaceae

In this family the thallus is eucarpic and usually polycentric, and the vegetative system may bear intercalary swellings, and septate **turbinate cells** (sometimes termed **spindle organs**). Many of its members are saprophytic on decaying plant remains, but some are parasitic on algae and on animal eggs.

Cladochytrium (Fig. 52)

Cladochytrium replicatum is a common representative, which is to be found in decaying pieces of aquatic vegetation, and can be distinguished from other chytrids by the bright orange globules found in the sporangia. It can frequently be isolated during the winter months if moribund aquatic vegetation is placed in a dish of water and 'baited' with boiled grass leaves. The bright orange sporangia which are visible under a dissecting microscope appear within about 5 days arising from an extensively branched hyaline rhizomycelium bearing two-celled intercalary swellings. The zoospores on release from the sporangium each contain a single orange globule and bear a single posterior flagellum. After swimming for a short time the zoospore attaches itself to the surface of the substratum and puts out usually a single germ tube which can penetrate the tissues of the host plant. Within the cell of the host the germ tube expands to form an elliptical or cylindrical swollen cell which is often later divided into two by a transverse septum. This swollen cell is termed a spindle organ or turbinate cell. The zoospore is uninucleate and during germination the single nucleus is transferred to the swollen turbinate cell which becomes a vegetative centre from which further growth takes place. Nuclear division is apparently confined to the turbinate cells. From the turbinate cell rhizoids are put out which in turn produce further turbinate cells (see Fig. 52B,D). Nuclei are transported through the rhizoidal system but are not resident there. The thallus so established branches profusely and the characteristic two- or three-celled turbinate cells and the fine rhizoidal system may be widely distributed in the host tissues. At certain points on the thallus spherical zoosporangia form either terminally or in an intercalary position. Sometimes one of the cells of a pair of turbinate cells swells and becomes transformed into a sporangium, and in culture both cells may be modified in this way. The spherical zoosporangium undergoes progressive nuclear division and becomes multinucleate. Meanwhile the contents of the sporangium acquire a bright orange colour, due to the accumulation of the globules

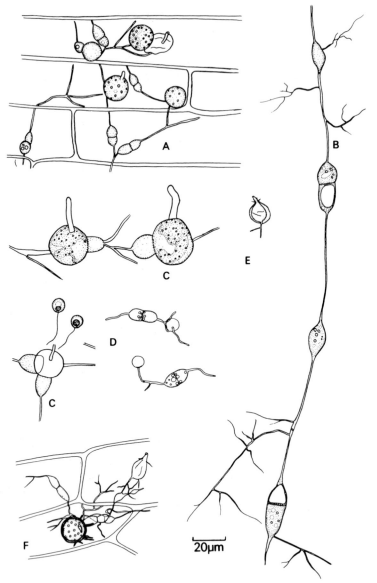

Fig. 52. *Cladochytrium replicatum*. A. Rhizomycelium within the epidermis of an aquatic plant bearing the two-celled hyaline turbinate cells and globose orange zoosporangia. B. Rhizomycelium and turbinate cells from a culture. C. Zoosporangia from a two-week-old culture. One zoosporangium has released zoospores, each of which contains a bright orange coloured globule. D. Germinating zoospores on boiled wheat leaves. The empty zoospore cysts are spherical. The germ tubes have expanded to form turbinate cells. E. A zoosporangium which has proliferated internally to form a second sporangium. F. Rhizomycelium within a boiled wheat leaf bearing a thick-walled, spiny resting sporangium.

containing the carotene lycopene, which are later found in the zoospores. Cleavage of the cytoplasm to form uninucleate zoospore initials follows. The zoospores escape through a narrow exit tube which penetrates to the exterior of the host plant and becomes mucilaginous at the tip. There is no operculum. Sometimes the zoosporangia may proliferate internally, a new zoosporangium being formed inside the wall of an empty one. Resting sporangia with thicker walls and a more hyaline cytoplasm are also formed either terminally or in an intercalary position on the rhizomycelium. In some cases the wall of the resting sporangium is reported to be smooth and in others spiny, but it has been suggested (Sparrow, 1960) that the two kinds of resting sporangia may belong to different species. However, studies by Willoughby (1962) of a number of single-spore isolates have shown that the presence or absence of spines is a variable character. The contents of the resting sporangia divide to form zoospores which also have a conspicuous orange granule, and escape by means of an exit tube as in the thin-walled zoosporangia. Whether the resting sporangia are formed as a result of a sexual process is not known. Pure cultures of *C. replicatum* have been studied by Willoughby (1962) and Goldstein (1960b). The fungus is heterotrophic for thiamine. Biotin, while not required absolutely, stimulates growth. Nitrate and sulphate were utilised. A number of different carbohydrates were utilised, and a limited amount of growth took place on cellulose.

Chytridiaceae

The Chytridiaceae and the Megachytriaceae are distinguished from the preceding families in the method by which their sporangia discharge zoospores, by the removal of a circular lid or operculum. In the Chytridiaceae the thallus is monocentric, eucarpic and epi- or endobiotic, whilst in the Megachytriaceae the thallus is polycentric. Most Chytridiaceae are parasitic or saprophytic on fresh-water and marine algae, fungi, pollen and protozoa. There is a remarkable parallel in thallus organisation to the inoperculate chytrids. The details of sexual reproduction are known in only a few species, and usually involves fusion of adjacent 'male' and 'female' thalli, as in *Zygorhizidium* and *Chytridium*.

Chytridium (Fig. 53)

This is a large genus with about forty species mostly parasitic on algae or on other fungi or protozoa. The sporangium is epibiotic, being formed by enlargement of the zoospore cyst, whilst the branched rhizoidal system penetrates the host. The thick-walled resting sporangium is endobiotic. *C. olla* is a parasite of oogonia and oospores of *Oedogonium*, but it can readily be grown in pure culture (Emerson, 1958). In culture the sporangia are globose, with granular pale-pink contents, and arise from an extensive rhizoidal system,

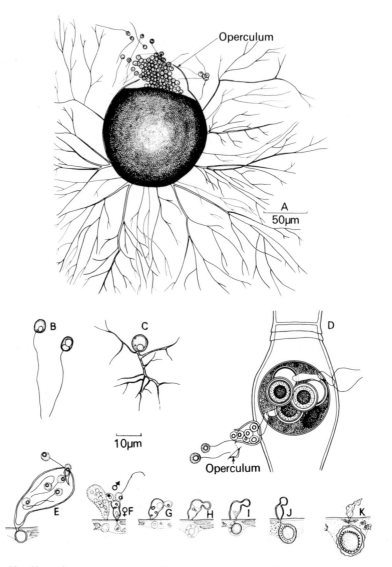

Fig. 53. *Chytridium* spp. A–D, *C. olla* E–K, *C. sexuale*. A. Mature zoosporangium releasing zoospores through a wide operculum. B. Zoospores. C. Young plant showing the enlargement of the zoospore cyst to form the zoosporangium. D. Oogonium of *Oedogonium* with an immature oospore killed by the parasite. The oospore contains five resting spores of *Chytridium*, two of which have germinated to form a sporangium. Note the operculum cast off from the mouth of the lower sporangium. E. Zoosporangium of *C. sexuale* attached to a *Vaucheria* filament. Zoospores are being released. F–K. Stages in the formation of a resting sporangium. F. An epibiotic female filament (female) bearing an attached encysted male thallus and a male motile cell. G, H, I, J. Stages in plasmogamy between male thallus and female thallus. The cytoplasm of the female thallus is gradually transferred from the epibiotic part to an endobiotic spherical cell, and in J the transfer is complete. K. Mature resting spore. The epibiotic part of the female thallus and the attached male thallus have collapsed. A, D to same scale; B, C, E–K to same scale. D, after de Bary (1887); E–K, after Koch (1951).

with a series of main axes branching into very fine extremities. When mature sporangia (about 3 days old) are flooded with water they dehisce by lifting off a large operculum and numerous posteriorly uniflagellate zoospores escape and swim for a few hours (Fig. 53A). On coming to rest, the flagellum is withdrawn, and the zoospore develops a rhizoidal system from one end. The zoospore cyst meanwhile enlarges to form the body of the zoosporangium (Fig. 53B,C). In this species resting spores have been described on the endobiotic rhizoidal system within the algal host, but the details of their formation are not known (Sparrow, 1973d). They are smooth and thick-walled, and after a long resting period they germinate to form a cylindrical germ tube which grows to the exterior of the *Oedogonium* oogonium and expands to form a pear-shaped epibiotic sporangium, resembling the normal epibiotic zoosporangia (Fig. 53D). Sexual fusion preceding the formation of resting spores has been described by Koch (1951) in *C. sexuale*. In this species which is parasitic on *Vaucheria* the motile cells may encyst on the host-cell wall and develop into a sporangial thallus, forming a further crop of zoospores from an epibiotic sporangium (Fig. 53E). Alternatively, the thalli may develop as 'female' thalli to which another zoospore may become attached, to form a 'male' thallus. Koch believes that 'the female thallus appears to be nothing more than a sporangial thallus with its exogenous development arrested at an early stage by the engagement of the male cells; however, there is no proof of this'. Later the encysted male cell empties its contents into the female thallus and the combined protoplasts move into an endobiotic swelling, increase in size and become surrounded by a thick warty wall (Fig. 53F–K). The germination of the resting spore has not been described.

Megachytriaceae

In this family most of the polycentric operculate chytrids are included. The thalli are epi- and endobiotic, and saprophytic in decaying remains of aquatic plants or occur in soil. The zoosporangia arise from terminal or intercalary swellings of the mycelium. Resting spores are also known, apparently formed asexually.

Nowakowskiella (Fig. 54)

Species of *Nowakowskiella* are widespread saprophytes in soil and in decaying aquatic plant debris, and can be obtained by 'baiting' aquatic plant remains in water with boiled grass leaves, cellophane and the like. *N. elegans* is often encountered in such cultures, and pure cultures can be obtained and grown on agar or on liquid culture media (Emerson, 1958). In boiled grass leaves the fungus forms an extensive rhizomycelium with intercalary swellings. Zoosporangia are formed terminally or in an intercalary position (see Fig. 54), and are globose or pear-shaped with a subsporangial swelling or apophysis, and

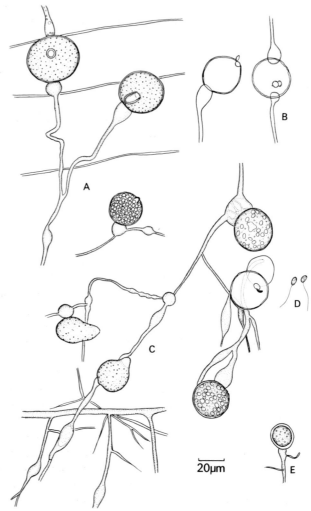

Fig. 54. *Nowakowskiella elegans*. A. Polycentric mycelium bearing zoosporangia. B. Empty zoosporangia showing opercula. C. Mycelium showing turbinate cells and zoosporangia. D. Zoospores from culture. E. Resting spore from culture.

granular or refractile hyaline contents. At maturity some sporangia develop a prominent beak, but in others this is not present. Sporangia dehisce by detaching an operculum (Fig. 54B,C) followed by escape of the zoospores which at first remain clumped together at the mouth of the sporangium. Thicker-walled yellowish resting sporangia have been described (Fig. 54E) especially when liquid cultures are grown at slightly reduced temperatures (Emerson, 1958). There is no evidence that these structures are formed following sexual fusion, and details of their germination are not known.

Goldstein (1961) has reported that the fungus requires thiamine, and can utilise nitrate, sulphate and a number of carbohydrates including cellulose, but cannot utilise starch.

In *N. profusa* three kinds of dehiscence have been described: **exo-operculate** in which the operculum breaks away to the outside of the sporangium; **endo-operculate**, in which the operculum remains within the sporangium; and **inoperculate** where the exit papilla opens without any clearly defined operculum (Chambers, Markus & Willoughby, 1967; Johnson, 1973). This variation in dehiscence in a single strain of the fungus adds emphasis to criticisms of the value of dehiscence as a primary criterion in classification.

BLASTOCLADIALES

The Blastocladiales are mostly saprophytes in soil, water, mud or inhabiting plant and animal debris, but one genus *Coelomomyces* is made up of obligate parasites of insects, usually mosquito larvae. This genus is also unusual in having a naked plasmodium-like thallus lacking rhizoids. Another unusual feature is that the life cycle is completed in two alternate hosts: sporophytic thalli occurring in mosquito larvae, and gametophytic thalli in a copepod. Attempts are being made to use *Coelomomyces* in the biological control of mosquitoes (Whisler *et al.*, 1975; Federici, 1977). In the remaining genera the thallus is eucarpic. The morphologically simpler forms such as *Blastocladiella* are monocentric, with a spherical or sac-like zoosporangium or resting sporangium arising directly or on a short one-celled stalk from a tuft of radiating rhizoids. These simpler types show considerable similarity to the monocentric chytrids, and in the vegetative state they may be difficult to distinguish from them. The more complex organisms such as *Allomyces* are polycentric, and the thallus is differentiated into a trunk-like portion bearing rhizoids below, branching above, often dichotomously, and bearing sporangia of various kinds at the tips of the branches. Studies of the composition of the wall of *Allomyces* by microchemical, X-ray diffraction and electron microscope techniques have demonstrated that it is composed of microfibrils of chitin, and that appreciable quantities of glucan, ash and protein are intimately associated with the walls. Chitin has also been demonstrated in the walls of *Blastocladiella* (Aronson, 1965). The zoospores of Blastocladiales have a single posterior flagellum of the whip type. The details of the structure of zoospores of *Blastocladiella* and *Allomyces* have been described on pp. 75–79. After swimming, the zoospore comes to rest, withdrawing its flagellum in *B. emersonii*. The flagellum is possibly wound in around the nuclear cap. In *Allomyces*, the zoospore cyst produces, at one point, a germ tube which branches to form the rhizoidal system. At the opposite pole, the zoospore cyst forms a wider germ tube which gives rise to branches bearing sporangia. The bipolar method of germination is another point of difference between the Blastocladiales and the Chytridiales, in which germination is typically unipolar. A number of distinct life history patterns are found in the group. In *Allomyces arbusculus*, for example, an isomorphic alternation of haploid gametophyte and diploid sporophyte has been demonstrated. In *A. neo-moni-liformis* (= *A. cystogenus*), there is no free-living sexual generation, but this stage is represented by a cyst (see below). In *A. anomalus*, the asexual stage only has been found in normal cultures, but experimental treatments may result in the development of sexual plants. These types of life cycle have also been found in other genera, such as *Blastocladiella*. A characteristic feature of the asexual plants of the Blastocladiales is the presence of resting sporangia with dark-brown pitted walls. The brown pigment in the resting sporangia of *Allomyces* is of the melanin group but γ-carotene is also present. Melanin has also been identified in the resting sporangia of *Blastocladiella*. The pits are inwardly directed conical pores in the wall. The inner ends of the pores abut against a smooth colourless inner layer of wall material surrounding the cytoplasm (Skucas, 1967, 1968). Such pigmented pitted resting sporangia are not found in the Chytridiales. The resting sporangia of *Allomyces* can remain viable for up to thirty years in dried soil.

The ease with which certain members of the group can be grown in culture has facilitated extensive studies of their nutrition and physiology, and the results of some of these investigations are discussed below.

Sparrow (1960, 1973b) has recognised three families, but of these we shall study only representatives of the Blastocladiaceae.

Blastocladiaceae

Allomyces (Figs. 55–57)

Species of *Allomyces* are found most frequently in mud or soil from the tropics or subtropics, and if dried samples of soil are placed in water and 'baited' with boiled hemp seeds, the baits may become colonised by zoospores. From such crude cultures, it is possible to pipette zoospores onto suitable agar media and to follow the complete life history of these fungi in the laboratory. Good growth occurs on a chemically ill-defined medium containing yeast extract, peptone and soluble starch (YPSS), but chemically defined media have also been used (see Youatt, 1973b). There is a requirement for thiamine and organic nitrogen in the form of amino acids. Sporangial development is stimulated by a deficiency in amino acids (Youatt *et al.*, 1971). Emerson (1941) has analysed soil samples from all over the world for species of *Allomyces* and has distinguished three types of life history, represented by three subgenera.

1. Eu-Allomyces

The *Eu-Allomyces* type of life history is exemplified by *A. arbusculus* and *A. macrogynus*. In *A. arbusculus* the outer wall of the brown pitted resting sporangia cracks open by a slit and the inner wall balloons outwards, then opens by one or more pores, to release about forty-eight posteriorly uniflagellate swarmers. The resting sporangia are formed on asexual, diploid plants and cytological studies have shown that the resting sporangia contain about twelve diploid nuclei, which undergo meiotic division during the early stages of germination (Wilson, 1952a). The cytoplasm cleaves around the forty-eight haploid nuclei to form the zoospores. Since meiosis occurs in the resting sporangia, they have been termed **meiosporangia**, and the haploid zoospores **meiospores**. The meiospores swim by movement of the trailing flagellum, and on coming to rest, germinate as described above to form a rhizoidal system and a trunk-like region which bears dichotomous branches. Repeated nuclear division occurs to form a coenocytic structure, and ring-like ingrowths from the walls of the trunk-region and branches extend inwards to form incomplete septa with a pore in the centre through which cytoplasmic connection can be seen (Fig. 55D). The haploid plants which develop from the meiospores are gametophytic or sexual plants. They are monoecious, and the tips of their branches swell to form paired sacs – the male and female gametangia. The male gametangia can be distinguished from the female by the presence of a bright orange pigment, γ-carotene, whilst the female gametangia are colourless. In *A. arbusculus* the male gametangium is subterminal, or hypogynous, beneath the terminal female gametangium, but in *A. macrogynus* the positions are reversed and the male gametangium is terminal or epigynous, above the subterminal female (Fig. 55E,I). The gametangia bear a number of colourless papillae on their walls which eventually dissolve to form pores. The contents of the gametangia differentiate into uninucleate gametes which differ in size and pigmentation. The female gametangium forms colourless swarmers, whilst the male gametangium releases smaller, more active orange-coloured swarmers. After escape through the papillae in the walls of the gametangia the gametes swim for a time and then pair off. A female gamete which fails to pair can germinate to form a new sexual plant. There is evidence that a hormone, sirenin, secreted by the female gametangia during gametogenesis and by the released female gametes stimulates a chemotactic response in male gametes (Machlis, 1958a,b, 1972; Carlile & Machlis, 1965a,b). The chemical structure of sirenin has been determined, and *d*- and *l*-forms have been synthesised. It is a bicyclic ketone probably derived from the parent hydrocarbon sesquicarene (Plattner & Rapoport, 1971). The paired gametes swim for a time, the zygotes showing two flagella. The cytoplasm of the two gametes becomes continuous, the zygote comes to rest, encysts, casts off the flagella, and nuclear fusion then follows. The zygote develops immediately into a diploid asexual plant resembling the sexual plants in general habit, but bearing two types of zoosporangia instead of gametangia. The first type is a thin-walled, papillate zoosporangium formed singly or in rows at the tips of the branches. Within these thin-walled sporangia the nuclei undergo mitosis and the cytoplasm cleaves around the nuclei to form diploid colourless zoospores which are released

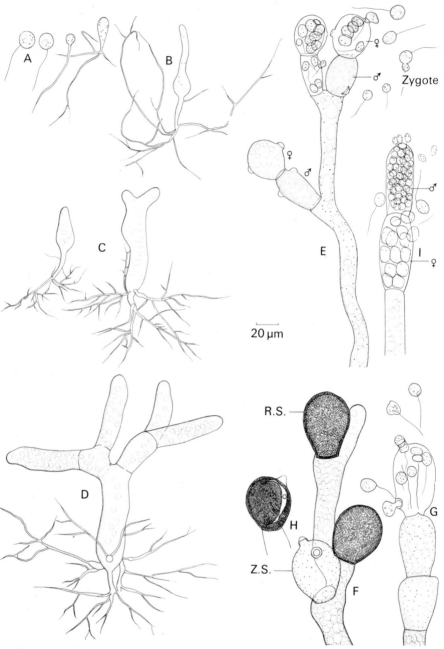

Fig. 55. *See opposite for legend.*

from the sporangia as plugs blocking the papillae in the wall dissolve to form circular pores through which the swarmers can escape. It has been suggested that the plugs might be composed of a pectic substance (Skucas, 1966) or a glycopeptide (Youatt, 1973a). The fine structure of zoospore cleavage has been studied by Barron & Hill (1974) who have shown that cleavage is brought about by the coalescence of cleavage vesicles, probably derived by blebbing from the zoosporangial membrane, induced by the availability of free water.

Since nuclear division in the thin-walled sporangia is mitotic they are termed **mitosporangia**, and the swarmers they release are **mitospores**. The mitospores, after a swimming phase, encyst and are capable of immediate germination to form a further diploid asexual plant resembling the plant which bore them. The second type of zoosporangium is the dark-brown, thick-walled, pitted resting sporangium, or meiosporangium, formed at the tips of the branches. Meiotic divisions within these sporangia result in the formation of the haploid meiospores, which develop to form sexual plants. The life cycle of a member of the subgenus *Eu-Allomyces* is an isomorphic alternation of generations, the haploid sexual plant or gametophyte forming gametes, and the resulting zygotes develop into diploid sporophytes on which the mito- and meio-sporangia are formed (Fig. 57). Comparisons of the nutrition and physiology of the two generations show no essential distinction between them up to the point of production of gametangia or sporangia.

Emerson & Wilson (1954) and Emerson (1954) have made cytological and genetical studies of a number of collections of *Allomyces*. Interspecific hybrids between *A. arbusculus* and *A. macrogynus* have been produced in the laboratory, and it has been shown that the fungus earlier described as *A. javanicus* is a naturally occurring hybrid between these two species. Cytological examination of the two parent species and of artificial and natural hybrids showed a great variation in chromosome number. In *A. arbusculus* the basic haploid chromosome number is 8, but strains with 16, 24 and 32 chromosomes have been found. In *A. macrogynus* the lowest haploid number encountered was 14, but strains with 28 and 56 chromosomes are also known. The demonstration that these two species each represent a polyploid series was the first to be made in the fungi.

The behaviour of the hybrid strains is of considerable interest. The parent species differ in the arrangement of the primary pairs of gametangia, *A.*

Fig. 55. A–H. *Allomyces arbusculus*. A. Zoospores (haploid meiospores). B. Young gametophytes, 1 day old. C. Young sporophytes, 18 hours old. D. Sporophyte, 30 hours old. Note the perforations visible in some of the septa. E. Gametangia at tips of branches of gametophyte. Note the disparity in the size of the gametes. The smaller male gametes are orange in colour whilst the larger female gametes are colourless. Compare the arrangement of gametangia with those of *A. macrogynus* in I. F. R.S. Resting sporangia (or meiosporangia) and Z.S. zoosporangia (or mitosporangia) on sporophyte. G. Release of mitospores from mitosporangia of sporophyte. H. Rupture of meiosporangium. I. *Allomyces macrogynus*. Gametophyte showing gametangia. The male gametangia are here terminal or epigynous.

arbusculus being hypogynous, whilst *A. macrogynus* is epigynous. Zygotes formed following fusion of gametes derived from the different parents, germinated and gave rise to sporophyte plants. The meiospores from the hybrid sporophytes had a low viability ($0\cdot1$–$3\cdot2\%$ as compared with a viability of about 63% for *A. arbusculus* meiospores), but some germinated to form gametophytes. The arrangement of the gametangia on these F_1 gametophytes showed a complete range from 100% epigyny, to 100% hypogyny. Also in certain gametophytes the ratio of male to female gametangia (normally about one in the two parents) was very high with less than 1 female per 1,000 male gametangia. It was concluded from these experiments that since intermediate gametangial arrangements are found in hybrid haploids this arrangement is not under the control of a single pair of non-duplicated allelic genes, but that a fairly large number of genes must be involved. Hybridisation in some way upsets the mechanism which controls the arrangement of gametangia in the parental species. By treating meiospores of *A. macrogynus* with DNA extracted from gametophytic cultures of *A. arbusculus*, Ojha & Turian (1971) have demonstrated that inversion of the normal gametangial arrangement in a proportion of the colonies occurs, i.e. instead of the normal epigynous arrangement, a proportion of the colonies from DNA-treated meiospores developed hypogynous antheridia. Similar inversions were also obtained in converse experiments. In an isolate of the naturally occurring hybrid *A. javanicus*, Ji & Dayal (1971) have shown that although copulation between anisogamous gametes results in the formation of sporophytic plants bearing thin-walled and thick-walled sporangia, the swarmers from the thick-walled sporangia rarely develop into gametophytic thalli, but into sporophytic thalli. This is not surprising for a hybrid, and is possibly due to a failure of meiosis in the thick-walled sporangia.

2. Cystogenes

A life cycle differing from *Eu-Allomyces* is found in *Allomyces moniliformis* and *A. neo-moniliformis*. There is no independent gametophyte generation, but this stage is probably represented by a cyst. The asexual plants resemble those of *Eu-Allomyces*, bearing both mitosporangia and brown thick-walled punctate meiosporangia. The mitosporangia are thin-walled, and bear a single terminal papilla, through which discharge of the posteriorly uniflagellate mitospores takes place. These swarmers germinate to form asexual plants again. The development of the meiosporangia is, however, very distinctive. The thin sporangial wall surrounding the sporangium breaks open irregularly and the brown pitted sporangium slips out (Fig. 56F). It is capable of germination within about 2–4 days after its release. Before germination the sporangium tends to become more round in shape, then its outer brown wall cracks open. Through the crack the thin inner wall expands and one to four discharge pores develop (Skucas, 1967, 1968). The contents of the resting sporangium

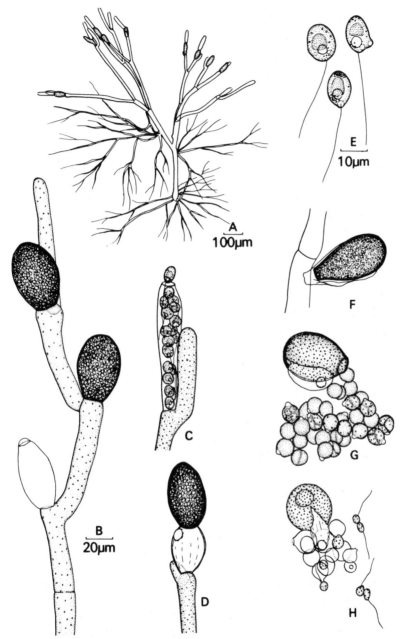

Fig. 56. *Allomyces neo-moniliformis*. A. Whole plant. B. Branch tips showing resting sporangia and zoosporangium (mitosporangium). C. Mitosporangium releasing mitospores. D. Branch tip showing a terminal resting sporangium and a subterminal empty mitosporangium. E. Mitospores. F. Resting sporangium escaping from its surrounding envelope. G. Germinating resting sporangium. A thin walled vesicle has ballooned out through the cracked wall. From the pore in the vesicle swarmers have escaped which have encysted immediately. H. Release of swarmers from the cysts. The swarmers pair to form biflagellate zygotes. B, C, D, E, G, H to same scale.

have meanwhile divided to form up to thirty amoeboid bodies which escape through the pores. There are reports that the amoeboid bodies bear a pair of posterior flagella (Emerson, 1941; Wilson, 1952b) or even as many as six (Teter, 1944; Wilson, 1952b), but flagella are not always present (McCranie, 1942; Teter, 1944). The movement of these bodies is very sluggish, and they rarely move far from their parent resting sporangium. They encyst quickly, frequently all clumped together (Fig. 56G,H). The period of encystment is short, often only a few hours, and then the contents of the cysts divide to form four colourless posteriorly uniflagellate swarmers, which escape through a single papilla on the cyst wall. These swarmers behave as isogametes and copulate to form biflagellate zygotes. There is evidence that the swarmers from one cyst are capable of copulation. The zygote swims for a time by movements of both flagella, then comes to rest and develops to form an asexual plant (Fig. 56A). The cytological details of the life cycle in *Cystogenes* have been worked out by Wilson (1952b). As in *Eu-Allomyces* the meiotic divisions occur in the resting sporangia. Before final cleavage of the cytoplasm to form swarmers the nuclei pair, being held together by a common nuclear cap. Cleavage thus results in the formation of binucleate cells, some of which bear two flagella, and some do not. It is these cells which form the cysts. During encystment a single mitotic division occurs in each cyst, and four separate haploid nuclei are formed, each with a distinct nuclear cap, and cleavage of the cytoplasm around these nuclei results in the formation of the four gametes. In the *Cystogenes* life cycle there is thus a diploid asexual generation, and the haploid phase is represented by the cyst and gametes. The life cycle is illustrated diagrammatically in Fig. 57.

3. Brachy-Allomyces

In certain isolates of *Allomyces* which have been placed in a 'form-species' *A. anomalus* there are neither sexual plants nor cysts. Asexual plants bear mito-sporangia and brown, resting sporangia. The spores from the resting sporangia develop directly to give asexual plants again. The cytological explanation proposed by Wilson (1952b) for this unusual behaviour is that due to complete or partial failure of chromosome pairing in the resting sporangia, meiosis does not occur and nuclear divisions are mitotic. Consequently the zoospores produced from resting sporangia are diploid, like their parent plants and, on germination, give rise to diploid asexual plants again. Similar failures in chromosome pairing were also encountered in the hybrids between *A. arbus-culus* and *A. macrogynus* leading to very low meiospore viability from certain crosses. In view of this it seemed possible that some of the forms of *A. anomalus* might have arisen through natural hybridisation. In a later study, Wilson & Flanagan (1968) showed that there were two distinct ways in which the life cycle of this fungus is maintained without a sexual phase. The first method, resulting from failure of meiosis in the resting sporangia, has already

Eu-Allomyces

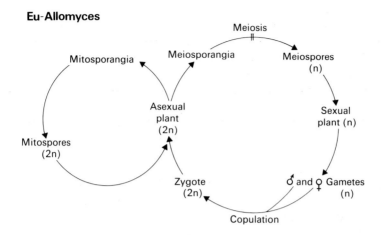

Cystogenes

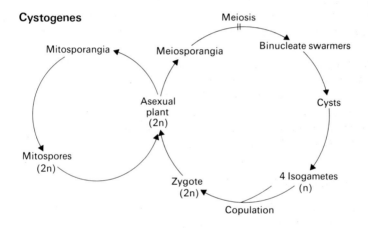

Brachy-Allomyces

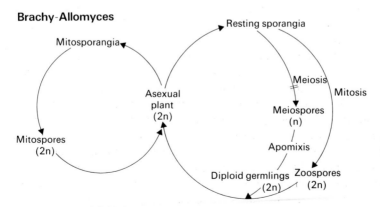

Fig. 57. Diagrammatic outlines of life cycles of *Allomyces*.

been described. Alternatively, in certain isolates, meiosis occurs in the resting sporangia, followed by apomixis, so that the germlings develop into sporophytic thalli. This is probably brought about by a doubling of the chromosomes in the germinating meiospore. By germinating resting sporangia in dilute K_2HPO_4, a small percentage of zoospores were produced which developed into gametophytes. From a study of the arrangement of the gametangia on these gametophytes, some were identified as *A. macrogynus* and some as *A. arbusculus*. No hybrids were found. Thus *A. anomalus* is not a single species, but represents sporophytic thalli of these two species in which the normal alternation of generations has been upset by cytological deviations.

Blastocladiella (Fig. 58)

Species of *Blastocladiella* have been isolated from soil and mud. Most of the isolates have been from soils from southerly latitudes, but two species have been found in Britain, one parasitic on the blue-green alga *Anabaena* (Canter & Willoughby, 1964). The form of the thallus is comparatively simple, resembling that of some monocentric chytrids. There is an extensive branched rhizoidal system which either arises directly from a sac-like sporangium, or from a cylindrical trunk-like region bearing a single sporangium at the tip. Different species of *Blastocladiella* have life cycles resembling those of the three subgenera of *Allomyces*, and Karling (1973) has proposed that *Blastocladiella* should similarly be divided into three subgenera: *Eucladiella*, corresponding to *Eu-Allomyces*; *Cystocladiella*, corresponding to *Cystogenes*; and *Blastocladiella*, corresponding to *Brachy-Allomyces*. In some species, there is an isomorphic alternation of generations, probably matching in essential features the *Eu-Allomyces* pattern, but cytological details are needed to confirm this. For example, in *Blastocladiella variabilis* two kinds of asexual plant are found. One bears thin-walled zoosporangia which release posteriorly uniflagellate swarmers. These swarmers may develop to form plants resembling their parents or may give rise to the second type of asexual plant bearing a thick-walled dark-brown sculptured resting sporangium within the terminal sac. The resting sporangial walls crack open on germination to allow papillae to protrude. The papillae dissolve to release posteriorly uniflagellate swarmers which, after swimming, germinate to form sexual plants of two kinds. In habit the sexual plants resemble the zoosporangial asexual plants, bearing a terminal club-shaped sac. About half of the sexual plants are colourless, and about half are orange-coloured. By analogy with *Allomyces* where the orange-coloured gametes are smaller or 'male', the orange-coloured plants of *Blastocladiella variabilis* are regarded as male. However in *Blastocladiella* there is no distinction in size between the gametes, in contrast with the anisogamy of *Eu-Allomyces*. The orange and colourless gametes pair to produce zygotes, which germinate directly to produce asexual plants. Because the orange and colourless plants were produced in approximately equal numbers from swarmers derived from resting sporangia it was believed that sexuality (i.e. colour) was probably genotypically determined at meiosis in the resting sporangia. A similar life cycle has been described for *B. stübenii*. In other species (e.g. *B. cystogena*) the life cycle is of the *Cystogenes* type. In this species there are no zoosporangial thalli but only resting sporangial thalli. The resting sporangia crack open to release posteriorly uniflagellate swarmers which encyst in an irregular mass. The cysts germinate to give rise to four uniflagellate isogamous gametes. These gametes pair to form a biflagellate zygote which develops into a plant bearing a resting sporangium. In some other species there is no clear evidence of sexual fusion. In *B. emersonii* (Fig. 58) the resting sporangial thallus contains a single globose, dark-brown, resting sporangium with a dimpled wall. After a resting period the wall cracks open, and one to four papillae protrude from which swarmers are released. The swarmers germinate to form two types of plant bearing thin-walled zoosporangia. About 98% of the swarmers give rise to plants bearing colourless sporangia, and about 2% to plants with orange-coloured sporangia due to the presence of γ-carotene. The proportion of these two types of plant is, however, variable. Naturally occurring mutants are known in which the swarmers from resting sporangia develop almost exclusively into orange plants. Also, by growing plants in the presence of the antibiotic cycloheximide the ratio can be adjusted to 25% orange to 75% colourless plants.

Cantino & Horenstein (1956) have speculated that the incidence of colourless and orange plants was controlled by the random distribution of a hypo-

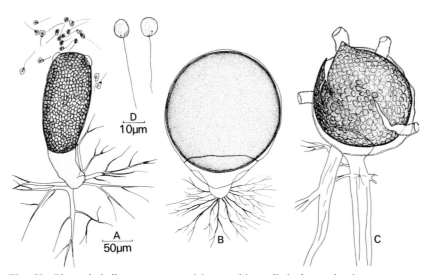

Fig. 58. *Blastocladiella emersonii*. A. Mature thin-walled plant releasing zoospores, 3-day-old plant. B. Immature resting sporangium, 3 days old. C. Germinating resting sporangium showing the cracked wall and four exit tubes. D. Zoospores from a thin-walled plant.

thetical cytoplasmic factor *gamma*. On cleavage of the cytoplasm inside the resting sporangium prior to the differentiation of swarmers some zoospore initials would have a high concentration of *gamma*, and would give rise to colourless plants, whilst others would have a low concentration of *gamma*, giving rise on germination to orange-coloured plants. When the zoospores derived from colourless, wild-type, orange-coloured, and mutant orange plants were stained with Nadi reagent (a 1% aqueous solution of *p*-amino-dimethylaniline hydrochloride mixed with an equal volume of 1% alcoholic *alpha*-naphthol) it was found that they contained a different distribution pattern of small blue-black particles. The swarmers from colourless plants had an average of 12·5 particles per spore, whilst the swarmers from the wild-type orange plants and mutant orange plants had averages of 7·5 and 8·0 particles per spore respectively. The distribution of these Nadi-positive particles thus corresponds with the predicted distribution of the hypothetical cytoplasmic factor *gamma*, although it should not too readily be assumed on the evidence so far available that *gamma* and the Nadi-positive particles are identical. It has been possible to isolate *gamma* particles from disrupted cells of *Blastocladiella*. They contain DNA, an RNA of low molecular weight, and the enzyme chitin synthetase (Myers & Cantino, 1974).

Despite careful and repeated examination there is no evidence of conventional sexual fusion between swarmers derived from orange and colourless plants. When swarmers from colourless and *mutant* orange plants are mixed together, however, transitory contact may be established by means of a cytoplasmic bridge, but the two cells eventually separate and nuclear fusion does not occur. Such contacts may occur several times, a colourless swarmer being visited in turn by more than one orange swarmer. There is some evidence that during such transitory contact exchange of cytoplasmic genetical determinants occurs. For example mutant zoospores show negligible viability on a purely synthetic medium, whilst the colourless zoospores possess good viability on the same medium. If zoospore suspensions of the two types are mixed, the number of viable plants derived from the suspension increases with the length of time the suspensions remain mixed. This and other evidence has led Cantino & Horenstein (1954) to the view that in the absence of ordinary gametic copulation in *Blastocladiella emersonii* its motile swarmers appear capable of unilateral transfer of certain cytoplasmic materials.

B. emersonii has a number of other unusual features. If zoospore suspensions are sprayed on to Peptone Yeast Glucose Agar the majority of plants which develop will be of the thin-walled, colourless type. On the same medium containing 10^{-2} M bicarbonate, resting sporangial plants with the thick, pitted brown walls will result. The addition of 40–80 mM KCl, NaCl or NH_4Cl, or exposure of cultures to ultra-violet light, will similarly induce the formation of resting sporangia (Horgen & Griffin, 1969). Thus, by means of a simple manipulation of the environment, it is possible to switch the metabolic activities of the fungus into one of two morphogenetic pathways. Cantino and his colleagues have analysed the differences in metabolism of plants grown in the presence and absence of bicarbonate and have shown that there are a number of important differences in the level of activity of certain enzymes (for reviews of literature see Cantino & Lovett, 1964; Cantino *et al.*, 1968). In the absence of bicarbonate, there is evidence for the operation of a Krebs cycle. However, in the presence of bicarbonate, part of the Krebs cycle is reversed, leading to alternative pathways of metabolism. In addition a polyphenol oxidase, absent in the thin-walled plant, replaces the normal cytochrome oxidase. There is also increased synthesis of melanin and of chitin in the presence of bicarbonate. Analysis of such changes will provide an insight at the enzymic level of differences between the two morphogenetic pathways.

The effect of bicarbonate can also be brought about by suitable levels of CO_2. However another unusual feature is that *B. emersonii* fixes CO_2 more rapidly in the light than in the dark. In the presence of CO_2, light-grown plants show a number of differences when compared with dark-grown controls. Illuminated plants take about 3 hours longer to mature, and are larger than dark-grown plants. They also have an increased rate of nuclear reproduction and a higher nucleic acid content. The most effective wavelengths for this increased CO_2 fixation (or lumisynthesis) lie between 400 and 500 nm, i.e. at the blue end of the spectrum. This suggests that the photoreceptor should be a yellowish substance. Attempts to identify the photoreceptor have, as yet, been unsuccessful, but it is known that it is not a carotenoid and not chlorophyll. Also the actual locus of the light-activated reaction has not yet been identified.

In *B. britannica* similarly, no sexual fusion has been reported (Cantino, 1970), but resting sporangial plants occur. The formation of resting sporangia in this species is stimulated by incubating cultures in white light. There is, however, no evidence of stimulation of formation of resting sporangia by bicarbonate.

MONOBLEPHARIDALES

This group is represented by three genera, *Monoblepharis*, *Gonapodya* and *Monoblepharella*. The first two are to be found on submerged twigs and fruits in fresh-water ponds and streams, especially in the spring, whilst *Monoblepharella* has been reported from hemp-seed baited water cultures from soil samples collected in the warmer parts of North and South America and from the West Indies. In all genera the thallus is eucarpic and filamentous. Studies of the wall of *Monoblepharella* show that it contains microfibrils of chitin. A characteristic feature is the frothy or alveolate appearance of the cytoplasm caused by the presence of numerous vacuoles often arranged in a regular fashion. Asexual reproduction is by posteriorly uniflagellate zoospores which are borne in terminal, cylindrical or flask-shaped sporangia. The zoospores have not yet received the same detailed examination as those of the Chytridiales and Blastocladiales, but the fine structure of the zoospore of *Mono-blepharella* has been studied (Fuller, 1966; Fuller & Reichle, 1968). The zoospore differs in a number of significant ways from those of the Blasto-cladiales. The nucleus is well separated from the kinetosome. Associated with the nucleus is a group of closely packed ribosomes, but at present there is some doubt whether this should be regarded as a nuclear cap. In addition to lipoid bodies, and several mitochondria, there is a characteristic laterally situated organelle termed the rumposome, whose function is unknown. It is inter-preted as being made up of a network of interconnecting tubules. Germina-tion may be bipolar or monopolar. Very little is known about the physiology of the group (Springer, 1945; Perrott, 1955, 1958; Aronson & Preston, 1960). Sexual reproduction is unusual for fungi in being oogamous with motile spermatozoids. After fertilisation has occurred, the egg may move to the mouth of the oogonium by amoeboid crawling in some species of *Mono-blepharis*, or propelled by movement of the spermatozoid flagellum in *Mono-blepharella* (Sparrow, 1939; Springer, 1945) and *Gonapodya* (Johns & Benja-min, 1954). Sparrow (1960) has separated the three genera into two families on the basis of this behavioural difference, the Monoblepharidaceae for *Mono-blepharis*, and the Gonapodyaceae for *Gonapodya* and *Monoblepharella*.

Monoblepharis (Fig. 59)

Species of *Monoblepharis* can be collected on waterlogged twigs on which the bark is still present, from quiet silt-free pools where the water is neutrally alkaline, i.e. pH 6·4 to 7·5. Twigs of birch, ash, elm and especially oak are suitable substrata, and although twigs collected at varying times throughout the year may yield growths of the fungus, there are two periods of vegetative growth, one in spring and another in autumn, with resting periods during the summer and winter months (Perrott, 1955). Low temperatures appear to favour development and good growths can be obtained on twigs incubated in dishes of distilled water at temperatures around 3 °C. The mycelium is delicate and vacuolate. The hyphae are multinucleate. During the formation of a sporangium a multinucleate tip is cut off by a septum. The cytoplasm cleaves around the nuclei to form zoospore initials which are at first angular, later pear-shaped. The ripe sporangium is cylindrical or club-shaped and may not be much wider than the hypha bearing it. A pore is formed at the tip of the sporangium through which the zoospores escape by amoeboid crawling. The free zoospores have a single posterior flagellum and swim away. On coming to rest the zoospore germinates to form a germ tube. The single nucleus of the zoospore divides and further nuclear divisions occur as the germ tube elon-gates.

Sexual reproduction can be induced by incubating twigs at room tempera-ture. In most species the sex organs may arise on the same plant which bears sporangia, although in two species *M. regignens* and *M. ovigera* no sexual stages have been reported. Whether these species are correctly classified in *Monoblepharis* is an open question, because it is possible that if they do reproduce sexually they might behave in the manner of *Monoblepharella*. In *Monoblepharis polymorpha* and related species the antheridia are epigynous. The antheridium is formed from the tip of the hypha which becomes cut off by a cross-wall. Beneath the antheridium the hypha becomes swollen, somewhat asymmetrically so that the antheridium is displaced into a lateral position. The swollen subterminal part becomes spherical and is then cut off by a basal septum to form the oogonium. In *M. sphaerica* and some other species the arrangement of the sex organs is the reverse of that in *M. polymorpha* the oogonium being terminal and the antheridium subterminal. In both groups of species the antheridium has often discharged sperm before the adjacent

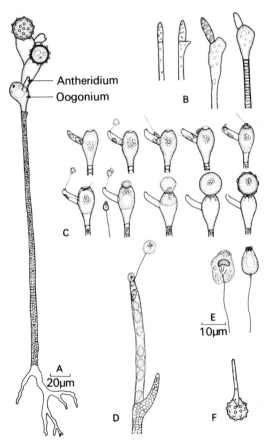

Fig. 59. *Monoblepharis polymorpha* (after Sparrow, 1960). A. Complete plant showing rhizoidal system, antheridia, oogonia and mature oospores. B. Stages in the development of gametangia. C. Stages in fertilisation. D. Discharging sporangium. E. Zoospores. F. Germinating oospore. A, B, C, D, F to same scale.

oogonium is ripe. The release of sperm resembles zoospore release, and each antheridium releases about four to eight posteriorly uniflagellate swarmers which resemble, but are somewhat smaller than, the zoospores. The oogonium contains a single spherical uninucleate oosphere, and when this is mature an apical receptive papilla on the oogonial wall breaks down. A spermatozoid approaching the receptive papilla of the oogonium becomes caught up in mucus and fusion with the oosphere then follows, the flagellum of the spermatozoid being absorbed within a few minutes. Following plasmogamy the oospore secretes a golden-brown membrane around itself and nuclear fusion later occurs. In some species, e.g. *M. sphaerica*, the oospore remains within the oogonium (endogenous) but in others, e.g. *M. polymorpha*, the oospore begins to move towards the mouth of the oogonium within a few minutes of fertilisation, and remains attached to it (exogenous) – Fig. 59. In the exogenous species nuclear fusion is delayed but finally fusions occur and the oospore becomes uninucleate. In some species the oospore wall remains smooth, but in others the wall may be ornamented by hemispherical warts or bullations. The oospore germinates after a resting period which coincides with frozen winter conditions or summer drought by producing a single hypha which branches to form a mycelium. The cytological details of the life cycle are not fully known but it seems likely that reduction division occurs during the germination of the over-wintered oospores (Laibach, 1927).

The Monoblepharidales are possibly related to the Blastocladiales, and the two groups have in common similar protoplasmic structure, uniflagellate swarmers with a well developed nuclear cap and bipolar zoospore germination (Sparrow, 1958). However, there is no parallel to the range of life cycles found in the Blastocladiales.

2: OOMYCETES

The Oomycetes are Mastigomycotina with a characteristic heterokont type of zoospore bearing two flagella, one of the whiplash type, and the other of the tinsel type. The Saprolegniales may have dimorphic zoospores, i.e. two kinds of zoospore may be present, one with apically attached flagella and the other with lateral attachment. In the Peronosporales, only laterally biflagellate zoospores occur. Most Oomycetes are aquatic, although some Saprolegniales and Peronosporales grow in soil, and some Peronosporales attack the shoots of terrestrial plants. Although the possession of zoospores implies a dependence on water for dispersal, in some Peronosporales causing diseases of plants, the sporangia are detached from the sporangiophore and dispersed by wind. Subsequent germination may be by means of zoospores or, as in some downy mildews, may be by means of germ tubes – i.e. zoospores may be lacking. The cell walls of Oomycetes are unusual for fungi, in that chitin is absent, but Lin & Aronson (1970) have claimed that in *Apodachlya* (Leptomitales) chitin and cellulose are present. In most Oomycetes only small amounts of cellulose are present, but the principal components are glucans with β-(1→3) and β-(1→6) glucosidic linkages (Aronson *et al.*, 1967). The form of the thallus varies. In the Lagenidiales (for the most part parasitic on algae, other aquatic fungi and microscopic aquatic animals), the thallus is holocarpic. In the remaining orders, the thallus is predominantly eucarpic, often composed of coarse coenocytic hyphae. In the Leptomitales, the hyphae are constricted at intervals and, at the point of constriction, a plug of the carbohydrate cellulin is present. For a fuller account of these groups, see Sparrow (1960, 1973c; Dick, 1973a). Sexual reproduction of the Oomycetes is oogamous, and results in the formation of one or more thick-walled resting spores or oospores. In the Lagenidiales, two holocarpic thalli of different sizes may fuse, but in the Saprolegniales, Peronosporales and Leptomitales, fusion is between a more or less globose oogonium containing one to several eggs, and an antheridium. It is now generally accepted that Oomycetes are diploid, having a haplobiontic type B life cycle with gametangial mitosis (see p. 91 and Dick & Win-Tin, 1973).

Oomycetes have been classified into four orders (Martin, 1961).

> Holocarpic or eucarpic; if latter, hyphae without constrictions or cellulin plugs, not arising from a well-defined basal cell; oogonium containing one to several oospores **Saprolegniales (p. 141)**

> Eucarpic; hyphae constricted, with cellulin plugs, arising from a well-defined basal cell; oogonium typically containing a single oospore **Leptomitales**

> h. Primarily aquatic; mostly parasitic on algae and water moulds; thallus holocarpic, slightly developed, simple or somewhat branched **Lagenidiales**

> h. Primarily terrestrial, living in soil or parasitic on vascular plants; zoosporangia, in latter case, often functioning as conidia **Peronosporales (p. 161)**

Dick (1976) has proposed that a new subdivision of the fungi, the Heterokontimycotina, should be erected for the Oomycetes. Possibly the group is related to heterokont algae, which have a similar life cycle.

SAPROLEGNIALES

This is the best-known group of aquatic fungi, often termed the water-moulds. Members of the group are abundant in wet soils, lake margins and fresh water, mainly as saprophytes on plant and animal debris, whilst some occur in brackish water. Most Saprolegniaceae thrive best in fresh water and are relatively intolerant of brackish water. A few species of *Saprolegnia* and *Achlya* are economically important as parasites of fish and their eggs (Scott &

O'Bier, 1962; Willoughby, 1968, 1969, 1970; Wilson, 1976; Willoughby & Pickering, 1977). *Aphanomyces euteiches* causes a root rot of peas and some other plants, whilst another species is a serious parasite of the crayfish *Astacus* (Unestam, 1965). Algae, fungi, rotifers and copepods may also be parasitised by members of the group and occasional epidemics of disease among zoo-plankton have been reported.

In the Ectrogellaceae, parasites of fresh-water and marine diatoms and Phaeophyceae, the thallus is holocarpic and endobiotic, resembling *Olpidium*. In the largest family (the Saprolegniaceae) the thallus is eucarpic and coenocy-tic, often forming a vigorous, coarse, stiff mycelium.

The zoospores are biflagellate and when the flagella are attached laterally the anterior flagellum is of the tinsel type and the posterior is of the whip type (Manton *et al.*, 1951). It has been suggested that the posterior flagellum is concerned with propulsion of the zoospore whilst the anterior tinsel-type flagellum functions as a rudder (McKeen, 1962). Two types of zoospore are found in some genera. The first motile stage is usually pear-shaped with two apically attached flagella. After swimming for a short time, the zoospore encysts, withdrawing its flagella. The cyst usually germinates to produce a second type of zoospore, bean-shaped, with laterally attached flagella. The second stage zoospore after swimming also encysts and on germination may produce a further zoospore of the second type. Repeated encystment and motility may occur and this phenomenon is sometimes called repeated emer-gence. Alternatively the cyst may germinate to form a germ tube from which the filamentous thallus develops. In some genera the behaviour of the zoo-spores appears to be modified (see below).

The sexual reproduction of the Saprolegniales is oogamous, with a large, usually spherical oogonium containing one to many eggs. Antheridial branches apply themselves to the wall of the oogonium, penetrating the wall by fertilisation tubes through which it is presumed that a single male nucleus is introduced into each egg.

Most of the Saprolegniaceae will grow readily in pure culture, even on purely synthetic media, and extensive studies of their physiology have been made (Cantino, 1950, 1955; Papavizas & Davey, 1960; Barksdale, 1962; Unestam & Gleason, 1968; Gleason *et al.*, 1970a,b; Faro, 1971; Gleason, 1972, 1973, 1976; Nolan & Lewis, 1974). Most species examined have no requirements for vitamins. Organic forms of sulphur such as cysteine, cystine, glutathione and methionine are preferred, and most species are unable to reduce sulphate. Organic nitrogen sources such as amino acids, peptone and casein are preferred to inorganic sources. Nitrate is generally not available although ammonium is utilised. Glucose, maltose, starch and glycogen are utilised by some species, but in others glucose is the only carbon source available. In liquid culture *Saprolegnia* can be maintained in the vegetative state indefinitely if supplied with organic nutrients in the form of broth. When the nutrients are replaced by water the hyphal tips quickly develop into zoosporangia. The formation of sexual organs can similarly be affected by manipulating the external conditions in some species, and the concentration of salts in the medium may play a decisive rôle. Deficiency of sulphur, phosphorus, calcium, potassium and magnesium may limit the formation of oogonia (Barksdale, 1962). In pure cultures of *Aphanomyces euteiches* oogonia are only produced in abundance when sulphur is supplied in the reduced form (Davey & Papavizas, 1962).

In the account which follows, only the Saprolegniaceae will be considered. Literature on the taxonomy of the group has been assembled by Dick (1973b) and on their ecology by Dick (1976).

Saprolegniaceae

Members of this group can readily be obtained from water, mud and soil by floating boiled hemp seed in dishes containing pond water or soil samples or waterlogged twigs covered with water (for details see Johnson, 1956). Within about 4 days the fungi can be recognised by their stiff, radiating, coarse hyphae bearing terminal sporangia, and cultures can be prepared by transferring hyphal tips or zoospores to maize-extract agar or other suitable media. About ten to twenty genera are recognised but the most commonly encountered are *Saprolegnia*, *Achlya*, *Aphanomyces*, *Dictyuchus* and *Thraustotheca*. In all the genera the hyphae are coenocytic, containing, within the wall, a peripheral layer of cytoplasm surrounding a continuous central vacuole. In the peripheral cytoplasm, streaming can be readily seen. Numerous nuclei are present unseparated by cell walls. Mitotic division is associated with the replication of paired centrioles and the development of an intranuclear mitotic spindle (Flanagan, 1970; Heath & Greenwood, 1970b, 1971; Heath, 1974, 1976). Filamentous mitochondria and fatty globules can also be observed in vegetative hyphae. Fine structure of vegetative hyphae has been studied in *Aphanomyces* (Shatla *et al.*, 1966; Hoch & Mitchell, 1972a), in *Saprolegnia* (Gay & Greenwood, 1966; Heath *et al.*, 1971) and in *Achlya* (Dargent *et al.*, 1973). In *Saprolegnia* the hyphal tip is tapering, and in an apical region 10–40 μm long, mitochondria and nuclei are virtually absent. This apical region is mainly filled with membrane-bounded vesicles which have been termed 'wall vesicles', present in maximum concentration in the apical 5 μm, but also present further behind the apex in the peripheral cytoplasm adjacent to the cell wall. The wall vesicles are presumed to contain precursors of wall material. The origin of the wall vesicles is not certain, but a Golgi apparatus is present, and it is possible that the vesicles arise from Golgi cisternae as in other Oomycetes such as *Pythium* and *Phytophthora* (see p. 60). The mitochondria are cylindrical, measuring up to 10 μm in length. They are most frequent some distance behind the hyphal apex, and are orientated parallel to the long axis of the hypha. They are sufficiently large to be seen in cytoplasmic streaming in living material.

A feature of many Saprolegniaceae, especially when grown in culture, is the formation of thick-walled enlarged terminal or intercalary portions of hyphae which become packed with dense cytoplasm and are cut off from the rest of the mycelium by septa. These structures, which may occur singly or in chains (see Fig. 60), are termed gemmae or chlamydospores, and their formation can be induced by manipulation of the culture conditions. Morphologically less distinct, but similar structures are frequently found in old cultures. Although it is known that gemmae cannot survive desiccation or prolonged freezing, they remain viable for long periods. They may function either as female gametangia or as zoosporangia, but more frequently germinate by means of a germ tube. Another feature of old cultures is the fragmentation of cylindrical pieces of mycelium cut off at each end by a septum.

ASEXUAL REPRODUCTION IN THE SAPROLEGNIACEAE

Saprolegnia (Fig. 60).

Species of *Saprolegnia* are common in soil and in fresh water, saprophytic on plant and animal remains but a number of species such as *S. ferax* and *S. parasitica* have been implicated in disease of fish and their eggs. These species have been experimentally inoculated into wounded fish, resulting in death within 24 hours (Tiffney, 1939a,b; Vishniac & Nigrelli, 1957; Scott & O'Bier, 1962). In nature *S. parasitica* may be associated with severe epidemics of disease among fish (Stuart & Fuller, 1968). The sporangia of *Saprolegnia* develop at the tips of the hyphae. The hyphal tip which is pointed in the vegetative condition, swells and becomes club-shaped, rounded at the tip and accumulates denser cytoplasm around the still clearly visible central vacuole. A septum develops at the base of the sporangium and it is at first convex with respect to the sporangial tip, i.e. it bulges into the sporangium. The hyphal tip contains numerous nuclei, and cleavage furrows separate the cytoplasm into uninucleate pieces, each of which differentiates into a zoospore. As the zoospores are cleaved out the central vacuole is no longer visible. The tip of the cylindrical sporangium contains clearer cytoplasm and a flattened protuberance develops at the end. As the sporangium ripens and the zoospores become fully differentiated they show limited movement and change of shape (Fig. 60b–d). Evidence of a build-up of pressure within the sporangium can be seen from the shape of the basal septum, which shortly before zoospore discharge becomes concave, i.e. is pushed into the lumen of the hypha beneath the sporangium. The septum is formed in the plane to convex position with respect to the hyphal tip, i.e. bulging into the sporangium, and very quickly the bulge becomes concave during maturation of the sporangium. After cleavage, the positive turgor pressure of the sporangium relative to the hypha is lost and the septum again bulges into the sporangium while the zoospores become fully differentiated. The sporangium undergoes a slight change in shape at this time and the tip of the sporangium breaks down at the clear

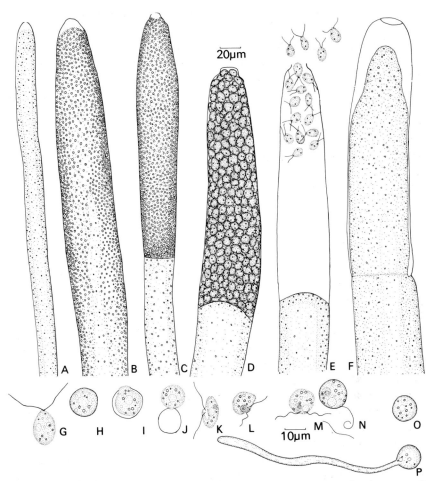

Fig. 60. *Saprolegnia*. A. Apex of vegetative hypha. B–D. Stages in the development of zoosporangia. E. Release of zoospores. F. Proliferation of zoosporangium. A second zoosporangium is developing within the empty one. G. Primary zoospore (first motile stage). H. Cyst formed at end of first motile stage (primary cyst). I–J. Germination of primary cyst to release a second motile stage (secondary zoospores). K–M. Secondary zoospores. N. Secondary zoospore at moment of encystment. Note the cast-off flagella. O. Secondary cyst. P. Secondary cyst germinating by means of a germ tube. A–F to same scale; G–P to same scale.

protuberant tip. The spores are released quickly, many zoospores escaping in a few seconds, moving as a column through the opening as though pushed out under pressure. The whole process of sporangium differentiation from the inception of the basal septum to the release of zoospores takes about 90 minutes. The zoospores leave the sporangium backwards, with the blunt posterior emerging first. The size of the zoospore is sometimes slightly greater than the diameter of the sporangial opening so that the zoospores are squeezed through it. If release of the zoospores is caused by a build-up of pressure, the difference in size of the zoospore and the opening may be important in preventing sudden loss of pressure. However, an occasional zoospore is sometimes left behind and swims about in the empty sporangium, finally making its way out backwards through the opening. Zoospores in partially empty sporangia orientate themselves in a linear fashion along the central axis of the sporangium as though repelled from the walls. Possibly electrostatic repulsion is involved, because the orientation can be disturbed if sporangia are placed in an electrical field. The zoospores are attracted towards the cathode (Borkowski, 1969).

A characteristic feature of *Saprolegnia* is that, following the discharge of the zoosporangium, growth is renewed from the septum at its base so that a new apex develops inside the old sporangial wall. This in turn may develop into a zoosporangium, discharging its spores through the old pore (Fig. 60F). The process may be repeated so that several empty zoosporangial walls may be found inside, or partially inside, each other.

The zoospores on release slowly revolve and eventually swim with the pointed end directed forwards. They are pear-shaped and bear two apically attached flagella (Fig. 60G). Each spore also contains a nucleus and a contractile vacuole. Electron micrographs of *Saprolegnia* zoospores have been provided by Manton *et al.* (1951). The zoospores swim for a time which may be less than a minute. However, the zoospores from a single sporangium show variation in their period of motility, the majority encysting within about a minute, but some remaining motile for over an hour. The zoospore then withdraws its flagella and encysts, i.e. the cytoplasm becomes surrounded by a distinct firm membrane (Fig. 60H) (Crump & Branton, 1966; Holloway & Heath, 1974, 1977a,b). Only the axonemes of the flagella are withdrawn: the flimmer hairs remain in a tuft at the surface of the cyst. Following a period of rest (2–3 hours in *S. dioica*) the cyst germinates to release a further zoospore (Fig. 60I–J). This zoospore differs in shape from the zoospore released from the sporangium, being bean-shaped (although amoeboid changes in shape occur) and bearing two laterally attached flagella which are inserted in a shallow groove running down one side of the zoospore (Fig. 60K–M). This laterally biflagellate bean-shaped zoospore is called the secondary zoospore or second motile stage, in contrast to the apically biflagellate primary zoospore or first motile stage. The secondary zoospore may swim vigorously for several hours before encysting. Salvin (1941) compared the rates of movement of primary and secondary zoospores in *Saprolegnia* and found that the secondary zoospores swam about three times more rapidly than the primary. He wrote 'as soon as there is a transition of the zoospore from a primary to a secondary type, not only is there a change in the general morphology, but in reality a transformation in the fundamental biochemical constitution'. Whilst this is possible, it is more likely that the reason for the more rapid rate of swimming of the secondary zoospore is that the lateral insertion of the flagella, with the tinsel flagellum pointing forwards, and the whip flagellum pointing backwards, makes possible a more effective method of propulsion. Movement of secondary zoospores is chemotactic and zoospores can be stimulated to aggregate about parts of animal bodies such as the leg of a fly. Fischer & Werner (1958a) have substituted this chemotactic stimulus by glass capillaries containing 10^{-1} M sodium chloride or potassium chloride and traces of amino acids. They have also shown (1958b) that encystment can be stimulated by the presence of traces of nicotinamide (10^{-7} M), but that if the cysts are washed free of this substance then a further motile stage may be formed. This phenomenon of repeated encystment and motility is termed **polyplanetism** or repeated emergence. At the end of the second motile stage, and any subsequent motile stages, the flagella are cast off, not withdrawn. The secondary cysts, instead of forming further zoospores, may germinate by means of a germ tube to give rise to vegetative filaments (Fig. 60P). Because there are two distinct motile stages in the asexual reproduction of *Saprolegnia*, it is said to be **diplanetic** (the zoospore is sometimes termed a planospore, another term for a motile spore). However, since as we have seen, there may be several periods of motility, and the term polyplanetic could equally well be applied. An alternative terminology has been suggested. Since the two motile stages differ morphologically, the term **dimorphic** has been proposed and may be preferable. Electron micrographs of the cysts formed at the end of the primary and secondary motile stages in *Saprolegnia* have shown that in this genus the *cysts* are also dimorphic. Primary cysts of *S. ferax* bear single or clustered spines $0.2–0.4$ μm long and about 10 nm in diameter. Secondary cysts bear double-headed recurved hooks, resembling boat hooks, each seated on a plaque of thicker wall material. In *S. ferax*, the hooks are single, but in *S. parasitica* they occur in tufts (Meier & Webster, 1954; Heath & Greenwood, 1970c) (Fig. 63). There is evidence that these hooks are effective in attaching the cysts to surfaces, but it is also possible that they may attach the cysts to the water meniscus. Trout and char, placed in a bath containing secondary zoospores of *Saprolegnia* for 10 minutes, followed by 1 hour in clear water, had an extremely high spore concentration attached to the skin (Willoughby & Pickering, 1977).

The fine structure of zoospore development in *Saprolegnia* has been studied by Gay & Greenwood (1966), Heath & Greenwood (1970c, 1971), Heath *et al.* (1970) and I. B. Heath (1976). Cleavage of the peripheral mass of cytoplasm surrounding the central vacuole is brought about by the development of a series of large vesicles, in contact with the tonoplast. The vesicles extend outwards until their membranes make contact with the plasmalemma, and in this way uninucleate zoospore initials are delimited. The fusion of the cleavage vesicles with the plasmalemma effectively makes the plasmalemma continuous with the tonoplast. This results in the loss by the sporangium of its semi-permeability and turgor. Vacuolar sap, i.e. the liquid originally contained within the tonoplast, is free to leak out of the sporangium. At the same time, the spore initials, which are now unrestrained and lying free from each other in the sporangium, are free to take up water, and swell.

During the development of zoosporangia, organelles termed 'bars' appear in the cytoplasm and persist in the cleaved zoospore initials. These structures are not found in vegetative hyphae. They are cylindrical, with hemispherical ends, up to 0.5 μm long and 0.1–0.15 μm wide, bounded by a single membrane, and contain a central bundle of parallel fibrils. The bars are frequently in contact with the plasmalemma in zoospores which have developed a wall, and it is believed that the bars contain material from which the cyst walls and spines are made. The flimmer hairs are believed to be derived from the endoplasmic reticulum. They appear as bundles of microtubules enclosed in square-ended, membrane-bounded vesicles in developing zoosporangia. The membranes of the vesicles are continuous with the endoplasmic reticulum and are studded with ribosomes. As the flagella of the zoospores develop, the flimmer hairs become attached to the flagellum sheath. At the onset of zoospore cleavage, the paired centrioles associated with the nuclei reorientate and develop into kinetosomes.

Achlya (Fig. 6)

Species of *Achlya* are also common in soil and in waterlogged plant debris such as twigs (Johnson, 1956). Certain species have also been reported as naturally occurring pathogens of fish. The zoosporangial development in *Achlya* is similar in all respects to that of *Saprolegnia*, and according to some earlier workers the zoospores develop flagella, but Johnson (1956) states that the question of primary zoospore flagellation is still unsolved. On discharge, the zoospores do not swim away, but cluster in a hollow ball at the mouth of the zoosporangium and encyst there (Fig. 61A). Partial fragmentation of the spore ball frequently occurs and may have ecological significance in spore dispersal prior to secondary zoospore formation. Unlike certain (but not all) *Saprolegnia* species, *Achlya* cysts are normally found at the bottom of culture dishes, and presumably at the water/bottom-mud interface in natural environments. Secondary zoospores do not develop simultaneously from the spore ball. Protoplasmic connections linking the zoospores together have been described. The first motile stage in *Achlya* is thus very brief. The primary cysts remain at the mouth of the sporangium for a few hours and then each cyst releases a secondary-type zoospore through a small pore (Fig. 61B,C). After swimming, the zoospore encysts and may germinate by a germ tube (Fig. 61E–F), or repeated emergence of secondary-type zoospores may occur. The cysts of *A. klebsiana* may remain viable for at least two months when stored aseptically at 5 °C (Reischer, 1951). When the zoosporangium of *Achlya* has released zoospores, growth is usually renewed laterally by the outpushing of a new hyphal apex just beneath the first sporangium (Fig. 61A).

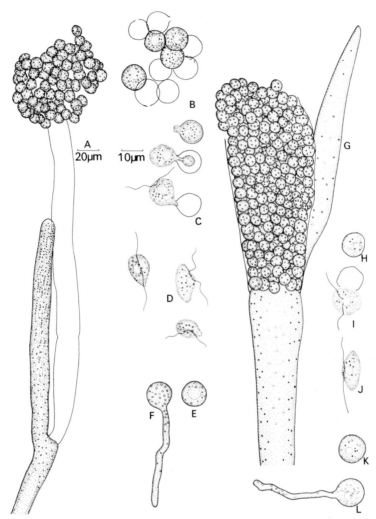

Fig. 61. A–F. *Achlya colorata.* A. Zoosporangium showing a clump of primary cysts at the mouth. Note the lateral proliferation of the hypha from beneath the old sporangium. B. Full and empty primary cysts. C. Stages in the release of secondary zoospores from a primary cyst. D. Secondary zoospores. E. Secondary cyst. F. Secondary cyst germinating by means of a germ tube. G–L. *Thraustotheca clavata.* G. Zoosporangium showing encystment within the sporangium. The primary cysts are being released through breakdown of the sporangial wall. H. Primary cyst. I. Primary cyst germinating to release a secondary-type zoospore (here the first motile stage). J. Secondary-type zoospore. K. Secondary cyst. L. Secondary cyst germinating by means of a germ tube. A and G to same scale; B–F and H–L to same scale.

Aphanomyces

Aphanomyces is distinguished from *Achlya* by its narrow, delicate mycelium and by its narrow sporangia containing a single row of spores. Species of this genus appear to be common on transient substrata such as insect exuviae in water (Dick, 1970). Keratinophilic species occur in soil (Seymour & Johnson, 1973). *Aphanomyces astaci* causes a disease of crayfish (Unestam, 1965), whilst *A. euteiches* is a pathogen of pea roots. Asexual reproduction in *Aphanomyces* is variable. In *A. euteiches*, Hoch & Mitchell (1972a,b) have shown that flagella do not develop on the primary spores. Protoplasts are cleaved out which move to the mouth of the sporangium and encyst. The zoospores which develop from the primary cysts are of the secondary type, but are the first truly motile stage. *Aphanomyces euteiches* is thus **monomorphic.** In *A. patersonii*, W. W. Scott (1956) has shown that the motility of the primary zoospore could be controlled by variation in temperature. Below 20 °C, encystment of the primary zoospores at the mouth of the sporangium occurred in the manner typical of the genus, but above this temperature the primary zoospores swam away and encysted at a distance from the zoosporangium. In this respect, its behaviour resembled that of a *Leptolegnia*. A similar phenomenon has been described in a *Saprolegnia*. Salvin (1941) has claimed that below 10 °C the primary zoospores did not behave in a normal manner, but in some instances they encysted almost immediately after emerging and formed a loose mass of encysted spores near the mouth of the sporangium.

Thraustotheca (Fig. 61)

In *T. clavata* the sporangia are broadly club-shaped, and there is no free-swimming primary zoosporic stage. Encystment occurs within the sporangia and the primary cysts are released by irregular rupture of the sporangial wall (Fig. 61G). After release, the angular cysts germinate to release bean-shaped zoospores with laterally attached flagella (Fig. 61I–J), and after a swimming stage, further encystment occurs, followed by germination by a germ tube (Fig. 61K–L), or by emergence of a further zoospore. The zoospores are thus monomorphic and polyplanetic.

Dictyuchus (Fig. 62)

In this genus, there is again no free-swimming primary zoospore stage. The entire zoosporangium may be deciduous, and detached zoosporangia are capable of forming zoospores. Zoospore initials are cleaved out but encystment occurs within the cylindrical sporangium. The cysts are tightly packed together and release their secondary zoospores independently through separate pores in the sporangial wall (Fig. 62A). When zoospore release is complete a network made up of the polygonal walls of the primary cysts is left behind.

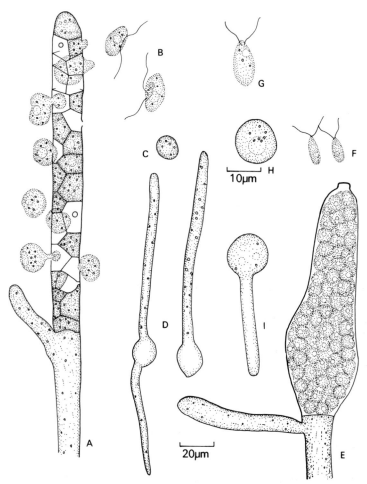

Fig. 62. A–D. *Dictyuchus sterile*. A. Zoosporangium showing cysts within the sporangium and the release of secondary-type zoospores through separate pores in the sporangium wall. Note the network of primary cyst membranes. B. Secondary-type zoospores. C. Secondary cyst. D. Germination of secondary cysts by means of germ tubes. E–I. *Pythiopsis cymosa*. E. Zoosporangium. F, G. Primary zoospores. H. Primary cyst. I. Primary cyst germinating by means of a germ tube. Secondary-type zoospores have not been described. A, B, C, E, F to same scale; G, H, I to same scale.

After swimming, the laterally biflagellate zoospore encysts (Fig. 62B,C). Electron micrographs have shown that the wall of the secondary cyst of *D. sterile* bears a series of long spines looking somewhat like the fruit of a horse-chestnut (Fig. 63D) (Meier & Webster, 1954; Heath *et al.*, 1970). Following the formation of the first zoosporangium a second may be formed immediately beneath it by the formation of a septum cutting off a subterminal segment of the original hypha, or growth may be renewed laterally to the first sporan-

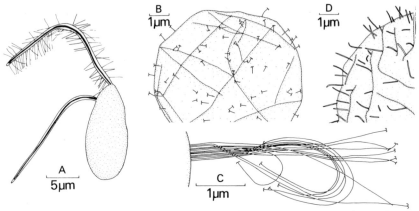

Fig. 63. Tracings from electron-micrographs of zoospores and cysts of Saprolegnia-ceae. A. *Saprolegnia ferax:* primary zoospore (after Manton *et al.*, 1951). B. *S. ferax:* secondary cyst. Empty cyst membrane showing two-headed hooks. C. *S. parasitica:* secondary cyst membrane showing tuft of two-headed hooks. D. *Dictyuchus sterile:* secondary cyst showing spiny projections.

gium. Commonly the entire sporangium in *Dictyuchus* may become detached from its parent hypha, but spore release proceeds normally.

Because there is only one motile stage in *Thraustotheca* and *Dictyuchus* (i.e. a zoospore of the secondary type), they are said to be monomorphic. How-ever, in *Pythiopsis cymosa* where there is again only one motile stage, the zoospore is of the primary type. After swimming, the zoospore encysts and then germinates directly (Fig. 62G–I).

APLANETIC FORMS

In certain cultures of Saprolegniaceae the zoosporangia produce zoospores which encyst within the sporangia but the cysts do not produce a motile stage. Instead, germ tubes are put out which penetrate the sporangial wall. It would appear then that there is no motile stage and such forms are said to be **aplanetic.** The aplanetic condition is occasionally found in staling cultures of *Saprolegnia*, *Achlya* and *Dictyuchus*. Some species produce sporangia only rarely and the genus *Aplanes* has been erected for these forms. In very clean cultural conditions, all have been shown to behave as *Achlya*. They are distinctive in other respects, and Dick (1973b) has assigned them to a separate group within the genus *Achlya*. Two species of Saprolegniaceae are not known to form sporangia at all. They are common in soil, and have been placed in a separate genus, *Aplanopsis*. Another genus, *Geolegnia*, forms sporangia con-taining multinucleate thick-walled aplanospores which never produce a flagel-late stage.

Variations in the method of zoospore release in well-known and less well-known species have brought the validity of generic boundaries into question. The genera of Saprolegniaceae were originally erected on the single criterion of zoospore release, but in the majority of cases other morphological features can also be used to reinforce generic concepts and allow the occurrence of, for example, dictyuchoid sporangial development in other genera.

SEXUAL REPRODUCTION IN SAPROLEGNIACEAE

In all the Saprolegniaceae sexual reproduction follows a similar course. Oogonia containing one to several eggs are fertilised by antheridial branches. Fertilisation is accomplished by the penetration of fertilisation tubes into the oogonia. The majority of species are homothallic (monoecious), that is a culture derived from a single zoospore will give rise to a mycelium forming both oogonia and antheridia. Some species are, however, heterothallic (dioecious) and sexual reproduction only occurs when two different strains are juxtaposed, one forming oogonia, the other antheridia. In some species the eggs develop apogamously (parthenogenetically) and in such forms ripe oogonia are found without any antheridia associated with them. There is a very extensive literature on the cytology of sexual reproduction (Dick, 1969, 1972; Dick & Win Tin, 1973). Light may inhibit the formation of oogonia (Szaniszlo, 1965).

The arrangement of oogonia and antheridia in monoecious forms is shown

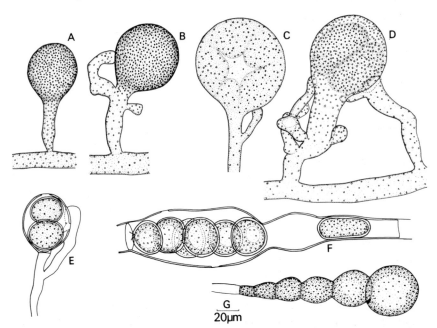

Fig. 64. *Saprolegnia litoralis.* A–D. Stages in the development of oogonia. C. Oogonium showing furrowed cytoplasm. D. Outline of two oospheres visible. E. Oogonium with two mature oospores. F. Intercalary oogonium lacking antheridia. The oospores have developed apogamously. G. Chain of gemmae.

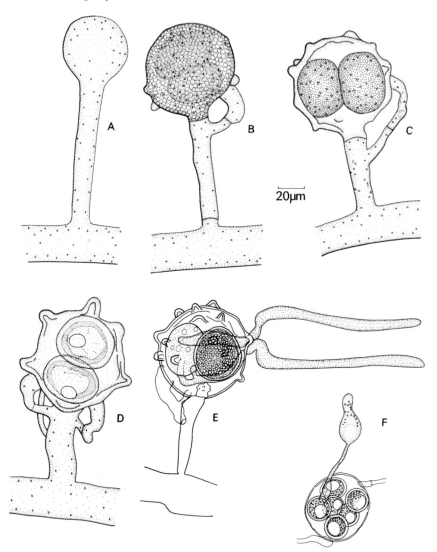

Fig. 65. A–E, *Achlya colorata*; F, *Thraustotheca clavata*. A–D. Stages in development of oogonia. E. Six-month-old oospores germinating after 40 hours in charcoal water. F. *T. clavata*: 6-month-old oospore germinating after 17 hours in charcoal water. The germ tube is terminated by a sporangium.

in Figs. 64 and 65, and in a dioecious form in Fig. 66. Where the antheridial branches arise from the stalk of the oogonium they are said to be **androgynous**. If they originate on the same hypha as the oogonium they are said to be **monoclinous**, and when they originate on different hyphae they are **diclinous**. The oogonial initial is multinucleate and as it enlarges nuclear divisions

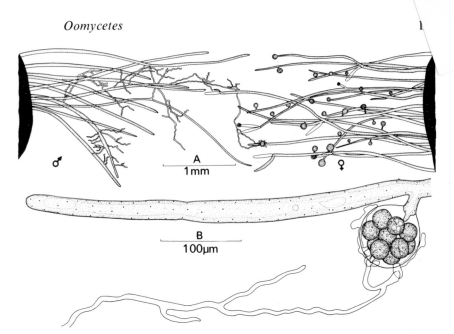

Fig. 66. *Achlya ambisexualis.* A. Male plant and female plant grown on hemp seeds and placed together in water for 4 days. Note the formation of antheridial branches on the male and oogonial initials on the female. B. Fertilisation, showing the diclinous origin of the antheridial branch.

continue. Eventually some of the nuclei degenerate leaving only those nuclei which are included in the eggs. From the central vacuole within the oogonium cleavage furrows radiate outwards to divide the cytoplasm into uninucleate portions which round off to form eggs. The number of eggs is variable. In *Aphanomyces* there is only a single egg, but in other genera there may be one to several or occasionally as many as thirty. The entire mass of cytoplasm within the oogonium is used up in the formation of eggs and there is no residual cytoplasm (periplasm) as in the related Peronosporales. The wall of the oogonium may be uniformly thick, but in some forms it shows uneven thickening, with thin areas or pits through which fertilisation tubes may penetrate (see Fig. 64E). In *Achlya colorata*, one of the commonest British species, the oogonial walls bear blunt rounded projections so that they appear somewhat spiny (Fig. 65D). A septum at the base of the oogonium cuts it off from the subtending hypha. The antheridia are also multinucleate. The antheridial branches grow chemotropically towards the oogonia and apply themselves to the oogonial wall. The tip of the antheridial branch is cut off by a septum, and the resulting antheridium puts out a fertilisation tube which penetrates the oogonial wall. Within the oogonium, the fertilisation tube may be branched and may contain several nuclei. After the tube has penetrated the oosphere wall, the male nucleus eventually fuses with the single egg nucleus. The fertilised egg (or oospore) now undergoes a series of changes. Its wall thickens and oil globules become obvious. They tend to aggregate in a characteristic way for a given species (see below). The eggs are rarely capable of immediate further development but seem to need a period of maturation, sometimes lasting weeks or months. Germination can be stimulated by transferring mature oogonia to freshly distilled water (preferably after shaking with charcoal and filtering), and occurs by means of a germ tube which grows out from the oospore, through the oogonial wall. Here it may continue growth as mycelium or may give rise to a sporangium (Fig. 65E,F).

The fine structure of the development of sex organs in Saprolegniaceae has been studied by Howard & Moore (1970), Gay *et al.* (1971), Ellzey (1974) and Beakes & Gay (1978a,b). Cleavage of the oospheres from the cytoplasm contained within the oogonium is brought about by the coalescence of specialised vesicles termed dense body vesicles, which finally fuse with the plasmalemma of the oogonium. Mature oospores contain a membrane-bound vacuole-like body, the **ooplast**, surrounded by cytoplasm containing various organelles. The ooplast contains particles in Brownian motion. The position of the ooplast is of taxonomic significance, and four types of oospore have been distinguished (Howard & Moore, 1970; Howard, 1971). *Centric* oospores have a central ooplast surrounded by one or two peripheral layers of small oil droplets. *Subcentric* oospores have two or three layers of small oil droplets on one side of the ooplast and one layer on the other. In *eccentric* oospores, there is one large oil globule in the cytoplasm which is disposed to one side, and with the ooplast in contact with the plasmalemma on the other side. *Expersate* oospores lack an ooplast and have one to several oil droplets entirely surrounded by protoplasm. This classification of oospore types is not accepted by all workers. Dick (1971) has claimed that all known species of Saprolegniaceae possess an ooplast and that the term expersate is therefore superfluous.

Evidence for meiotic divisions within the gametangia of Saprolegniaceae is of several kinds.

(a) Cytological observations by light microscopy have indicated a reduction in nuclear size during the nuclear divisions in antheridia and oogonia. Counts of chromosomes and arrangements of chromosomes are also indicative of meiosis.

(b) Ultrastructural evidence for meiosis has been obtained in the demonstration of synaptonemal complexes in developing oogonia and antheridia. Centriolar organisation characteristic of meiotic division has also been found. Meiosis is intranuclear, i.e. the nuclear envelope does not break down.

(c) Microspectrophotometric evidence in *Saprolegnia*, and also in the related fungus *Apodachlya*, a member of the Leptomitales, has been obtained by Bryant & Howard (1969) and Howard & Bryant (1971), supporting the occurrence of gametangial meiosis. The DNA content of antheridial and oogonial nuclei prior to the final nuclear divisions was 4C, but this was reduced to 1C in male and female gamete nuclei, whilst the DNA content of nuclei from vegetative hyphae was 2C.

(*d*) Genetical evidence based largely on the inheritance of mating type is most readily interpreted on the basis of a life cycle with gametangial meiosis (Mullins & Raper, 1965; Barksdale, 1966). Crosses were made between male and female strains of *Achlya* (see below) and oospores were germinated. From any one oospore, the progeny which developed were almost invariably either male or female, never both. If meiosis occurred at oospore germination, then, assuming that all products of meiosis survived, both male and female strains would be expected. Since this result is not obtained, it is concluded that the oospore gives rise to diploid progeny on germination.

There is evidence that the cytology of some Oomycetes is not straightforward. For example, in some forms with apogamous oogonial development, there is a failure of meiosis. In others, there is evidence of polyploidy (Dick & Win-Tin, 1973).

Studies on heterothallic species of *Achlya* (J. R. Raper, 1952, 1954, 1957), have shown that the sequence of events in sexual reproduction is co-ordinated by hormones secreted by the male and female plants. If isolates of *Achlya bisexualis*, *A. ambisexualis* and *A. heterosexualis* made from water or mud are grown singly on hemp seed in water, reproduction is entirely asexual, and no sexual organs are distinguishable. When certain of the isolates are grown together in the same dish, however, within 2–3 days it becomes apparent that one strain is forming oogonia, and the other antheridia. Evidently the presence of both strains in the same dish causes a mutual stimulation for the production of sex organs. It is not necessary for the two strains to be in direct contact but they can be held apart in the water or separated by a membrane of cellophane or by agar, and development of antheridial and oogonial initials occurs as the mycelia of the two strains approach each other. This suggests that a diffusible substance (or substances) may be responsible for the phenomenon. The first observable reaction as the compatible hyphae approach each other is the production of fine lateral branches behind the advancing tips of the male hyphae. These are antheridial branches. By growing male plants in water in which female plants had been growing Raper showed that the vegetative female plants initiated the development of antheridial branches on the male. When female plants were grown in water in which vegetative male plants had previously been growing there was no noticeable reaction on the part of the female, i.e. no oogonial initials were produced. The rôle of the vegetative female plants as initiators of the sequence of events in sexual reproduction was ingeniously confirmed by experiments in micro-aquaria consisting of several consecutive chambers through which water flowed by means of small siphons. In adjacent chambers male and female plants were placed alternately so that water from a male plant would flow over a female plant and so on. If a female plant was placed in the first chamber the male in the second reacted by developing antheridial hyphae. If, however, a male plant was placed in the first chamber and a female in the second, the female failed to respond, and the male plant in the third chamber was the first to react. Raper postulated that the development of the antheridial branches was in response to the secretion of a hormone, Hormone A, secreted by the vegetative females. By further experiments of this kind Raper showed that the later steps in the sexual process were also regulated by means of diffusible substances. He postulated that the antheridial branches secreted a second substance, Hormone B, which resulted in the formation of oogonial initials on the female plants, the next observable step following the formation of antheridial hyphae. The oogonial initials in their turn secreted a further substance, Hormone C, which stimulated the antheridial initials to grow towards the oogonial initials and also resulted in the antheridia being delimited. Having made contact with the oogonial initials the antheridial branches secreted Hormone D which resulted in the formation of a septum cutting off the oogonium from its stalk and the formation of oospheres. If the interchange of diffusible material was prevented by removing male plants from close proximity to the female the sexual process could be interrupted. If the direct contact of antheridia and oogonia is prevented by a barrier of cellophane fertilisation does not occur. The outlines of the proposed scheme are shown in table 3.

The original scheme necessitated four hormones, but later work suggested that Hormone A might be a complex of at least four hormones, some of which were secreted by male plants. Certain of the hormones of the 'A complex'

Table 3. *Effects of hormones on sexual reactions in* Achlya ambisexualis

Hormone	Produced by	Affecting	Specific action
A	Vegetative hyphae	Vegetative hyphae	Induces formation of antheridial branches
B	Antheridial branches	Vegetative hyphae	Initiates formation of oogonial initials
C	Oogonial initials	Antheridial branches	(1) Attracts antheridial branches (2) Induces thigmotropic response and delimitation of antheridia
D	Antheridia	Oogonial initials	Induces delimitation of oogonium by formation of basal wall

After Raper (1939).

augmented the hormonal activity of the vegetative female, and another in bited it. Separation of the different hormones of the 'A complex' can achieved by chemical means. Purification, and concentration and assay of Hormone A have been achieved. The purified hormone of *A. bisexualis* is a colourless crystal having the empirical formula $C_{29}H_{42}O_5$, to which the name antheridiol has been given (McMorris & Barksdale, 1967). The substance has been characterised chemically, and synthesised (see Barksdale, 1969; Machlis, 1972). Antheridiol is a sterol, with a stigmasterol skeleton. Whether the augmentation of the effect of Hormone A is due to further hormones now seems doubtful, since later work has shown that available carbon and nitrogen sources such as amino acids influence the expression of Hormone A (Barksdale, 1970). It has been shown that the effect of Hormone A in inducing antheridial branch formation is correlated with production of cellulase by responsive male strains. Presumably the cellulase brings about softening of the walls of the male hyphae to allow lateral branches to develop (Thomas & Mullins, 1969; Mullins, 1973). It has been suggested that the hormone affects morphogenesis by turning on the synthesis of mRNA which is then translated into protein required for the differentiation of antheridial initials (Kane *et al.*, 1973; Silver & Horgen, 1974).

Barksdale (1963b) has shown that if Hormone A is adsorbed on to small particles of the inert plastic polystyrene, the particles not only induce initiation of antheridial initials but also directional growth, and she has therefore questioned the necessity to postulate a hormone distinct from A which is exclusively concerned with directional growth.

Three chemically similar compounds have now been isolated from a hermaphroditic strain of *A. heterosexualis* which show Hormone B activity, i.e. are capable of inducing oogonial initials in female strains (McMorris *et al.*, 1975). The substances, which have been named oogoniol-1, -2 and -3, are steroids. Oogoniols probably arise by hydroxylation of β- or γ-sitosterol which are abundant plant sterols.

The hormonal control of sexual reproduction is not confined to heterothallic forms: there is evidence that in homothallic species there is similar coordination. The fact that it is possible to initiate sexual reactions between homothallic and heterothallic species of *Achlya* shows that some of the hormones are probably common to more than one species although there is also evidence of some degree of specificity of the hormones of different species (Raper, 1950; Barksdale, 1960, 1965). Partial or sexual reactions may be stimulated between closely related species (Couch, 1926; Salvin, 1942; Raper, 1950).

One further interesting phenomenon which has been disc to the heterothallic Achlyas is relative sexuality. If a numb *bisexualis* and *A. ambisexualis* from widely separated sourc possible combinations it is found that certain strains show either as male or female depending on the particular partne apposed. Other strains remain invariably male or invariably female and these are referred to as true males or strong males, etc. The strains can be arranged in a series with strong males and strong females at the extremes, and intermediate strains whose reaction may be either male or female depending on the strength of their mate. Interspecific responses between strains of *A. bisexualis* and *A. ambisexualis* are also possible and the behaviour of certain of these strains may also show reversibility. Further, some of the strains which appear heterothallic at room temperature are homothallic at lower temperatures. Similar findings have been reported for *Dictyuchus monosporus*, i.e. a given strain may behave as a heterothallic male or female, or as a homothallic strain, depending on the nature of its partner, or on culture conditions (Sherwood, 1969, 1971). Barksdale (1960) has postulated that the heterothallic forms may have been derived from homothallic ones. She has argued that the most notable difference between strong males and strong females is in their production and response to Hormone A. Very little of this substance is found in male cultures, and male cultures are also much more sensitive in their response to the hormone than female cultures. Another important difference is in the uptake of Hormone A. Certain strains appear capable of absorbing Hormone A much more readily than others, and it is these strains which have the ability to absorb the hormone that produce antheridial branches during conjugation with other thalli (Barksdale, 1963a). If one assumes that heterothallic forms have been derived from homothallic this might have occurred by mutations leading to increased sensitivity to A and hence to maleness. Mutations leading to extracellular accumulation of A should lead to increasing femaleness.

For a general discussion of mating hormones in *Achlya*, see van den Ende (1976).

Germination of the oospores of *A. ambisexualis* results in the formation of a multinucleate germ tube which develops into a germ sporangium if transferred to water, or to an extensive coenocytic mycelium in the presence of nutrients. This mycelium can be induced to form zoosporangia when transferred to water. From zoosporangia of either source single zoospore cultures can be obtained which can be mated with the parental male or female strains. All zoospores or germ tubes derived from a single oospore gave the same result in regard to their sexual interaction, and appeared to have the same tendency. These findings suggest that nuclear division on oospore germination is probably not meiotic, and thus are consistent with the idea that the life cycle is diploid (Mullins & Raper, 1965). Confirmation of these results, implying meiosis during gamete differentiation, has also been obtained with *A. bisexualis* (Barksdale, 1966).

RELATIONSHIPS

The Saprolegniales are probably closely related to the Peronosporales, both groups having walls containing cellulose, similar zoospore structure and ⟨o⟩gamous reproduction. The Leptomitales is another group of aquatic fungi showing some similarities with the Saprolegniales, but they are possibly more closely related with the Peronosporales.

PERONOSPORALES

Many fungi in this group are parasites of higher plants, and may cause diseases of economic importance such as late blight of potato, downy mildew of the grape-vine and blue-mould of tobacco. Some, particularly the Peronosporaceae and Albuginaceae, are biotrophic (obligate) parasites, but some members of the Pythiaceae are facultative (necrotrophic) parasites, and probably survive saprophytically, in soil, mud or on decaying vegetation.

The mycelium is coenocytic but delicate. The composition of the walls of species of *Pythium* and *Phytophthora* has been studied (Aronson *et al.*, 1967; Hunsley, 1973; Sietsma *et al.*, 1975). They are fibrillar in organisation, with a complex chemical structure consisting of polysaccharide, protein and lipid. Glucans constitute about 90% of the wall. Cellulose, a β-($1\rightarrow4$) linked glucan, may make up a relatively small proportion of this glucan (up to about 36%), but in some forms, e.g. *Pythium butleri*, cellulose is replaced by a β-($1\rightarrow2$) linked glucan, whilst in *Pythium acanthicum*, β-($1\rightarrow3$) and β-($1\rightarrow6$) linked glucans are present. A feature of Oomycete cell walls already noted is the presence of the amino-acid hydroxyproline. The fine structure of the hyphal apex has already been described (p. 60). In many of the biotrophic parasites, haustoria are formed within the host cell, and they vary in form from minute spherical or cylindrical ingrowths to more extensively branched or lobed structures. Asexual reproduction is by means of sporangia which vary greatly in shape. In certain species of *Pythium*, the sporangia are merely inflated lobes of mycelium, but more usually they are spherical or pear-shaped. In the Peronosporaceae they arise on differentiated and often characteristically branched sporangiophores, whilst in the Albuginaceae they occur in chains. The sporangia of foliar pathogens are detached from the sporangiophores and are dispersed by wind. In the water- and soil-inhabiting forms, the sporangia usually give rise to zoospores. Whilst zoospores are found in certain pathogens such as *Phytophthora*, *Albugo* and in *Plasmopara*, sporangial germination by means of a germ tube may also occur, especially in *Peronospora* where zoospore formation from sporangia does not occur, so that the term conidium is sometimes used for the asexual propagule. The zoospore is laterally biflagellate, with one flagellum of the whip type and one of the tinsel type (Colhoun, 1966; Reichle, 1969). It thus corresponds to the secondary type of zoospore found in the Saprolegniales with which this group may be related.

Sexual reproduction is oogamous. Each oogonium contains a single egg (except in *Pythium multisporum* where there are several). The antheridial and oogonial initials are commonly multinucleate at their inception and further nuclear divisions may occur during development. In many forms there is only one functional male and female nucleus, but in others multiple fusions occur.

A feature of the oogonia of the Peronosporales is that the residual cytoplasm left after the differentiation of the central oosphere persists as the periplasm which may play a part in the deposition of wall material around the fertilised oospore. There is evidence that meiosis occurs during the development of the gametangia, implying that the soma is diploid. Oospore germination may be by means of a germ tube or by zoospores. Most species are homothallic, but heterothallism and relative sexuality have been reported.

Nutritional and physiological studies have been made on members of the Pythiaceae which can be grown in artificial culture (for references see Cantino, 1950, 1955; Cantino & Turian, 1959; Fothergill & Hide, 1962; Fothergill & Child, 1964). Most species of *Pythium* and *Phytophthora* are heterotrophic for thiamine. It has also been shown that sterols are essential for asexual and sexual reproduction, and it seems likely that members of the Pythiaceae are unable to synthesise sterols. They also have a requirement for calcium which is perhaps unusual for fungi (Hendrix, 1970; Elliott, 1972). Sulphur requirements can be met in most forms by sulphate, but organic forms of sulphur are also utilised. Inorganic nitrogen compounds such at nitrates and ammonium salts may support growth, but species differ in their ability to utilise these sources. A range of sugars and sugar alcohols and other carbohydrates serve as sources of carbon.

Waterhouse (1973b), who has given a general account of the taxonomy of the Peronosporales, distinguishes three families.

Key to the families of Peronosporales

Nonobligate parasites or saprobes: sporangiophores or conidiophores usually undifferentiated from the mycelium, branched, indeterminate, resuming growth after the production of a sporangium or conidium either from below or within the previous empty sporangium; periplasm a thin layer or absent; haustoria absent or branched
Pythiaceae (p. 163)

Obligate parasites of plants with branched tree-like sporangiophores or conidiophores of determinate growth (no subsporangial regrowth), differentiated from the mycelium, emerging singly or in tufts from the epidermis usually via the stomata, producing sporangia or conidia singly at the branch tips; periplasm persistent and conspicuous; haustoria varied, usually branched
Peronosporaceae (p. 179)

Obligate parasites of plants with unbranched, clavate sporangiophores, each bearing a basipetal chain of deciduous sporangia in dense subepidermal clusters, forming on the host white or creamish sori, eventually erupting with the shedding of the sporangia; oogonial periplasm persistent and conspicuous; haustoria knoblike
Albuginaceae (p. 186)

Pythiaceae

Representatives of the Pythiaceae can be found in water and in soil growing saprophytically on plant and animal debris, or as parasites of animals (e.g. *Zoophagus*, parasitic on rotifers) and plants. Although about six genera can be distinguished (Middleton, 1952; Sparrow, 1960; Waterhouse, 1973b) we shall consider only *Pythium* and *Phytophthora*.

Pythium (Figs. 68–71)

Pythium grows in water and soil as a saprophyte, but under suitable conditions, e.g. where seedlings are grown crowded together in poorly drained soil, it can become parasitic causing diseases such as pre-emergence killing, damping-off and foot-rot. Damping-off of cress (*Lepidium sativum*) can be demonstrated by sowing seeds densely on heavy garden soil which is kept liberally watered. Within five to seven days some of the seedlings may show brown lesions at the base of the hypocotyl, and the hypocotyl and cotyledons may become water-soaked and flaccid. In this condition the seedling collapses. A collapsed seedling may come into contact with other seedlings and so spread the disease (Fig. 67A). The host cells separate from each other easily due to the breakdown of the middle lamella, probably brought about by pectic and possibly cellulolytic enzymes secreted by the *Pythium*. The enzymes diffuse from the hyphal tips, so that softening of the host tissue actually occurs in advance of the mycelium. Pure culture studies *in vitro* suggest that species of

Fig. 67. A. Pot of cress (*Lepidium sativum*) one week old, showing symptoms of damping-off caused by *Pythium* spp. B. Symptoms of late blight caused by *Phytophthora infestans* on a terminal leaflet of potato.

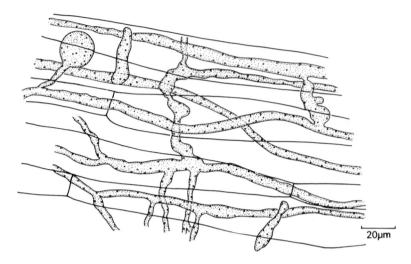

20µm

Fig. 68. *Pythium* mycelium in the rotting tissue of a cress seedling hypocotyl. Note the absence of haustoria and the spherical sporangium initial.

Pythium may also secrete heat-stable substances which are toxic to plants. Within the host the mycelium is coarse and coenocytic, with granular cytoplasmic contents (Fig. 68). At first there are no septa, but later cross-walls may cut off empty portions of hyphae. Thick-walled chlamydospores may also be formed. There are no haustoria. Several species are known to cause damping-off, e.g. *P. debaryanum* and, perhaps more frequently, *P. ultimum*. *Pythium aphanidermatum* is associated with stem-rot and damping-off of cucumber, and the fungus may also cause rotting of mature cucumbers. *P. mamillatum* causes damping-off of mustard and beet seedlings, and is also associated with root-rot in *Viola*. A taxonomic account of *Pythium* has been given by Middleton (1943), and keys to species and original descriptions by Waterhouse (1967, 1968). There are about ninety species.

  Asexual reproduction: the mycelium within the host tissue, or in culture, usually produces sporangia, but their form varies. In some species, e.g. *P. gracile*, the sporangia are filamentous and are scarcely distinguishable from vegetative hyphae. In *P. aphanidermatum* the sporangia are formed from inflated lobed hyphae (Fig. 69B). In many species however, e.g. *P. debaryanum*, the sporangia are globose (Fig. 69A). A terminal or intercalary portion of a hypha enlarges and assumes spherical shape, and becomes cut off from the mycelium by a cross-wall. The sporangia contain numerous nuclei. The cleavage of the cytoplasm to form zoospores begins within the sporangium, but is completed within a thin-walled vesicle which is extruded from the sporangium. Within the sporangium, cleavage vesicles coalesce to separate the cytoplasm into uninucleate pieces. Even before the extrusion of the thin-walled vesicle, membrane-bound packets of mastigonemes are present within

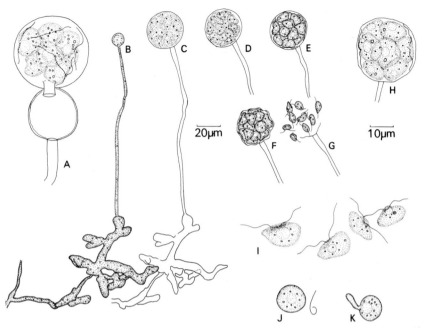

Fig. 69. Sporangia and zoospores of *Pythium*. A. *P. debaryanum*. Spherical sporangium with short tube and a vesicle containing zoospores. B–K. *P. aphanidermatum*. B. Lobed sporangium showing a long tube and the vesicle beginning to expand. C–G. Further stages in the enlargement of the vesicle, and differentiation of zoospores. Note the transfer of cytoplasm from the sporangium to the vesicle in C. The stages illustrated in B–G took place in 25 minutes. H. Enlarged vesicle showing the zoospores. Flagella are also visible. I. Zoospores. J. Encystment of zoospore showing a cast-off flagellum. K. Germination of a zoospore cyst. B–G to same scale; A, H, I, J, K to same scale.

the cytoplasm of the sporangium. In *P. middletonii* (sometimes referred to as *P. proliferum* (Fig. 70)), the sporangium is extended into an apical papilla capped by a basin-shaped mass of fibrillar material which is lamellate in structure (Lunney & Bland, 1976). Shortly before sporangial discharge, there is an accumulation of cleavage vesicles behind the apical cap, and at the periphery of the cytoplasm close to the sporangium wall. The cleavage vesicles around the sporangial cytoplasm discharge their contents to form a loose, fibrous interface between the cytoplasm and the sporangial wall. Discharge of the sporangium occurs by the inflation of the thin-walled vesicle at the tip of the papilla, and the partially differentiated zoospore mass is extruded into it. The movement of the cytoplasm from the sporangium into the vesicle is probably the result of several forces including the elastic contraction of the sporangium wall and possibly surface energy (Webster & Dennis, 1967). Lunney & Bland (1976) have also suggested that the fibrillar material extruded from the cleavage vesicles at the periphery of the sporangium wall may imbibe water, resulting in a build-up of pressure between the cytoplasm and the

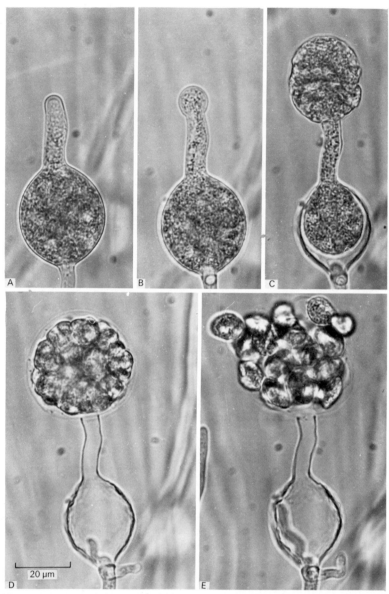

Fig. 70. *Pythium middletonii*: stages in zoosporangial discharge. A. Sporangium shortly before discharge. Note the thickened tip of the papilla containing clear cytoplasm. B. Inflation of the vesicle begins. C. Cytoplasm retreating from the sporangium. Note the shrinkage in sporangium diameter as compared with A. D. Zoospores differentiated within the vesicle, with flagella visible between the vesicle wall and the zoospores. E. Zoospores escaping following rupture of the vesicle wall. The whole process of discharge takes about 20 minutes.

sporangium wall. The vesicle enlarges as cytoplasm from the sporangium is transferred to it, and during the next few minutes the cytoplasm cleaves into about eight to twenty uninucleate zoospores which jostle about inside the sporangium, and cause the thin wall of the vesicle to bulge irregularly (Fig. 70). Finally, about 20 minutes after the inflation of the vesicle, its wall breaks down and the zoospores swim away. They are broadly bean-shaped with two laterally attached flagella. The zoospores of several species of *Pythium* have been shown to move chemotactically towards roots. After swimming for a time the flagella are cast off, the spore encysts and germination takes place by means of a germ tube (Fig. 69K). Repeated emergence has also been reported. In some forms, e.g. *P. ultimum* var. *ultimum*, zoospore formation from sporangia does not occur; the sporangia germinate directly to form a germ tube. However, Drechsler (1960) has described a variety, *P. ultimum* var. *sporangiferum*, in which zoospores develop from the zoosporangia. Sporangia showing direct germination are sometimes referred to as conidia. Zoospores are formed in *P. ultimum* var. *ultimum* when the oospore germinates, although direct germination of the oospores has also been reported (Fig. 71D,E). Sporangial proliferation occurs in certain species, e.g. *P. middletonii* and *P. undulatum*. In *P. ultimum*, the sporangia may survive in soil, whether moist or air dry, for several months, and are stimulated to germinate within a few hours by sugar-containing exudates from seed coats. Germination is by means of germ tubes which grow very rapidly so that an adjacent host may be penetrated within 24 hours (Stanghellini & Hancock, 1971a,b).

Sexual reproduction: the formation of oogonia and antheridia occurs readily in cultures derived from single zoospores, and it is therefore probable that most species of *Pythium* are homothallic. However, certain species fail to form oospores in culture, and it has been shown that several of them, e.g. *P. sylvaticum*, *P. heterothallicum* and *P. splendens*, are heterothallic, although homothallic isolates are sometimes encountered (Campbell & Hendrix, 1967; Papa *et al.*, 1967; van der Plaats-Niterink, 1968, 1969; Pratt & Green, 1971, 1973). In crosses between strongly male and strongly female strains of *P. sylvaticum*, Pratt & Green (1973) obtained a preponderance of male progeny from germinating oospores, but female, neutral and bisexual progeny were also obtained in lower numbers. The genetic studies suggest that complex or multiple genetic systems govern sexuality in *P. sylvaticum*. Oogonia arise as terminal or intercalary spherical swellings which become cut off from the adjacent mycelium by cross-wall formation. In some species, e.g. *P. mamillatum*, the oogonial wall is folded into long projections (Fig. 71B). The antheridia arise as club-shaped swollen hyphal tips, often as branches of the oogonial stalk, or sometimes from separate hyphae. In some species, e.g. *P. ultimum*, there is typically only a single antheridium to each oogonium, whilst in others, e.g. *P. debaryanum*, there may be several (Fig. 71A). The young oogonium is multinucleate and the cytoplasm within it differentiates into a multinucleate central mass, the ooplasm, from which the egg develops, and a peripheral

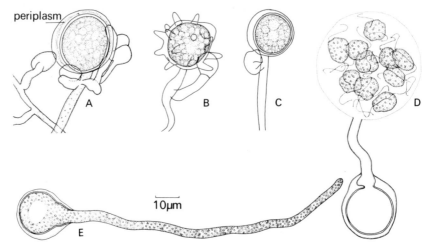

Fig. 71. Oogonia and oospores of *Pythium*. A. *P. debaryanum*. Note that there are several antheridia. B. *P. mamillatum*. Oogonium showing spiny outgrowths of oogonial wall. C. *P. ultimum*. D, E. *P. ultimum* germinating oospores (after Drechsler, 1960).

mass, the periplasm, also containing several nuclei. The periplasm does not contribute to the formation of the egg. There are conflicting accounts of the type of nuclear division which occurs. Many early workers, assuming that meiosis occurred on germination of the oospore and that the vegetative filaments were therefore haploid, have described only mitoses during the development of the egg. However Sansome (1961, 1963) has claimed that meiosis occurs in the oogonia and antheridia of *P. debaryanum*. In *P. debary-anum* Sansome has described young oogonia with more numerous nuclei than older ones, due to degeneration of certain nuclei. One to eight nuclei survive and enlarge, undergoing meiosis. At a later stage of division up to thirty-two nuclei were counted. In this and other species only one egg nucleus survives in the centre of the ooplasm, while the remaining nuclei degenerate. The nuclei of the periplasm also degenerate. The young antheridia are multinucleate, but all nuclei except one degenerate. The surviving nucleus undergoes meiosis so that older antheridia have four nuclei. The antheridia are applied to the oogonial wall and penetrate it by means of a fertilisation tube. Following penetration only three nuclei were counted in the antheridium, suggesting that one had entered the oogonium. Later still, empty antheridia were found, and it is presumed that the three remaining nuclei enter the oogonium and join the nuclei degenerating in the periplasm. Fusion between a single antheridial and egg nucleus has been described. The fertilised egg secretes a double wall, and a globule of reserve material appears in the cytoplasm. Material derived from the periplasm may also be deposited on the outside of the egg. Such oospores may need a period of rest (after-ripening) of several weeks before they are capable of germinating. Germination may be by means of a germ tube, or by the formation of a vesicle in which zoospores are differentiated (Fig. 71D,E), or in some forms the developing oospore produces a short germ tube ending in a sporangium.

Pythium can live saprophytically, and can also survive in air-dry soil for several years. It is more common in cultivated than in natural soils, and appears to be intolerant of acid soils. As saprophytes, species of *Pythium* are important primary colonisers of virgin substrata, and probably gain initial advantage in colonisation by virtue of their rapid growth rate. They do not, however, compete well with other fungi which have already colonised a substrate and appear to be rather intolerant of antibiotics.

The control of diseases caused by *Pythium* is obviously rendered difficult by its ability to survive saprophytically in soil. Its wide host range means that it is not possible to control diseases by means of crop rotation. The effects of disease can be minimised by improving drainage and avoiding overcrowding of seedlings. In the greenhouse or nursery some measure of control can be achieved by partial sterilisation using steam or formalin to destroy the fungus, and recolonisation of the treated soil by *Pythium* is slow.

Phytophthora (Figs. 72–74)

The name *Phytophthora* is derived from the Greek (*phyton*, plant; *phthora*, destruction); and some species are certainly destructive parasites. The best-known is *P. infestans*, the cause of late blight of potatoes. This fungus is confined to Solanaceous hosts but others have a much wider range. For example *P. cactorum* has been recorded from over forty families of flowering plants causing a variety of diseases such as damping-off, root-rot and fruit-rot. Other important pathogens are *P. erythroseptica* associated with a pink-rot of potato tubers, *P. fragariae* the cause of red-core of strawberries, and *P. palmivora* causing pod-rot and canker of cacao. *P. cinnamomi* is a serious pathogen of woody hosts, including conifers. A taxonomic account of the genus has been given by Tucker (1931); an invaluable general account by Hickman (1958). Waterhouse (1956) has compiled a collection of original descriptions of species, and provided a key (1963). About forty species are known. Whilst some species may live saprophytically as water-moulds, the majority of pathogenic forms probably have no prolonged free-living saprophytic existence, but survive in the soil in the form of oospores, or within diseased host tissue. However, almost all the pathogenic forms can be isolated from their hosts at the appropriate season and can be grown in pure culture. Selective media, often incorporating antibiotics such as pimaricin, have been devised for the isolation of *Phytophthora* (Tsao, 1970). Most species form a non-septate mycelium producing branches at right angles, often constricted at their point of origin. Septa may be present in old cultures. Within the host, the mycelium is intercellular, but haustoria are formed which penetrate the host

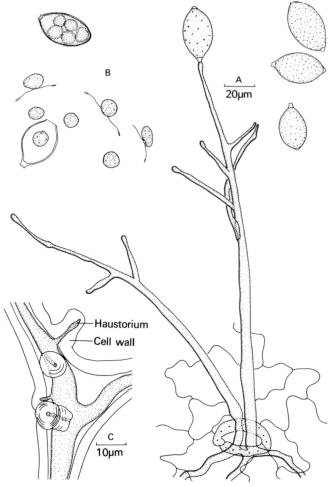

Fig. 72. *Phytophthora infestans.* A. Sporangiophores penetrating a stoma of a potato leaf. B. Sporangial contents dividing and releasing zoospores. C. Intercellular mycelium from a potato tuber showing the finger-like haustoria penetrating the cell walls. Note the thickening of the cell wall around the haustorium.

cells. In *P. infestans* within potato tubers, the haustoria are finger-like protuberances which may be in part surrounded by thickenings of host wall material (Fig. 72C). Electron micrographs of haustoria in potato leaves show that the haustoria are not surrounded by host cell wall. They are, however, surrounded by an encapsulation whose origin and function are not certain (Ehrlich & Ehrlich, 1966, 1971). Nuclei have been reported in haustoria in potato leaf cells (Siva & Shaw, 1969). Sporangia are usually pear-shaped and arise on well-differentiated simple or branched sporangiophores. On the host plant, the sporangiophores may emerge through the stomata, as in *P. infestans*

(Fig. 72A). The first sporangium is terminal, but the hypha bearing it may push it to one side and proceed to the formation of further sporangia. The mature sporangium has a terminal papilla, and it is separated from the sporangiophore by a thickening of wall material which forms a basal plug. The fine structure of sporangia in *Phytophthora* spp. has been studied by several workers (Chapman & Vujičić, 1965; Hemmes & Hohl, 1969; Williams & Webster, 1970; Elsner *et al.*, 1970; Christen & Hohl, 1972). Cleavage of the cytoplasmic contents to form zoospore initials is brought about by cleavage vesicles which are derived from the Golgi apparatus. Flagella are visible early in development.

In terrestrial forms, the sporangia are detached, possibly aided by hygroscopic twisting of the sporangiophore on drying, and are dispersed by wind before germinating, but in aquatic forms zoospore release commonly occurs whilst the sporangia are still attached. In these aquatic species, internal proliferation of sporangia may occur.

The germination of sporangia may occur by two alternative methods. *Indirect germination* by the production of zoospores can be induced in many species by suspending zoosporangia in chilled water or soil extract. *Direct germination* by the production of a germ tube is induced by higher temperatures. For example, in *P. infestans*, below 15 °C uninucleate zoospores are produced, whilst above 20 °C multinucleate germ tubes arise. With increasing age, sporangia lose their capacity to produce zoospores. Direct germination is preceded by resorption of the flagella, formed inside the sporangia. In *Phytophthora*, when zoospores are formed, they are typically differentiated *within* the sporangium, and not in a vesicle as in *Pythium*, but this distinction is not absolute. An evanescent vesicle may be formed from the papilla and the differentiated zoospores pass into it before release, but in many cases the zoospores are released directly from the sporangium by the breakdown of the papilla. The uninucleate laterally biflagellate zoospores swim for a time, often attracted chemotactically to host tissue, then usually encyst and germinate by means of a germ tube, but repeated emergence has occasionally been reported. In *P. cactorum*, sporangia have been preserved for several months under moderately dry conditions. On return to suitable conditions, such a sporangium may germinate by the formation of a vegetative hypha, or a further sporangium. Thick-walled spherical chlamydospores have also been described.

Sexual reproduction in *Phytophthora* is similar in outline to that of *Pythium* and oospore formation is similarly stimulated by sterols. Two distinct types of antheridial arrangement are found. In *P. cactorum* and a number of other species, antheridia are attached laterally to the oogonium and are described as **paragynous** (Fig. 73G–I). In some other species including *P. erythroseptica* and *P. infestans*, the oogonia, during their development, penetrate and grow through the antheridium. The oogonial hypha emerges above the antheridium and inflates to form a spherical oogonium, with the antheridium persisting as

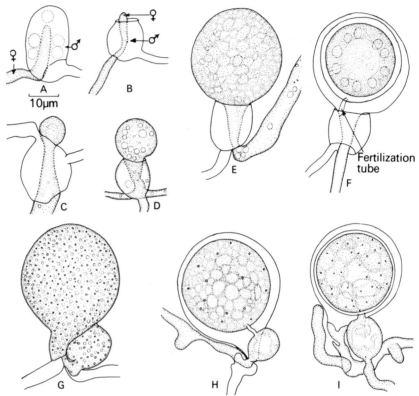

Fig. 73. Oogonial development in *Phytophthora*. A–F. Stages in development of *P. erythroseptica*. G–I. Stages in development of *P. cactorum*.

a collar about its base (Fig. 73A–F). This arrangement of the antheridium is termed **amphigynous**. Hemmes & Bartnicki-Garcia (1975) have confirmed by means of thin sections and transmission electron microscopy that penetration of the antheridium by the oogonium occurs in *P. capsici*, thus refuting the claims of Stephenson & Erwin (1972) that the antheridium encircled the oogonium during development. Both the oogonia and the antheridia are multinucleate, but as the oosphere matures, only a single nucleus remains at the centre, while the remaining nuclei are included in the periplasm. Fertilisation tubes have been observed and a single nucleus is introduced from the antheridium (Fig. 74). Fusion between the egg nucleus and the antheridial nucleus is delayed until the oospore wall is mature. This wall has a thin exospore derived from the periplasm composed of pectic substances, and a thicker endospore of cellulose, protein and possibly other reserve substances. Within, the cytoplasm contains a large oil globule. Such oospores do not germinate immediately, but undergo a period of maturation lasting several weeks or months. Before germination, nuclear fusion will have occurred and

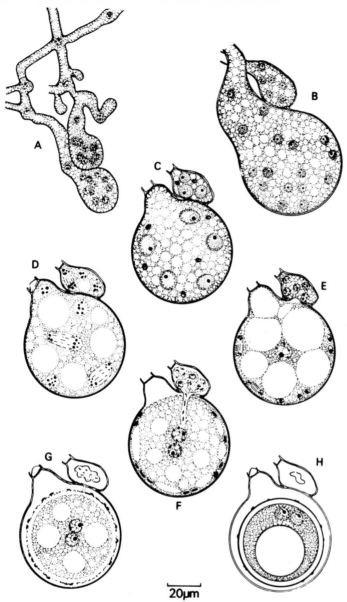

Fig. 74. *Phytophthora cactorum.* Development of oogonium, antheridium and oospore. A. Initials of oogonium and antheridium. B. Oogonium and antheridium grown to full size: the oogonium has about twenty-four nuclei and the antheridium about nine. C. Development of a septum at the base of each, and degeneration of some nuclei in each until the oogonium has eight or nine nuclei and the antheridium four or five. D. A simultaneous division of the surviving nuclei in oogonium and antheridium. The protoplasm has large vacuoles. E. Separation of ooplasm from periplasm. Nuclei in division in periplasm prior to degeneration. Oogonium presses into antheridium. F. Entry of one antheridial nucleus by a fertilisation tube. The protoplasm and remaining nuclei of the antheridium degenerate. G. Development of oospore wall. H. Oospore enters on its dormant period with exospore formed from dead periplasm, endospore (of cellulose, protein, etc.) deposited upon it, and paired nuclei in association but not yet fused. A–H are composite drawings of eight stages in sequence (after Blackwell, 1943a).

the fusion nucleus will have divided several times. The egg swells and the exospore is stretched. The oil globule breaks up into a number of smaller globules and there are signs of digestion of the endospore from within. A germ tube penetrates the exospore and the oogonial wall, and sporangia may be formed at the tip. There is now evidence that in *P. cactorum, P. capsici* and *P. drechsleri*, meiosis occurs in the process of gamete formation (Sansome, 1963, 1976; Galindo & Zentmyer, 1967; Mortimer & Shaw, 1975). Genetical studies on the inheritance of a requirement for the amino-acid methionine in *P. cactorum* are entirely consistent with the supposition that the zoospores are diploid and that meiosis occurs during gametogenesis (Elliott & MacIntyre, 1973). However, for some species, the genetical evidence has been interpreted as favouring the view that the zoospores are haploid and that meiosis occurs on oospore germination. Day (1974) has written 'Where meiosis occurs in *Phytophthora* is still a vexing question . . . It may be that, in some Oomycetes, meiosis occurs in the gametangia, whereas in others it occurs at oospore germination.'

Oospores may survive in soil for long periods. In any population of oospores there is considerable physiological heterogeneity, and no one set of environmental conditions will stimulate all the oospores to germinate (Blackwell, 1943a,b).

Some species of *Phytophthora*, e.g. *P. cactorum* and *P. erythroseptica*, are homothallic and form oospores readily in cultures derived from single uninucleate zoospores. Others, e.g. *P. infestans, P. palmivora*, are heterothallic and normally form oospores only when different isolates are paired together, i.e. they are self-incompatible. In both these species, it has been shown that there are two *compatibility types*, A_1 and A_2, controlled by two alleles at a single locus. Only when isolates of the A_1 type are paired with isolates of the A_2 type are oospores formed. However, it is also known that each compatibility type is potentially bisexual, i.e. a given strain is capable of forming antheridia in certain pairings, or oogonia in others. The bisexual nature of each isolate is confirmed by the observation that selfing can be induced by pairing, not only between strains of the same species, but in interspecific pairings. For example, single strains from both compatibility types of the heterothallic fungus *P. palmivora* can be induced to form sex organs when separated by a cellophane membrane from a culture of the homothallic *P. heveae*. True interspecific matings with one species contributing an antheridial hypha and the other an oogonium have also been demonstrated in other species. *Phytophthora infestans* shows relative sexuality. Whether a given strain will behave as male or as female in a cross depends in part on the degree of 'maleness' or 'femaleness' of its partner, and also on nutrition. Starved hyphae of *P. infestans* function as males, whilst better-nourished hyphae form oogonia (Galindo & Gallegly, 1960; Savage *et al.*, 1968; Brasier, 1972). These findings, reminiscent of the situation in heterothallic species of *Achlya* (p. 157) suggest that sexual reproduction is under hormonal or chemical control.

The two compatibility types in heterothallic *Phytophthora* species are not equally common. In Mexico, believed to be near the centre of origin of the genus *Solanum* to which the potato belongs, both compatibility types of *P. infestans* have been found to be equally abundant in nature, and oospores develop in potato leaves. In North America, where native members of the Solanaceae are few, only the A_1 compatibility type has been found, and the same is true for Europe (Gallegly & Galindo, 1958).

We have discussed evidence suggesting hormonal or chemical control of sexual reproduction in heterothallic species of *Phytophthora*. It is also known that unrelated fungi can induce oospore development in A_2 isolates. Chance contamination of an A_2 isolate of *P. palmivora* by the common soil fungus *Trichoderma viride* resulted in oospore formation. Subsequent experiments showed that the '*Trichoderma* effect' worked with A_2 isolates of other hetero-thallic species. Moreover, several different species of *Trichoderma* are capable of inducing sexual reproduction. Direct physical contact between hyphae of the *Trichoderma* and *Phytophthora* is not necessary. Oospores develop in *Phytophthora* cultures placed over *Trichoderma* in such a way that volatile substances produced by the latter can be absorbed. The effect is probably non-specific, because exposure to vapour of the fungicide chloroneb induces oospore development in A_2 isolates of *P. capsici*, whilst mechanical injury or injury from hydrogen peroxide induces oospore development in *P. cinnamonii* and *P. cryptogea*. It has been suggested that the *Trichoderma* effect may be of ecological significance. If the distribution of compatibility types is such that the chance of sexual reproduction is remote, it might be to the advantage of a *Phytophthora* colonising a root system from the soil, in competition with *Trichoderma*, to form oospores as a defence mechanism against antibiotics produced by this fungus (Brasier, 1975a,b). It is not understood why volatile antibiotics or fungicides should have the same effect on A_2 strains as association with A_1 strains, but it is possible that the stimulus switches off a self-sterility mechanism. The reasons why A_1 strains fail to respond to such stimuli are also unknown.

LATE BLIGHT OF POTATO

Late blight of potato caused by *P. infestans* is a notorious disease. In the period between 1845 and 1847 it resulted in famine among the working-class population of Ireland who had come to depend on potatoes as their major source of food. 'The cost of the potato famine in terms of life and health of the people is very difficult to estimate. We have certain fixed points. The population of Ireland in 1841 was 8,175,124. In 1851 it was 6,552,385, a diminution of 1,622,739' (Salaman, 1949). Indirectly, however, the famine provided a stimulus to the study of plant pathology. The history of the Irish famine has been ably documented by Large (1958) and Woodham-Smith (1962). The status of the disease has been assessed by Cox & Large (1960). It is found wherever the crop is grown, but it is not the major cause of lost production because certain virus diseases are more injurious. There is now an immense literature to the disease (Moore, 1959).

The fungus overwinters in tubers infected during the previous season, and a very low proportion of such tubers give rise to infected shoots. In experimental plots the proportion of infected plants developing from naturally and artificially infected tubers was found to be less than 1% (Hirst, 1955; Hirst & Stedman, 1960a,b). Nevertheless such infected shoots form foci within the crop from which the disease spreads. Sporangia formed on the diseased shoots are blown to healthy leaves, and there germinate either by the formation of germ tubes or zoospores. Zoospore production is favoured by temperatures between 9 and 15 °C. After swimming for a time the zoospores encyst and then form germ tubes which usually penetrate the epidermal walls of the potato leaf, or occasionally enter the stomata. An appressorium is formed at the tip of the germ tube, attaching the zoospore cyst firmly to the leaf, and penetration of the cell wall is probably achieved by both mechanical and enzymatic action. Penetration can occur within 2 hours. Within the leaf tissues mycelium develops which grows out radially from the point of penetration. The resulting lesion had a dark green water-soaked appearance associated with tissue disintegration, possibly aided by toxin-secretion. Such lesions (see Fig. 67B) are visible within about 3–5 days of infection under suitable conditions of temperature and humidity. Around the margin of the advancing lesion, especially on the lower surface of the leaf, a zone of sporulation is found with sporangiophores emerging through the stomata. Sporulation is most prolific during periods of high humidity and commonly occurs at night following the deposition of dew. In potato crops as the leaf canopy closes over between the rows to cover the soil a humid microclimate is established which may result in extensive sporulation. As the foliage dries during the morning the sporangiophores undergo hygroscopic twisting which results in the flicking off of sporangia. Thus the concentration of sporangia in the air usually shows a characteristic diurnal fluctuation, with a peak at about 10 a.m. Sporangia are sensitive to drying, and may be incapable of germination if exposed for 5 minutes to a relative humidity of 95%, in the absence of a water film on the surface of the sporangium. Possibly dispersal on water droplets is necessary for long-range spread (Warren & Colhoun, 1975).

The destructive action of *P. infestans* is directly associated with the killing of the foliage, which results in a reduction in the photosynthetic area of the potato plant and hence a reduction in the weight of tubers. When about 75% of the leaf tissue has been destroyed, further increase in weight of the crop ceases (Cox & Large, 1960). Thus an early epidemic of blight can result in a lowering of yield. This reduction tends to be offset by the fact that such epidemics are more common in rainy cool seasons which are conducive to high crop yields.

A more sinister cause of crop losses is occasioned by the fact that tubers can be infected by sporangia falling on to them, either during growth or lifting.

Such infected tubers may rot in storage, and the diseased tissue is often invaded by secondary bacterial and fungal saprophytes.

Control of potato blight is achieved by several means:

1. Spraying

By spraying with suitable fungicides, which prevent sporangial germination, epidemic spread of the disease can be delayed. This can result in a prolongation of photosynthetic activity of the potato foliage and hence an increase in yield. Various types of fungicide have been developed, and among the most common is one containing copper sulphate and calcium oxide (Bordeaux mixture). However, the damage caused by the copper on the potato leaves and by tractor wheels of the spraying apparatus to the potato haulms may cause a greater reduction in the potential crop yield, than the blight which might have resulted had the crop not been sprayed. To avoid unnecessary spraying and to ensure that timely spray applications are made it has proved possible to provide, for certain countries, forecasts of the incidence of potato blight epidemics. Analysis of the incidence of blight epidemics in south Devon (England) by Beaumont (1947) established that a 'temperature–humidity rule' controlled the relationship between blight epidemics and weather. After a certain date (which varied with the locality) blight followed within 15–22 days, a period during which the minimum temperature was not less than 10 °C and the relative humidity was over 75% for two consecutive days. The warm humid weather during this period provides conditions suitable for sporulation and the initiation of new infections. Modified in the light of experience, this relationship, coupled with field observations on the first appearance of blight lesions in the field, has enabled the Plant Pathology Laboratory of the Ministry of Agriculture to issue accurate forecasts of the likelihood of potato blight epidemics (Cox & Large, 1960; Bourke, 1970), and thus to increase the potential value of spraying, make substantial savings and minimise pollution of soil by fungicides.

2. Haulm destruction

The danger of infection of tubers by sporangia falling on to them from foliage at lifting time can be minimised by ensuring that all the foliage is destroyed beforehand. This is achieved by spraying the foliage 2–3 weeks before lifting time with such sprays as sulphuric acid, copper sulphate, tar acid compounds or sodium chlorate. The ridging of potato tubers also helps to protect the tubers from infection. Although sporangia may survive in soil for several weeks they do not penetrate deeply into it.

3. Use of disease-resistant varieties

A world-wide search for species of *Solanum* showed that a number had natural resistance to *P. infestans*. One species which has proved to be an important source of resistance is *S. demissum*. Although this species is valueless in itself for commercial cultivation it is possible to cross it with *S. tuberosum* and some of the hybrids are resistant to the disease. *Solanum demissum* contains at least four major genes for resistance (R_1, R_2, R_3 and R_4), together with a number of minor genes which determine the degree of susceptibility in susceptible varieties (Black, 1952). The four genes may be absent from a particular host strain (O), present singly (e.g. R_1) in pairs (e.g. R_1, R_3), in threes (e.g. R_1, R_2, R_3), or all together (e.g. R_1, R_2, R_3, R_4) so that sixteen host genotypes are possible representing different combinations of R genes. The identification of the R gene complex was dependent on the discovery that the fungus itself exists in a number of strains or physiologic races. On the assumption that the fungus itself carries genes which correspond to, and enable it to overcome the effect of, a host R gene, sixteen races of the fungus should theoretically be demonstrable. If the corresponding genes of the fungus are termed 1, 2, 3 and 4 then the different races can be labelled (0), (1), (2), etc., (1,2) (1,3) etc., (1,2,3), (1,3,4) etc., 1,2,3,4. By 1953, thirteen of the sixteen had been identified. The prevalent race commonly found, however, was Race 4. By 1969, eleven R genes had been recognised in Britain (Malcolmson, 1969). Resistance based on a small number of genes of major effect has been termed *oligogenic resistance* (Day, 1974).

In addition to the major genes for resistance in potato, a large number of genes probably also exist which, although individually of small effect, may, if present together, contribute to resistance. Resistance of this type is known as *polygenic resistance* or *field resistance*, and some potato breeding programmes aim at producing varieties possessing it. This is probably preferable in the long run because one danger of relying on oligogenic resistance is that variants within the pathogen population are more likely to be selected which can attack the potato varieties which possess this type of resistance (van der Plank, 1971; Day, 1974).

The origin of the physiologic races is difficult to determine. If sexual reproduction is a rare phenomenon, then the opportunity which this presents for genetical recombination must be small. However, it is possible that some of the races arise by mutation followed by selection on a susceptible host (Fincham & Day, 1971). Another possibility is that the mycelium of *P. infestans* may be heterokaryotic, carrying nuclei of more than one race (Graham, 1955). Yet another possibility is that parasexual recombination is involved. By mixing sporangia of two different races, new races with a different pattern of virulence towards potato varieties have been obtained after several cycles of inoculation (Malcolmson, 1970; Denward, 1970; Day, 1974). There is also some evidence to suggest that the pathogenicity of a strain of the parasite can be increased by continued passage through resistant hosts (see Buxton, 1960).

Within 1–2 days of infection, the tissues of a resistant host undergo necrosis so rapidly that sporulation of the fungus does not occur. Because of this rapid reaction of the host tissue, further growth of the fungus becomes impossible. Such a reaction is sometimes termed hypersensitivity. According to Müller (1959), the function of the *R* genes is to accelerate this host reaction. When potato tuber tissues are inoculated with an avirulent race of *P. infestans*, they respond by secreting antifungal substances (phytoalexins). Two of the phytoalexins formed by resistant tubers are rishitin and phytuberin. Rishitin (originally isolated from the potato variety Rishiri) is a bicyclic sesquiterpene. Tomiyama *et al.* (1968) showed that R_1 tuber tissue inoculated with an avirulent race of *P. infestans* produced over 270 times the amount of rishitin than when inoculated with a virulent race. It is possible that the *R* genes of the potato determine the ability of host tissue to respond to avirulent races of *P. infestans* (Day, 1974).

Peronosporaceae

The Peronosporaceae are in nature biotrophic parasites of higher plants and are responsible for a group of diseases known collectively as downy mildews. None has yet been grown in pure culture, but callus cultures of host tissue will support growth of some pathogens, e.g. *Peronospora farinosa* on sugar beet (Ingram & Joachim, 1971). Most species are confined to particular host families or host genera. The mycelium in the host tissues is coenocytic and intercellular, with haustoria of various types penetrating the host cells (Fraymouth, 1956). The sporangia arise on well-differentiated branched sporangiophores which grow out from stomata. They are disseminated by wind. In some genera, germination by means of biflagellate zoospores has been reported, but in most the sporangia germinate directly, by means of a germ tube. The term conidium is sometimes used for sporangia showing direct germination.

Representative genera of the family are *Peronospora*, *Plasmopara* and *Bremia*.

Peronospora (Figs. 75–77)

A number of diseases of economic importance are caused by species of *Peronospora*. *Peronospora destructor* causes a serious disease of onions and shallots whilst *P. farinosa* causes downy mildew of sugar beet, beetroot and spinach, but can also be found on weeds such as *Atriplex* and *Chenopodium*. *Peronospora tabacina* causes blue-mould of tobacco. This name refers to the bluish-purple colour of the sporangia which is a feature of many species of *Peronospora*. *Peronospora parasitica* attacks members of the Cruciferae. Although many specific names have been applied to forms of this fungus on

Fig. 75. A. Inflorescence of *Capsella bursa-pastoris* infected with *Peronospora parasitica*. The white 'fur' is due to development of sporangiophores. B. Stem of *Capsella* infected with *Albugo candida*. Note the distortion and hypertrophy of the infected stem. Compare the diameter of the infected stem with that of the uninfected stem behind it. The white blisters have not yet ruptured.

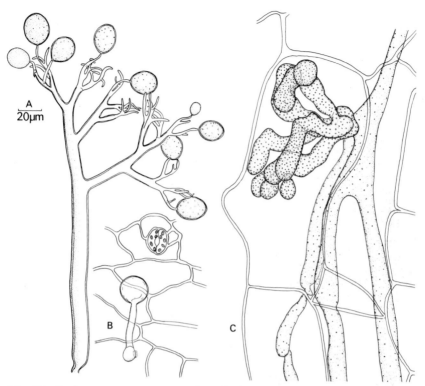

Fig. 76. *Peronospora parasitica* on *Capsella bursa-pastoris*. A. Sporangiophore. B. Sporangium germinating by means of a germ tube. C. L.S. host stem showing intercellular mycelium and coarse lobed haustoria.

different host genera, it is now customary to regard them all as belonging to a single species (Yerkes & Shaw, 1959). Turnips, swede, cauliflower, brussels sprouts and wallflowers (*Cheiranthus*) are commonly attached, and the fungus is frequently found on Shepherd's purse (*Capsella bursa-pastoris*). Diseased plants are detected by their swollen and distorted stems bearing a white 'fur' of sporangiophores (Fig. 75A). On leaves the fungus is associated with yellowish patches on the upper surface and the formation of white sporangiophores beneath. Sections of the diseased tissue show a wide coenocytic intercellular mycelium and branched lobed coarse haustoria in certain host cells (Fig. 76C). According to Fraymouth (1956) the haustoria of the Peronosporales do not actually penetrate the host protoplast. In plasmolysis experiments following staining of the protoplast with Neutral Red, she was able to show that the protoplast withdrew from the haustorium leaving it entirely free. Following penetration of the host cell reactions are set up between the host protoplasm and the invading fungus. The haustorium becomes ensheathed by a layer of callose which is often visible as a thickened collar around the base of the

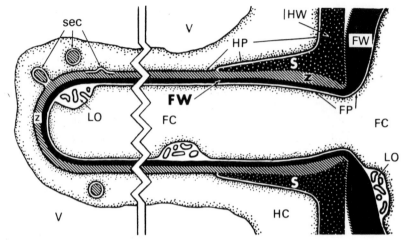

Fig. 77. *Peronospora manshurica*. Diagram of host–parasite interface in haustorial region. Fungal cytoplasm (FC) is bounded by the fungal plasma membrane (FP), lomasomes (LO), and the fungal cell wall (FW) in both the intercellular hypha (right) and the haustorium (centre). The relative positions of the host cell vacuole (V), host cytoplasm (HC) and host plasma membrane (HP) are indicated. The host cell wall (HW) terminates in a sheath (S). The zone of apposition (Z) separates the haustorium from the host plasma membrane. Invaginations of the host plasma membrane and vesicular host cytoplasm are considered evidence for host secretory activity (sec) (after Peyton & Bowen, 1963).

haustorium. Fraymouth's findings have been confirmed and extended by Peyton & Bowen (1963), who studied thin sections under the electron microscope of leaf tissue of soya bean infected with *P. manshurica*. Their interpretation of the haustorium is shown in Fig. 77. At the point of penetration of the host cell, the haustorium is surrounded by a sheath of host wall material. It is enclosed for the whole of its length by a further sheath, the zone of apposition, possibly also derived from the host. The plasma membrane of the haustorium invaginates at certain points to form lomasomes. Secretory bodies are apparently formed in the host cell around the haustoria, and the concentration of host ribosomes is also apparently increased. The sporangiophores emerge singly or in groups from stomata. There is a stout main axis which branches dichotomously to bear the oval sporangia at the tips of incurved branches (Fig. 76). Detachment of the sporangia is possibly caused by hygroscopic twisting of the sporangiophores related to changes in humidity. In *P. tabacina*, however, it has been suggested that changes in turgor of the sporangiophores occur which parallel changes in the water content of the tobacco leaf. It has also been claimed for this fungus that the sporangia are discharged actively by energy applied at the point of attachment of the sporangia. Violent sporangial discharge occurs also in *Sclerospora philippinensis*, a downy mildew of maize in the Philippines. Here, a flat septum is formed between the sporangium and the sporangiophore, and separation is brought about by the rounding-off of the turgid cells on either side of the septum, so that the sporangia bounce off for a distance of 1–2 mm. In *P. parasitica* the sporangia, on alighting on a suitable host, germinate by the formation of a germ tube, and not by zoospores. The germ tube penetrates the wall of the epidermis (Fig. 76B).

Oospores of *P. parasitica* are embedded in senescent tissues and are found throughout the season. There is evidence that some strains of the fungus are heterothallic, whilst others are homothallic (McMeekin, 1960). Both the antheridium and oogonium are at first multinucleate. Nuclear division precedes fertilisation, and chromosome reduction occurs in the oogonium and in the antheridium (Sansome & Sansome, 1974). The single functional gamete nuclei do not immediately fuse, and fusion is delayed until the oospore wall is partly formed.

The wall of the oospore in *P. parasitica* is very tough and it is difficult to induce germination. In *P. destructor* and some other species, germination occurs by means of a germ tube but in *P. tabacina* zoospores have been described. It is probable that oospores overwinter in the soil and give rise to infection in subsequent seasons. The oospores of *P. destructor* have been germinated after twenty-five years, but it has not proved possible to infect onions from such germinating oospores. Possibly in this case the disease is carried over by means of systemic infection of onion bulbs (McKay, 1957). Although it has been found possible to grow *P. parasitica* in callus culture (Guttenberg & Schmoller, 1958) it is very unlikely that the fungus grows saprophytically under natural conditions.

Plasmopara (Fig. 78)

Although downy mildews caused by species of *Plasmopara* are rarely serious in Britain, *P. viticola* is potentially a very destructive pathogen of the grapevine. The disease which was endemic in America and not particularly destructive there was probably imported into France during the nineteenth century with disastrous results. Historically the disease is of great interest because experiments by Millardet to control the disease in France led to the formulation of Bordeaux mixtures which proved effective not only against this fungus but against *Phytophthora infestans*, and a number of other important foliar pathogens.

Plasmopara nivea is occasionally reported in Britain on Umbelliferous crops such as carrot and parsnip, and is also found on *Aegopodium podagraria*. *Plasmopara pygmaea* is found on yellowish patches on the leaves of *Anemone nemorosa*, whilst *P. pusilla* is similarly associated with *Geranium pratense*. The haustoria of *Plasmopara* are knob-like; the sporangiophores are branched monopodially and the sporangia are hyaline (Fig. 78). Two types of sporangial germination have been reported. In *P. pygmaea* there are no zoospores but the entire sporangial contents escape and later produce a germ

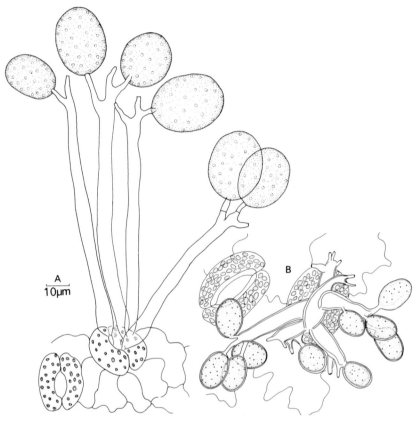

Fig. 78. *Plasmopara.* A. *P. pusilla*; sporangiophores on *Geranium pratense.* B. *P. pygmaea*; sporangiophores on *Anemone nemorosa.*

tube. In other species the sporangia germinate by means of zoospores which penetrate the host stomata. Oospore germination in *P. viticola* is also by means of zoospores.

Bremia (Fig. 79)

Bremia lactucae causes downy mildew of lettuce (*Lactuca sativa*) and strains of it can also be found on a number of Composite hosts such as *Sonchus* and *Senecio.* Cross-inoculation experiments using sporangia from these hosts have failed to result in infection of lettuce and it seems that the fungus exists in a number of host-specific strains (*formae speciales*). Although wild species of *Lactuca* can carry strains capable of infecting lettuce, these hosts are not sufficiently common to provide a serious source of infection. The disease can be troublesome both in lettuce grown in the open and under frames, and in market gardens there may be sufficient overlap in the growing time of lettuce

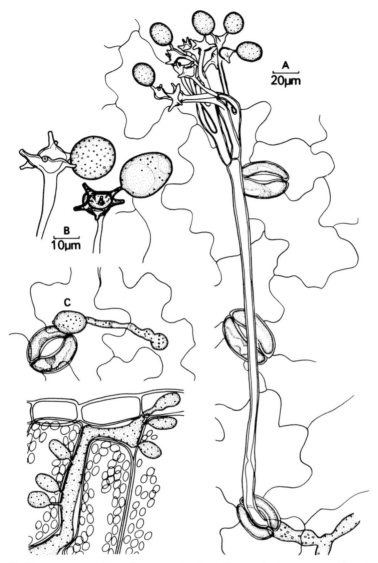

Fig. 79. *Bremia lactucae* from *Senecio vulgaris*. A. Sporangiophore protruding through a stoma. B. Sporangiophore apex. C. Sporangium germinating by means of a germ tube. D. Cells of epidermis and palisade mesophyll, showing intercellular mycelium and haustoria. A, C, D to same scale.

crops for the disease to be carried over from one sowing to the next. The damage to the crop caused by *Bremia* may not in itself be severe, but infected plants are prone to infection by the more serious grey mould, *Botrytis cinerea*. Systemic infection can occur. The intercellular mycelium is often coarse, and the haustoria are sac-shaped, often several in each host cell (Fig. 79D). The sporangiophores emerge singly or in small groups through the stomata and branch dichotomously. The tips of the branches expand to form a cup-shaped disk bearing short cylindrical sterigmata at the margin, and occasionally at the centre, and from these the hyaline sporangia arise (Fig. 79A,B). Germination of the sporangia is usually by means of a germ tube which penetrates an epidermal cell directly (Fig. 79C) or through a stoma, but zoospore formation has also been reported but has not been confirmed. Oogonia are apparently rare on lettuce (Tommerup *et al.*, 1974; Ingram *et al.*, 1975).

Albuginaceae (Figs. 80–82)

This family has only a single genus, *Albugo*, with about thirty species of biotrophic parasites of flowering plants, causing diseases known as white blisters or white rusts. The commonest British species is *A. candida* (= *Cystopus candidus*) causing white blister of crucifers such as cabbage, turnip, swede, horse-radish, etc., but particularly common on Shepherd's purse (*Capsella bursa-pastoris*) (Fig. 75B). There is some degree of physiological specialisation in the races of this fungus on different host genera. Another less common species is *A. tragopogi* causing white blisters of salsify (*Tragopogon porrifolius*), goatbeard (*T. pratensis*) and *Senecio squalidus*.

In *A. candida* on shepherd's purse, diseased plants may be detected by the distorted stems (Fig. 75B) and the shining white raised blisters on the stem, leaves and fruits before the host epidermis is ruptured. Later, when the epidermis is burst open, a white powdery pustule is visible. The distortion is possibly associated with changes in auxin level of diseased as compared with healthy tissues. The host plant may be infected simultaneously with *Peronospora parasitica*, but the two fungi are easily distinguishable microscopically both on the structure of the sporangiophores and by their different haustoria. In *Albugo* the mycelium in the host tissues is intercellular and only small spherical haustoria are found (Fig. 80), which contrast sharply with the coarsely lobed haustoria of *P. parasitica*. The fine structure of the haustorium of *Albugo candida* has been studied by Berlin & Bowen (1964) (see Fig. 81). The haustoria are spherical or somewhat flattened and about 4 μm in diameter, connected to the intercellular mycelium by a narrow stalk about 0·5 μm wide. Within the plasma membrane of the haustorium lomasomes, a system of unit-membranes and tubules, apparently derived from the plasma membrane, are more numerous than in the intercellular hyphae. The cytoplasm of the haustorial head is densely packed with mitochondria, ribosomes, endoplasmic reticulum and occasional lipoidal inclusions, but nuclei have not been observed. Since nuclei of *Albugo* are about 2·5 μm in diameter they may be unable to penetrate the constriction which joins the haustorium to the intercellular hypha. The base of the haustorium is surrounded by a collar-like sheath which is an extension of the host cell wall, but this wall does not normally completely surround the haustorium. Between the haustorium and the host

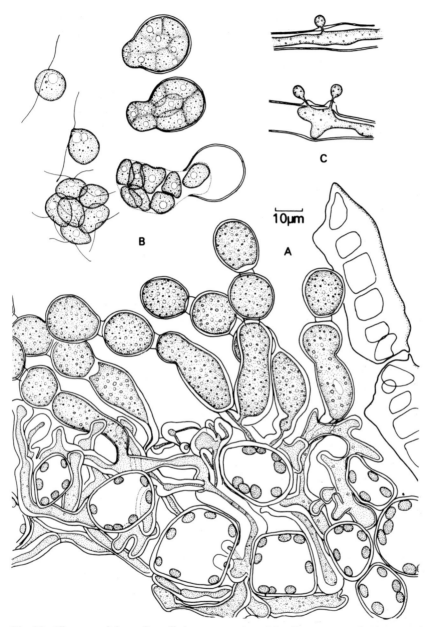

Fig. 80. *Albugo candida* on *Capsella bursa-pastoris*. A. Mycelium, sporangiophores and chains of sporangia formed beneath the ruptured epidermis (right). B. Germination of sporangia showing the release of eight biflagellate zoospores. The stages illustrated took place within 2 minutes. C. Haustoria.

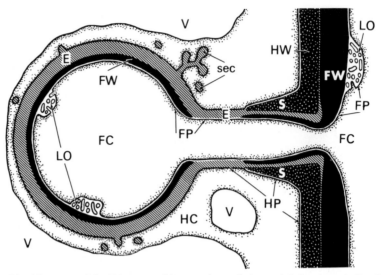

Fig. 81. *Albugo candida.* Diagram of haustorium of *A. candida* in mesophyll cell of *Raphanus.* Fungal cytoplasm (FC) is bounded by the fungal plasma membrane (FP) with its lomasomes (LO) and the fungal cell wall (FW) in both the intercellular hypha (right) and the haustorial head (left). The relative positions of the host cell vacuole (v), host cytoplasm (HC) and the host plasma membrane (HP) are indicated. The host cell wall (HW) terminates in a collar-like sheath (S). The encapsulation (E) separates the haustorium from the host. Note the discontinuity of the fungal cell wall in the stalk proximal to the haustorial head. Invaginations of the host plasma membrane and vesicular host cytoplasm are suggested as evidence for host secretory activity (sec) (after Berlin & Bowen, 1964).

plasma membrane is an encapsulation. Host cytoplasm reacts to infection by an increase in the number of ribosomes and Golgi complexes. In the vicinity of the haustorium the host cytoplasm contains numerous vesicular and tubular elements not found in uninfected cells. These structures have been interpreted as evidence of a secretory process induced in the host cell by the presence of the parasite. The intercellular mycelium masses beneath the host epidermis to form a palisade of cylindrical or skittle-shaped sporangiophores which give rise to chains of spherical sporangia in basipetal succession – i.e. new sporangia are formed at the base of the chain. The pressure of the developing chains of sporangia raises the host epidermis and finally ruptures it. The sporangia are then visible externally as a white powdery mass, dispersed by wind. If sporangia alight on a suitable host leaf they are capable of germinating within a few hours in films of water to form biflagellate zoospores, about eight per sporangium (Fig. 80B). After swimming for a time a zoospore encysts and then forms a germ tube which penetrates the host epidermis. A further crop of sporangia may be formed within 10 days. Infection may be localised or systemic. Oogonia grow in the intercellular spaces of the stem and leaves. Both the antheridium and the oogonium are multinucleate at their inception, and during development two further nuclear divisions occur so that the oogonium may contain over 200 nuclei. However there is only one functional male and one functional female nucleus. In the oogonium all the nuclei except one migrate to the periphery and are included in the periplasm. Following nuclear fusion a thin membrane first develops around the oospore. Division of the zygote nucleus takes place, and is repeated, so that at maturity the oospore may contain as many as thirty-two nuclei. Sansome & Sansome (1974) report that meiosis occurs within the gametangia. They have also suggested that *A. candida* is heterothallic. The high incidence of oospores of *Albugo* in *Capsella* stems simultaneously infected with *Peronospora parasitica* may result from some stimulus to sexual reproduction produced by the *Peronospora* which stimulates self-fertilisation in the *Albugo*, a situation analogous to the *Trichoderma*-induced sexual reproduction by heterothallic species of *Phytophthora*. The mature oospore is surrounded by a brown warty exospore, thrown into folds (Fig. 82A). Germination of the oospores only takes place after a resting period of several months. Under suitable conditions the outer wall of the oospore bursts and the endospore is extruded as a thin spherical vesicle which may be sessile or formed at the end of a wide cylindrical tube. Within the thin vesicle 40–60 zoospores are differentiated and are released on its breakdown (Fig. 82B,C).

The cytology of oospore development in some other species of *Albugo* differs from that of *A. candida*. In *A. bliti* a parasite of *Portulaca* in North America and Europe, the oogonia and antheridia are also multinucleate and two nuclear divisions take place during their development. Numerous male

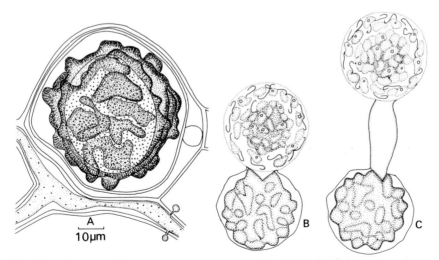

Fig. 82. *Albugo candida* oospores. A. Oogonium and oospore from *Capsella* leaf. B, C. Two methods of oospore germination (after Vanterpool, 1959).

nuclei fuse with numerous female nuclei and the fusion nuclei pass the winter without further change. In *A. tragopogi* a multinucleate oospore develops, and again there are two nuclear divisions involved in the development of the egg and the antheridium, but finally there is a single nuclear fusion between one male and one female nucleus. The fusion nucleus undergoes repeated division so that the overwintering oospore is multinucleate.

2

Zygomycotina

The Zygomycotina is an assemblage of fungi which reproduce asexually by non-motile aplanospores. The spores are contained in sporangia which may be violently projected, but more usually spores are passively dispersed by wind, rain or animals. Sexual reproduction is by gametangial copulation which is typically isogamous, and results in the formation of a **zygospore**. The mycelial organisation is coenocytic, and the cell wall contains chitin.

Two classes are included, the Zygomycetes and the Trichomycetes. The Zygomycetes comprise two orders, the Mucorales and Entomophthorales. Mucorales are ubiquitous in soil and dung, mostly as saprophytes, although a few are parasitic on plants and animals. Entomophthorales include a number of insect parasites, but some saprophytic forms exist. The Trichomycetes is a group of uncertain affinity, and are mostly parasitic in the guts of arthropods, e.g. millipedes and insect larvae. This group will not be considered further. For references see Lichtwardt (1973a,b; 1976).

1: ZYGOMYCETES

MUCORALES

Unlike the fungi previously considered the Mucorales do not possess motile zoospores, but reproduce asexually by non-motile spores carried passively by wind, or dispersed by rain-splash, insects or other animals. In many forms numerous spores are contained in globose sporangia surrounding a central core or columella. Some also possess few-spored sporangia, termed **sporangiola**, dispersed as a unit. Others may reproduce by means of unicellular propagules termed **conidia**. It is believed that conidia may have evolved several times within different groups of Mucorales from forms with monosporous sporangiola. A distinction between sporangiospores and conidiospores is that, before germination of sporangiospores, a new wall, eventually continuous with the germ tube, is laid down within the original spore wall (Ekundayo, 1966), whilst in conidia there is no new wall layer laid down (see Fig. 105 and Dykstra, 1974). Sporangia are also known in which there is no columella, or where the spores are arranged in a row inside a cylindrical sac termed a **merosporangium**. Sexual reproduction is by conjugation, the fusion of usually equal gametangia derived from branch tips, to form a warty zygospore.

The Mucorales are widely distributed in soil, and are mostly saprophytic. Some may cause spoilage of food. A few are weak parasites of fruits; some are parasitic on other fungi, and some cause diseases of animals including man. A number of species have been used in fermentations for the production of alcohol. Extensive studies of their nutrition and physiology have been made (for references see Foster, 1949; Lilly & Barnett, 1951; Cochrane, 1958; Cantino & Turian, 1959). A wide variety of sugars can be used, and whilst starch can be decomposed by some species, cellulose is not utilised by most. Under anaerobic conditions, ethyl alcohol and numerous organic acids are produced. Many Mucoraceae need an external supply of vitamins for growth in synthetic culture. Thiamine is a common requirement, and since *Phycomyces* needs an external supply of this substance, the amount of growth of the fungus can be used as a sensitive assay for the concentration of thiamine. Carotene synthesis by cultures of certain Mucorales may prove to be an economic proposition. Zycha *et al.* (1969) has given a general account of the taxonomy of Mucorales. Different authors have adopted different systems of classification. Hesseltine & Ellis (1973) recognise fourteen families, but in the Key below, Martin (1961) distinguishes nine.

1 Sporangia all columellate and alike   2
  Columellate sporangia lacking or, if present, accompanied by sporangioles or conidia   3

2 Sporangial membrane thin, fugacious; sporangiospores liberated by
  breaking or dissolution of sporangial wall; suspensors rarely tong-
  like   **Mucoraceae (p. 193)**
  Sporangial wall densely cutinised above; sporangium violently discharged or passively separated as a unit from sporangiophore; suspensors always tong-like   **Pilobolaceae (p. 213)**

3 Columellate sporangia usually present, accompanied by few-spored
  subglubose sporangioles in which spores are never formed in linear
  series, or by conidia   4
  Columellate sporangia never present; sporangioles one-spored, con-
  idium-like, or bearing spores in linear series, or modified non-columellate sporangia functioning as dissemilules   5

4 Sporangioles borne in clusters on lower part of sporangiophore, the
  latter usually tipped by a columellate sporangium or a spine
                                                    Thamnidiaceae (p. 220)
  Sporangioles or conidia never borne on same sporangiophores as
  columellate sporangia; zygospores with tong-like suspensors
                                                    Choanephoraceae (p. 224)

5 Merosporangia borne on tips of sporangiophores, at first cylindrical,

dividing into a single row of sporangiospores, simulating a chain of conidia **Piptocephalidaceae (p. 227)**
Merosporangia lacking, or not clearly defined 6

6 Conidia or very short merosporangia borne unilaterally on special branches (sporocladia) or sporophores, in comb-like or fan-like aggregates **Kickxellaceae**
Non-columnellate sporangia or conidia present, but latter not borne on sporocladia 7

7 Conidia present borne on spicules on inflated tips of sporophores; sporangia lacking **Cunninghamellaceae (p. 229)**
Conidia lacking; sporangia non-columellate 8

8 Sporangia borne freely on surface, entire sporangium sometimes modified to function as a disseminule; zygospores in hyphal matrix but not forming a sporocarp **Mortierellaceae (p. 230)**

9 Sporocarp present, sclerotium-like, enclosing sporangia, zygospores or azygospores **Endogonaceae (p. 233)**

Representatives of all these families with the exception of the Kickxellaceae will be discussed.

Mucoraceae

Members of this family are abundant in soil, on dung, and on moist fresh organic matter in contact with soil. For the most part they are saprophytic, and play an important role in the early colonisation of substrata in soil. Sometimes, however, they can behave as parasites of plant tissues, e.g. *Rhizopus stolonifer* can cause a rot of sweet potatoes or fruit such as apples, strawberries and tomatoes. They are also known to cause fungal diseases of animals and man (mucormycosis), and such conditions seem to be particularly frequent in patients suffering from diabetes, leukaemia and cancer. Lesions may be localised in the brain, lungs or other organs, or may be disseminated, e.g. at various points in the vascular system. Species of *Rhizopus* and *Mucor* are reported from human lesions, and these genera together with species of *Absidia* are associated with mucormycosis in domestic animals. Occasionally *Rhizopus* and *Mucor* cause spoilage of bread and other food. Certain species, e.g. *Mucor rouxii,* are also used industrially to break down starch to sugar before fermentation by yeasts to alcohols, since yeasts do not contain amylolytic enzymes for the initial breakdown.

The mycelium first established by a germinating spore on a solid substrate is coarse, coenocytic and richly branched; the branches usually taper to fine points (Fig. 83). Later, septa may appear. Thick-walled mycelial segments or chlamydospores may be cut off by such septa, and in certain species, e.g. *Mucor racemosus,* the presence of chlamydospores may be a useful diagnostic

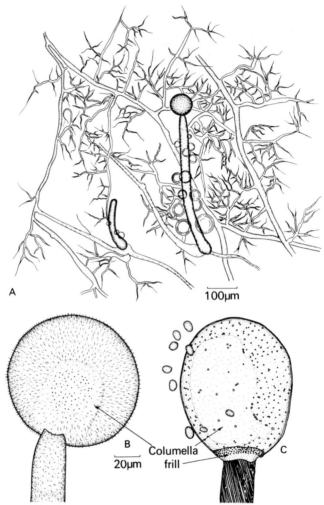

Fig. 83. *Mucor mucedo.* A. Mycelium and young sporangiophores, with globules of liquid attached. B. Immature sporangium with the columella visible through the sporangial wall. C. Dehisced sporangium showing the columella, the frill representing the remains of the sporangial wall, and sporangiospores.

feature (Fig. 84A). In anaerobic liquid culture, especially in the presence of CO_2, *Mucor* may grow in a yeast-like instead of a filamentous form (Fig. 85) but reverts to filamentous growth in the presence of O_2 (Bartnicki-Garcia & Nickerson, 1962a, b; Lara & Bartnicki-Garcia, 1974). The cell walls of Mucoraceae are complex chemically. Whilst chitin microfibrils are present, chitosan may be the most abundant component. Other polysaccharides such as gluco-samine and galactose, poly-D-glucuronides, polyphosphates, proteins, lipids, purines, pyrimidines, magnesium and calcium have also been detected (Bart-

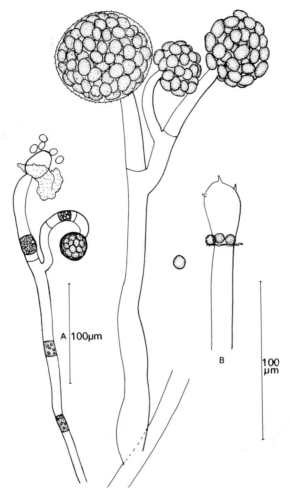

Fig. 84. A. *Mucor racemosus*: branched sporangiophore containing chlamydospores.
The remnants of the sporangial wall in the upper dehisced sporangium are clearly
visible. B. *M. plumbeus*: branched sporangiophore. The sporangial wall has disappeared in two sporangia. Note the spiny process on the columella.

nicki-Garcia & Nickerson, 1962c; Bartnicki-Garcia *et al.*, 1968; Bartnicki-
Garcia & Reyes, 1968; Jones *et al.*, 1968; Kreger, 1954). Comparison of the
structure and composition of yeast-like and filamentous cells of *Mucor rouxii*
show that the yeast-like cells have much thicker walls. They also have a
mannose content about five times as great as that of filamentous cell walls.

The synthesis of chitin microfibrils may be correlated with the presence of
multivesicular bodies, which seem to arise from the endoplasmic reticulum.
Such structures have been described from sporangiophores of *Phycomyces*

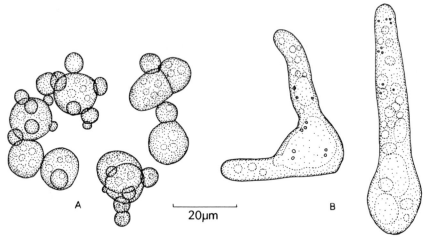

Fig. 85. *Mucor rouxii.* A. Yeast-like growth in liquid medium under anaerobic conditions 24 hours after inoculation with spores. B. Filamentous growth from spores in liquid medium under aerobic conditions 4 hours after inoculation.

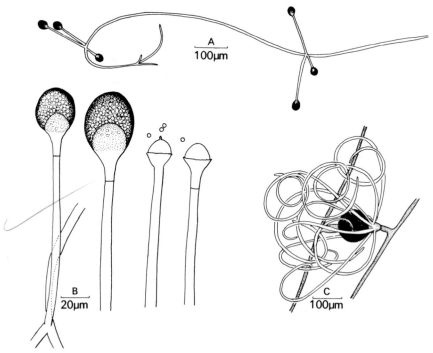

Fig. 86. *Absidia glauca.* A. Habit showing whorls of pear-shaped sporangia. B. Intact and dehisced sporangia. Note the single spine-like projection on certain columellae. C. Zygospore showing the arching suspensor appendages.

and from a number of other fungi with chitin walls (Peat & Banbury, 1967; Marchant *et al.*, 1967).

The cytoplasm often shows rapid streaming, and phase contrast micro-scopy shows granules being carried in the stream, mitochondria which are short at the hyphal tips but longer and filamentous in older parts, sap vacuoles and nuclei. The nuclei are often irregular in shape. Conventional mitosis and spindle formation have not been seen, and the nuclei appear to divide by constriction, although chromosomes can be demonstrated (Cutter, 1942a,b; Robinow, 1957a,b, 1962). Electron micrographs of the mycelium of *Rhizopus* reveal no especially distinctive features (Hawker & Abbott, 1963a).

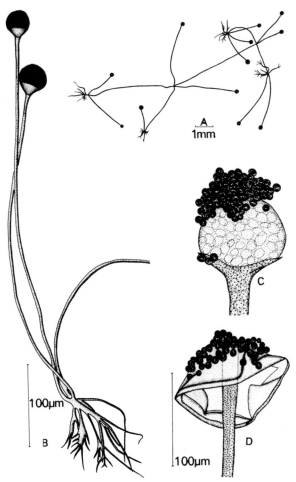

Fig. 87. *Rhizopus stolonifer*. A. Habit sketch, showing stolon-like branches which develop rhizoids and tufts of sporangiophores. B. Two sporangiophores showing basal rhizoids. C. Columella and attached spores. D. Invaginated columella.

ASEXUAL REPRODUCTION

Asexual reproduction is by non-motile sporangiospores contained in globose or pear-shaped sporangia. The sporangia may be borne singly at the tip of a sporangiophore or may occur on a branched sporangiophore. In some genera, e.g. *Absidia* (Fig. 86), the sporangia may be arranged in whorls on aerial branches, and in many species of *Rhizopus* the sporangiophores arise in groups from a clump of rhizoids (Fig. 87). The sporangiophores are commonly phototropic, and numerous studies on the phototropism of the large sporangiophores of *Phycomyces* have been made (Shropshire, 1963; Castle, 1966; Bergman *et al.*, 1969; Cerdá-Olmedo, 1974).

The sporangiophore of this fungus, in view of its large size, may not be typical of the Mucoraceae. Nevertheless, it has been the subject of elegant studies by physiologists interested in sensory perception. 'The sporangiophore of the fungus *Phycomyces* is a gigantic, single-celled, erect, cylindrical aerial hypha. It is sensitive to at least four distinct stimuli: light, gravity, stretch, and some unknown stimulus by which it avoids solid objects. These stimuli control a common output, the growth rate, producing either temporal changes in

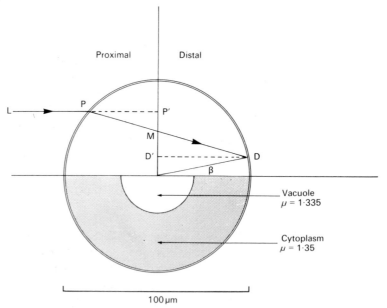

Fig. 88. Sporangiophore of *Phycomyces* as a cylindrical lens. Upper portion: light ray L impinges from the left and is refracted at the first surface. The ratio of 'path length' of the light ray in the proximal part of the sporangiophore to the path length in the distal part is PM/DM. The maximum value of the angle β is about 20°. Lower portion: sporangiophore in section to show the peripheral layer of cytoplasm surrounding the central vacuole. The values are estimates of the refractive index of cytoplasm and vacuolar sap. Diagram modified from Bergman *et al.* (1969).

growth rate or tropic responses' (Bergman *et al.*, 1969). It may reach 15–20 cm in height, and for much of its length is about 100 μm in diameter. The wall, about 0·6 μm thick, encloses a peripheral layer of cytoplasm about 30 μm thick surrounding a central vacuole about 20 μm in radius (Fig. 88). The globose sporangium measures about 500 μm in diameter.

It has been suggested (Fig. 88) that the sporangiophore functions as a cylindrical lens focussing unilateral light on the distal wall of the sporangiophore, and that this would result in more intense illumination of the distal side of the sporangiophore. If, as a result of the more intense illumination of the distal half of the sporangiophore, the wall became more plastic in that region, the turgor pressure of the sporangiophore would cause more rapid stretching there, causing the sporangiophore to bend towards the light.

It has been argued that if the sporangiophore is surrounded by a liquid of higher refractive index than its contents, the sporangiophore should function as a diverging lens and bend *away* from the light. This has been shown to be true when sporangiophores are immersed in liquid paraffin with a refractive index of 1·47. It has also been possible to prepare liquids with a 'neutral refractive index', i.e. liquids within which the sporangiophores fail to respond either positively or negatively to unilateral light. Using light of 440 nm wavelength, the neutral refractive index of a mixture of fluorocarbon oils was found to be 1·295, appreciably below that of either the cytoplasm or the vacuole. The idea that light is normally focussed to greater intensity on the distal side of the sporangiophore, and that this somehow results in more rapid growth of the distal wall, receives support from experiments in which a very narrow beam of light is made to fall on the edge of a sporangiophore, resulting in more rapid growth of the illuminated region. Examination of Fig. 88 shows that the path length of a ray of light passing through the distal half of a sporangiophore is longer than that in the proximal half and, provided the attenuation of the light rays as they pass through the proximal half is not severe, the total amount of light energy absorbed in the distal half may well be greater. However, as Bergman *et al.* have argued, it is, perhaps, unlikely that the way in which the sporangiophore evaluates the dissymmetry of the light is related to this primitive explanation.

A central problem in studies of photoresponses is the nature of the photoreceptor or receptors, and a clue to their possible nature can be obtained by studying the action spectrum of the response over a range of light wavelengths. The phototropic curvature of *Phycomyces* sporangiophores has a similar action spectrum to the growth response of the vegetative mycelium which is also stimulated by light. There are several clearly defined peaks, at 485, 455, 385 and 280 nm, for the most part in the blue part of the spectrum. Whether these four peaks correspond to absorption by one or by more than one pigment in *Phycomyces* is not known. Its photoreceptor has not yet been identified, although it has been suggested that β-carotene or a flavoprotein may be involved. The discovery of carotene-deficient mutants with less than

1/1000th of the β-carotene content of wild-type strains, which are still fully photosensitive, suggests that the bulk of the β-carotene is not the effective pigment. Both β-carotene and flavoproteins are components of the cytoplasm. For a fuller discussion, see Carlile (1965).

As the sporangiophore of *Phycomyces* develops it rotates. Castle (1942) followed the growth and rotation of the sporangiophore by attaching *Lycopodium* spores as markers. The displacement of the markers was then followed. His findings are illustrated in Fig. 89. After a period of apical growth of the tubular sporangiophore (Stage I) the sporangium appears as a terminal swelling and growth ceases. During this period (Stage II) growth is limited to sporangial enlargement. In the next period (Stage III) no further enlargement of the sporangium occurs, and elongation is at a standstill. During Stages IVA and IVB further elongation occurs and growth is mainly localised in a zone somewhat below the sporangium. During Stage I the tip of the sporangio-

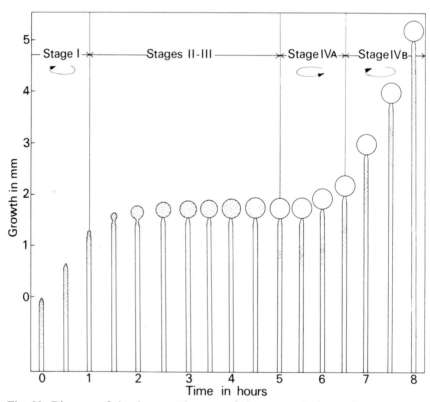

Fig. 89. Diagram of developmental stages of the sporangiophore of *Phycomyces*. Regions in which growth is taking place are stippled. The rotary component of growth is indicated. During Stage I the axis of growth is directed sinistrally, in Stages II and III growth is unoriented. In Stage IVA dextral spiralling occurs and in Stage IVB sinistral spiralling again takes place (after Castle, 1942).

phore rotates clockwise (as seen from above looking down) through a maximum angle of about 90°. There is no rotary movement during Stages II and III. However when the sporangium is completed and elongation recommences in Stage IVA, rotation is again detected so that the markers move upwards spirally. The direction of rotation is now *anti-clockwise* (as seen from above) and the direction of spiral growth is *dextral* (right-handed) instead of *sinistral* (left-handed) as in Stage I. During this stage, which lasts about 1 hour, markers attached to the growth zone may make up to two complete revolutions around the axis. Finally, during Stage IVB the direction of rotation reverses once more and becomes sinistral.

The reasons for the spiral growth are far from clear. It is known that the chitin microfibrils which make up the wall of the sporangiophore show a right-handed or Z spiral orientation. One possible explanation is that the laying down of the fibrils in this way is responsible for the rotation. A second is that the extension due to turgor of a cylinder whose walls are composed of spirally arranged fibrils would naturally result in a passive rotation. The phenomenon of spiral growth is not peculiar to *Phycomyces* but occurs during the extension of various cylindrical plant cells. For a fuller discussion of the problems involved see Castle (1953), and Roelofsen (1950, 1959).

The mechanical properties of the sporangiophore of *Phycomyces* change during development. During Stage II, when no elongation of the sporangiophore is taking place, the sporangiophore shows elastic deformation when small loads are applied to it, i.e. the fractional extension in length is directly proportional to the applied load, and on removal of the load, the sporangiophore returns to its original length. During Stage IV, when elongation of the sporangiophore is renewed, although the sporangiophore increases in length in response to applied loads, upon unloading the cell wall behaves quite unelastically, and the sporangiophore does not return to its original length. These differences suggest that the cell wall in the growing zone beneath the sporangium undergoes a physical change as the sporangiophore matures (Ahlquist & Gamow, 1973). There is now good evidence that the spores secrete a substance or substances which control elongation of the sporangiophore. In experiments in which mature sporangia were removed, it was found that growth of the sporangiophore stopped. Replacement of the detached sporangium with a substitute sporangium, with a suspension of spores, or with a drop of supernatant liquid from a centrifuged spore suspension, resulted in resumption of growth. Another effect of the removal of a ripe sporangium is that branching is induced in the sporangiophore, an effect which has been likened to apical dominance exerted by the terminal buds of Angiosperms (Goodell, 1971).

Sporangiophores develop as coarser, blunt-tipped, aerial hyphae, which grow away from the substratum. The tip expands to form the sporangium initial containing numerous nuclei which continue to divide. A dome-shaped septum is laid down cutting off a distal portion which will contain the spores from a central cylindrical or subglobose spore-free core, the columella. It should be stressed that the columella is curved from its inception and does not arise by the arching of a flat septum into the sporangium. Cleavage planes separate the nuclei within the sporangium and finally the spores are cleaved out. They may be uninucleate or multinucleate according to the species, e.g. *M. hiemalis* and *Absidia glauca* have uninucleate spores whilst *Phycomyces blakesleeanus, Rhizopus stolonifer* and *Syzygites megalocarpus* have multinucleate spores (Sjöwall, 1945; Robinow, 1957a,b; Hawker & Abbott, 1963b). The number of spores formed is very variable. On nutrient-poor media minute sporangia containing very few spores may be formed, but in *Phycomyces blakesleeanus* the number may be as high as 50,000–100,000 per sporangium (Ingold & Zoberi, 1963).

The fine structure of developing sporangia of *Gilbertella* (Choanephoraceae) has been studied by Bracker (1966). The cleavage of the sporangial cytoplasm to form spores is accomplished by the fusion of cleavage vesicles associated with endoplasmic reticulum which is continuous with the nuclear envelope.

The sporangial wall often darkens and may develop a spiny surface due to the formation of crystals. Despite the apparent similarity in sporangial structure in members of the Mucoraceae at least two distinct mechanisms are involved in spore liberation. In many of the commonest species of *Mucor* (e.g. *M. hiemalis*) the sporangium becomes converted at maturity into a 'sporangial drop'. The sporangial wall dissolves and the spores absorb water so that the tip of the sporangiophore bears a drop of liquid containing spores adhering to the columella. The remnants of the sporangial wall can often be seen as a frill at the base of the columella. In large sporangia e.g. *M. plasmaticus* and *M. mucedo* and in *Phycomyces* the spores are embedded in mucilage. The sporangial wall does not break open spontaneously, but only when it is touched; the slimy contents then exude. Such sticky spore masses are not readily detached by wind or by mechanical agitation, but possibly by insects or rain-splash, or after drying. In other species, e.g. *M. plumbeus,* the sporangial wall breaks into pieces. Here air currents or mechanical agitation readily liberate spores. In this species the columella may terminate in one or more finger-like or spiny projections (Fig. 84B) and in *Absidia* the columella may also bear a single nipple-like projection (Fig. 86). In *Rhizopus stolonifer* the columella is large, and as the sporangium dries the columella collapses so that it appears like an inverted pudding bowl balanced on the end of a stick represented by the stiff sporangiophore (Ingold & Zoberi, 1963) – see Fig. 87. Associated with these changes in columella shape the sporangium wall breaks up into many fragments and the dry spores can escape in wind currents.

SEXUAL REPRODUCTION

The Mucoraceae reproduce sexually by a process of conjugation resulting in the formation of zygospores. Some species are homothallic, zygospores being

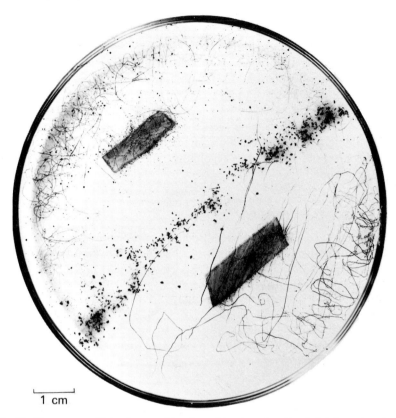

1 cm

Fig. 90. *Phycomyces blakesleeanus.* Ten-day-old culture in a Petri dish inoculated with two compatible strains. The black zone between the two inoculum blocks contains numerous zygospores.

formed in cultures derived from a single sporangiospore (e.g. *Rhizopus sexualis, Syzygites megalocarpus, Zygorhynchus mölleri* and *Absidia spinosa*). However, the majority of species are heterothallic, and only form zygospores when compatible strains are mated together. If the appropriate strains are inoculated at opposite sides of a Petri dish the mycelia grow out and where they meet in the centre of the dish a line of zygospores develops (Fig. 90). The two compatible strains rarely differ in any regular way from each other, although there may be slight differences in growth rate and carotene content. Because it was not possible to designate one strain as male and the other as female, Blakeslee labelled one strain as ($+$) and the other as ($-$). The two compatible strains can be said to differ in *mating type*. The morphological events preceding zygospore formation are sufficiently similar to allow a general description of the process, although there are some morphological features peculiar to certain genera, and also differences in cytology.

When two compatible strains approach each other, three reactions can be distinguished (Burgeff, 1924; Mesland *et al.*, 1974):

(*a*) a 'telemorphotic reaction' which involves the induction of aerial **zygophore** formation. The zygophores are club-shaped and often coloured yellow due to their content of *β*-carotene.

(*b*) a 'zygotropic reaction', in which directed growth of zygophores of (+) and (−) mating partners towards each other is observed.

(*c*) a 'thigmomotropic reaction', involving the events which occur after contact of the respective zygophores, such as gametangial fusion and septation.

Zygophore induction

Evidence has been obtained (see van den Ende & Stegwee, 1971; Gooday, 1973, 1974a; van den Ende, 1976) that zygophores in both (+) and (−) strains are induced by trisporic acids B + C. Trisporic acids (see Fig. 91) are oxidised, unsaturated derivatives of trimethyl cyclohexane.

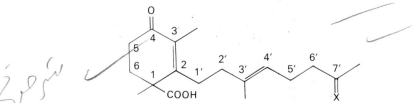

Fig. 91. Trisporic acid **B**, X = O: trisporic acid **C**, X = H.OH.

In heterothallic Mucorales, trisporic acids are only produced in appreciable quantities when (+) and (−) cultures are in continuous diffusion contact with each other, either on solid media or in well-aerated liquid cultures, although small quantities of trisporic acid have occasionally been detected in unmixed cultures. The greatly increased synthesis in mated cultures is brought about by the production and secretion of inducers, which have been shown to be precursors of trisporic acids (Werkman & van den Ende, 1973). A diagrammatic representation of the sequence of events is shown below (Gooday, 1973).

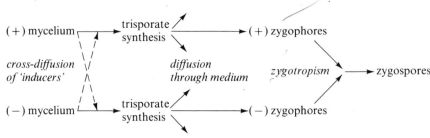

The specific inducers of trisporic acid biosynthesis have been identified. Unmated cultures of (+) and (−) strains produce small quantities of 'prohormones' that are converted to trisporic acid (= trisporate) by the (−) and (+) strains respectively. The major (+) prohormone is methyl-4-OH-trisporate, and the major (−) prohormones are trisporol and *gem*-dimethyl compounds (Bu'Lock *et al.*, 1974a,b).

Trisporic acids have several effects:

(a) Greatly enhanced secretion of the specific precursors of trisporic acid, and also of β-carotene, which is itself also a precursor. Thus, in a mated culture, trisporic acid has a positive feed-back effect on its own synthesis by stimulating the rate of formation of precursor compounds (Werkman & van den Ende, 1973).

(b) Zygophore formation is induced.

(c) Sporangiophore formation is repressed.

The effects of trisporic acid in inducing carotene synthesis and zygophore formation are non-specific; for example, trisporic acid produced by *Choanephora (Blakeslea) trispora* can induce zygophore development in other Mucorales such as *Mucor mucedo*. It is also known that trisporic acid is involved in the sexual response of some homothallic Mucorales such as *Zygorhynchus mölleri*, *Mucor genevensis* and *Syzygites megalocarpus* (Werkman & van den Ende, 1974). The common nature of the hormones of homothallic and heterothallic species could also be inferred from earlier observations of attempted matings between such forms, either at the interspecific or intergeneric level.

Zygotropic reaction

When (+) and (−) zygophores are near, a directed outgrowth towards each other occurs, and it has been suggested that this is a chemotropic response to a volatile stimulus. Banbury (1955) has claimed that, as well as the mutual attraction between zygophores of opposite sign, there is mutual repulsion between zygophores of like sign, but this needs confirmation. Mesland *et al.* (1974) have shown that the *vegetative mycelium* of *Mucor mucedo*, even when not in diffusion contact with vegetative mycelium of the opposite strain, could produce volatile substances which induced zygophore development and the zygotropic reaction. There was no evidence that the vegetative mycelia contained trisporic acids, which are, in any case, non-volatile. It now seems likely that the volatile substances are identical with the specific precursors characteristic of each mating type. Thus these volatile compounds may have a dual role, to enhance (i.e. de-repress) trisporic acid synthesis in the opposite mating type, and to mediate the zygotropic response.

Thigmotropic reaction

When compatible zygophores make contact they develop into progametangia. The tip of each progametangium becomes cut off by a septum to separate a distal multinucleate gametangium from the subterminal suspensor (see Fig. 92).

The walls separating the two gametangia break down so that the numerous nuclei from each cell become surrounded by a common cytoplasm. The fusion cell, or zygospore, swells and develops a dark warty outer layer. After a resting period the zygospore may germinate by developing a germ sporangium, usually on an unbranched sporangiophore, terminated by a columellate sporangium of the usual type. In some cases vegetative mycelium develops from the germinating zygospore. The precise conditions for zygospore germination are imperfectly known, and the development of reliable techniques for germination would greatly facilitate genetical analysis of Mucoraceae. There have been numerous accounts of the cytology of zygospore formation, and among more recent accounts the investigations of Cutter (1942a,b),

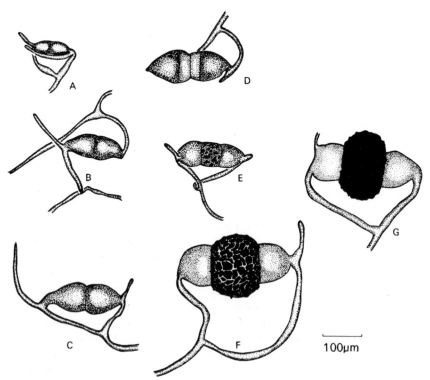

100µm

Fig. 92. *Rhizopus sexualis.* A–G. Successive stages in the formation of zygospores. The fungus is homothallic.

Rhizopus stoloiFer heterothelic

Sjöwall (1945, 1946) and Laane (1974) may be cited. Four main types of nuclear behaviour can be distinguished:

1. In *Mucor hiemalis, Absidia spinosa* and some other members of the Mucoraceae all the nuclei fuse in pairs within a few days, then quickly undergo meiosis so that the mature zygospore contains only haploid nuclei.

2. In *Rhizopus stolonifer* and *Absidia glauca* some of the nuclei entering the zygospore do not pair, but degenerate. The remainder fuse in pairs but do not undergo meiosis immediately. Meiosis is delayed until germination of the zygospore.

3. In *Phycomyces blakesleeanus* the nuclei continue to divide in the young zygospore, so increasing in number. They then become associated in groups each containing several nuclei, with occasional single nuclei. Before germination some of the nuclei pair, and in the germ sporangium are found diploid nuclei and also haploid nuclei some of which are probably meiotic products; others may represent the scattered solitary nuclei which failed to pair. This assortment of nuclei in the germ sporangium is reflected in the mating-type reactions of the spores it contains (see below).

4. In *Syzygites megalocarpus* nuclear divisions continue in the young zygospore, but nuclear fusion and meiosis apparently do not occur. This fungus can therefore be described as *amictic* (Burnett, 1965).

The results of various workers (Burgeff, 1914; Köhler, 1935) suggest that there is a single mating type locus with two alternative alleles, + and −, which segregate at meiosis. However there are a number of anomalous results for which a full cytological explanation is still awaited.

The fine structure of zygospore development has been studied in the homothallic *Rhizopus sexualis* (Hawker & Gooday, 1967, 1968; Hawker & Beckett, 1971). Following contact of the tips of the two zygophores (Fig. 92A–C), their walls become adherent and flattened, and each cell becomes distended to form a progametangium. In each progametangium, an oblique septum, concave to the developing zygospore, develops as a flange-shaped diaphragm, gradually extending centripetally until the septum is complete. In section, the septum appears as a narrow-based wedge continuous at its base with the primary wall of the progametangium, narrowing at the inner edge where it is apparently derived by the coalescence of a series of vesicles. When the septum is complete, it divides the terminal gametangia from the suspensors. However, cytoplasmic continuity between the suspensor and the gametangium persists through a series of plasmodesmata, which probably enable food materials to continue to flow into the developing zygospore from the surrounding mycelium. Numerous nuclei congregate on the flattened walls separating the tips of the two gametangia. It has been estimated that there may be over 150 nuclei in a pair of progametangia, but the number may rise to over 300 in a pair of completely delimited gametangia, reflecting divisions of gametangial nuclei.

The breakdown of the fusion wall separating the gametangia is associated with an accumulation of vesicles which are present in large numbers in the neighbourhood of the dissolving wall. They are presumed to function in the transport of hydrolytic enzymes to the wall and of degraded wall material from it. The fusion wall is entirely dissolved, leaving no detectable solid residues. Once the cytoplasmic contents of the two gametangia are continuous, the nuclei become arranged in the periphery of the cytoplasm. In *R. sexualis*, it is probable that most of the gametangial nuclei fuse in pairs immediately, and that the fusion nuclei then quickly divide.

The development of the walls surrounding the mature zygospore is as follows. After the formation of the septa separating gametangia from suspensors, but before the fusion wall is completely dissolved, the primary outer wall of the young zygospore becomes thicker, possibly as a result of gelatinisation. Beneath this original wall, the warts (which will eventually ornament the wall of the mature zygospore) are initiated as widely separated patches shaped like inverted saucers (Fig. 93C). The cytoplasm fills the domes of the 'saucers' and also balloons out between them, enveloped by the plasmalemma. As the

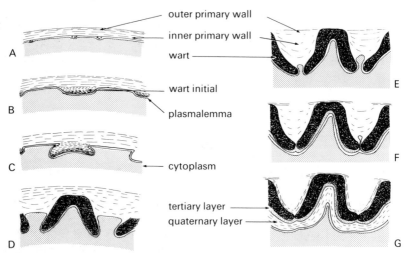

Fig. 93. Development of lateral wall of zygospore of *Rhizopus sexualis* (diagrammatic, after Hawker & Beckett, 1971). A. Primary wall before inflation of zygospore, showing thin electron-dense outer layer, thicker less electron-dense inner one, and scattered lomasome-like bodies. B. Blocks of secondary material (wart initials) developing locally on inner surface of primary wall. C. Wart initials growing by deposition of secondary material at the rims to give saucer-shaped pigmented masses. D. Warts becoming flower-pot shaped by further growth at rims, inner layer of primary wall becoming gelatinous and swollen. Note pockets of cytoplasm between warts. E. Rims of warts nearly touching, inner layer of primary wall showing stress lines, pockets of cytoplasm between warts much reduced. F. Edges of warts touching, warts lined with tertiary smoothing layer, outer layer of primary wall torn. G. Thick stratified impermeable layer of quaternary material laid down inside smoothing layer, inner gelatinous layer of primary wall has collapsed as a horny skin enveloping the warts.

zygospore continues to enlarge, the saucers change in shape and size to resemble inverted flower-pots which, by the addition of new material at their rims, increase in size until they are contiguous. Before the bases of the warts become contiguous, the developing zygospore is permeable to fixatives, but when contact is complete, fixatives can no longer penetrate. This explains why cytological studies of zygospore development have proved difficult. Eventually, the tips of the warts become pushed through the original primary wall. At least three other wall layers are laid down beneath the original primary wall (Fig. 93G). The darkening of the wall is probably due to the deposition of melanin.

The sculpturing of the zygospore wall of other members of the Mucorales, as seen by scanning electron microscopy, shows different patterns, ranging from circular or conical warts, to branched stellate warts, and the pattern may have some taxonomic significance, i.e. information on zygospore sculpturing may assist in indicating relationships within Mucorales (Schipper *et al.*, 1975).

In *Mucor mucedo*, the wall of the zygospore is rich in sporopollenin, which is also known to occur in the walls of pollen grains. Sporopollenin is extremely resistant to degradation, and may enable zygospores to remain dormant but undamaged in the soil for long periods. Sporopollenin is formed by oxidative polymerisation of β-carotene, and this may explain the high content of this pigment in developing zygospores (Gooday *et al.*, 1973). However, β-carotene and sporopollenin appear to be absent from the zygospore walls of *R. sexualis* (Hocking, 1963).

Zygospore germination

Blakeslee (1906) defined four kinds of zygospore germination in relation to the distribution of mating type in the spores of the germ sporangium.

1. Pure germinations, in which all germ sporangiospores were of one mating type, i.e. all + or all −. *Mucor mucedo, M. hiemalis* and *Phycomyces blakesleeanus* behave in this way (Gauger, 1965; Hocking, 1967). A possible cytological explanation of this finding is that only one diploid nucleus survives in the zygospore and that when this nucleus undergoes meiosis only one of the four haploid daughter nuclei survives which then divides to give nuclei which are all alike – i.e. all + or all −. Some support for this view is found in the results of Köhler (1935) who made crosses involving two pairs of alleles in *M. mucedo*, but found that only one of the four possible genotypes was found in any one germ sporangium. Sjöwall (1945) reported degeneration of nuclei in zygospores during the resting state in *M. hiemalis*. In *P. blakesleeanus*, a cross between a heterokaryotic (−) strain with two separate gene markers for colour, white and red, and a (+) strain with a yellow marker, yielded zygospores. Only one germ-sporangium out of thirty-two contained all three colours. 'It is concluded that, however many nuclei enter the zygospore, only a

pair of nuclei, rarely more, become parents of the progeny. The other nuclei are presumably used as food and energy reserves' (Cerdá-Olmedo, 1974.)

2. Pure germinations in which all the spores were homothallic. *Mucor genevensis, Zygorhynchus dangeardi* and *Syzygites megalocarpus* behave in this way.

3. Mixed germinations in *Phycomyces nitens* where the same germ sporangia sometimes contained $+$, $-$ and homothallic (i.e. self-fertile) spores. Cutter's findings that diploid nuclei enter the germ sporangia may be the explanation of the presence of homothallic spores.

4. Mixed germinations were postulated in which $+$ and $-$ spores, but not homothallic spores, occurred together in the same sporangium. Blakeslee himself did not observe this type of behaviour, but it has since been discovered by Gauger (1961) in *Rhizopus stolonifer*. In an analysis of 33 zygospore germinations he found 19 germ sporangia to contain all $+$ spores, 9 yielded all $-$ spores, and 5 germ sporangia yielded mixed $+$ and $-$ spores. The pure germinations would presumably be explained cytologically in a similar way to *M. mucedo*, whilst for the mixed germinations it would be necessary only to postulate the survival of more than one meiotic product so that both mating types were represented. In some germ sporangia 'neuter' spores were found, i.e. spores which on germination yielded mycelia which failed to mate both with tester $+$ and $-$ strains. The ability to mate was restored in time in certain isolates. Loss of ability to mate seems to be a fairly common phenomena in laboratory cultures after repeated subculturing, but an explanation of the effect is awaited.

In *Phycomyces* and *Absidia* the suspensors may bear appendages which arch over the zygospore. In *Phycomyces* the appendages are dark black and are forked, whilst in *Absidia* they are unbranched but curved inwards or coiled (see Figs. 86, 94). The function of such appendages is unknown: possibly they assist in attaching zygospores to passing animals. The forked appendage tips of *Phycomyces* bear a drop of liquid, and Burgeff (1925) regarded them as hydathodes. In the homothallic species *A. spinosa* the appendages arise on only one suspensor. The suspensors of *Zygorhynchus* also differ in size, one being appreciably larger than the other (Fig. 95). Forms in which a distinction can be made between the two suspensors are said to be **heterogametangic**.

Hybridisation experiments have been conducted between different species and genera of Mucoraceae. In some cases imperfect zygospores are formed. Attempted copulation has also been observed between homothallic and heterothallic strains (Burgeff, 1924).

An unusual type of mating behaviour has been discovered in *Mucor pusillus* which is predominantly heterothallic but in which homothallic strains are known. It has been possible to induce a $(+)$ strain to mutate to a $(-)$ strain, and also to a homothallic strain by γ-irradiation (Nielsen, 1978).

In some Mucorales, gametangial copulation may fail to take place normally, and one or both gametangia may give rise parthenogenetically to a

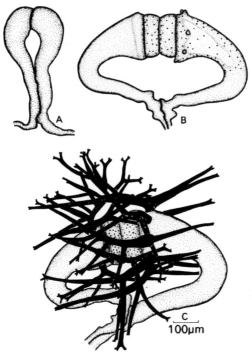

Fig. 94. *Phycomyces blakesleeanus*. A–C. Stages in zygospore formation. The fungus is heterothallic. Note the dichotomous suspensor appendages.

structure morphologically similar to the zygospore, termed an **azygospore**. Azygospores therefore appear as warty spherical structures born on a single suspensor-like cell, or occasionally on a sporangiophore. They are formed regularly in cultures of *Mucor bainieri* and *M. azygospora* (Benjamin & Mehrotra, 1963) and in certain isolates of *M. hiemalis* (Gauger, 1966, 1975). Gauger's azygosporic isolates of *M. hiemalis* were derived from spores from germ-sporangia which developed from germinating normal zygospores. If the azygosporic strains are subcultured, either from single sporangiospores, or by mass transfer, they show a tendency to 'break down' to strains of (+) or (−) mating type of normal appearance. Gauger has suggested that azygosporic strains of *M. hiemalis* are typically diploid, and heterozygous for mating type – i.e. the diploid nucleus carries both (+) and (−) mating type alleles. He has suggested that the breakdown to the normal (+) or (−) mating type condition is brought about by somatic (i.e. non-meiotic) reduction leading to aneuploid intermediates, and finally to haploids.

Notes on the characteristic features of some common genera of the Mucoraceae are given below.

Mucor (Figs. 83, 84)

Cosmopolitan, and widespread in soil and on dung and other organic substrata. The sporangia are globose and borne on branched and unbranched sporangiophores. Most species are heterothallic and isogamous, i.e. the gametangia and suspensors are equal in size. Amongst the most common species from soil are *M. hiemalis, M. racemosus* and *M. spinosus. Mucor mucedo* is common on dung. Schipper (1978) has given a key to all accepted species.

Zygorhynchus (Fig. 95)

Mostly reported from soil, often from considerable depth (Hesseltine *et al.*, 1959). All species are homothallic and have heterogametangic zygospores. The sporangiophores are usually branched and the columella is often broader than high.

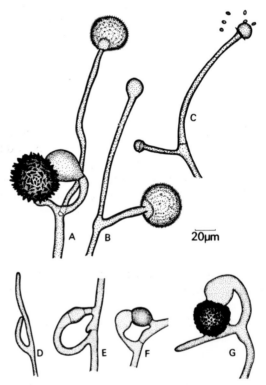

Fig. 95. *Zygorhynchus molleri.* A. Zygospore and sporangium. B. Young sporangiophores. C. Dehisced sporangia. D–G. Stages in zygospore formation. Note that the fungus is homothallic and the suspensors unequal.

Rhizopus (Figs. 87, 92)

Occurs not only in soil but on fruit, other foods, all kinds of decaying materials and as a laboratory contaminant. *Rhizopus stolonifer* is often found on ripe bananas, especially if they are incubated in a moist atmosphere. The characteristic features are the presence of rhizoids at the base of the sporangiophores (which may grow in clusters), and the stoloniferous habit. An aerial hypha grows out and where it touches on the substratum it bears rhizoids and sporangiophores. Growth in this manner is repeated. Most species are heterothallic but *R. sexualis* is homothallic and forms zygospores freely within 2 days in the laboratory.

Absidia (Fig. 86)

The characteristic features are the pear-shaped sporangia produced in partial whorls at intervals along stolon-like branches. These branches produce rhizoids at intervals but not opposite the sporangiophores. The zygospores are surrounded by curved unbranched suspensor appendages which may arise from one or both suspensors. Most species are heterothallic but *A. spinosa* is homothallic. *Absidia glauca, A. orchidis* and *A. spinosa* are amongst the most commonly isolated species.

Phycomyces (Fig. 94)

Benjamin & Hesseltine (1959) recognise three species but the two best-known are *P. nitens* and *P. blakesleeanus*. The spores of *P. nitens* are larger than those of *P. blakesleeanus*, but it is likely that many workers confused the two before Burgeff (1925) pointed out the difference. Much of the literature on *P. nitens* probably refers to *P. blakesleeanus*. Neither species is particularly common; but a likely habitat is on fatty products and on empty oil casks. Bread and dung are other recorded substrata.

Syzygites (Fig. 96)

S. megalocarpus (= *Sporodinia grandis*, Hesseltine, 1957) is found on decaying sporophores of various toadstools, especially *Boletus, Lactarius* and *Russula*. It grows readily in culture and is homothallic (Davis, 1967). The sporangiophores are dichotomous and the sporangia thin-walled.

Pilobolaceae

There are two common genera in this family, *Pilobolus* and *Pilaira*. Both grow on the dung of herbivores occurring early in the succession of fungi which fruit regularly on such substrata. In both genera the sporangium dehisces by a

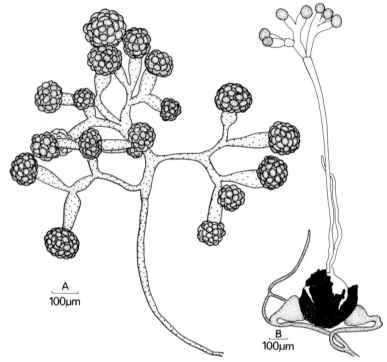

Fig. 96. *Syzygites megalocarpus*. A. Sporangiophore. B. Germinating zygospore. The fungus is homothallic.

transverse crack running around the base, and when this bursts a mucilaginous secretion is exuded so that the spores themselves are not released at this stage. In *Pilobolus* the sporangiophores are swollen and the sporangia are shot away violently by a jet of liquid, whilst in *Pilaira* the sporangiophore is cylindrical and the sporangium becomes converted into a sporangial drop. A monograph of the family has been given by Grove (see Buller, 1934).

Pilobolus (Figs. 97–100)

The generic name means literally the hat-thrower, referring to the sporangial discharge. If fresh horse-dung is brought into the laboratory and incubated in a glass dish on a window-sill, after a preliminary phase of fruiting of *Mucor* lasting about 4–7 days the characteristic bulbous sporangiophores of *Pilobolus* appear (Fig. 97). Common species are *P. kleinii, P. longipes* and *P. crystallinus*. A full account of the development and discharge of the sporangium has been given by Buller (1934). The sporangia of *Pilobolus* becomes attached to the vegetation and are eaten by herbivores: in the gut the spores are released and germinate. In the faeces the spores develop into mycelium

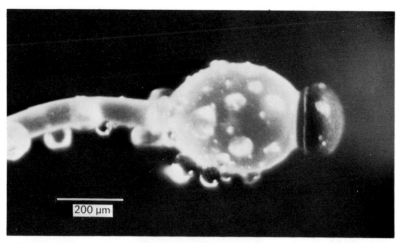

Fig. 97. *Pilobolus* sp. Sporangiophore growing on dung. Note the flattened black sporangium, with a line of dehiscence around its base and the swollen subsporangial vesicle bearing drops of liquid.

which after about 4 days forms at the surface of the dung trophocysts, swollen segments coloured yellow by carotene (Fig. 98). Sporangiophores develop from the trophocysts in a regular daily sequence, and the stage of development can be correlated fairly closely with the time of day. During the late afternoon the sporangiophore grows away from the trophocyst towards the light and during the night its tip enlarges to become the sporangium. The swelling of the subsporangial vesicle takes place mainly between midnight and early morning. Young sporangiophores, even before their sporangia are differentiated, are highly phototropic, and the clear tip of the developing sporangiophore is the sensitive region. Despite the bright yellow colour of the trophocysts and young sporangiophores, due to their carotene content, studies of the light response of young sporangiophores to light of different wavelength suggest that the photoreceptor is more likely to be a flavin than a carotenoid (Page & Curry, 1966).

The fully developed sporangiophores are also highly phototropic. Light projected along the axis of the sporangiophore is brought to a focus at a point beneath the swollen vesicle. In this region is an accumulation of carotene-rich cytoplasm which glows orange when illuminated. When light falls asymmetrically on the sporangiophore it is focused on to the back of the subsporangial vesicle near its base, and some stimulus is probably transmitted to the cylindrical part of the sporangiophore resulting in more rapid growth on the side away from the light. Curvature of the whole sporangiophore thus occurs until it is orientated symmetrically with respect to the light (see Fig. 99). The structure of the sporangium differs in a number of ways from that of the Mucoraceae. The sporangium is flattened, its wall is dark black, shiny, tough and unwet-

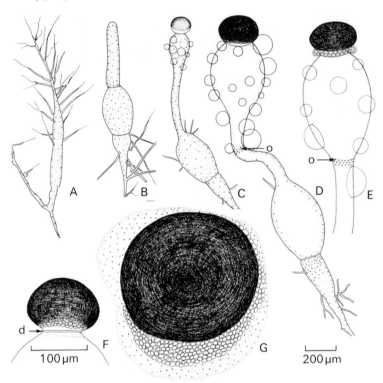

Fig. 98. *Pilobolus kleinii* asexual reproduction. A. Developing trophocyst, becoming distended by carotene-rich cytoplasm. B. Trophocyst with immature sporangiophore. The clear tip of the sporangiophore is light-sensitive. C. Trophocyst bearing developing sporangium: the upper part of the sporangium is beginning to darken. Globules of liquid accumulate on the sporangiophore. 9.00 p.m. D. Trophocyst with sporangium which has not yet dehisced. 9.00 a.m. The arrow (o) points to a carotene-rich band of cytoplasm or ocellus. E. Sporangiophore bearing a sporangium which has dehisced near its base. Spores have extruded and are held in place by a ring of mucilage. 11.30 a.m. F. Sporangium showing dehiscence line at its base (d). G. Discharged sporangium surrounded by dried-out vesicular sap. The spores are enclosed in mucilage. A–E to same scale: F, G to same scale.

table. At the base of the sporangium is a conical columella which is separated from the spores by a pad of mucilage. During the late morning the sporangium cracks open by a suture running around the base, just above the columella. The spores are, however, not released at this time because they are prevented from escaping by the mucilaginous pad which protrudes through the crack in the sporangium wall as a ring of mucilage (Fig. 98E,F).

The subsporangial vesicle is turgid: it contains liquid under pressure. It has been estimated that the osmotic pressure of the liquid is of the order of 5·5 bars. Drops of excreted liquid commonly adhere to the sporangiophore. Eventually, usually about midday, the sporangial vesicle explodes at a line of

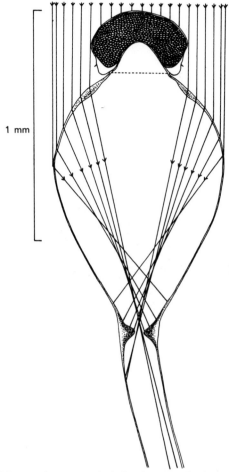

1 mm

Fig. 99. *Pilobolus kleinii*. Diagrammatic L.S. of sporangiophore showing the path of light rays which are brought to a focus beneath the subsporangial vesicle. The sporangiophore illustrated is orientated symmetrically with respect to the incident light. Note the mucilaginous ring extruded at the base of the sporangium (after Buller, 1934).

weakness just beneath the columella. Due to the elasticity of the vesicle wall the liquid contents are squirted out, projecting the entire sporangium forwards in the direction of the light. Photographs of the jet show that it is at first cylindrical, but eventually breaks up into fine droplets (Page, 1964). The velocity of projection varies between wide limits in *P. kleinii*, 4·7–27·5 m/second with a mean of 10·8 m/second (Page & Kennedy, 1964). The sporangia can be projected vertically upwards for as much as 2 m and horizontally for up to 2·5 m. On striking herbage surrounding the dung the sporangia become attached in such a way that the mucilaginous ring adheres, with the black sporangium wall facing outwards. Buller has suggested that the reason why

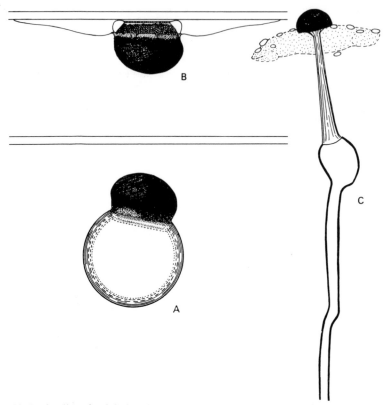

Fig. 100. Projectiles of *Pilobolus* (diagrammatic). A. Sporangium with adherent drop of sporangiophore sap about to strike an obstacle. B. Sporangium after striking the obstacle. The sporangiophore sap has flowed round the sporangium which has turned outwards so that the mucilage ring adheres to the surface of the obstacle (after Buller, 1934). C. Sporangiophore projecting a sporangium. Note the jet of liquid and the bending of the narrow base of the sporangiophore under the recoil of the discharge (after Page, 1964).

the *Pilobolus* sporangium adheres in this way is that the projectile consists of a drop of liquid attached to the sporangium (Fig. 100A). When the projectile strikes an object the liquid flows around the sporangium, but because the sporangium wall is unwettable and because the base of the sporangium is surrounded by the wettable mucilaginous ring, the sporangium turns round in the drop of liquid so that its wall faces outwards (Fig. 100B). The non-wettable nature of the sporangial wall may be related to the presence on its surface of hollow, blunt-tipped spines as seen by electron microscopy. Crystals of unidentified composition may be grouped around the spines (Bland & Charles, 1972). As the drop of liquid dries, the mucilage becomes more firmly attached, and dried sporangia are extremely difficult to detach from vegetation. The spores of *Pilobolus* are therefore not released at the time of sporangial discharge, but only after the sporangia have been eaten by an animal, when they are released into the gut.

So far as is known, all members of the Pilobolaceae are heterothallic (Hesseltine & Ellis, 1973).

The physiology of *Pilobolus* shows a number of interesting features possibly correlated with its coprophilous habit. The spores germinate best above pH 6·5, and can be induced to germinate by treatment with alkaline pancreatin. Mycelial growth is best above pH 7·0. Growth on synthetic media is stimulated by the addition of thiazole, hemin and coprogen, an organo-iron compound produced by various fungi and bacteria. Ammonia stimulates sporangial production and in dual culture *Mucor plumbeus* can release sufficient ammonia to enhance sporulation (Page 1952, 1959, 1960; Hesseltine *et al.*, 1953; Pidacks *et al.*, 1953; Lyr, 1953).

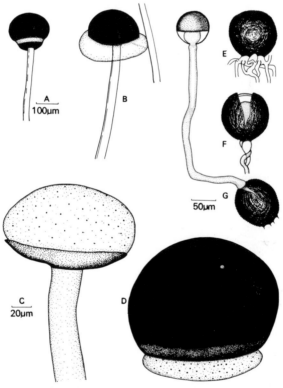

Fig. 101. *Pilaira anomala*. A. Sporangiophore from rabbit dung showing the rupture at the base of the sporangium. B. Sporangium with extruded mucilage ring adhering to an adjacent hypha. C. Columella after sporangium has been detached. D. Detached sporangium showing basal mucilage ring. E. Zygospore. F, G. Stages in zygospore germination (E, F, G after Brefeld, 1881).

Pilaira (Fig. 101)

Pilaira anomala is found on dung of various herbivorous animals such as horse and rabbit. The structure of the sporangium closely resembles that of *Pilobolus*, in that the spores are separated from the columella by a mucilaginous ring which extrudes from the base of the sporangium. There is, however, no subsporangial vesicle and sporangial release is non-violent. The sporangiophores are phototropic, and when they are mature, they elongate quite rapidly (H. J. Fletcher, 1969, 1973). Their development essentially resembles that of *Phycomyces*. The mucilaginous ring, on making contact with adjacent herbage, becomes firmly attached to it, and the sporangium is detached from the columella. In a moist atmosphere, the mucilaginous ring may absorb water and swell considerably so that a large sporangial drop is formed (Ingold & Zoberi, 1963).

P. anomala forms zygospores resembling those of *Pilobolus*. On germination, a sporangium is produced.

Thamnidiaceae

In this family two kinds of asexual reproductive structure are found; columellate sporangia of the *Mucor* type, and smaller, usually non-columellate sporangia, termed sporangiola, often borne in whorls or at the tips of branches. The branches bearing the sporangiola may be borne laterally on the columellate sporangiophores or may arise separately. In some cases the branch system bearing the sporangiola is terminated by a spine. There are relatively few spores in the sporangiola and in *Chaetocladium* there may be a single spore only. Hesseltine & Ellis (1973) recognise seven genera, but we shall consider only *Thamnidium*, *Helicostylum* and *Chaetocladium*.

Thamnidium (Fig. 102)

The commonest species is *T. elegans* which grows on dung, in soil and has been reported from meat in cold storage. In culture large terminal columellate sporangia are produced with dichotomous lateral branches bearing fewer-spored non-columellate sporangiola. The sporangiola may also be borne on separate branch systems. Low temperature and light induce the formation of sporangia as opposed to sporangiola. During the development of the sporangiophores, spiral growth occurs as in *Phycomyces* (Lythgoe, 1961, 1962). Electron microscope studies of the development of sporangia and sporangiola show that they develop in essentially the same way (J. Fletcher, 1973a,b). At maturity, the columellate sporangia become converted into sticky sporangial drops not easily detached by wind currents or mechanical agitation. The sporangiola are, however, easily detached by such treatment and in wind tunnel experiments become attached to slide traps, whilst the sporangia do not. Change from damp to dry air leads to an increase in sporangiole liberation (Zoberi, 1961; Ingold & Zoberi, 1963). *Thamnidium elegans* is heterothallic and forms zygospores resembling those of *Mucor* or *Rhizopus*, but they

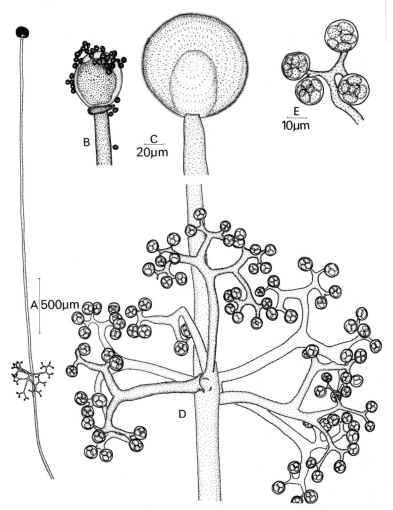

Fig. 102. *Thamnidium elegans.* A. Sporangiophore showing terminal sporangium and lateral branches bearing sporangiola. B. Dehisced terminal sporangium showing columella and spores. C. Immature terminal sporangium showing columella. D. Branches bearing sporangiola. E. Sporangiola. Note the absence of a columella. B, C, D to same scale.

are produced best at low temperatures such as 6–7 °C and not at 20 °C (Hesseltine & Anderson, 1956).

Helicostylum (Fig. 103)

In this genus, some of the sporangiolum-bearing branches end in terminal

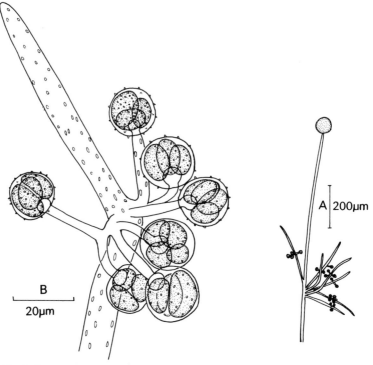

Fig. 103. *Helicostylum fresenii.* A. Habit sketch to show terminal sporangium and lateral sporangioles. B. Branch ending in a terminal spine and bearing lateral sporangiola. Note that the sporangiola are columellate.

spines, but a *Mucor*-type sporangium is borne at the tip of the sporangiophore. *H. fresenii* illustrates this situation well. This fungus, which has been isolated from frozen meat, was earlier known as *Chaetostylum fresenii* (Hesseltine & Anderson, 1957; Lythgoe, 1958). In the wind tunnel, the sporangia and sporangiola behave as in *T. elegans.*

Chaetocladium (Fig. 104)

In *Chaetocladium,* there are no *Mucor*-like sporangia. Sporangiola, each containing a single spore, are borne on branches which end in spines. Such monosporous sporangiola are sometimes termed conidia. However, as pointed out by Dykstra (1974), a distinction can be made between sporangiospores, monosporous sporangiola and conidia in Mucorales (see Fig. 105). Sporangiospores, at the onset of germination, synthesise a new wall layer inside the primary wall. The new wall layer is continuous with the wall of the germ tube. The spore wall may fracture on germination (Fig. 105C,D). The spore contained within a monosporous sporangiolum behaves in essentially

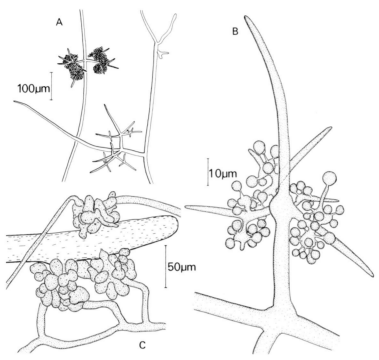

Fig. 104. *Chaetocladium brefeldii.* A. Habit sketch to show branches ending in spines and bearing lateral sporangiola. B. Branch showing spine and sporangiola. C. Hypha of *Pilaira anomala* bearing bladder-like outgrowths following parasitism by *Chaetocladium.*

the same way, and synthesises a new wall at the onset of germination. During germination, the sporangiolum wall either fractures in advance of the fracture of the spore wall (Fig. 105G) or, in some cases, is separated from the spore as a unit. The germ tube is continuous with the newly synthesised wall. During the first stage of germination of conidia, the wall expands, but the outer layers do not fracture as in germinating sporangiospores, and the only fracture is at the point of germ tube emergence (Fig. 105J,K). In *Chaetocladium,* the distinction between the wall of the spore and the sporangiolum is sometimes visible on germination. There are two common species, *C. jonesii* and *C. brefeldii,* both parasitic on other members of the Mucorales. They often occur on *Mucor* or *Pilaira* on dung. At the point of attachment to the host, there are numerous bladder-like outgrowths, which contain nuclei of both the host and the parasite (Burgeff, 1924). The parasites can, however, be cultured in the absence of a host (Hesseltine & Anderson, 1957). Both species are heterothallic. *C. brefeldii* is heterogametangic, forming zygospores resembling *Zygorhynchus.* Burgeff (1920, 1924) has claimed that a given strain of *Chaetocladium* can only parasitise one of the two mating type strains of heterothallic

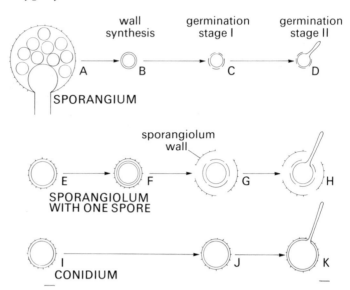

Fig. 105. Diagrammatic representation of morphological changes occurring during germination in a sporangiospore, a single-spored sporangiolum and a conidium (after Dykstra, 1974). The single lines in the diagram each represent a distinct wall layer.

Mucors. He has made the interesting suggestion that the parasitic habit of fungi such as *Chaetocladium* may have originated from attempted copulation with other members of the Mucorales.

Choanephoraceae

In this family, as in the Thamnidiaceae, both sporangia and sporangiola occur. The sporangia are usually columellate and often hang downwards. They contain dark brown sporangiospores with a striate epispore and bristle-like appendages. The sporangiola contain one to a few spores of similar construction. The zygospores also have striate walls. Three genera are recognised, *Choanephora*, *Blakeslea* and *Gilbertella*. *Blakeslea* is treated by some authorities as a synonym of *Choanephora* (Hesseltine, 1953, 1960; Poitras, 1955; Hesseltine & Benjamin, 1957; Hesseltine & Ellis, 1973).

Choanephora and *Blakeslea* (Figs. 106, 107)

Species of *Choanephora* are found in warmer soils. *C. cucurbitarum* causes a rot of cucumbers and related fruits and is also commonly isolated from decaying flowers of various kinds. *Blakeslea trispora*, which has been isolated from cowpeas, tobacco and cucumber leaves, forms in culture two kinds of asexual reproductive structure: nodding columellate or non-columellate sporangia with brown, faintly striate spores which usually bear bristle-like appen-

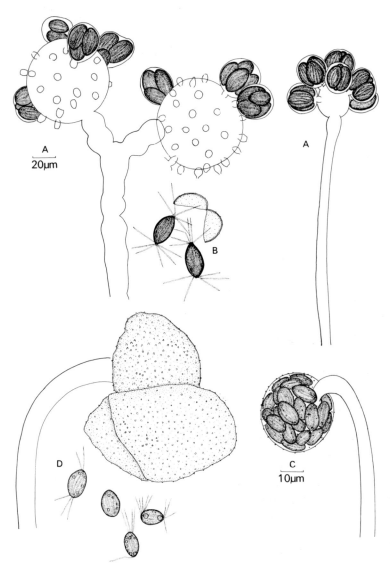

Fig. 106. *Blakeslea trispora.* A. Sporangiophores with globose vesicles bearing sporangiola containing three or four spores. B. A dehisced sporangiolum showing two spores. Note the striate epispore and mucilaginous appendages, and that the sporangiolum splits into two halves. C. Sporangiophore bearing a drooping sporangium. No columella was observed but columellate sporangia have been described. D. Dehisced sporangium also lacking a columella. Note the split sporangial wall and the sporangiospores with striate epispore and mucilaginous appendages.

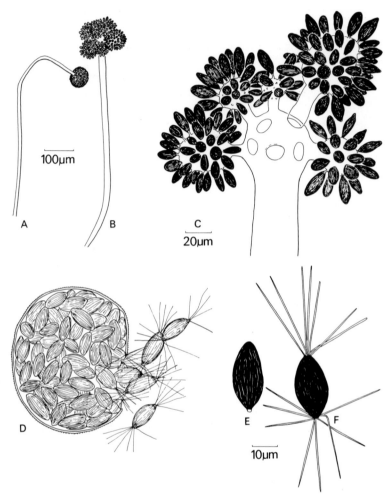

Fig. 107. *Choanephora cucurbitarum.* A. Sporangiophore with drooping sporangium. B. Conidiophore and conidia. C. Apex of conidiophore showing vesicles and conidia. D. Dehisced sporangium showing striate spores with appendages. E. Conidium. F. Sporangiospore. C and D to same scale.

dages, and non-columellate sporangioles borne in large numbers on globose vesicles. The sporangioles contain from two to five, but typically three, distinctly striate, dark brown spores also with bristle-like appendages. In *C. cucurbitarum*, two similar structures can be recognised, but here the sporangiole contains only a single spore (Poitras, 1955), showing the characteristic striations. Such monosporous sporangiola are sometimes termed conidia. The sporangiola of *B. trispora* are readily detached by wind and break open in water like the two halves of a bivalve shell to release the spores. The function of the appendages is not known, but the 'conidia' are carried by insects from one plant to another. In the related genus *Gilbertella,* there are no sporangioles, but the sporangia split open to release the striate appendaged spores. *Gilbertella persicaria* is a parasite of peaches and tomatoes.

The Choanephoraceae have been the subject of physiological investigations (Hesseltine, 1961). In *C. cucurbitarum,* it has been shown that the optimum temperature for sporangiolum production is near 25 °C. A temperature of 31 °C is unfavourable for sporangiolum formation but stimulates the production of large sporangia. High relative humidity inhibits conidial formation but stimulates sporangial formation at 20 °C. All species studied are heterothallic, and interspecific crosses may also result in the formation of zygospores. An interesting phenomenon in connection with intra- and interspecific crosses is that the production of β-carotene is markedly enhanced when $(+)$ and $(-)$ strains are mated on liquid media, as compared with production from either strain grown singly. This method of producing β-carotene may prove to be commercially valuable. The discovery that carotene production could be stimulated by an acid fraction of culture filtrates from mixed cultures of *B. trispora* led to the discovery of trisporic acid as the sex hormone of Mucorales.

Piptocephalidaceae

A characteristic feature of this family is that asexual reproduction occurs by means of cylindrical sporangia containing typically a single row of sporangiospores. Such sporangia are termed **merosporangia** and are formed in groups on inflated vesicles or at the tips of dichotomous branches (Benjamin, 1966). Benjamin (1959) and Hesseltine & Ellis (1973) include only *Piptocephalis* and *Syncephalis,* parasites of other fungi, in this family. The saprophytic *Syncephalastrum* is sometimes classified in a separate family, the Syncephastraceae.

Syncephalastrum (Fig. 108)

S. racemosum grows in soil and on dung. In culture its growth is similar to *Mucor,* and it forms aerial branches terminating in club-shaped or spherical enlargements known as vesicles. The vesicles are multinucleate and bud out all over their surface to form cylindrical outgrowths, the merosporangial primordia. Into these outgrowths one or perhaps several nuclei pass, and nuclear division continues. The cytoplasm in the merosporangium cleaves into a single row of five to ten sporangiospores each with one to three nuclei. The cleavage process is substantially similar to that of other Mucorales (Fletcher, 1972). The sporangial wall shrinks at maturity so that the spores appear in chains reminiscent of an *Aspergillus.* Occasionally the merospores may lie in more than a single row. The spore heads remain dry and entire rows of spores are detached by wind (Ingold & Zoberi, 1963). *Syncephalastrum racemosum* is heterothallic and forms zygospores resembling those of the Mucoraceae.

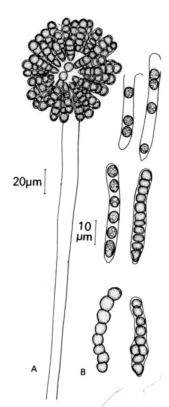

20μm

10 μm

A

B

Fig. 108. *Syncephalastrum racemosum*. A. Sporangiophore bearing a vesicle and numerous merosporangia. B. Merosporangia and merospores.

Piptocephalis (Fig. 109)

Most species of *Piptocephalis* parasitise the mycelium of other members of the Mucorales, but *P. xenophila* attacks various Ascomycetes and Fungi Imperfecti. A characteristic habitat for *P. freseniana* is dung at the end of the phase of fruiting of *Mucor*. If situated near to the hypha of a suitable host the cylindrical spore of *Piptocephalis* germinates laterally (not from the ends) forming one or more germ tubes which make contact with the host hypha and form enlarged appressoria (Fig. 109D). Beneath the appressorium fine haustoria develop. The mycelium of the parasite then develops externally to the host and forms stolon-like branches which develop further appressoria and haustoria. Merosporangia are formed at the tips of dichotomously branched aerial hyphae, arising from specialised swellings or head-cells. The merospores in *Piptocephalis* usually contain one or occasionally two nuclei. At maturity the merosporangia behave in two distinct ways. In *Piptocephalis virginiana* the spore chains remain dry and entire chains as well as head-cells may be dispersed by air currents, although single spores are also dispersed. In

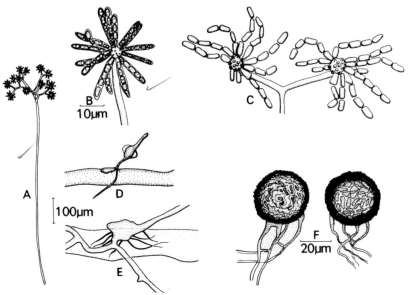

Fig. 109. *Piptocephalis virginiana*. A. Habit sketch to show dichotomous sporangiophore. B. Head cell and intact merosporangia. C. Head cells showing breakdown of merosporangia to form chains of spores. D. Spore germination and formation of appressorium on a host hypha. E. Appressorium and branched haustorium on host hypha. The parasite mycelium is branched and extending to other host hyphae. F. Zygospores. The fungus is homothallic.

P. freseniana all the spore chains in any one head become involved in the formation of a sticky spore drop, and the entire spore drop is dispersed by wind (Ingold & Zoberi, 1963). Most species are homothallic (Leadbeater & Mercer, 1956, 1957a,b; Benjamin, 1959). Zygospores are usually formed within the agar in cultures. The mature zygospore is a spherical dark brown sculptured globose cell held between two tong-shaped suspensors.

Cultures of *Piptocephalis* in the absence of a host grow very poorly (Manocha, 1975). On suitable media spores will germinate and give rise to a limited mycelium producing dwarf sporophores. The spores so formed do not germinate in the absence of a host but can infect a suitable fungus. There is evidence that spore germination and chemotropic growth of the germ tubes are stimulated by fungal secretions (Berry & Barnett, 1957). The effects of *Piptocephalis* spp. on the growth rate of their hosts are variable (Curtis *et al.*, 1978). In some cases, the rate of growth of dual cultures was not significantly different from that of uninfected controls, in others it was reduced, whilst in others it was enhanced. These effects are temperature dependent.

Cunninghamellaceae

In this family, asexual reproduction is entirely by means of conidia: sporangia and sporangiola are not formed. For this reason, the genera included here have been separated from the Choanephoraceae with which they were formerly classified. Hesseltine & Ellis (1973) recognise four genera, but we shall study only *Cunninghamella*.

Cunninghamella (Fig. 110)

Species of *Cunninghamella* are found in soil in the warmer regions of the world. The asexual conidia are hyaline and clustered on globose vesicles on branched or unbranched conidiophores. Although the conidia are possibly to be interpreted as one-spored sporangiola, there is no evidence of this from their structure. Although the original wall of the conidium is two-layered, there is no evidence of synthesis of a new wall layer at or before germination, and the germ tube is continuous with a wall layer whose appearance is consistent with its being a chemically changed part of the original wall rather than an entirely new wall formed *de novo* (Hawker *et al.*, 1970; Dykstra, 1974). In some species, e.g. *C. echinulata* and *C. elegans*, the conidia are spiny, but in others they are smooth. Zygospores are of the *Mucor* type.

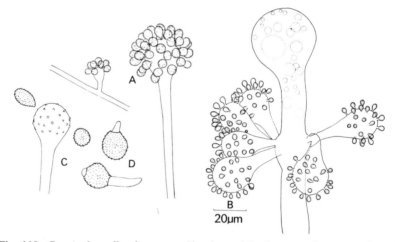

Fig. 110. *Cunninghamella elegans*. A. Simple conidiophores. B. Immature branched conidiophore showing developing conidia. C. Mature conidiophore showing scars of attachment of conidia. D. Germinating conidia, 5 hours.

Mortierellaceae

The characteristic feature of this family is that the sporangia lack columellae. In some species, stalked globose detachable one-celled spores, or **stylospores,** occur, possibly representing modified sporangia. In the most frequently encountered genus, *Mortierella,* zygospores are commonly heterogametangic and may be naked or enclosed in a weft of mycelium. Hesseltine & Ellis (1973) recognise four genera. The family has been monographed by Linnemann in Zycha *et al.* (1969).

Mortierella (Figs. 111, 112)

About eighty species of *Mortierella* are known, occurring widely in soil and on plant and animal remains in contact with soil. These fungi can be isolated readily on nutrient-poor media which prevent the growth of more vigorous moulds. Some species, e.g. *M. wolfii*, are associated with mycotic abortion in cattle, and can be isolated from the placenta and foetal stomach contents (Ainsworth & Austwick, 1973). The mycelium is fine and often shows a characteristic series of fan-like zones. The sporangia are borne on branched or unbranched, usually tapering, sporangiophores. The sporangium wall is deli-

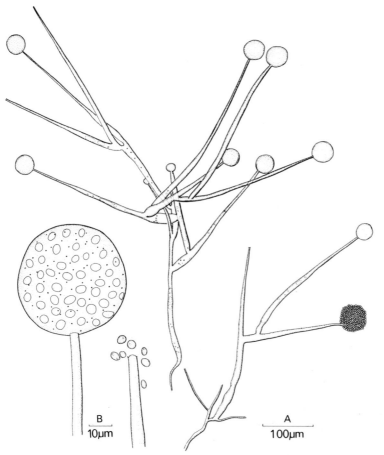

Fig. 111. *Mortierella* sp. A. Branched sporangiophores. B. Intact sporangium and tip of sporangiophore after sporangium dehiscence. Note the absence of a columella.

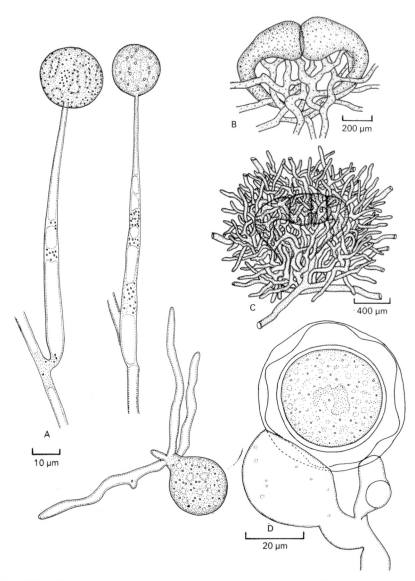

Fig. 112. A. *Mortierella zonata* stylospores, one germinating. B, C. *Mortierella rostafins-kii* (after Brefeld, 1876). B. Developing zygospore. C. Older zygospore surrounded by weft of hyphae. D. *Mortierella epigama* zygospore showing unequal suspensors arising from a common branch.

cate and may collapse around the spores. There is no columella. Frequently the entire sporangium is detached. In a number of species, there may be only two or three spores per sporangium. Some species bear stalked globose unicellular stylospores which are easily detached and can germinate to form new mycelium, and in some species, e.g. *M. stylospora* and *M. zonata*, only stylospores are present; true sporangia are lacking.

The zygospores of some Mortierellas may be surrounded by an investment of sterile hyphae (see Fig. 112B,C). Whilst some species are homothallic, *M. parvispora* is heterothallic and heterogametangic, and its zygospores are not surrounded by sterile hyphae. *Mortierella marburgensis* is also heterothallic. Here the suspensors are at first equal, but later one increases in size whilst the other loses most of its cytoplasm (Gams & Williams, 1963; Williams *et al.*, 1965; Gams *et al.*, 1972; Kuhlman, 1972, 1975; Chien *et al.*, 1974).

Endogonaceae

This family is included in the Mucorales because of the presence of zygospores following conjugation. The zygospores may occur in clusters, surrounded by wefts of sterile hyphae, and such structures are termed **sporocarps**. Sporocarps are also known which contain azygospores or chlamydospores, which resemble zygospores but are not formed as a result of conjugation. Sporocarps vary in size from about 1 mm in diameter (e.g. in *Glomus mosseae*, Fig. 113A,B) to 20 mm or more in some species of *Endogone*.

The family is of great interest because it contains mycorrhizal associates of a wide range of herbaceous and woody hosts, from all groups of vascular plants, including many crop plants of economic importance (Mosse, 1973; Sanders *et al.*, 1976). The earliest known vascular plants, the Rhynie fossils or Psilophytales, of Devonian age, some 400 million years ago had mycorrhizal infections of their roots which are similar to modern infections caused by Endogonaceae.

The taxonomy of the group presents difficulties, because early workers failed to appreciate the distinction between zygosporic, azygosporic and chlamydosporic forms. However, Gerdemann & Trappe (1974) have separated seven genera. The genus *Endogone* is restricted to forms with true zygospores. True zygospores are distinguished by the fact that they are attached to two suspensors, whilst chlamydospores are subtended by a single hypha (Fig. 113B). Many of the experiments on the effects of mycorrhizal infection have been carried out using *Glomus*, a genus in which chlamydospores occur freely in soil, or in solid sporocarps. In other genera such as *Glaziella*, the sporocarps are hollow. Asexual reproduction by means of sporangia is uncommon, but is found in *Modicella*. Growth of most Endogonaceae in pure culture has not yet proved possible, but it is possible to extract sporocarps and chlamydospores from soil by wet-sieving and surface-sterilising them. Surface-sterilised spores can be used to infect aseptically germinated seedlings grown in sterile soil. The

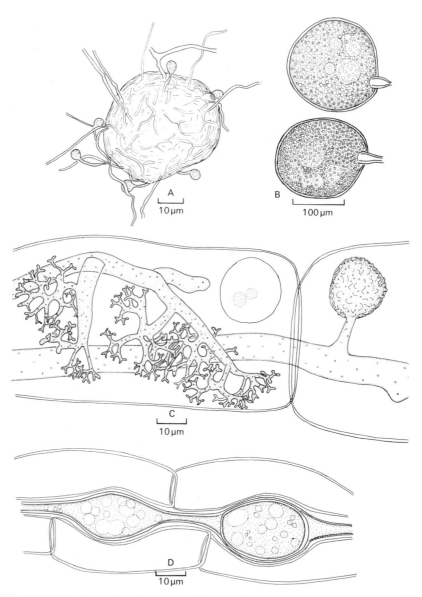

Fig. 113. Vesicular–arbuscular mycorrhiza. A. *Glomus mosseae* sporocarp: note the presence of naked chlamydospores on the external hyphae. B. Chlamydospores dissected from the interior of a sporocarp: note that they are borne on a single subtending hyphae. C. *Glomus mosseae* infection of onion roots. The cell to the left contains a nucleus with two nucleoli and a branched haustorium or arbuscule. In the cell to the right, the arbuscule has degenerated. D. Vesicles from roots of *Arum maculatum*.

chlamydospores germinate in close proximity to roots, and develop germ tubes which penetrate the epidermal cells. Two distinct types of mycorrhizal association involving Endogonaceae are known, sheathing mycorrhizas, and endotrophic mycorrhizas of the vesicular–arbuscular type (see below). Sheathing mycorrhizas of *Pinus* and *Pseudotsuga* have been found to be associated with *Endogone lactiflua,* a sporocarpic form with true zygospores (Fassi *et al.*, 1969). This is, however, rather unusual because sheathing mycorrhizas are normally associated with Basidiomycete infections (Harley, 1969). The vesicular–arbuscular type of mycorrhizal association, associated with infections by *Glomus, Gigaspora, Acaulospora* and other genera is an extremely common type of root infection. Infected root systems bear coarse angular hyphae projecting into the soil. On this extramatrical mycelium, chlamydospores or sporocarps may be attached. The chlamydospores are usually large, often more than 150 μm in diameter, so that, by selecting sieves of appropriate mesh, it is possible to sieve them out from soil. They are often balloon-shaped, thick-walled, and range in colour from white to yellow, honey-coloured to brown (Mosse & Bowen, 1968). Within the root, coarse non-septate hyphae grow between the cells, penetrating the cells by richly branched haustoria (Fig. 113C). The plasmalemma of the host cell is folded round the haustorium, so that the surface area of the interface between the fungus and its host is large (Cox & Sanders, 1974). The branched haustoria have been termed **arbuscules**. Later, the fine tips of the branches may undergo digestion to form irregular globose bodies to which the term sporangiole has been, unfortunately, misapplied (Fig. 113C). At intervals within the root system, either inside or between the cells, large terminal or intercalary thick-walled **vesicles** may develop (Fig. 113D).

Comparisons between the dry weights of plants infected with vesicular–arbuscular mycorrhizas and uninfected plants have shown that, in soils low in available phosphate, infection results in more rapid growth and increased phosphate uptake. The increased uptake is not due to the fact that the fungus has access to a source of phosphorus unavailable to the root, but rather that the fungus can explore a larger volume of soil. The level of phosphate in soil is often limiting to plant growth, and its rate of diffusion in soil is far lower than the rate at which it is translocated along the fungal hyphae.

The absence of asexual reproduction in most Endogonaceae, coupled with their widespread distribution in soil, is at first sight puzzling. However, it has been found that the guts of rodents often contain the characteristic spores, and it is possible that rodents bring about dispersal (Silver-Dowding, 1955). The spores of *Glomus macrocarpus* have been germinated after passage through a rodent gut (Trappe & Maser, 1976).

ENTOMOPHTHORALES

Many Entomophthorales are parasites of insects and other animals, including man, whilst some parasitise desmids or fern prothalli. Others are saprophytic in dung or soil. The cells are uninucleate or coenocytic, with chitinous walls. In culture or in their insect hosts, their mycelia tend to break up into segments (hyphal bodies). Asexual reproduction is by means of forcibly discharged uninucleate or multinucleate conidia, and on germination such conidia may sometimes develop secondary conidia. Sexual reproduction is by isogamous or anisogamous conjugation between multinucleate gametangia, to give a thick-walled zygospore. Azygospores may also be formed without conjugation.

Ideas on the classification of the group vary. Martin (1961) separated three families, but Waterhouse (1973c) recognises only one, Entomophthoraceae, with six genera. The Zoopagaceae of Martin are now treated as a separate order, Zoopagales. This order, which we shall not consider further, includes parasites of soil amoebae and nematodes, which reproduce by non-violently discharged conidia (see Duddington, 1973).

Basidiobolus (Figs. 114–116)

If a frog is captured and placed in a jar with a little water the dung can be filtered off. When the damp filter paper is placed in the lid of a Petri dish containing a suitable medium (e.g. 1% peptone agar) conidia of *B. ranarum* will be projected on to the agar surface and within a few days the coarse septate mycelium will become visible. *Basidiobolus* is present in the gut of frogs in the form of spherical uninucleate cells up to 20µm in diameter (Levisohn, 1927). These cells are voided with the faeces and under dry conditions are capable of surviving for several months. Under moist warm conditions the spherical cells germinate to give a coarse septate mycelium from which conidiophores develop. The cytoplasm in the mycelium moves towards the hyphal apex so that only a few terminal segments contain cytoplasm whilst the older segments are empty. The cytoplasm-filled mycelial segments are termed hyphal bodies. The conidiophores are phototropic and resemble the sporangiophores of *Pilobolus*, but bear a colourless pear-shaped conidium. A conical columella projects into the conidium. Beneath is a swollen subconidial vesicle containing liquid under pressure. A line of weakness can be detected around the base of the vesicle, and when this ruptures the conidium and vesicle fly forwards for a distance of 1–2 cm. The elastic upper portion of the vesicle contracts and the sap within it squirts out backwards, so that the projectile behaves as a minute rocket. During its flight the conidium and the rocket motor (i.e. the vesicle) are often separated although the two parts may also remain attached. Such conidia may germinate to form secondary conidia of the same type, or may germinate directly to form a septate mycelium. The conidia are not very resistant to drying, and they are eaten by beetles. Within the gut of the beetle the conidium remains unchanged, but when beetles which have eaten *Basidiobolus* conidia are ingested by frogs the conidia are released,

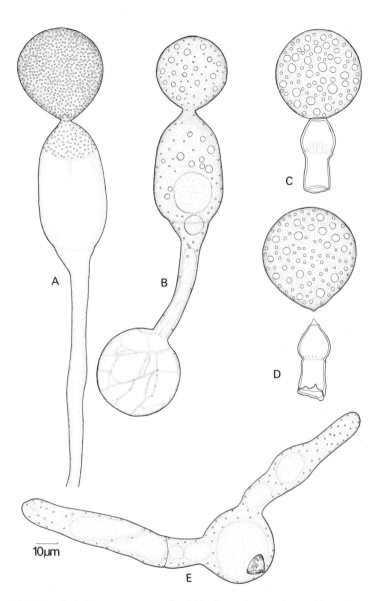

Fig. 114. *Basidiobolus ranarum*. A. Conidiophore from culture. Note the conical columella and the swollen vesicle with the line of weakness around its base. B. Conidium germinating to produce a secondary conidiophore. C. Discharged conidium with remnant of vesicle attached. D. Discharged conidium separated from the remnant of the vesicle. E. Conidium germinating to form a septate mycelium.

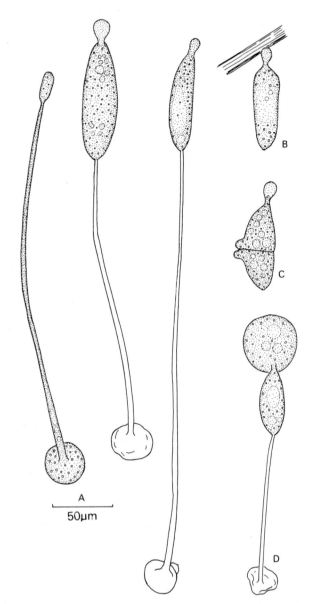

Fig. 115. *Basidiobolus ranarum*. A. Germination of primary conidia to form slender conidiophores bearing secondary, non-propulsive adhesive conidia. B. Attachment of secondary conidium to a hair by a terminal adhesive pad. C. Septate secondary conidium, germinating. D. Secondary conidium enlarging to form a globose conidium.

following digestion, into the gut of the frog. The conidial contents divide into numerous spherical cells and it is from these cells that the mycelium develops later. There is no evidence that the fungus harms either the frog or the beetle. A second type of asexual reproduction has been reported by Drechsler (1956). The globose conidia on germination may give rise to germ tubes, to further globose conidia of the same type or to elongate sausage-shaped uninucleate or binucleate secondary conidia with an adhesive pad at the distal end (Fig. 115). The contents of the secondary adhesive conidia may divide by transverse and longitudinal division to form a multicellular structure. These bodies have been found attached to the bristles of mites, and it is possible that frogs ingest secondary conidia.

In culture it has been found that light, especially blue light of wavelength 440–480 nm, stimulates conidial development and discharge. The effect of light is to stimulate aerial growth from hyphal bodies within the medium, and the aerial hyphae which develop in the light become modified as conidiophores (Callaghan, 1969a,b). The fate of discharged conidia in culture is largely determined by the nutrient level, light and the pH of the substratum on which they land. At low nutrient levels, e.g. water agar, most conidia germinate by producing a secondary conidiophore which discharges a further conidium. On nutrient-rich agar, the majority of conidia give rise to vegetative mycelia (Callaghan, 1969c; 1974).

Zygospores are formed following conjugation. Development can readily be followed in 4- to 5-day-old cultures derived from a single conidium. Zygospore development appears to occur most readily in the dark, and under these conditions the hyphal bodies become bicellular, prior to developing into zygospores (Callaghan, 1969b). On either side of a septum, beak-like projections develop. The tips of these branches become cut off by septa, and the subterminal cells beneath them fuse to form the zygospore which has a wrinkled thick wall when mature (Fig. 116). The cytology of zygospore formation has been described by several workers (see Woycicki, 1927). The vegetative cells of *Basidiobolus* are uninucleate and a single nucleus migrates into each beak-like projection. There each nucleus divides mitotically. One daughter nucleus is cut off by a septum in the terminal cell of the beak, and later disintegrates. The other nucleus resulting from division migrates back into the parent cell. Following this one of the parent cells enlarges to several times the volume of the adjacent cell and a pore is formed connecting the two cells through the original septum separating them. The nucleus from the smaller cell passes through the pore and lies close to the nucleus of the larger cell. Nuclear fusion may occur directly or after a further division, following which one daughter nucleus from each pair fuses, whilst the others disintegrate. Reduction division occurs within the mature zygospore to give four haploid nuclei, of which three usually degenerate. On germination the zygospore may form a germ tube, or may form a conidiophore terminated by a globose conidium (Drechsler, 1956). *Basidiobolus ranarum* is somewhat un-

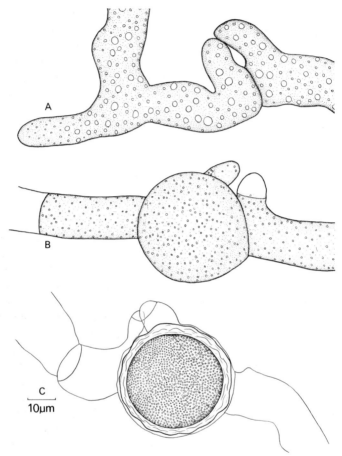

Fig. 116. *Basidiobolus ranarum.* A–C. Successive stages in the formation of zygospores. A. Progametangia. B. Young zygospore. C. Mature zygospore.

usual in that its mycelium becomes segmented into uninucleate segments. Its nuclei are also unusually large, up to 25 μm, and this fact has led to several investigations of its cytology (e.g. Robinow, 1963; Tanaka, 1970; Sun & Bowen, 1972). The number of chromosomes has been estimated to be as high as 900, and it has been suggested that the nucleus may be polyploid.

There are occasional reports of the isolation of *Basidiobolus* from man and domestic animals. The isolates pathogenic to man have been identified as *B. meristosporus* (Greer & Friedman, 1966). Benjamin (1962) has given a key to the species.

Entomophthora (Figs. 117–121)

The name *Entomophthora* means insect-destroyer. About 150 species are known, mostly as insect parasites (MacLeod, 1963; MacLeod & Müller-Kögler, 1973; Waterhouse, 1975). Whilst some appear to be rather specific in their host range, others are capable of attacking a wide range of hosts. A number of species have been grown in culture. One of the easiest to maintain is *E. coronata,* a fungus which has been referred to under various names, e.g. *Delacroixia coronata, Conidiobolus villosus, C. coronata* (see Srinivasan & Thirumalachar, 1964; Tyrell & MacLeod, 1972; Waterhouse, 1973c). This fungus is a parasite of aphids and termites, but is also a common saprophyte of plant debris. Infection of termites can occur by penetration of germ tubes through the exoskeleton, or via the oesophagus following ingestion of germinated conidia (Yendol & Paschke, 1965). It has also been isolated from nasal granulomatous lesions in horses (Emmons & Bridges, 1961). Mycotoxins, possibly proteinaceous, have been isolated from culture filtrates of *E. coronata* (Prasertphon & Tanada, 1961). The toxins induce intoxication, damage to blood cells or early death in several insects when injected into the haemocoel. In culture, the fungus grows rapidly, forming a septate mycelium which, within 2–3 days, forms numerous phototropic conidiophores (Fig. 117), which shoot off conidia onto the lid of the Petri-dish. Conidial discharge takes place both in the light and in the dark, but is enhanced by light (Callaghan, 1969a). The terminal conidium is separated from the conidiophore by a hemispherical columella which bulges into the spore. The columella is double-walled and the spore is violently projected up to 4 cm from the tip of the conidiophore by the pushing outwards, due to turgor, of the wall of the conidium which was at first folded inwards over the columella. After discharge, this part of the spore wall can be seen projecting outwards as a conical papilla. The behaviour of the conidia on germination depends on pH, availability of light, and nutrients. If the conidium falls on a medium containing readily available nutrients, it germinates by means of a germ tube, but on nutrient-poor media such as water agar, it may develop into a secondary conidiophore, forming a slightly smaller conidium. The secondary conidiophore develops from the illuminated side of a primary conidium, and the conidiophore which develops is phototropically orientated, but not very precisely (Page & Humber, 1973). In addition to the smooth globose conidia, more pointed spiny conidia are also formed and appear to be projected in a similar way. They may occasionally arise as secondary conidia on germination of the smooth conidia. The precise conditions under which spiny conidia are formed are not known, nor is it known whether they are physiologically different from the smooth conidia, although Martin (1925) has called them resting spores. Structures resembling zygospores have also been described (Kevorkian, 1937; Emmons & Bridges, 1961). *Entomophthora coronata* can grow well in a purely synthetic medium containing mineral salts, arginine hydrochloride and glucose, but the introduction of peptone results in more rapid growth. It appears to be autotrophic for vitamins.

Another well-known species of *Entomophthora* is *E. muscae.* This fungus is

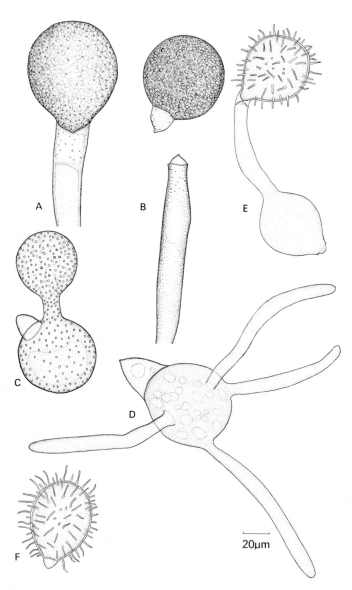

Fig. 117. *Entomophthora coronata*. A. Conidiophore with attached conidium. B. Apex of conidiophore and conidium after discharge. C. Conidium germinating to produce a secondary conidium. D. Conidium germinating to form several germ tubes. E. Conidium germinating to form a spiny spore. F. Detached spiny spore.

Fig. 118. A. *Entomophthora muscae*. Dead fly attached to a window-pane, surrounded by discharged conidia. B. *Entomophthora americana*. Dead blow-fly attached to a leaf. Note the three bands of conidiophores penetrating between the abdominal segments, and the conidia on the leaf surface.

a parasite of the house-flies and other insects, and the disease is apparently more frequent in wet weather. Diseased flies can be found occasionally attached to the glass of a window-pane surrounded by a white halo about 2 cm in diameter made up of discharged conidia (Fig. 118A). The fly shows a distended abdomen with white bands of conidiophores projecting between the segments of the exoskeleton. The condidiophores are unbranched and multinucleate and arise from the coenocytic mycelium which plugs the body of the dead fly. The conidia are also multinucleate (Fig. 119B). They are projected by a forwardly directed jet of cytoplasm from the elastic conidiophores. Recently discharged conidia have a drop of cytoplasm around them. The cytoplasmic coating may act as a protective agent against desiccation. If the conidium impinges on the body of a fly it secretes an appressorium, or adhesive pad, which attaches the conidium firmly to the cuticle. Penetration of the cuticle is probably brought about by mechanical means. A few hours after infection tri-radiate fissures can be seen in the cuticle beneath attached conidia. If the cuticle in such a region is examined from the inside a thin-walled bladder-like expansion can be seen above the tri-radiate fissure. From this cell mycelial branches develop. The hyphae grow towards the fatty tissues, and as these are consumed the hyphae break up into rounded cells termed hyphal bodies, which are carried by the circulation to all parts of the body (Schweizer, 1947) (Fig. 119C). About a week after infection the flies die, often crawling to the top of a grass stem and clasping it or adhering to walls or window-panes by the proboscis. The hyphal bodies then grow out into coenocytic hyphae which penetrate between the abdominal segments and develop into conidiophores. The primary conidia remain viable for only 3–5 days. If they fail to penetrate a fly the primary conidia may produce secondary conidia within 12 hours. The secondary conidia are formed at the tips of short conidiophores and are discharged by a different mechanism, by the rounding-off of a two-ply columella as in *E. coronata*. The secondary conidia may germinate by a germ tube or may produce tertiary conidia.

Within the body of the dead fly multinucleate spherical resting bodies are formed asexually, and it is presumably from such cells that infection begins each year (Goldstein, 1923). Their germination is said to be stimulated by the action of chitin-decomposing bacteria. *Entomophthora muscae* has been grown in pure culture on extracts of animal tissue but only if the media are sterilised cold (i.e. chemically). Growth is markedly stimulated by the presence of animal fat and by glucosamine, a breakdown product of chitin (Schweizer, 1947). Successful cultures have also been established on a medium containing wheat grain extract, peptone, yeast extract and glycerine (Srinivasan *et al.*, 1964).

In many other species of *Entomophthora* the conidiophores are branched, and this is well shown by *E. americana* (Figs. 118B, 120), a fungus commonly found on blow-flies in the autumn, especially around corpses of dead animals or stinkhorns, where blow-flies congregate. In wet weather severe epidemics of

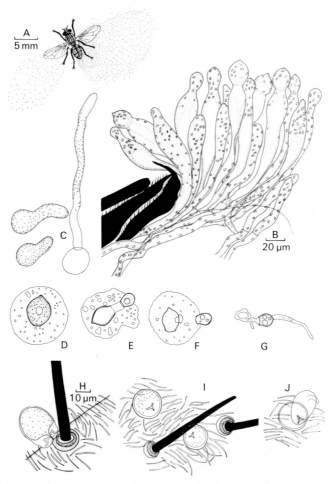

Fig. 119. *Entomophthora muscae*. A. House-fly adhering to window-pane, surrounded by halo of conidia. B. L.S. house-fly showing palisade of unbranched conidiophores projecting between segments of exoskeleton. The conidiophores are multinucleate. C. Hyphal bodies from body of recently dead fly. The hyphal bodies are extending to form conidiophores. D. Conidium immediately after discharge surrounded by cytoplasm from the conidiophore. E–F. Germination of primary conidia to form secondary conidia within 12 hours of discharge. Note the septum cutting off the secondary conidium in E, and the rounding-off of the septum on discharge of the secondary conidium. G. Germination of a secondary conidium by two germ tubes. H. Attachment of primary conidium to integument of a fly. Note the thickened appressorium and the narrow point of penetration. I. Two primary conidia attached to integument, and penetrating it by a tri-radiate fissure. J. View of penetration from within the integument. Note the bladder-like expansion inside the tri-radiate fissure. B–G to same scale; H–J to same scale.

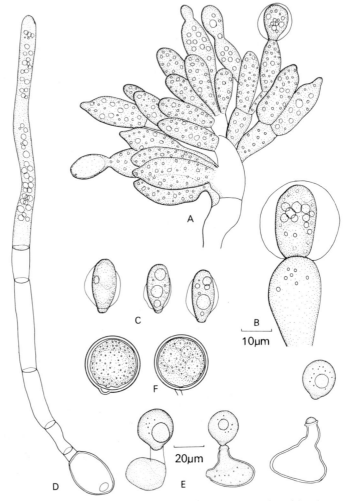

Fig. 120. *Entomophthora americana* from blow-fly. A. Branched conidiophore. B. Single conidiophore and conidium. C. Conidia after discharge. D. Conidium germinating by germ tube. E. Conidia germinating to produce secondary conidia. F. Spherical resting bodies from dead fly.

blow-fly populations may occur, severely reducing their numbers. The dead flies are often attached to the adjacent plants by filamentous rhizoid-like hyphae. The conidiophores form yellowish pustules between the abdominal segments and the branched tips bear conidia. The wall of the conidium is double and the two layers are frequently separated from each other by liquid (Fig. 120A–C). These conidia are projected for several centimetres from the host and, on germination, may form germ tubes, or may produce secondary conidia which are projected by the rounding-off of a two-ply columella.

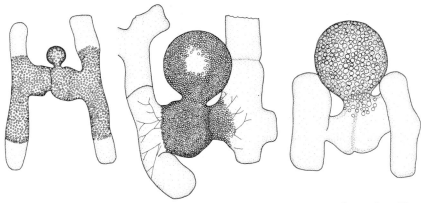

Fig. 121. *Entomophthora sepulchralis.* Three stages in zygospore formation. Two hyphal bodies conjugate and the zygospore arises as a bud from the fusion cell (after Thaxter, 1888).

Within the dry body of the dead fly numerous smooth hyaline thick-walled resting spores are formed.

In some species of *Entomophthora,* e.g. *E. sepulchralis,* resting spores (zygospores) are believed to be formed following conjugation between hyphal bodies (see Fig. 121). Riddle (1906) has claimed that the resting spores of *E. americana* also form zygospores by fusion of hyphal bodies but because doubt has been cast on these observations (Krenner, 1961) it would be valuable to have them confirmed.

The resting spores of many parasitic Entomophthoras do not germinate readily in the laboratory, but it seems likely that they remain viable for one or possibly two years. In *E. virulenta* 2–5% of the resting spores (azygospores) are ready to germinate immediately on formation. Germination is stimulated by soaking the resting spores in water, and no other special treatment, such as exposure to chinin-splitting bacteria appears to be necessary (Hall & Halfhill, 1959).

Interest in the parasitic Entomophthoras has been aroused by the possibility of using them in the biological control of insect pathogens. A number of them have been grown in artificial culture (for references see Müller-Kögler, 1959, 1965; Gustafsson, 1965, 1969). Many have complex requirements and will not grow easily in synthetic media, but can be grown on animal substrata such as meat, fish, coagulated egg-yolk, and milk, or on tissue-culture medium supplemented by bovine serum albumen. On tissue-culture media, *E. egressa,* a parasite of the spruce budworm, can grow and multiply in a wall-less protoplast state. Similar protoplasts have been found in the haemocoels of infected larvae (Tyrell, 1977). Unfortunately, however, attempts to use such cultures to induce epidemics in natural populations have not been successful.

3

Ascomycotina
(Ascomycetes)

This is the largest class of fungi, containing some 15,000 species, although it is likely that many remain to be described. They occur in a wide variety of habitats: in soil, dung, in marine and in fresh water; as saprophytes of plant and animal remains; as animal and plant pathogens. Their characteristic feature is that the sexually produced spores (sometimes called the 'perfect' spores) are borne in a sac or ascus, typically containing eight spores (ascospores) which are explosively ejected. The vegetative structure consists either of single cells (as in yeasts) or septate filaments, each segment often containing several nuclei. The cell walls of filamentous Ascomycetes contain a microfibrillar skeleton of chitin, and in addition various other compounds have been detected such as other amino-sugars, protein, mannose and glucose. It is important to realise that the cell wall is not a functionally inert coating, but may contain surface enzymes (Sussman, 1957; Aronson, 1965). Electron micrographs show that the septum is perforated by a pore so that cytoplasmic continuity between adjacent segments occurs. The pore is also wide enough to allow mitochondria and nuclei to pass through (Fig. 122, Shatkin & Tatum, 1959; Moore, 1965). Where several nuclei occur in a single cell or mycelium, they are not always genetically identical. Differences between nuclei may arise by mutation or by anastomosis of hyphae of different genotype followed by nuclear migration. A cell (or mycelium) which contains nuclei of more than one genotype is said to be a heterokaryon (or heterokaryotic) and the phenomenon of heterokaryosis has been demonstrated in numerous Ascomycetes and in Basidiomycetes and Fungi Imperfecti (Davis, 1966). Its significance is potentially profound:

1. ADAPTABILITY

The ratio of the different kinds of nuclei in a heterokaryon may vary during the growth of a colony, or in relation to the changing nutrient content of the substratum. Jinks (1952) showed that *Penicillium cyclopium*, one of the Fungi Imperfecti (Deuteromycotina), can occur in nature as a heterokaryon. In pure culture the heterokaryon gave rise to two distinguishable homokaryotic components by virtue of the fact that individual conidia are uninucleate. The growth of the component homokaryons was inferior to that of the heterokaryon on semi-natural media over a range of concentrations (Fig. 123). More-

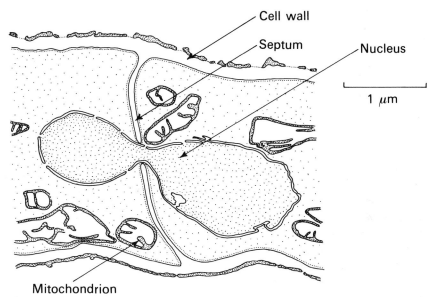

Fig. 122. *Neurospora crassa*. L.S. hypha showing the perforated transverse septum through which a nucleus is passing (based on an electron micrograph by Shatkin & Tatum, 1959).

over, using the proportion of the different kinds of conidia as a means of estimation, it was shown that the proportion of nuclei in the heterokaryon varied with the concentration of apple pulp in the medium used (as shown in table 4).

Table 4. *Nuclear ratios in a heterokaryon of* Penicillium cyclopium *in mixtures of apple pulp and minimal medium*

Medium		% of 'A' nuclei in
10% apple	Minimal	heterokaryon
40	60	12·82
35	65	15·15
30	70	13·70
25	75	15·87
20	80	15·38
15	85	17·86
10	90	20·41
5	95	29·41
0	100	57·47

After Jinks (1952).

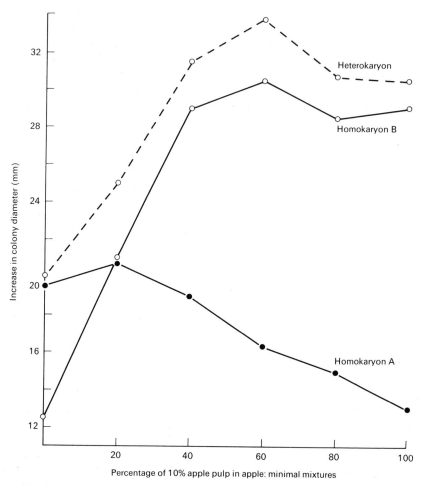

Fig. 123. Growth of two component homokaryons (**A** and **B**) and a heterokaryon of *Penicillium* containing **A** and **B** nuclei on a medium of varying concentration of apple pulp. Note that the growth of the heterokaryon is superior to that of either homokaryon (after Jinks, 1952).

2. COMPLEMENTATION

The observation that heterokaryons may make better growth than their component homokaryons has also been made in other fungi. Using biochemical mutants of *Neurospora crassa* (i.e. mutants produced by irradiation which differ from the wild type in their inability to synthesise from the ingredients of a minimal medium a particular substance, and which therefore need to be supplied with an exogenous supply of this substance) Beadle & Coonradt (1944) showed that two different mutants, individually incapable of growth on minimal medium were able to grow if inoculated simultaneously on to this medium. One of the mutants lacked the ability to synthesise *p*-amino-benzoic acid, and the other nicotinic acid. Since the genes controlling the synthesis of these two substances are non-allelic the *p*-amino-benzoic acid-less mutant would have an intact gene for nicotinic acid synthesis, whilst the nicotinic-acid-less mutant would have unimpaired ability to synthesise *p*-amino-benzoic acid. Once anastomosis of the mutant hyphae had occurred, the two kinds of nuclei would exist in a common cytoplasm which would then have the necessary biochemical facilities for synthesis of both the deficiencies in the original mutants. Each mutant therefore *complements* the deficiency of the other. In other experiments with the same fungus it has been found that growth of heterokaryons synthesised from complementary homokaryons was maximal over a wide range of nuclear ratios of the component homokaryons (Pittenger *et al.*, 1955; Pittenger & Atwood, 1956; Klein, 1960). Mutants of this type are often termed *auxotrophs* (auxotrophic) to distinguish them from wild-type strains which are *prototrophic*. Heterokaryons between auxotrophic mutants are termed forced heterokaryons (Caten & Jinks, 1966).

3. GENETICAL VARIABILITY

The occurrence within a single mycelium of nuclei of differing genotype represents a store of genetical material analogous to the heterozygous condition of diploid organisms.

4. BREAKDOWN OF HETEROKARYONS

A heterokaryotic mycelium may give rise to uninucleate or homokaryotic conidia, or to homokaryotic hyphal tips. Examples of multinucleate but homokaryotic conidia are found in certain species of *Aspergillus*. In *A. tamarii* (non-ascocarpic) there are typically three or four nuclei per conidium. Wild-type strains have green conidia, but a mutant is known with white conidia. Heterokaryons between the wild-type and mutant strains give rise to conidial heads bearing chains of green conidia and chains of white conidia, but with no intermediate colours. Single spore cultures derived from white or green chains breed true. It is thus inferred that although the conidiophore is heterokaryotic, each phialide (mother cell at the base of the spore chain) receives only a single nucleus, either mutant or wild-type (Fig. 124). A different situation is found in *A. carbonarius*, which has two to five nuclei per spore. In this fungus heterokaryotic conidiophores, formed between the black-spored wild-type strain and a pale brown mutant, bear spores intermediate in colour between these two strains, and when single spores from the heterokaryotic heads are cultured they often give rise to mixed colonies of the wild-type and mutant strains. From this it can be inferred that the spores themselves are heterokaryotic and arise from phialides which contained both kinds of nuclei. The inference that the phialides are multinucleate can be confirmed cytologically (Yuill, 1950).

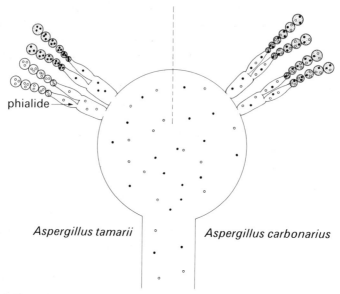

phialide

Aspergillus tamarii   *Aspergillus carbonarius*

Fig. 124. Heterokaryosis in *Aspergillus* spp. Left: Situation in *A. tamarii* where a heterokaryotic conidiophore formed between a dark-spored wild-type and a pale-spored mutant strain gives rise to separate kinds of spore chain, because each phialide receives a single nucleus from which the nuclei of the spore chain are all derived. The spores are thus homokaryotic. Right: Situation in *A. carbonarius* where a similar heterokaryotic conidiophore gives rise to conidia which are heterokaryotic because the phialides contain several nuclei.

5. RECOMBINATION

An important consequence of heterokaryosis is that recombination can occur even in homothallic fungi and in Fungi Imperfecti. In homothallic fungi such as *Emericella nidulans* or *Sordaria fimicola* heterokaryon formation may result in the presence of genetically distinct nuclei in the ascus initial. Nuclear fusion and meiosis associated with sexual reproduction of a conventional kind may thus give rise to recombinants between the two genotypes, so that a degree of outbreeding is possible even in homothallic organisms (Olive, 1963a). In Fungi Imperfecti, that is in fungi which are not known to reproduce sexually, recombination has been discovered in the progeny derived from heterokaryons. The recombination process is of a novel kind, not involving meiotic segregation, and has been termed parasexual recombination (Pontecorvo, 1956; Roper, 1966).

In *E. nidulans*, in which the conidia are uninucleate, the phenomenon was detected in heterokaryons between two strains marked by having spore colours different from the wild type, i.e. white or yellow as distinct from the green wild-type conidia, and in addition biochemical markers such as a requirement for adenine or lysine, which are not required by wild-type strains.

The genes for white and yellow spores are non-allelic and are recessive. On minimal medium on which the wild-type strain can grow, neither of the two mutant strains could grow singly. When inoculated simultaneously, however, a heterokaryon was formed and was able to grow because each mutant strain could complement the nutritional deficiency of the other, i.e. a strain requiring adenine would not require lysine and vice versa. When the heterokaryon formed conidia, in addition to the yellow and white conidia of the two parental strains, green conidia were also occasionally formed. The green conidia were diploid and were slightly larger than haploid wild-type conidia. The diploid condition is believed to have arisen by rare fusion between two different mutant nuclei. If such diploid conidia are inoculated into cultures, the resulting colony may give rise to conidia of four types: the two original mutant strains, further diploids and, rarely, conidia which when plated out show recombination of the properties of the two parental mutants. For example when one parent had white conidia and a requirement for lysine, and the other had yellow conidia and a requirement of adenine, recombinant strains isolated from diploid colonies had yellow conidia and a requirement for lysine, or white conidia and a requirement for adenine. It is believed that recombination is not the result of meiosis, but occurs during mitosis in diploid nuclei and that subsequent to mitotic recombination the diploid nuclei become haploid by progressive occasional loss of a chromosome. Pontecorvo (1956) believes that the events in parasexual recombination form part of a cycle which he has termed the parasexual cycle. The steps in the cycle are:

(*a*) Fusion of two unlike nuclei in a heterokaryon.
(*b*) Multiplication of the resulting diploid heterozygous nucleus side by side with the parent haploid nuclei in a heterokaryotic condition.
(*c*) Eventual sorting-out of a homokaryotic diploid mycelium which may become established as a strain.
(*d*) Mitotic crossing-over occurring during multiplication of the diploid nuclei.
(*e*) Vegetative haploidisation of the diploid nuclei.

Parasexual recombination has been demonstrated in several Ascomycetes, Basidiomycetes and Fungi Imperfecti. The implications of parasexual recombination are great:

1. It is possible to study the genetics of asexual organisms. In the case of *E. nidulans* it has proved possible to compare genetical maps based on conventional sexual reproduction and on parasexual recombination, and there is a close degree of correspondence.

2. It is possible to 'breed' asexual organisms such as members of the Fungi Imperfecti used in fermentations and so to combine, in one strain, desirable properties from different strains.

3. The occurrence of parasexual recombination explains how variation can

occur in Fungi Imperfecti, e.g. how new strains of pathogenic fungi can appear (Parmeter *et al.*, 1963).

In assessing the likely significance of heterokaryosis and parasexual recombination in nature it should be borne in mind, however, that many laboratory studies have been based on forced heterokaryons (complementary auxotrophs) which can only grow if they can form heterokaryons. Secondly, numerous studies have shown that heterokaryons are only formed readily between strains which have a closely similar genetical background (e.g. mutants derived from a common wild-type strain). This restricts the possibility of gene flow between strains which differ appreciably in genotype (Caten & Jinks, 1966).

For a discussion of heterokaryon incompatibility in *E. nidulans*, see Croft & Jinks (1977), and in *Neurospora*, Perkins & Barry (1977).

NUCLEAR MIGRATION IN ASCOMYCETES

As shown in Fig. 122 there is evidence that nuclei can migrate through septal pores. Indirect evidence of nuclear migration has also been obtained in experiments with mycelia of opposite mating type in heterothallic ascomycetes. *Gelasinospora tetrasperma* forms flask-shaped ascocarps (perithecia) containing asci which have ascospores of two sizes. Cultures started from the larger ascospores are self-fertile, but those derived from small ascospores are of two distinct mating types, 'A' and 'a', and perithecia are formed only when mycelia of the two kinds are mated together. If an 'A' mycelium is allowed to fill a Petri-dish culture a small block of 'a' mycelium can be added and within a

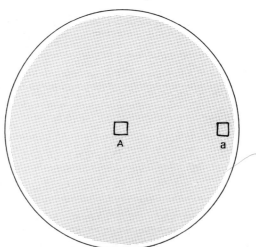

Fig. 125. Nuclear migration in *Gelasinospora tetrasperma*. A. Original inoculum of 'A' mating type. a. point at which a block of mycelium of the compatible 'a' mating type is later added after the 'A' mycelium has colonised the whole plate. The migration of 'a' nuclei can be followed by removing plugs at intervals of time at various distances from the point of inoculation.

few days perithecia appear at some distance from the point of inoculation of the 'a' block. If small plugs of agar are removed from the dish at intervals after addition of the 'a' block, and transferred to fresh agar plates, those plugs which have received 'a' nuclei will develop perithecia (Fig. 125). In this way estimates of the rate of nuclear migration of the order of 10·5 mm/hour have been made, about two to three times the rate of growth of the mycelium (Dowding & Bakerspigel, 1954). This rate is probably an underestimate and direct observation and photography give rates of 40 mm/hour (Dowding, 1958). The actual mechanism of nuclear migration probably depends on cytoplasmic streaming (Snider, 1965). However, in the Basidiomycete *Schizophyllum commune*, microtubules attached to the nuclear envelope are believed to be concerned in nuclear migration (Raudaskoski, 1972a). In some cases, nuclear migration is unilateral, i.e. there is no reciprocal exchange of nuclei. This phenomenon is discussed below.

SOMATIC NUCLEAR DIVISION   *Mitosis*

Nuclear division in Ascomycetes (and in related Fungi Imperfecti) is probably similar in essentials to that of most eukaryotes, but the small size of the nuclei and chromosomes is close to the limits of resolution of the light microscope, and this has led to some uncertainty of interpretation. In *Fusarium oxysporum* (a member of the Fungi Imperfecti because no perfect state is yet known: some other species of *Fusarium* have perfect states belonging to the Hypocreaceae, p. 337), Aist & Williams (1972) have shown that mitosis takes 5–6 minutes. As shown in Fig. 126, mitosis is intranuclear, i.e. the nuclear membrane does not break down prior to separation of sister chromatids. Breakdown of the nucleolus is followed by the division of the spindle pole body, an organelle situated outside the nuclear membrane in a small invagination. The two daughter spindle pole bodies move to opposite poles of the nucleus and, from the inside of the nuclear envelope opposite the spindle pole bodies, microtubules appear and ultimately stretch across the nucleus from one spindle pole body to the other. The microtubules form a parallel bundle, making up the spindle, which is thus enclosed within the nuclear membrane, and is not formed outside it as in many other eukaryotes (Pickett-Heaps, 1969). The chromosomes are not arranged on a metaphase plate, but are attached to the spindle microtubules by means of kinetochores at different points along the length of the spindle. During telophase, the nucleus elongates. The nuclear membrane constricts near the middle, separating the two daughter nuclei. Some $5\frac{1}{2}$ minutes from the beginning of nuclear division, new nucleoli reappear in the enlarging daughter nuclei. Day (1972) has reviewed various models which have been proposed for somatic nuclear division, and their consequences for fungal genetics.

MATING BEHAVIOUR

Ascomycetes may be homothallic or heterothallic. The basis for heterothal-

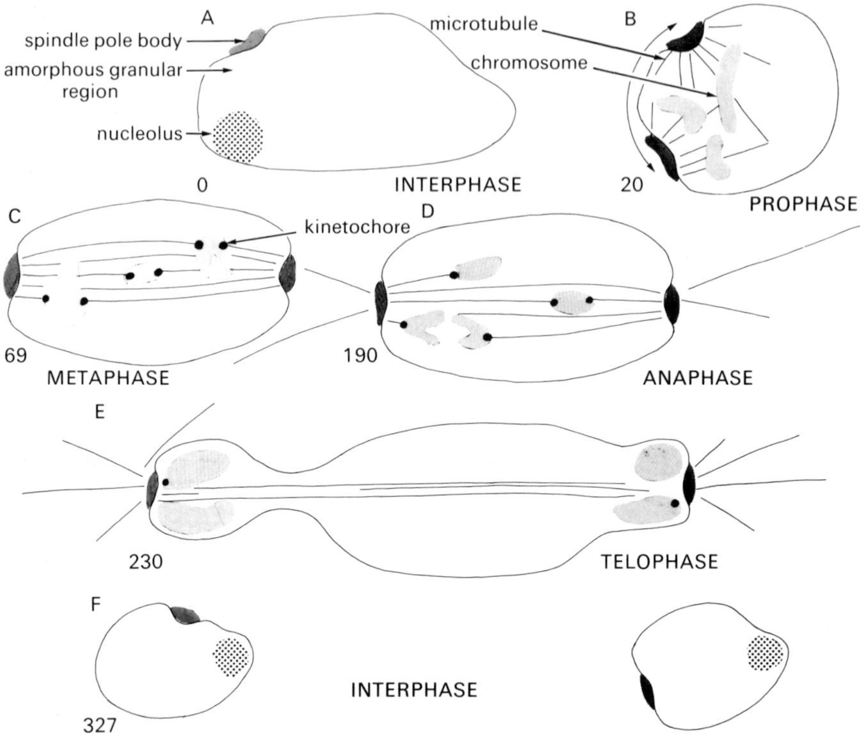

Fig. 126. Diagrammatic representation of somatic nuclear division in *Fusarium oxysporum* (modified from Aist & Williams, 1972). The numbers given at the left of each figure represent the time in seconds from the beginning of mitosis. A. Interphase: nucleus with intact nucleolus. B. Prophase: the nucleolus has broken down and chromosomes are visible. Half spindles have developed opposite the depressions in the nuclear envelope occupied by the spindle pole bodies. C. Metaphase: separate chromosomes are distinguishable, attached at different levels along the longitudinal axis of the spindle. Sister kinetochores are located at opposite sides of the chromosomes. D. Anaphase: non-synchronous separation of sister chromatids and their migration toward the spindle poles. E. Telophase: elongation of the entire nucleus is followed by constriction of the still-intact nuclear envelope. F. Interphase: nucleolus reappears.

lism is typically a single gene with two alleles, 'A' and 'a', and because of segregation during the meiotic division which precedes ascospore formation, the eight ascospores normally present in an ascus will include four of one mating type and four of the other. Since, in these forms, incompatibility results when identical alleles are present at the locus controlling sexual reproduction, this type of incompatibility is termed *homogenic* (Esser, 1971). In certain species with four-spored asci, e.g. *Neurospora tetrasperma, Podospora anserina*, the ascospores are binucleate, and commonly contain nuclei of both mating types. Such ascospores on germination would give rise to fully fertile mycelia, and it would appear that the fungus is homothallic. However, occasionally uninucleate ascospores are formed, and on germination the resulting mycelium is not fertile: ascocarps are only formed in 50% of matings between such mycelia. Since basically these fungi are heterothallic in their mating behaviour, the term *secondary homothallism* is used to describe the behaviour of their binucleate ascospores (Whitehouse, 1949a,b).

Sex organs are formed in some Ascomycetes. The female organ or **ascogonium** is commonly a coiled, multinucleate cell, sometimes surmounted by a receptive trichogyne. The male organs may take the form of slender branches, antheridia, or minute unicellular **spermatia** incapable of germination, or **microconidia** (oidia) which, whilst capable of fulfilling a sexual role are nevertheless capable of germination. In some heterothallic species, e.g. *Neurospora crassa*, ascogonia and microconidia are formed on a single strain, but they are self-incompatible – i.e. the microconidia are unable to fertilise the ascogonia on the same mycelium. Only when a fertilising element from the opposite mating type is brought into contact with the ascogonium does fertilisation occur. This paradoxical phenomenon is found in numerous Ascomycetes. Each mating type possesses both kinds of sex organs, and is morphologically indistinguishable from the opposite mating type. Whitehouse (1949a) has used the term *physiological heterothallism* to describe this type of behaviour, and since there are two alleles involved it has also been termed *two-allelomorph physiological heterothallism*.

In many Ascomycetes there are no recognisable sex organs; fusion takes place between ordinary hyphae. In heterothallic forms this can only happen where heterokaryosis is possible between hyphae of opposite mating type.

DEVELOPMENT OF ASCI

In yeasts and related fungi the ascus may arise directly from a single cell, but in most other Ascomycetes the ascus develops from a specialised hypha, the **ascogenous hypha**, which in turn develops from an ascogonium. The ascogenous hypha is multinucleate, and its tip is recurved to form a **crozier** (shepherd's crook). Within the ascogenous hypha nuclear division occurs simultaneously. Two septa at the tip of the crozier cut off a penultimate binucleate cell (Fig. 127C) destined to become an ascus. The terminal cell of the crozier curves round and fuses with the ascogenous hypha behind the penultimate cell, and this region of the ascogenous hypha may grow on to form a new crozier in which the same sequence of events is repeated. Repeated proliferation of the tip of the crozier can result in a cluster of asci. In the ascus initial the two nuclei fuse and the fusion nucleus undergoes meiosis to form four haploid daughter nuclei (Fig. 127D,E) (Olive, 1965). These nuclei then undergo a mitotic division so that eight haploid nuclei result. During these nuclear divisions the ascus is elongating, and the plane of the division is parallel to the length of the ascus. Cytoplasm is cleaved out around each nucleus to form an ascospore. In some forms the eight nuclei divide further so that each ascospore is binucleate. Where the ascospores are multicellular there

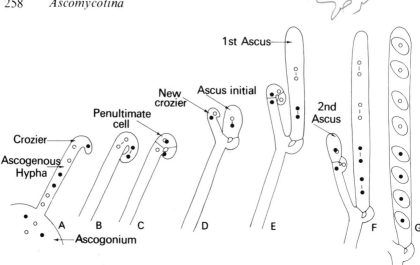

Fig. 127. Diagrammatic representation of cytological features of ascus development. A. Ascogenous hypha with a crozier at its tip developing from an ascogonium. B. Conjugate division of the two nuclei in the crozier. C. Two septa have cut off a binucleate penultimate cell. These two nuclei fuse to form a diploid fusion nucleus. The uninucleate terminal segment has fused with the ascogenous hypha. D. Penultimate cell enlarges to become the ascus within which the fusion nucleus begins to divide meiotically. A new crozier is developing from beneath the ascus. E. Second division of meiosis has occurred in the young ascus. The behaviour of the second crozier repeats that of the first. F. Mitotic division of the four haploid nuclei in the ascus. G. Ascospores formed.

are repeated nuclear divisions. In some forms more than eight ascospores are formed, or the eight ascospores may break up into part-spores. Studies of the fine structure of asci during cleavage of the ascospores have shown that a system of double membranes continuous with the endoplasmic reticulum extends from the envelope of the fusion nucleus in developing asci (Fig. 128). The double membrane forms a cylindrical envelope lining the young ascus. The lining layer is termed the ascus vesicle or ascospore membrane. The ascospores are cut out from the cytoplasm within the ascus by invagination of the double membrane. Between the two layers forming the membrane, the spore wall is secreted, and the inner membrane forms the plasma membrane of the ascospore (Delay, 1966; Carroll, 1967; Reeves, 1967; Oso, 1969; Wells, 1970, 1972).

It is important to note that because the division which follows the four-nucleate stage is mitotic and because the division plane is usually parallel to the length of the ascus, adjacent pairs of spores starting from the tip of an ascus are normally sister spores which are thus genetically identical. Rare exceptions to this situation are occasionally found where the division planes are oblique, or for other reasons. A neat illustration of the normal arrangement of sister spores is provided by crossing experiments between black-spored Ascomy-

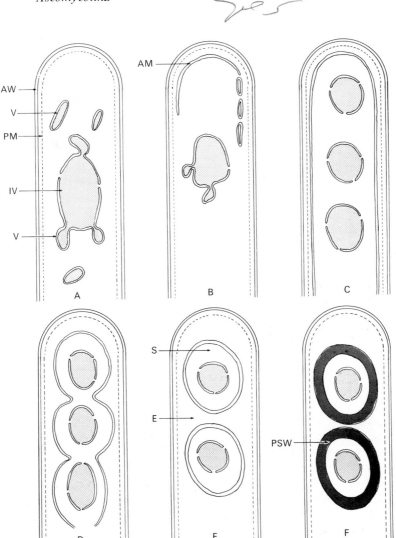

Fig. 128. Ascus development in *Ascobolus* (after Oso, 1969). A. Young ascus showing the formation of membrane-bounded vesicles (V) from the nucleus (N). The ascus wall (AW) is lined by the plasmalemma (PM). B. Appearance of the ascospore membrane (AM) at the tip of the ascus, and the arrangement of vesicles along the periphery of the ascus. C. Ascospore membrane now in the form of a peripheral tube open at the lower end. The diploid nucleus has divided. D. Invagination of the ascospore membrane between the haploid nuclei. E. Young ascospores (S) delimited by the ascospore membrane from the epiplasm (E). F. Separation of the two layers of the ascospore membrane due to the formation of the primary spore wall (PSW) between them.

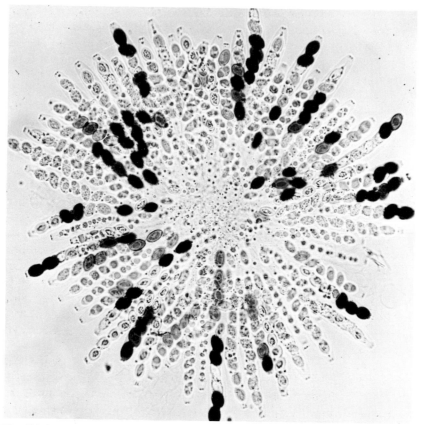

Fig. 129. *Sordaria fimicola.* Squash preparation from a hybrid perithecium. Most ripe asci contain four black and four white ascospores.

cetes and mutant derivatives with colourless spores. Hybrid asci with four black and four white ascospores are formed and, as shown in Fig. 129, the majority of asci show either alternating pairs of black and pairs of white, or groups of four black then four white spores. Alternating pairs result from second-division segregation of the gene for spore colour, whilst the four–four patterns result from first-division segregation.

The actual form of the mature ascus is very variable. In forms with non-explosive ascospore release the ascus is often a globose sac, but in the majority of Ascomycetes the ascus is cylindrical, and the spores are expelled from the ascus explosively. It is thought that the explosive release follows increased turgor, caused by water uptake. In the young ascus, after the spores have been cut out, epiplasm (residual cytoplasm rich in polysaccharides such as glycogen) remains lining the ascus wall surrounding a large central vacuole containing ascus sap, within which the ascospores are suspended. Conversion of the polysaccharide to sugars of smaller molecular weight is believed to bring about an increased osmotic concentration of the ascus sap, which is followed by increased water uptake by the ascus, and the resulting increase in turgor causes the ascus to stretch. Ingold (1939) has estimated that the osmotic pressure of the ascus sap in *Ascobolus furfuraceus* (= *A. stercorarius*) is about 10–13 bars and for *Sordaria fimicola* 10–30 bars (Ingold, 1966a). In the latter fungus glucose is the most abundant sugar present in ascus sap, but the amount present represents only a negligible fraction of the total solutes; the balance being made up by salts. In many cases the asci are surrounded by packing tissue in the form of paraphyses, pseudoparaphyses and other asci, so that they cannot expand laterally but are forced to elongate. In the cup-fungi or Discomycetes the elongation of the asci raises their tips above the general level of the hymenium. The ascus tips are often phototropic and when the increased pressure causes the ascus tip to burst the spores are shot out in a drop of liquid, the ascus sap. In this group large numbers of asci may be discharged simultaneously, so that a cloud of ascospores is visible. This phenomenon is known as puffing. In some Discomycetes (e.g. the Pezizales) the ascus tip is surmounted by a cap or operculum which is blown aside or actually blown off the tip of the ascus by the force of the explosion. However in other Discomycetes (e.g. the Helotiales) the ascus tip is perforated by a pore, and there is no operculum. These two types of asci are respectively termed operculate and inoperculate, and the presence or absence of an operculum is an important feature of classification.

In the flask-fungi (often loosely termed the Pyrenomycetes) the asci are enclosed in a cavity which opens to the exterior through a narrow pore, the ostiole. As an ascus ripens it elongates and takes up a position inside the ostiole, often gripped in position by a lining layer of hairs, periphyses. In this case the asci discharge their spores singly and puffing does not happen.

The asci of Pyrenomycetes are never operculate. In many groups, the ascus tip has a distinctive apical apparatus, the detailed structure of which may be of importance in classification (Chadefaud, 1973). Staining of the ascus apex by a variety of stains has led to claims of apical structures of great complexity and diversity. However, electron micrographs have, in general, failed to substantiate the presence of such complex structures (Greenhalgh & Evans, 1967; Reeves, 1971; Griffiths, 1973; Beckett & Crawford, 1973). In many Pyrenomycetes, there is an apical ring or annulus (Fig. 130), which may stain blue with iodine (for example, in *Xylaria* and its allies), or may fail to stain (for example, in *Sordaria*). When the ascus explodes, the apical ring is everted (Fig. 130B) and is believed to grip the ascospores as they are ejected.

The wall of the ascus may be single (**unitunicate**) or double (**bitunicate**). This is an important taxonomic feature. Bitunicate asci are characteristic of the Loculoascomycetes (Luttrell, 1973). Chadefaud (1960) has claimed that ascus walls are generally double, but this is not supported by electron microscopy. With certain fixation procedures prior to embedding for electron microscopy, layers of different density may be distinguishable in unitunicate asci, but the

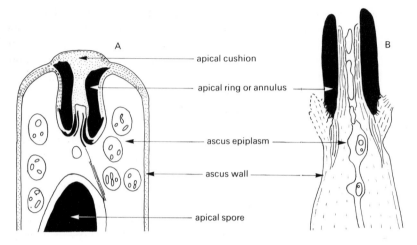

Fig. 130. *Xylaria longipes*: fine structure of ascus apex (after Beckett & Crawford, 1973). A. L.S. undischarged ascus showing the apical ring. B. L.S. discharged ascus showing the eversion of the apical ring.

layering is probably due to differences in the orientation of the microfibrils making up the ascus wall (Reeves, 1971). In bitunicate asci, there are two wall layers which physically separate from each other. The outer layer is termed the **ectoascus** or **ectotunica**, and the inner layer the **endoascus** or **endotunica**. Development of a bitunicate ascus takes place in two stages prior to ascospore formation. The first stage involves the growth of the ascus initial and the expansion of the ascus mother cell. In the second stage, a secondary wall layer, i.e. the endotunica, is laid down within the primary wall, i.e. the ectotunica. Both layers are made up of microfibrils embedded in an amorphous matrix. The two layers differ only in the arrangement of the microfibrils (Reynolds, 1971).

The behaviour of the bitunicate type of ascus during discharge has been described as the Jack-in-a-box mechanism. The outer wall is relatively rigid and inextensible. As the ascus expands, the outer wall ruptures laterally or apically (see Figs. 214–217), and the inner wall then stretches before the ascus explodes. In some forms, e.g. *Cochliobolus sativus*, the endoascus may break down and apparently plays no part in spore release (Shoemaker, 1955). This statement requires confirmation because in *C. cymbopogonis*, El Shafie & Webster (1980) have shown that the endoascus is incomplete, forming a thimble-like cap which sits over the recurved tips of the coiled ascospores as they pass through the pseudothecial neck.

The explosive release of the ascospores appears to throw all the spores out simultaneously. In fact it is likely that in most cases the spores are spatially separated from each other as they are constricted on passing through the ascus pore. This has been neatly demonstrated by spinning a transparent disk over the surface of a culture of *Sordaria* discharging spores (Ingold & Hadland, 1959a). The ascus contents are laid out on the disk in the order in which they are released. Various patterns of spore clumping and separation are visible, and although in many asci the eight spores are well separated from each other there is also a tendency for spores to stick together. From measurements of the length of the ascospore deposit and the velocity of rotation of the disk calculations of the velocity of ascospore discharge have been made. A value for the minimum initial velocity of 736 cm/second was obtained by this method.* By a separate method, a value of 1078 cm/second was obtained. The actual time taken for an ascus to discharge was estimated by the rotating disk method to be 0·000024 second.*

When ascospores stick together they are discharged further than single-spored projectiles. In many coprophilous Ascomycetes (e.g. *Ascobolus*, *Saccobolus*, *Podospora*) the spores may be attached together by mucilaginous secretions so that the spores may be projected for distances of 30 cm in *Ascobolus immersus* and *Podospora fimicola*. The distances to which individual ascospores are discharged vary. Single ascospores are commonly discharged for about 1–2 cm. Where puffing of the asci occurs the distance to which the spores are projected may be increased.

In some Ascomycetes the asci do not discharge their spores violently, and in such cases the asci are often globose instead of cylindrical. The Endomycetales and Eurotiales have asci of this type. In *Ceratocystis* the asci break down to produce a mass of sticky spores which ooze out as a drop from the tip of a cylindrical neck which surmounts the ascocarp. Breakdown of asci within the fruit-body is also found in *Chaetomium*. In Ascomycetes with subterranean fruit-bodies, e.g. the Tuberales, the ascospores are again usually not discharged violently, but are dispersed when the fruit-bodies are eaten by rodents.

TYPES OF FRUIT-BODY

In yeasts and related fungi (Endomycetales) the asci are not enclosed by hyphae, but in most Ascomycetes they are surrounded by hyphae to form an ascocarp. The form of the ascocarp is very varied. In *Gymnoascus* there is a loose open network of hyphae. In those species of *Aspergillus* and *Penicillium* which possess ascocarps and in the Erysiphales the asci are enclosed in a globose fructification with no special opening to the outside. Such ascocarps are termed **cleistocarps** (closed fruits) or **cleistothecia**. In the cup-fungi, e.g. Pezizales and Helotiales, the asci are borne in open saucer-shaped ascocarps, and at maturity the tips of the asci are freely exposed. Such fruit-bodies are termed **apothecia**. The Pyrenomycetes (Sphaeriales and Hypocreales) have **perithecia**, flask-shaped fruit-bodies opening by a pore or ostiole. The perithecial wall is formed from sterile cells derived from hyphae which surrounded

---

* In the original paper, there is an arithmetical error, and the value of 366 cm/second is given. The time for the ascus to empty is also incorrectly stated to be 0·000048 second.

the ascogonium during development. The ascus in these groups is unitunicate. The Loculoascomycetes (e.g. the Pleosporales) have ascocarps which bear a superficial resemblance to perithecia, but differ in details of development, and contain bitunicate asci. Such fruit-bodies are properly termed **pseudothecia**, but the term perithecium is sometimes applied loosely to include them.

Ascocarps may arise singly or are often clustered together. In many Pyreno-mycetes the perithecia are seated on a mass of tissue termed a **stroma** (e.g. *Nectria cinnabarina*) or may be embedded in it with only the ostioles visible at the surface (see for example the Xylariaceae, and the Clavicipitaceae).

For a general account of fruit-body form in Ascomycetes, see Booth (1966b).

CONIDIA OF ASCOMYCETES

Whilst some Ascomycetes reproduce by means of ascospores only, many have one or more conidial states. The recognition that Ascomycetes are pleomorphic we owe to the brothers Tulasne who in their monumental *Selecta Fungorum Carpologia* (1851–1855; translated into English in 1931) described the association of perfect (i.e. ascosporic) and imperfect (conidial) stages of many Ascomycetes, and illustrated their findings by exquisite figures unrivalled in their accuracy and detail. However, evidence of association can be misleading and the pure culture techniques exploited by Brefeld and later mycologists have provided conclusive evidence linking perfect and imperfect states. The type of conidial apparatus is sometimes a guide to relationships (Tubaki, 1958). For example the conidia of many Eurotiales and Hypocreaceae are phialospores, whilst there is a general similarity between the conidia of Erysiphales. However, such generalisations are dangerous because it is known that morphologically similar conidia may belong to quite distinct groups of fungi (compare the *Monilia* conidia of *Neurospora crassa*, a member of the Sphaeriales (Fig. 170D) with those of *Sclerotinia fructigena*, a member of the Helotiales (Fig. 193D). It is often found that conidia have different means of dispersal from ascospores. For example the conidia of *Nectria cinnabarina* are dispersed by rain-splash, whilst the ascospores are wind-dispersed. The conditions under which ascospores and conidia are produced may also be quite different. Conidia tend to be produced fairly soon after a new host or substratum has become infected, whilst ascospores may develop rather later. The viability of ascospores and conidia is usually different: conidia are relatively short-lived.

CLASSIFICATION OF ASCOMYCOTINA (ASCOMYCETES)

Different authorities hold widely differing views about the classification of Ascomycetes. The problem has been discussed by Miller (1949); Luttrell (1951); von Arx & Müller (1954); Müller & von Arx (1962). For the purpose of this account, the system proposed by Ainsworth (1973) will be followed.

Key to classes of Ascomycotina

1 Ascocarps and ascogenous hyphae lacking; thallus mycelial or yeast-like **Hemiascomycetes (p. 265)**

Ascocarps and ascogenous hyphae present; thallus mycelial 2

2 Asci bitunicate; ascocarp an ascostroma
Loculoascomycetes (p. 389)

Asci typically unitunicate; if bitunicate, ascocarp an apothecium 3

3 Asci evanescent, scattered within the astomous ascocarp which is typically a cleistothecium; ascospores aseptate
Plectomycetes (p. 283)

Asci regularly arranged within the ascocarp as a basal or peripheral layer 4

4 Exoparasites of arthropods; thallus reduced; ascocarp a perithecium: asci inoperculate **Laboulbeniomycetes***

Not exoparasites of arthropods 5

5 Ascocarp typically a perithecium which is usually ostiolate (if astomous, asci not evanescent); asci inoperculate with an apical pore or slit **Pyrenomycetes (p. 312)**

Ascocarp an apothecium or a modified apothecium, frequently macrocarpic, epigean or hypogean; asci inoperculate or operculate
Discomycetes (p. 358)

1: HEMIASCOMYCETES

The feature which distinguishes the Hemiascomycetes from other Ascomycetes is the absence of an ascocarp, i.e. an investment of sterile cells surrounding the asci. The asci are formed singly, usually following karyogamy, and are not borne on ascogenous hyphae. Three orders are included (Martin, 1961): the Protomycetales, Endomycetales and Taphrinales. The Protomycetales are parasitic on higher plants and produce spores in a spore-sac or synascus which is regarded as the equivalent of several asci. This group will not be considered further (see Kramer, 1973). The Endomycetales include a number of yeasts and related mycelial forms and are mostly saprophytic. The Taphrinales are parasitic on vascular plants, causing a variety of diseases such as leaf curl, witches' broom diseases, and diseases of fruits. The distinction between these two orders is set out below.

(*a*) Zygote a single cell transformed directly into an ascus; mycelium sometimes lacking. Mostly saprobic **Endomycetales**

(*b*) Hyphae bearing terminal chlamydospores or ascogenous cells, each of which produces a single ascus, usually forming a continuous hymenium-like layer on often modified tissues of host. Parasitic on vascular plants
Taphrinales

* This group is not considered further (see Benjamin, 1973).

ENDOMYCETALES

This group of fungi, sometimes also known as the Saccharomycetales, is important economically because it includes ascospore-forming yeasts such as *Saccharomyces* and *Schizosaccharomyces*, used in alcoholic fermentations and in bread-making. These two genera usually do not form true mycelium but exist as single cells which reproduce by budding or by division. There are, however, related forms which possess a true mycelium, e.g. *Eremascus*. In numerous yeast-like organisms ascospores have not been discovered. Some of these forms are possibly 'imperfect' yeasts, i.e. yeasts which have lost the power to reproduce sexually, or may represent haploid strains of heterothallic yeasts. These asporogenous yeasts include such genera as *Cryptococcus*, *Torulopsis*, *Pityrosporum* and *Candida* and some of them are important human and animal pathogens. A taxonomic account of yeasts has been given by Lodder (1970) and Kreger-van Rij (1973), and Keys by Barnett & Pankhurst (1974). More general accounts of the biology of yeasts will be found in Ingram (1955); Cook (1958); Reiff *et al.* (1960); Phaff *et al.* (1966); Rose & Harrison (1969–1970). The details of classification vary from one author to another. Kreger-van Rij (1973) recognises four families, but we shall only refer to two, the Endomycetaceae and Saccharomycetaceae.

Endomycetaceae

This group includes *Schizosaccharomyces*, *Endomyces* and *Eremascus*. For a recent account of the taxonomy of the family, see Redhead & Malloch (1977).

Schizosaccharomyces (Fig. 131)

Four species of *Schizosaccharomyces* are known (Slooff, 1970). *S. japonicus* var. *versatilis* isolated from canned grape juice is of interest because, as well as growing like a yeast, it can also form a mycelium (Wickerham & Duprat, 1945). The two best-known species are *S. pombe* and *S. octosporus*. *S. pombe* is the fermenting agent of African millet beer (pombe), arak in Java, and it has also been isolated from sugar molasses and grape juice. It has been widely used in genetical and cell-biology studies (Leupold, 1970; Flores de Cunha, 1970; Mitchison, 1970; Gutz *et al.*, 1974). Both species grow well on liquid or on solid media such as malt extract agar, and develop ripe asci within 3 days at 25 °C. Individual cells are globose to cylindrical, uninucleate and haploid. Cell division is preceded by intranuclear mitosis, towards the end of which the nucleus constricts and becomes dumb-bell shaped (Conti & Naylor, 1959; McCully & Robinow, 1971). The division of the cell into two daughter cells is brought about by the centripetal development of a septum which cuts the cytoplasm into two (Johnson *et al.*, 1973). The two sister cells may remain attached for a time, or may separate by breakdown of a layer of material in the middle of the septum. Because cell replication occurs by division, *Schizosac-charomyces* is termed a fission yeast. In contrast, the cells of *Saccharomyces* reproduce by budding. Ascus formation in *Schizosaccharomyces* is preceded by copulation. *S. octosporus* is homothallic, and quite often adjacent sister cells may fuse together. *S. pombe* may be homothallic or heterothallic (Leupold, 1970; Gutz & Doe, 1975). When cells of opposite mating type of this fungus are grown together in liquid culture, a strong sexual agglutination occurs, and this becomes obvious as the cells clump together, leading to flocculation of the culture (Egel, 1971; Calleja & Johnson, 1971). The cytology of this process has been reinvestigated by Widra & Delamater (1955) with the light microscope, and by Conti & Naylor (1960) and Yoo *et al.* (1973) using the electron microscope. Two cells come into contact by a portion of the cell wall. Often the fusing cells are sister cells formed by division. A pore is formed in the centre of the attachment area and this widens and elongates to form a conjugation canal (Fig. 131B). During this process the nuclei, one from each cell, migrate towards each other and fuse. Vacuoles may appear in the young ascus following nuclear fusion. The fused nucleus elongates and may reach half the length of the ascus, and then divides by constriction, the nuclear membrane remaining intact during division. The two daughter nuclei migrate to opposite ends of the ascus and then divide further. These divisions constitute meiosis, and a single mitosis follows so that eight haploid nuclei result, and eight ascospores are finally differentiated. The ascospores are released by breakdown of the ascus wall. Four-spored asci are also common. The life cycle of *Schizosaccharomyces* is thus interpreted as being based on haploid vegetative cells, which fuse to form asci, the only diploid cells. Meiosis in the ascus

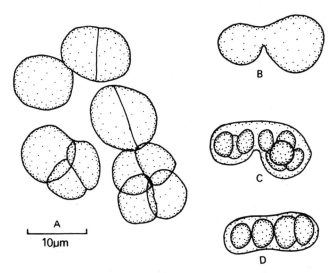

Fig. 131. *Schizosaccharomyces octosporus*. A. Vegetative cells showing transverse division. B. Copulation. C. Eight-spored ascus. D. Four-spored ascus.

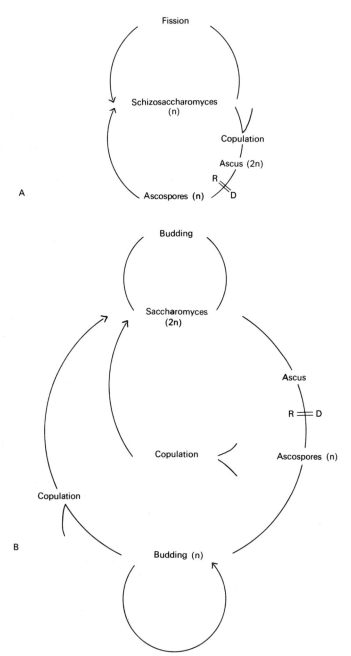

Fig. 132. Life cycles of *Schizosaccharomyces* and *Saccharomyces*.

restores the haploid condition (Fig. 132A). Some variation in this pattern may occur. For example, in *S. japonicus* var. *versatilis* and *S. pombe*, limited division of the zygote may occur in the diploid state before ascospore formation takes place (Suminoe & Dukmo, 1963). It is possible to select diploid strains.

Physical and chemical analyses of the cell walls of *Schizosaccharomyces* show that they are principally composed of two types of glucan and a galactomannan. By growing cells of *S. octosporus* in the presence of magnesium sulphate and 2-de-oxy-*d*-glucose (which appears to interfere with the phosphorylation of sugars in glucan formation), it has been possible to induce the development of wall-free protoplasts which undergo budding (Berliner, 1971). The walls of the spores give a blue reaction with iodine. Growth of the cell wall occurs at one end of the cell only, at the end opposite from the division scar (MacLean, 1964). The cells of *Schizosaccharomyces* show a moderate tolerance of alcohol (about 5–7%), an important property of yeasts used in fermentations (Gray, 1941; Ingram, 1955). In contrast, many wine yeasts can tolerate much higher concentrations of alcohol.

Endomyces (Fig. 133)

This genus represents a mycelial parallel to *Schizosaccharomyces* forming a mycelium which fragments into segments (see Fig. 133). The best-known species is *E. geotrichum* (*Galactomyces geotrichum* of Redhead & Malloch, 1977). In the conidial, non-ascosporic state, this fungus is known as *Geotrichum candidum*, a ubiquitous mould which is common in soil, dairy products, sewage and other substrata. Some isolates are saprophytic or pathogenic in the gut or lungs of humans (Carmichael, 1957), whilst others are pathogenic to plants, e.g. fruits of lemon, tomato and melon (Butler, 1960; Butler *et al.*, 1965). The fungus grows readily in culture, forming broad hyphae with finer lateral branches. The vegetative cells contain 1–4 nuclei.

In its mycelial or conidial state, *E. geotrichum* has been the subject of investigations of its fine structure (e.g. Steele & Fraser, 1973a,b), morphogenesis and growth kinetics (for references see Robinson & Smith, 1976). Branching is of two kinds, pseudodichotomous near the apex, and lateral immediately behind a septum. It is from such lateral branches that conidia develop (Fig. 133A,B). Conidiophores are difficult to differentiate from vegetative hyphae. Prior to conidium formation, apical growth of a hypha ceases, then septa are laid down in the tip region. The septa are two-ply, and separation of the two layers making up the septa leads to the disarticulation of the terminal part of a hypha into cylindrical segments termed **arthrospores** or **arthroconidia** (Cole & Kendrick, 1969c; Kendrick, 1971b). The development of abundant conidia gives cultures a slimy appearance. The perfect state of *E. geotrichum* is less often encountered (Butler & Petersen, 1972). It is homothallic, but mutates to give self-sterile isolates which will mate to form asci in

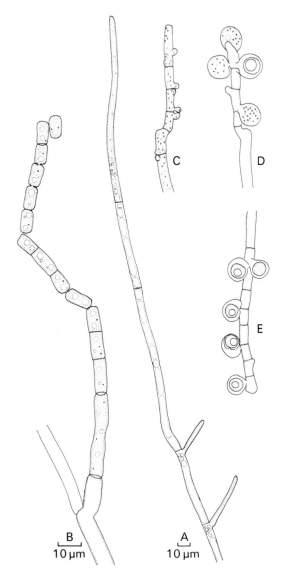

Fig. 133. *Endomyces geotrichum.* A. Vegetative hyphal apex. The two lateral branches near the base are developing conidiophores. B. Conidiophore showing the development and separation of arthroconidia. C. Gametangia developing as lateral bulges of the hyphae on either side of a septum. D. Fusion of gametangia to form asci. In one ascus, a single ascospore is differentiated. E. Mature asci each containing a single ascospore.

certain pairings. In homothallic forms, gametangia arise in pairs on either side of a septum, on the broad main hyphae or on short side branches (Fig. 133C–E). Fusion of the gametangia gives rise to a globose fusion cell which becomes transformed directly into an ascus. The ascus contains only a single ascospore, which has two walls, a smooth inner layer and a furrowed outer layer. In heterothallic strains, the ascus develops by conjugation of the tips of short lateral branches. These develop single ascospores, which on germination develop into homothallic strains. The ascospores contain 1–2 nuclei. Whether or not meiosis occurs is not yet known. Butler & Petersen (1972) have suggested that *Geotrichum candidum* may represent a complex of asexual fungi, for which there may be more than one perfect state. *Endomyces reessii* forms conidia virtually indistinguishable from those of *E. geotrichum*. *Endomyces magnusii*, originally isolated from the slime flux of trees, has four-spored asci formed by conjugation of lateral branches.

Eremascus (Fig. 134)

Two species of *Eremascus* are known. *E. albus* and *E. fertilis* (Fig. 134) are associated with sugary substrata such as mouldy jam, but several collections of *E. albus* have been made from powdered mustard. Harrold (1950) has

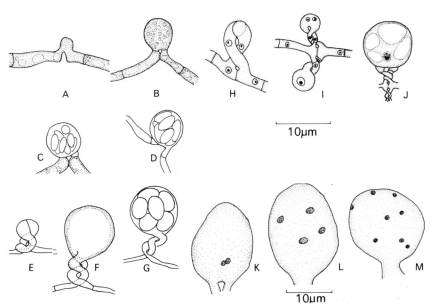

Fig. 134. *Eremascus*. A–D. *E. fertilis,* stages in development of asci. E–G. *E. albus,* stages in development of asci. Note the coiling of the gametangia and the globose ascospores of *E. albus*. H–M. *E. albus,* nuclear behaviour during ascus formation (after Harrold, 1950). H. Uninucleate gametangia. I. Plasmogamy and karyogamy. K–M. Nuclear divisions preceding ascospore formation.

shown that both fungi grow best on media with a high sugar content (e.g. 40%
sucrose), and do not grow well in a saturated atmosphere. The mature
.nycelium consists of uninucleate segments. Both species are homothallic. On
either side of a septum short gametangial branches arise, swollen at their tips,
and coiling around each other in *E. albus*. The gametangial tips in *E. albus* are
usually uninucleate and, following breakdown of the wall separating the tips
of adjacent gametangia, nuclear fusion occurs. This is followed by meiosis and
mitosis (Delamater *et al.*, 1953), so that eight nuclei result, each one being
surrounded by cytoplasm to form the uninucleate ascospores (Fig. 134M). The
ascospores are dispersed passively following breakdown of the ascus wall. On
germination a multinucleate germ tube first forms, but as septa appear the
uninucleate condition is established.

Saccharomycetaceae

Saccharomyces (Figs. 135–137)

About forty species of *Saccharomyces* have been distinguished (van der Walt,
1970a), but the best known is *S. cerevisiae*, strains of which are used in the
fermentation of certain beers and wines, and in baking. *S. cerevisiae* is found
in nature on ripe fruit. Grape wines are often made by spontaneous fermen-

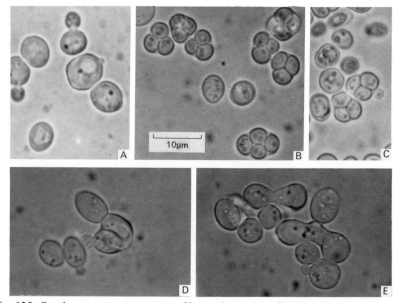

Fig. 135. *Saccharomyces cerevisiae*. A. Vegetative yeast cells (diploid) showing buds
and bud scars. Note the prominent vacuole. B. Yeast asci mostly containing four
spores, sometimes with only three spores in focus. C. Ascus showing a budding
ascospore. D. Ascus in which two spores have fused together and are budding. E. Two
ascospores fusing (top left), and two fused ascospores forming a diploid bud (right).

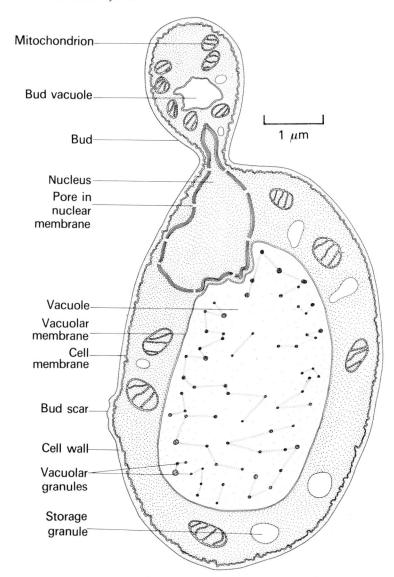

Mitochondrion

Bud vacuole

Bud

Nucleus
Pore in
nuclear
membrane

Vacuole
Vacuolar
membrane
Cell
membrane

Bud scar

Cell wall

Vacuolar
granules

Storage
granule

1 μm

Fig. 136. *Saccharomyces cerevisiae*. Diagrammatic representation of a section of a budding yeast cell as seen under an electron microscope.

tation by yeasts growing on the surface of grapes. Because of the economic significance of yeasts, there is an extensive literature on their cytology, genetics, ecology, nutrition and physiology, and on the technology of yeast (see Rose & Harrison, 1969–1970). The cells of *S. cerevisiae* are elliptical and about 6–8 × 5–6 μm. In suitable conditions, they multiply by budding (Figs. 135A, 136). The relatively small size makes the interpretation of their internal structure difficult at the magnification possible using the light microscope. However, following studies using transmission electron microscopy of thin sections, and examination of carbon-replicas from freeze-fractured material, supplemented by chemical analyses of wall preparations and fractionated cell contents, a much clearer understanding of yeast cell structure has emerged (Matile *et al.*, 1969). In cells with thick walls, three wall layers have been distinguished, but these are less obvious in thin-walled cells. The three layers differ mainly in their chemical composition. The outer layer consists mainly of mannan–protein and some chitin. The middle layer is largely composed of glucan, whilst the innermost layer contains protein–glucan. Some lipid and phosphate is also present. The glucan (yeast cellulose), which may make up to 30% of the dry weight of the cell wall, is composed of polymers of $\beta(1\rightarrow3)$-linked and $\beta(1\rightarrow6)$-linked glucose residues. The mannan (yeast gum), which equals the amount of glucan, is a branched polymer of mannose with an $\alpha(1\rightarrow6)$-linked backbone, and $\alpha(1\rightarrow2)$- and $\alpha(1\rightarrow3)$-linked side chains. Chitin, a polymer of $\beta(1\rightarrow4)$-linked N-acetyl glucosamine, is present in small quantities, and is especially abundant in the region of the bud scars (see below). Protein makes up about 7% of the dry weight of the wall. Some of this protein may be in the form of enzymes such as invertase and other hydrolases which have been demonstrated in wall preparations. The 'skeletal' material of the wall is made up of a random array of microfibrils of chitin, glucan and mannan.

The cell membrane or plasmalemma is of the usual unit-membrane type, but is unusual in containing a series of shallow, elongated pits or invaginations as seen in freeze-etched material. Other inclusions in the cell resemble those of other eukaryotic cells: an endoplasmic reticulum, ribosomes, mitochondria, lipid granules (sphaerosomes), Golgi apparatus, and a nucleus enclosed in a perforated nuclear membrane. The centre of a mature yeast cell contains a prominent large vacuole limited by a single membrane, the tonoplast. The vacuole contains a watery substance, and granules of polymetaphosphate and lipid. The mitochondria are very variable in shape, depending partly on the conditions under which the yeast is growing. They may be spherical, rod-like, thread-like, unbranched or branched. Under anaerobic conditions, the mitochondria may decrease in size and degenerate into tiny spherical or rod-shaped structures lacking cristae. Respiratory-deficient mutants, which form smaller colonies than wild-type forms (petite mutants), possess fewer mitochondria.

The nucleus is difficult to see in living yeast cells. In budding yeast cells, it is to be found between the vacuole and the bud. It consists of a cup-shaped nucleolus and dome-shaped nucleoplasm. Mitotic division is *intranuclear*, i.e. the nuclear membrane remains intact during nuclear division. An intranuclear spindle made up of microtubules stretches between a pair of spindle pole bodies (spindle plaques) on opposite sides of the dividing nucleus (Moens & Rapport, 1971). Vegetative cells of *S. cerevisiae* are generally diploid. There are conflicting reports on the chromosome number of the diploid yeast cell. Cytological studies suggest that the diploid number is eight (McClary *et al.*, 1957; Ganesan, 1959), but this does not correspond with the fourteen linkage groups postulated (Mortimer & Hawthorne, 1966). Later estimates suggest that there may be seventeen linkage groups (Sherman & Lawrence, 1974). Polyploid yeasts are also known.

When the yeast cell buds, its nucleus appears to divide by constriction, and the nuclear envelope does not break down. A portion of the constricted nucleus enters the bud along with other organelles. The cytoplasmic connection is closed by the laying down of wall material. Eventually the bud separates from the parent cell, but its previous point of attachment is distinguishable as a birth scar.

The point of origin of buds is at the poles of the yeast cell. Possibly this is connected with the fact that internal fluid pressure in an ellipsoidal cell is exerted maximally against the cell wall at the points of maximum curvature. The bud scars can be demonstrated by fluorescence microscopy using the fluorescent stain primuline. The bud scars remain visible as circular, craterlike, raised scars, and as many as twenty-three have been seen on a single cell. Calculations based on the average surface area of a yeast cell show that the maximum number of scars which could be accommodated would be about 100, suggesting that yeast cells are not capable of unlimited budding. The origin of a bud is presumably associated with the loosening of the bonds linking the molecules making up the wall so that the wall at that point becomes more plastic. At the optimum temperature for growth, 30 °C, the time taken for a complete cell cycle is about 100 minutes. For a fuller discussion of the biochemistry of morphogenesis in yeast, see Bartnicki-Garcia & McMurrough (1971), Hartwell (1974) and Cabib (1975).

S. cerevisiae can be induced to form ascospores by suitable treatment, and is therefore termed an **ascosporogenous** yeast, in contrast to **asporogenous** yeasts in which ascospores have not been observed. Ascospore development can be induced by growing the yeast on a nutrient-rich presporulation medium containing an assimilable sugar, a suitable nitrogen source for good growth, and vitamins of the B group. Such a medium results in well-grown cells which will sporulate on transfer to a sporulation medium. Sporulation occurs best on media in which budding is inhibited. Low concentrations of an assimilable sugar are necessary to provide energy for the sporulation process. It has been shown that sodium or potassium acetate ($0.1–1.0\%$ w/v) stimulates sporulation (Fowell, 1969; Haber & Halvorson, 1975).

Following treatment with a suitable sporulation medium, the diploid yeast cells develop directly into asci within 12–24 hours. The cytoplasm differentiates into four thick-walled spherical spores, although the number of spores may be fewer (see Fig. 135D). The nuclear divisions which precede spore formation are meiotic, a fact which has been confirmed by genetical studies. Electron microscope studies have shown that, as in mitotic division, the

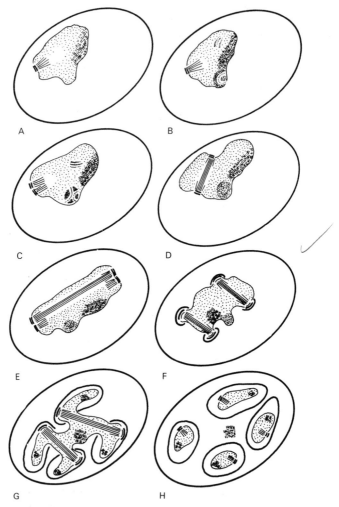

Fig. 137. *Saccharomyces cerevisiae.* Diagrammatic summary of the processes of meiosis and ascospore delimitation (from Beckett *et al.*, 1974). A–D. Spindle pole body replicates and the two spindle pole bodies move to opposite poles of the nucleus: nuclear membrane remains intact. E–F. Further replication of spindle pole bodies and rearrangement: nuclear membrane still intact. New membranes, the ascospore delimiting membranes, form outside the spindle pole bodies. G–H. Envelopment by the ascospore delimiting membranes of the lobes of the dividing nucleus results in the formation of haploid uninucleate ascospores.

nuclear membrane remains intact during meiosis, so that the divisions to form the four haploid daughter nuclei all take place within the original nuclear membrane. This type of nuclear division has been termed **uninuclear** (Moens & Rapport, 1971; Illingworth *et al.*, 1973; Beckett *et al.*, 1974), see Fig. 137.

The spores of yeast are richer in carbohydrate (glucan, mannan and trehalose) and lipid, but poorer in protein and amino-acid content, than vegetative yeast cells. Studies on the metabolic changes which accompany ascospore formation have been reviewed by Fowell (1969), Tingle *et al.* (1973) and Haber & Halvorson (1975).

If ascospores are dissected from the asci using a micromanipulator and allowed to germinate in a nutrient medium they form haploid buds which are often smaller and rounder than the diploid yeast cell, and single spore cultures can be maintained indefinitely in the haploid state. The diploid state may be re-established in several ways:

1. By fusion of ascospores: this may occur inside or outside the ascus. The walls separating ascospores break down, or short conjugation tubes develop which bring the cytoplasm of the two spores into contact. Nuclear fusion follows and the zygote develops diploid buds (Fig. 135).
2. Haploid cells fuse to form diploid cells.
3. Fusion may take place between an ascospore and a haploid cell.

Many strains of *S. cerevisiae* are heterothallic, and the ascospores are of two mating types. Mating type specificity is controlled by a single gene which exists in two allelic states *a* and α, and segregation at the meiosis preceding ascospore formation results in two *a* and two α ascospores. Fusion normally occurs only between cells of differing mating type, and this has been termed legitimate copulation. Such fusions result in diploid cells which readily form asci with viable ascospores (Lindegren & Lindegren, 1943). There are, however, exceptions to the fusion of cells or nuclei of opposite mating type:

1. In haploid colonies devised from single ascospores mutation from *a* to α and from α to *a* may occur, followed by copulation (Ahmad, 1965).
2. In haploid colonies derived from single ascospores two-spored asci have been reported: the ascospores are relatively non-viable. It is believed that such asci arise from diploid cells derived from fusion of cells of the same mating type (illegitimate copulation; Lindegren, 1949).
3. Spontaneous diploidisation may occur following the first mitotic division of a germinating spore by fusion of the two sister nuclei (Winge & Laustsen, 1937).
4. In *S. chevalieri* and in hybrids between this yeast and *S. cerevisiae* a gene *D* for diploidisation may be present. The presence of this gene permits diploidisation to occur in haploid progeny of either mating type. Diploid yeast cells may be heterozygous for the *D* gene (i.e. *Dd*), and then may give rise to asci from which two ascospores will give rise to diploid colonies directly and two will not (Winge & Roberts, 1949). For a fuller discussion of the genetics of mating behaviour in yeasts, see Fowell (1969), Mortimer & Hawthorne (1969).

The mechanism by which haploid yeast cells of opposite mating type fuse together is by a binding process involving the walls of the two cells. A clue to the mechanism is afforded by the behaviour of the yeast *Hansenula wingei*. When haploid cultures of opposite mating type are mixed, intense agglutination occurs, a process which has been termed *sexual agglutination*. This is interpreted as a pre-conjugation phenomenon ensuring intimate contact between cells of opposite mating type. The agglutinated cells cohere so tightly that individual cells assume a polygonal shape. After agglutination, conjugation occurs. The agglutinating factor has been extracted in cell-free suspensions, and has been shown to be a heterogeneous complex of protein and mannan (Fowell, 1969). In most species of *Saccharomyces*, only weak sexual agglutination reactions have been found. A sex-specific factor produced by mating-type α cells of *S. cerevisiae* has also been partially purified which inhibits budding in cells of the opposite mating type, and induces them to elongate to form copulatory processes (Yanagishima, 1969; Duntze *et al.*, 1970; Herman, 1971).

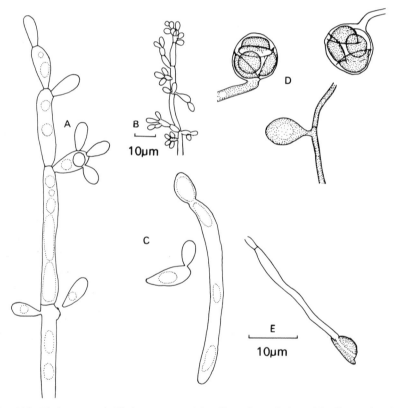

Fig. 138. *Endomycopsis fibuligera*. A, B. Mycelium from 3-day-old culture showing blastospore formation. C. Blastospores germinating by germ tube, or budding to form a further blastospore. D. A young ascus and two mature asci containing four hat-shaped ascospores. E. Germinating ascospore. A, C, D, E to same scale.

Endomycopsis (Fig. 138)

Endomycopsis is a mycelial form which reproduces by buds (blastospores) and also forms asci parthenogenetically or following isogamous fusion. About ten species are known (Kreger-van Rij, 1970). *Endomycopsis fibuligera* grows in flour, bread and macaroni, and produces an active extracellular diastase, an unusual property of yeasts (Wickerham *et al.*, 1944). In culture, it may form budding yeast cells and septate branched hyphae which produce blastospores laterally and terminally (Fig. 138). Arthrospore formation has also been demonstrated (Müller, 1964). Ascus formation can be induced by growing the yeast for a few days on malt extract agar and transferring to distilled water. The asci are mostly four-spored, and the spores are hat-shaped (Fig. 138D), having a flange-like extension of the wall. *Endomycopsis fasciculata* is an ambrosia fungus lining the galleries of ambrosia beetles and serving as a source of food for the larvae (Batra, 1963). The fungus is homothallic and forms two-spored asci. *Endomycopsis scolyti*, associated with bark beetles of the genus *Scolytus*, parasitic on conifers, is heterothallic, and conjugation only occurs when yeast cells of both types are mixed. Conjugation results in the formation of diploid cells which give rise to asci containing one to four hat-shaped ascospores (Phaff & Yoneyama, 1961). There are a number of mycelial forms which form blastospores resembling *Endomycopsis*, but in which asci have not been seen. One of the best-known is *Candida albicans*, the cause of various diseases in man such as thrush (Gentles & La Touche, 1969). Other species of *Candida* have been identified as strains of heterothallic species of *Endomycopsis* (Wickerham & Burton, 1954).

TAPHRINALES

The Taphrinales are parasitic on flowering plants and ferns, causing a wide variety of disorders. Two families, Taphrinaceae and Protomycetaceae, are recognised by some authors, but others have treated the latter group as a separate order, Protomycetales (Kramer, 1973; Reddy & Kramer, 1975). This group, which will not be considered further, includes *Protomyces* which causes infected hosts to form blister-like galls on leaves and petioles. Common species are *P. macrosporus* on *Aegapodium* and *P. inundatus* on *Apium*.

The Taphrinaceae includes only a single genus *Taphrina*, formerly named *Exoascus*.

Taphrina (Figs. 139, 140)

About 100 species are known, mostly parasitic on Amentiferae and Rosaceae (Mix, 1949), causing diseases of three main kinds:

1. Leaf curl or leaf blister diseases, e.g. *T. deformans*, the cause of peach leaf

curl; *T. tosquinetii*, the cause of leaf blister of alder, and *T. populina*, the cause of yellow leaf blister of poplar.

2. Diseases of branches in which the infected plant undergoes repeated branching to form dense tufts of twigs called witches' brooms, e.g. *T. betulina* causing witches' brooms of birch. Similar twig proliferation is also caused by

Figure 139. *Taphrina deformans.* Peach leaf showing leaf curl.

mites. *T. instititiae* causes witches' broom of plum and damsons, and *T. cerasi* witches' brooms and leaf curl of cherry.

3. Diseases of fruits, e.g. *T. pruni* which causes the condition known as pocket plums in which the fruit is wrinkled and shrivelled and has a cavity in the centre in place of a stone.

T. deformans. Peach leaf curl is common on leaves and twigs on peach and almond, especially after a cool moist spring. Towards the end of May, infected peach leaves show raised reddish puckered blisters which eventually

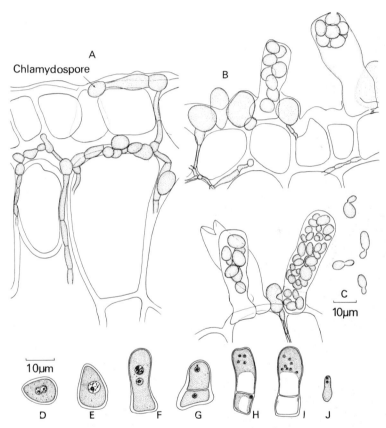

Fig. 140. *Taphrina deformans*. A. T.S. peach leaf showing intercellular mycelium and subcuticular chlamydospores. B. T.S. peach leaf showing chlamydospores and asci, containing eight ascospores. C. T.S. leaf showing a dehisced ascus, an eight-spored ascus and an ascus in which the ascospores are budding. Ascospores budding outside the ascus are also shown. D–J. Cytology of ascus formation (after Martin, 1940). D, E. Fusion of nuclei in chlamydospore. F. Elongating ascogenous cell containing two nuclei formed by mitosis from the fusion nucleus. The upper nucleus has begun to divide meiotically. G. Uninucleate ascus with uninucleate basal cell. H, I. Four- and eight-nucleate asci. J. Binucleate germ tube in germinating ascospore.

acquire a waxy bloom (Fig. 139). Sections of leaves in this condition show an extensive septate mycelium growing between the cells of the mesophyll and between the cuticle and the epidermis, where the hyphae end in swollen chlamydospores (see Fig. 140). The interface between the parasitic fungus and the host takes the form of contact between their walls. No specialised haustoria have been found in *T. deformans*, although they have been reported in some other species (Syrop, 1975a,b; Marte & Gargiulo, 1972). Cytological studies (Martin, 1940; Kramer, 1961; Caporali, 1964) show that the segments of the mycelium and the young chlamydospores are binucleate. In the chlamydospore, the two nuclei fuse and the diploid nucleus divides mitotically. The upper of the two daughter nuclei then undergoes meiosis followed by a mitosis, so that eight nuclei result, which form the nuclei of the eight ascospores. The lower daughter nucleus remains in the lower part of the chlamydospore and is often separated from the other nucleus by a cross-wall. During these nuclear divisions, the wall of the chlamydospore has stretched to form an ascus. Delimitation of the ascospores occurs at the eight-nucleate stage. The individual nuclei become enclosed by double delimiting membranes which arise by invagination of the plasmalemma of the developing ascus (Syrop & Beckett, 1972). Within the ascus, the ascospores may bud so that ripe asci may contain numerous buds (see Fig. 140c). The asci form a palisade-like layer above the epidermis, and it is their presence which gives the leaf its waxy bloom. The ascospores or buds are projected from the asci which often opens by a characteristic slit (Fig. 140c). Following treatment with potassium hydroxide, the ascus wall of *T. populina* appears double, and Schneider (1956) has compared it with the bitunicate asci of other Ascomycetes. Yarwood (1941) has shown that there is a diurnal cycle of ascus development and discharge in *T. deformans*. Nuclear fusion takes place during the afternoon or evening; the nuclear divisions are complete by about 5 a.m. and the spores appear mature by 8 a.m. However, maximum spore discharge does not occur until about 8 p.m. Outside the ascus the spores or conidia may continue budding and the fungus can be grown saprophytically in a yeast-like manner in agar or in liquid culture. Young leaves can be infected from such budding cells, and it has been shown that a culture derived from a single spore can cause infection resulting in the formation of asci, so that *T. deformans* is homothallic. In this respect it differs from some other species, e.g. *T. epiphylla* where fusion of buds, presumably of different mating type, is necessary before infection can occur (Wieben, 1927). In *T. deformans* the binucleate condition is established at the first nuclear division of a bud placed on a peach leaf; the two daughter nuclei remain associated in the germ tube which penetrates the cuticle (Fig. 140J).

The distortion of the host tissue is associated with division and hypertrophy of the cells of the palisade mesophyll. In liquid cultures of *T. deformans*, especially on media containing tryptophane, considerable quantities of indole acetic acid have been demonstrated (Crady & Wolf, 1959; Sommer, 1961).

Comparison of healthy and infected leaf tissue shows that infected leaves contain increased cytokinin activity, increased indole-3-acetic acid and tryptophane content (Sziráki *et al.*, 1975; Kern & Naef-Roth, 1975).

The fungus overwinters in two ways:

1. Mycelium persists throughout the winter in the cortex of infected twigs. Towards the end of the winter, the mycelium penetrates buds beneath the infected area.
2. Ascospores and conidia survive on the surface of twigs and between bud scales during autumn and winter. Between November and March, the spores and conidia develop thick walls and, in spring, as the peach buds open, the conidia produce germ tubes which penetrate the young leaves (Caporali, 1964).

Control of peach leaf curl is achieved by spraying with a suitable fungicide before bud-break in spring (Burchill *et al.*, 1976).

2: PLECTOMYCETES

In this group (which is really an assemblage of unrelated fungi) are included Ascomycetes with ascocarps which are rudimentary, or consist of a loose investment of hyphae or are globose cleistocarps, i.e. closed ascocarps not opening by an ostiole. We shall consider two orders, the Erysiphales and the Eurotiales. The two groups are not closely related. The Erysiphales are biotrophic parasites of higher plants. The ascocarps contain one to several oval to club-shaped asci which discharge their ascospores violently. The Eurotiales are mostly saprophytic. The ascocarps are very variable in form, but the asci are small and globose. The ascospores are not violently discharged.

ERYSIPHALES

The Erysiphales is an important group of plant pathogens (Spencer, 1978). Yarwood (1973, 1978) recognises two families: Perisporiaceae, the dark mildews, and Erysiphaceae, the powdery mildews or white mildews. The Perisporiaceae occur in warm, humid, tropical forest on adult leaves. The powdery mildews are biotrophic parasites of Angiosperms, and their common name is derived from the mealy appearance of the conidia on infected foliage. Diseases caused by the Erysiphales are of economic significance – e.g. cereal and grass mildew caused by *Erysiphe graminis*, apple mildew caused by *Podosphaera leucotricha*, American gooseberry mildew caused by *Sphaerotheca mors-uvae*. The mycelium of the Erysiphales is usually superficial, with haustoria often confined to the epidermal cells. Chains of conidia arise in basipetal succession from a mother cell on the mycelium. Later ascocarps (cleistothecia) may be formed. The cleistothecia are brown globose bodies and have no ostiole. They may contain one to several asci which discharge their spores explosively.

There is a voluminous literature on ascocarp development (Luttrell, 1951; Gordon, 1966). Development probably follows the same pattern in most species (Fig. 141). Two branches of the superficial mycelium come into contact. The terminal cells are at first uninucleate. One cell encircles the other, and the encircled cell enlarges. Following breakdown of the walls separating them, or by the development of a conjugation tube a nucleus is transferred to the larger central cell. Fusion of the two nuclei has been reported, but accounts differ in this respect. Many early authors have interpreted the central enlarged cell as an ascogonium, and the encircling cell as an antheridium, and have claimed that the ascogonium initiates the development of asci, but Gordon, although claiming nuclear fusion in the central cell, finds no evidence that it plays a functional role in ascus development. He therefore terms the central cell a pseudoascogonium and the encircling cell (and cells derived from it) the pseudoantheridium. In Gordon's account (which is unfortunately a composite account based on the development of four species), further development is as follows. The pseudoantheridial cells which completely encircle the pseudoascogonium divide to form the peripheral cells of the ascocarp. The outer

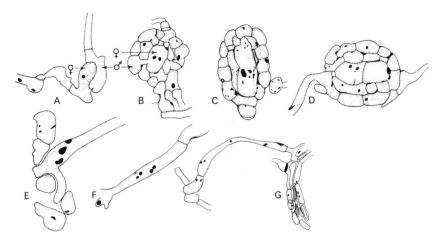

Fig. 141. Ascocarp development in various Erysiphaceae (after Gordon, 1966). A. Contact between pseudoascogonium (female) and pseudoantheridium (male) in *Erysiphe cichoraceacum*. B. Conjugation between a cell of the pseudoantheridum (male) and the pseudoascogonium (female) in *E. cichoraceacum*. Note that the pseudoascogonium is binucleate. C. Binucleate pseudoascogonium of *Microsphaera diffusa*. The pseudoascogonium is surrounded by two layers of pseudoantheridial cells. D. Extension of a mother cell of the pseudoantheridium to form a receptive hypha in *E. cichoraceacum*. E. Tip of receptive hypha in *E. cichoraceacum* making contact and encircling a cell of the superficial mycelium. F. Pairs of nuclei in the receptive hyphae of *M. diffusa*. G. Receptive hypha of *E. cichoraceacum* showing its connection with a cell of the peridium, derived from the pseudoascogonium.

cells (which we may call mother cells) develop two- to five-celled hyphae, receptive hyphae, with uninucleate segments. The tips of the receptive hyphae make contact with vegetative hyphae on the surface of the host. A nucleus from the vegetative hypha apparently migrates into the receptive hypha and pairs with the nucleus already present. One of the paired nuclei, thought to be the nucleus from the vegetative hypha, divides, and a daughter nucleus migrates through the septal pore into the next cell of the receptive hypha. This process is repeated until a daughter nucleus pairs with the nucleus of the peripheral cell which gave rise to the receptive hypha (i.e. mother cell). This is followed by deliquescence of the terminal part of the receptive hypha. The binucleate mother cell now enlarges and becomes multinucleate and divides. Inner cells formed by division from the enlarged mother cell become binucleate whilst the outer cells remain uninucleate. Many of the outer uninucleate cells produce new receptive hyphae and the whole process of anastomosis and nuclear migration is repeated.

During the events described above the two nuclei in the pseudoascogonium fuse and the enlarged nucleus divides. Following cell wall formation between the daughter nuclei the pseudoascogonium becomes a series of three to five uninucleate cells which at first contain denser cytoplasm than the surrounding pseudoparenchymatous cells, but later this distinction is no longer obvious. Occasional binucleate cells have been found in the three- to five-celled pseudoascogonium, but it is possible that such cells were dividing. The immature ascocarp consists of a pseudoparenchymatous centrum composed largely of binucleate cells derived from the mother cells of the pseudoantheridium intermixed with some uninucleate cells, and surrounded by a peridium, some four to six cell layers thick, which become darkly pigmented. Uninucleate and binucleate cells above the middle part of the centrum lyse (break down). Karyogamy (i.e. nuclear fusion) occurs within certain of the binucleate cells, more or less isolated from surrounding cells by lysis. These cells then enlarge to form asci. The asci appear to have the ability to absorb the uninucleate and binucleate cells of the centrum. Eventually the asci (or in some cases a single ascus) occupy almost the entire centrum. Meiosis of the fusion nucleis in developing asci is usually delayed until the centrum cells have all been absorbed.

In addition to the 'receptive hyphae' the cleistothecia of all Erysiphaceae bear thick-walled hyphae (appendages) which may be branched or unbranched, and often have a highly distinctive appearance. The type of appendage has proved a useful aid to classification. Another criterion is whether there is only one ascus in each ascocarp or several, and using such criteria eight genera have been distinguished (Blumer, 1967; Yarwood, 1973, 1978).

Erysiphe (Figs. 142–146)

Diseases caused by *Erysiphe* include grass and corn mildew (*E. graminis*),

Figure 142. *Erysiphe graminis*. A. Conidial pustules on wheat leaf. B. Cleistothecia on wheat leaf sheath. About twice natural size.

mildew of cucurbits and other plants (*E. communis* = *E. cichoracearum*) and pea and clover mildew (*E. polygoni*). The last-named is especially common in Britain on clover, *Polygonum* and on *Heracleum sphondylium* but it has an exceptionally wide range of other hosts. *E. graminis* can be collected throughout the year in the conidial state on numerous grasses and cereals. A number of host-specific forms (or *formae speciales*) of *E. graminis* exist, e.g. *E. graminis* f. sp. *tritici* infects wheat (*Triticum*) but not barley, whilst *E. graminis* f. sp. *hordei* infects barley (*Hordeum*) but not wheat. It is now known that hybrids between *formae speciales* can occur, especially if the hosts themselves will hybridise. For example, it has been shown that hybridisation between the three *E. graminis* f. sp. *agropyri*, *tritici* and *secalis* can occur, resulting in viable ascospores. The three host genera *Agropyron* (couch grass), *Triticum* (wheat) and *Secale* (rye) can also hybridise (Hiura, 1978). The cleistothecia are commonly found on cereals in the summer and autumn. The appearance of conidial pustules on wheat leaves is shown in Fig. 142. The pustules are white

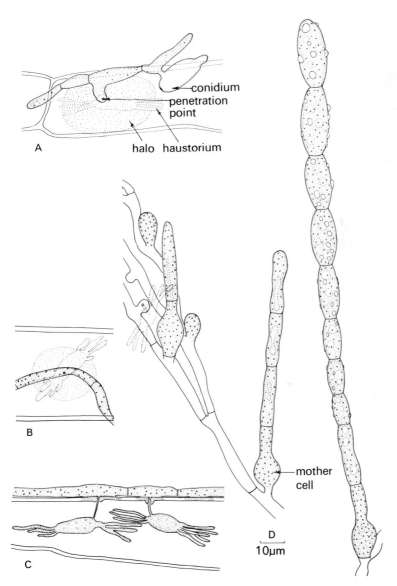

Fig. 143. *Erysiphe graminis.* A. Two-day-old germinating conidium on wheat leaf, showing penetration point, surrounded by a 'halo' (stippled). A haustorium has developed beneath the penetration point. B. Penetration from an established mycelium. C. Section of an epidermal cell showing two penetration points and two haustoria. Note the thickening of the epidermal cell beneath the penetration point. D. Mycelium and conidiophores, showing the swollen flask-shaped mother cell.

to pale brown in colour. Conidia on wheat leaves germinate within 1–2 days to form short germ tubes (Fig. 143A). The germ tube attaches itself to the epidermis by a pad or appressorium and, beneath the point of attachment, a fine infection tube penetrates the host cell wall. Penetration is probably a two-stage process (Edwards & Allen, 1970; Stanbridge *et al.*, 1971; Ellingboe, 1972). The cuticle and epidermal cell wall are subjected to enzymic degradation, and evidence of this can be seen in the 'halo' surrounding the actual point of penetration of the cell wall which is stained by cotton-blue. Following enzymic softening of the outer wall, mechanical penetration of the inner wall takes place. Opposite the point of penetration, the host wall may bear a thickened papilla (Fig. 143C). It is possible that some types of genetic resistance to infection possessed by certain cereals are conferred by repression of the fungal wall-degrading enzymes. The thickened wall papilla has different staining properties around points of unsuccessful penetration as compared with points of successful penetration. Around unsuccessful penetrations, the papilla stains with bromo-phenol blue, and the staining reaction extends throughout the affected cell to adjacent mesophyll cells (Edwards, 1970; Lin & Edwards, 1974). Where infection of a suitable host takes place, the infection tube enlarges within the epidermal cell to form an elongate uninucleate haustorium with finger-like projections developing from opposite ends (Fig. 143C). Such haustoria are typical of *E. graminis*. In most other members of the Erysiphales, the haustoria appear as simple globose bodies (Fig. 146B), but in fact they are also lobed. The lobes may be folded backward over the body of the haustorium (Perera & Gay, 1976; Bushnell & Gay, 1978). The whole of the haustorium, i.e. the body and the finger-like lobes, is enclosed in a **sheath** (Fig. 144A). The lobes may be individually enclosed, or several lobes may be enclosed in a common extension of the sheath. The body of the haustorium contains a single nucleus, mitochondria and unidentified vesicles, whilst the most prominent features of the haustorial lobes are their mitochondria and branched system of tubules forming the endoplasmic reticulum. Below its point of penetration of the host cell wall the neck of the haustorium is surrounded by a collar which differs in structure from the host wall and may represent a deposit of material on it. The sheath surrounding the haustorium is probably a modified invagination of the host plasmalemma (ectoplast), so that the haustorium does not lie freely in the host cytoplasm but is bounded by a membrane continuous with the plasmalemma. Immediately outside the wall of the haustorium there is a wide **sheath matrix**, itself bounded by a **sheath membrane** contiguous with the tonoplast of the host cell. Thus the key interface between fungus and the host is the sheath membrane. At various points the sheath membrane shows tubular invaginations, and these presumably represent the points at which interchange of material between the host cell and the pathogen are taking place. In the region of the haustorial neck, the sheath surrounding the neck is very tightly attached, and thickened to form two neck-bands, composed of distinctive material. It is believed that these

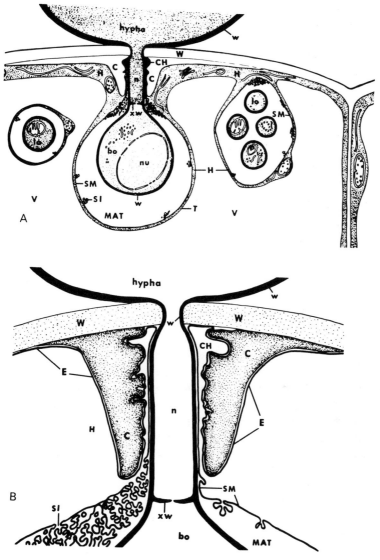

Fig. 144. Interpretation of fine structure of haustorium of *Erysiphe graminis* (Bracker, 1968). A. Section of host leaf at point of penetration. The body of the haustorium (bo) containing a single nucleus (nu) lies immediately beneath the point of penetration. The body of the haustorium is enclosed in a sheath with extensive matrix (MAT). The sheath membrane (SM) is in contact with the host tonoplast (T). The sheath membrane bears invaginations (SI). A single lobe of another haustorium enclosed in an extension of the sheath is shown to the left of the diagram, and four lobes enclosed in a common sheath to the right. B. Enlargement of neck of haustorium. Note the thickened collar (C) deposited on the host cell wall (W). The sheath membrane (SM) is continuous with the host ectoplast (E). (XW), crosswall or septum; (CH), channel; (H), host cytoplasm.

neck-bands constitute an impervious seal ensuring that materials entering the haustorium from the host cell do so through the sheath membrane surrounding the main body and lobes of the haustorium (Bracker, 1968; Bushnell, 1972; Bracker & Littlefield, 1973; Bushnell & Gay, 1978). Many of the features of the haustorium of *E. graminis* have also been found in *E. cichoracearum* (McKeen *et al.*, 1966).

In a susceptible host, the infected cell remains alive. Following successful establishment in an epidermal cell, the superficial mycelium develops branches, further appressoria and haustoria (Slesinski & Ellingboe, 1969; Ellingboe, 1972) (Fig. 143B,D), and within 7–10 days begins to develop conidia. If, however, a non-susceptible grass host is infected, the host cell undergoes rapid necrosis and adjacent cells may also die, thus restricting further development of the pathogen. Infection of races of wheat and barley bred for resistance to mildew differs in several respects from infection of susceptible hosts. Resistance genes confer resistance in several other ways:

1. The proportion of conidia which germinate and produce haustoria is reduced.
2. Haustorial development is delayed and the size of the haustoria is reduced.
3. Sporulation, i.e. development of conidia, is suppressed (Masri & Ellingboe, 1966).

The conidia develop from a flask-shaped mother cell within which nuclear division occurs. The mother cell elongates away from the host leaf and a cross-wall cuts off the hyphal tip. Further cross-walls develop so that a chain of cells is formed, increasing in length at its base – i.e. by further divisions of the mother cell. Each conidium is uninucleate. The segments become swollen and barrel-shaped, and are detached by wind. This type of conidial apparatus is usually assigned to the form-genus *Oidium* of the Fungi Imperfecti, but Yarwood (1978) supports the view that *Acrosporium* is the correct name for the form-genus. Similar conidia are found in most members of the Erysiphales. Interpretation of the mother cell as a rudimentary phialide is not generally accepted. Hughes (1953) refers to these conidia as meristem arthrospores and not phialospores. The fine structure of conidia of *E. cichoracearum* has been studied by McKeen *et al.* (1967). It has been estimated that the time taken for conidium development of *E. graminis* is about 3 hours. Only the terminal spores of the chain are released. These spores differ from more proximal spores in that they are attached to adjacent spores by a smooth central cushion. Spore discharge is aided by wind causing agitation of the leaves, and is inhibited by surface wetness of the leaf. There is a general diurnal periodicity of spore release, with the maximum spore release in the early afternoon, but this is probably related to conditions for spore separation rather than to periodicity in spore formation (Hammett & Manners, 1971, 1973, 1974; Plumb & Tanner, 1972; Ward & Manners, 1974).

The conidia of *Erysiphe* are unusual in their ability to germinate at low humidities, even at zero relative humidity. This has been attributed to the fact that they have a water content as high as 70%, compared with about 10% for other representative air-borne fungus spores (Somers & Horsfall, 1966). It has also been postulated that the large lipid reserves of the conidia provide a substrate which releases relatively large amounts of water on respiration (McKeen, 1970). Other reserves within conidia are fibrosin bodies which are disk-like bodies containing carbohydrate. Possibly the ability to germinate at low humidity is related to the fact that powdery mildews are abundant in hot dry seasons (Schnathorst, 1965). It is of interest that free water may actually inhibit infection by some powdery mildews, e.g. *Erysiphe graminis* (Manners & Hossain, 1963), *Sphaerotheca pannosa* (Perera & Wheeler, 1975; Butt, 1978).

The cleistothecia of *E. graminis* are dark brown and globose, and grow on the basal leaves and leaf-sheaths of cereals nestling in a dense mass of mycelium (Figs. 142B; 145). In many species of *Erysiphe*, the cleistothecial wall also bears unbranched dark appendages with free ends (see Fig. 146 of *E. polygoni*). Each cleistothecium has a wall made up of several layers of cells, surrounding a number of asci. There is no ostiole, and it is for this reason that the ascocarps are termed cleistothecia, although the term perithecium is also used. They crack open by swelling of the contents and the asci discharge their spores. Ascospores of *E. graminis* formed in the current season are capable of

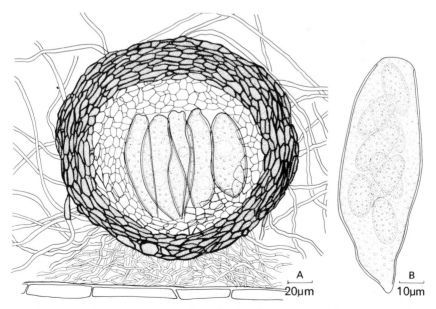

Fig. 145. *Erysiphe graminis*. A. Section of cleistothecium containing several asci. B. Ascus.

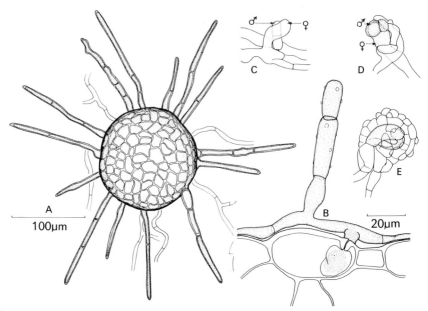

Fig. 146. *Erysiphe polygoni*. Cleistothecium showing dark, free-ended equatorial appendages, and the superficial mycelium anchoring the cleistothecium to the host leaf. B. T.S. host leaf showing simple haustorium, superficial mycelium and conidial chain. C. Contact between pseudoascogonium (female) and pseudoantheridium (male). D. Pseudoascogonium becoming surrounded by cells of pseudoantheridium. E. Pseudoascogonium almost completely enclosed.

infection, but they can also survive for up to thirteen years (Moseman & Powers, 1957). *E. graminis* is heterothallic, and this is probably true of some other Erysiphaceae (Yarwood, 1978).

Infection of cereals by *E. graminis* results in increased respiration of the host, and a decrease in photosynthesis, and this may mean that an infected leaf is unable to export carbohydrate. Areas of healthy leaf adjacent to mildew pustules produce carbohydrate which is translocated towards the pustule. These effects are ultimately reflected in the reduced weight of shoots, roots and ears (Brooks, 1972a; Jenkyn & Bainbridge, 1978).

Chemical control of diseases caused by Erysiphales can be achieved by dusting or spraying host plants with a fungicide containing sulphur. Some systemic fungicides can be used, e.g. ethirimol. The value of using systemic fungicides is that they give protection against mildew throughout the growing period. Increase in yields of barley of 100% have been obtained using ethirimol. The quality of the grain is also improved (Bent, 1970, 1978). Control can also be achieved by selecting and breeding resistant host varieties (McIntosh, 1978). The inheritance of mildew resistance is complex. Seven loci for resistance have been located in *E. graminis hordei*, at least five of them located on one barley chromosome. At least seventeen alleles conditioning resistance are known, so that the potential range of combinations of resistance in the host is very large (Moseman, 1966; Wolfe, 1972).

Sphaerotheca (Fig. 147)

Common species are *S. macularis* var. *fuliginea* (= *S. fuliginea*), on dandelions and other Compositae; *S. macularis* (= *S. humuli*), hop mildew; *S. mors-uvae*,

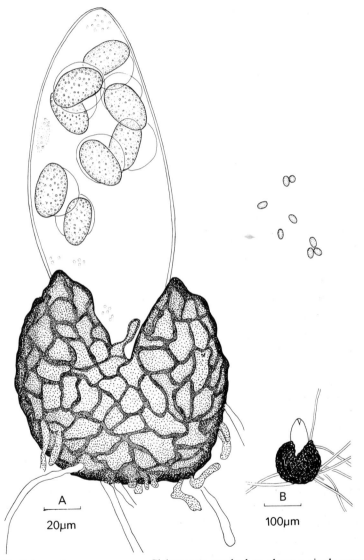

Fig. 147. *Sphaerotheca pannosa*. A. Cleistocarp crushed to show a single ascus. B. Cleistocarp showing discharged ascospores.

American gooseberry mildew (on gooseberries and blackcurrants); and *S. pannosa*, rose mildew.

The cleistothecial structure of *Sphaerotheca* closely resembles that of *Erysiphe*. The appendages are simple, but instead of containing several asci, each ascocarp contains only one (Fig. 147). *S. macularis* var. *fuliginea* is homothallic. The fine structure of developing and mature cleistothecia of *S. mors-uvae* has been studied by Martin *et al.* (1976). They have shown that the darker melanised cells forming the peridium are, like those of vegetative hyphae, uninucleate. Most of the inner cells of a cleistocarp are binucleate, suggesting that they may have arisen from a binucleate ascogonial fusion cell. Another interesting discovery is that fibrosin bodies, previously reported from conidia, are also present in the ascospores.

The mycelium and conidia of *S. pannosa* are common on leaves and shoots of cultivated and wild roses. Cleistothecia are formed on twigs, embedded in a dense mycelial felt. Overwintering is not only by means of ascospores, but as mycelium within dormant buds. Price (1970) has concluded that in *S. pannosa* the cleistocarps do not play any significant rôle in perennation. In the case of *S. mors-uvae* from blackcurrants, less than one in a thousand overwintered cleistocarps seem to be functional. Loss in viability is associated with degeneration of asci and ascospores. Cleistocarps overwintered on the soil surface are invaded by chitinolytic micro-organisms which may play an important rôle in degrading the walls of the cleistocarp prior to cracking open as the ascus expands (Jackson & Wheeler, 1974; Jackson & Gay, 1976).

Podosphaera (Fig. 148A)

The cleistothecia of *Podosphaera* contain a single ascus and bear characteristic flattened dichotomously branched appendages (Fig. 148A). *P. leucotricha* is the cause of apple mildew, and the mycelium and conidia are visible in spring on the expanding foliage and young shoots, probably developing from a perennating mycelium. Cleistothecia are formed on the young branches. Woodward (1927) has reported that the asci themselves are projected from the perithecia. The ascocarp gapes open as the asci expand by absorbing water, but the wall is elastic and eventually the asci are thrown out by the snapping together of the 'jaws' of the cleistothecium, for a distance of several centimetres. If the ascus alights in water it continues to expand and explodes, shooting out its ascospores. A similar double discharge has been reported for *Erysiphe graminis* (Ingold, 1939). Another common species of *Podosphaera* is *P. clandestina* (= *P. oxyacanthae*) on hawthorn. The incidence of disease is associated with hedge-clipping during the summer when conidia are abundant in the air. Clipping removes the apical buds, thus destroying apical dominance and leading to the enlargement of axillary buds. The fungus can penetrate these lateral buds and overwinter there. Effective control can be achieved by avoiding clipping, or restricting it to winter (Khairi & Preece, 1978a,b).

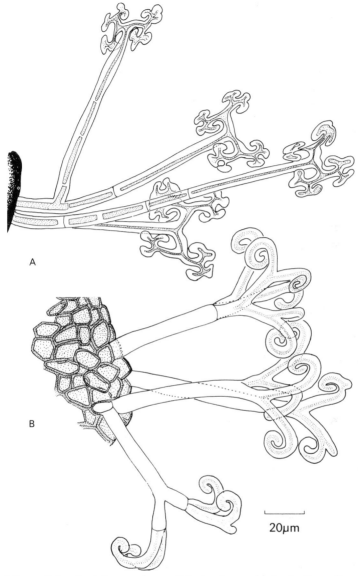

Fig. 148. A. Cleistothecial appendages of *Podosphaera clandestina* from *Crataegus*. B. Cleistothecial appendages of *Uncinula bicornis* from *Acer*.

Microsphaera has similar appendages to *Podosphaera*, but differs in having more than one ascus in each cleistothecium. *Microsphaera alphitoides* is the cause of oak mildew, common on sucker shoots and seedlings. The cleistothecial stage is rare, and seems only to be found during hot summers.

Uncinula (Fig. 148B)

The cleistothecia of *Uncinula* have several asci, and appendages which may be branched but with recurved tips (Figs. 147B). *U. bicornis* (= *U. aceris*) forms cleistothecia on the underside of sycamore leaves. *U. necator* is the cause of vine mildew. When first discovered the conidial state only was known and the fungus was called *Oidium tuckeri*. It threatened the wine industry of France, but experiments on controlling the disease led to the discovery of the efficacy of sulphur-containing fungicides.

Phyllactinia (Figs. 148–150)

P. guttata (= *P. corylea*) grows on hazel leaves forming cleistothecia on the

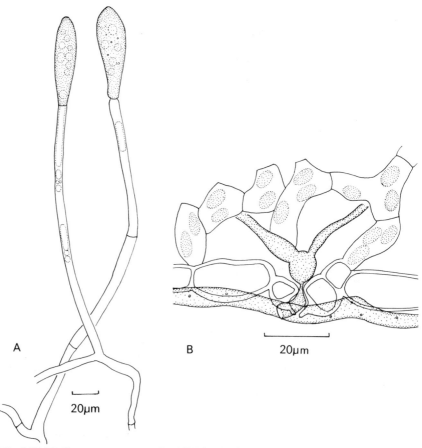

Fig. 149. *Phyllactinia guttata*. A. Conidiophores showing the single terminal conidium. B. T.S. leaf of *Corylus avellana* showing penetration of stoma in lower epidermis, and extension of the mycelium into the mesophyll.

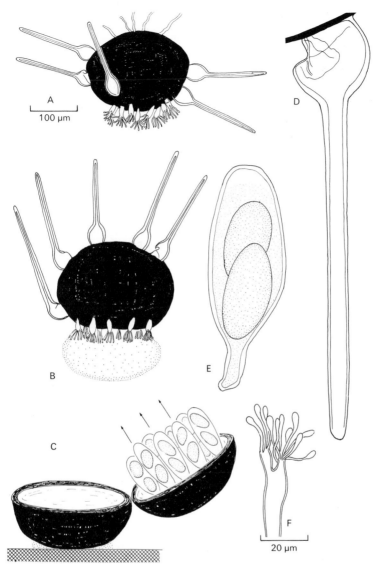

Fig. 150. *Phyllactinia guttata*. A. Cleistocarp on lower side of hazel (*Corylus*) leaf. The radiating bulbous appendages are horizontal. The branched secretory appendages which crown the cleistocarp are on the morphologically upper side, i.e. the side nearest the apices of the ascus. B. Position of cleistocarp during its fall from the host leaf. The bulbous appendages are now folded to form flight vanes, ensuring that the sticky mass of mucilage faces downwards. C. Diagrammatic representation of an opened cleistocarp. The cleistocarp is shown attached by mucilage to a surface. The cleistocarp has opened by a circumscissile equatorial line of weakness, and has hinged back so that the apices of the asci now point outwards. Arrows indicate direction of ascospore discharge. D. A bulbous appendage, showing the differential thickening of the wall of the bulb. Collapse of the thinner walls results in movement of the appendage. E. Two-spored ascus. F. Branched secretory appendage. A, B to same scale. D, E, F to same scale.

lower leaf surface in late summer and autumn. As well as the superficial mycelium there is penetration of the mesophyll (Fig. 149B). The conidia are unusual in being formed singly, and are club-shaped (Fig. 149A). Conidia of this type are classified in the form-genus *Ovulariopsis*. In a damp chamber chains of up to four conidia formed in basipetal succession as in other Erysiphales have been observed. The cleistothecia bear two types of appendage, an equatorial group of radiating unbranched appendages with bulbous bases, and a crown of repeatedly branched appendages which secrete mucilage (Fig. 150). The base of the bulbous appendage is thick-walled above and thin-walled below. On drying the appendage bends towards the leaf surface as the thin part buckles inwards, and the pressure of the appendage-tip levers the cleistothecium free from the superficial mycelium. The bulbous appendages now function as 'flights' or vanes, and the cleistocarp plummets downwards like a shuttlecock, orientated during its fall with the sticky mucilage on the lower side. The blob of mucilage between the apical crown of branched appendages helps to stick the cleistothecium onto twigs and leaves. The asci usually contain only two spores (Fig. 150E). Overwintered cleistothecia open by means of an equatorial line of dehiscence, and the base of the fruit-body hinges back to place the asci in a suitable position for ascus discharge to occur (Cullum & Webster, 1977; Webster, 1979).

EUROTIALES

The Eurotiales include fungi of great economic importance, such as *Aspergillus* and *Penicillium*, some species of which cause spoilage of food and textiles, whilst others are used in fermentation. The Gymnoascaceae have aroused interest because of their relationship to skin pathogens of animals and man. Whilst the asci in *Aspergillus* are completely enclosed by a well-defined envelope of sterile hyphae or **peridium**, those of *Gymnoascus* are only partially enclosed by loose hyphae. In *Byssochlamys* the asci, although arising in clumps from ascogenous hyphae, are not enclosed. The inclusion of this genus in the Eurotiales (Kuehn, 1958; Fennell, 1973) is not accepted by all mycologists, and some would place it in the Endomycetales. However, its conidial structures are similar to those of some other Eurotiales, and the genus is, in some respects, intermediate between the Endomycetales and Eurotiales.

A general account of the taxonomy of the Eurotiales has been given by Fennell (1973) who recognises nine families. Of these, we shall consider only two, Gymnoascaceae and Eurotiaceae. The general features shown by all members of the Eurotiales are: ascocarps lacking ostioles and paraphyses; asci irregularly distributed throughout the ascocarp and not arranged in a bundle, produced from fertile hyphae which ramify throughout the centrum of the ascocarp, typically eight-spored, sessile, thin-walled, quickly evanescent; ascospores unicellular, lacking germ pores or germ slits. Many Eurotiales have conspicuous and characteristic imperfect states.

Gymnoascaceae

About fifteen genera have been grouped here (Benjamin, 1956; Kuehn, 1958; Apinis, 1964). Most species occur in soil, and fruit on animal substrata such as dung, feathers, wool, skin and bones. Some are skin pathogens (dermatophytes) of animals including man (Kuehn *et al.*, 1964; Padhye & Carmichael, 1971; Ainsworth & Austwick, 1973; Ajello, 1977). The perfect states of many of these dermatophytes belong to the genera *Nannizzia* and *Arthroderma*, with imperfect states belonging to the form-genera *Microsporum* and *Trichophyton*. The general term 'ringworm' has been used to describe the symptoms of the diseases which they cause. The ascocarp usually consists of a loose mesh of hyphae surrounding the asci, or, rarely, the asci may be naked.

Gymnoascus (Figs. 151–152)

Accounts of this genus have been given by Kuehn (1959), Orr *et al.* (1963) and Apinis (1964). *Gymnoascus reessii* forms minute reddish-brown globose ascocarps on animal substrata, sacking etc., consisting of branched, recurved, thick-walled hyphae loosely enclosing a mass of asci (Fig. 151). Ascocarp

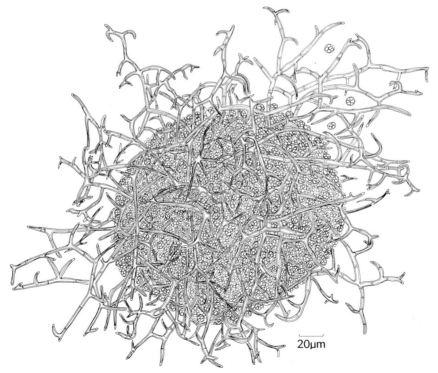

20µm

Fig. 151. *Gymnoascus reessii.* Ascocarp showing branched peridial hyphae and asci.

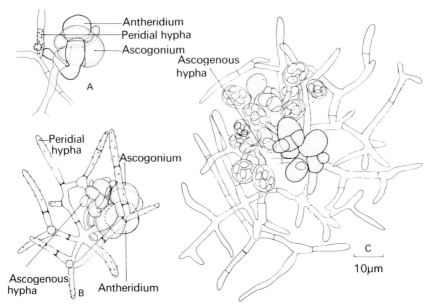

Fig. 152. *Gymnoascus reessii*: development of ascocarp. A. Antheridium and asco-gonium. B. Ascogonium showing development of ascogenous hyphae. The peridial envelope is also developing. C. Young ascocarp showing asci at the tips of ascogenous hyphae.

development in culture begins from paired gametangia which arise from the same or from different hyphae. The antheridium is club-shaped and the ascogonium coils around it (Fig. 152A). The ascogonium becomes septate and its cells give rise to ascogenous hyphae (see Fig. 152B) whose tips develop into croziers. Asci develop from the penultimate cells of the croziers (Kuehn, 1956). The branched peridial hyphae arise from vegetative hyphae in the region of the gametangium (Fig. 152B,C). The asci do not discharge violently; the ascus wall disappears and the spores escape through the loose envelope. There is no conidial stage.

Eurotiaceae

This family, as circumscribed by Fennell (1973), includes several genera with ascocarps ranging in form from a cluster of unenclosed asci in *Byssochlamys*, with soft hyphal aggregates enclosing asci, to hard sclerotium-like fructifica-tions of certain species of *Eupenicillium*. The conidial states of Eurotiaceae are generally phialidic, and include such important and well-known genera as *Aspergillus* and *Penicillium*. Here the conidia are developed within a special-ised cell termed the phialide. A phialide has been defined (Kendrick, 1971a) as:

a conidiogenous cell in which at least the first conidium initial is produced within an apical extension of the cell, but is liberated sooner or later by the rupture or dissolution of the upper wall of the parent cell. Thereafter, from a fixed conidiogenous locus, a basipetal succession of enteroblastic conidia is produced, each clad in a newly laid-down wall to which the wall of the conidiogenous cell does not contribute. . . . The length of the phialide does not change during the production of a succession of conidia . . .

Conidia produced from phialides (phialidic conidia) may be termed **phialo-conidia**. Phialoconidia are found in several groups of Ascomycetes as well as in the Eurotiales, e.g. the Hypocreaceae (p. 337) and in certain genera of the Fungi Imperfecti. The development of phialoconidia in *Aspergillus niger* is illustrated in Fig. 153. The phialides of *A. niger* are arranged in groups upon special cells termed **metulae** which are borne on a globose vesicle. Young phialides are somewhat club-shaped in outline. In *A. niger*, the phialides and

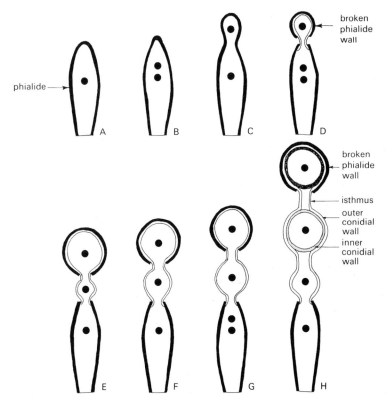

Fig. 153. *Aspergillus niger*: diagrams illustrating phialoconidium ontogeny (modified from Subramanian, 1971). A. Young phialide. B. Mitosis in phialide. C. Conidium initial with daughter nucleus. D. Breakage of phialide wall; formation of new wall layer surrounding conidial cytoplasm; mitosis continuing in phialide. E, F. Extrusion of second phialospore. Note that the phialide has not increased in length. G. Further mitosis in phialide. Note the isthmus connecting successive conidia. H. Three-spored chain. Note the formation of an inner conidial wall within the outer.

phialoconidia are uninucleate, but in some other species of *Aspergillus* they may be multinucleate. In *A. niger*, the single nucleus divides, and the tip of the phialide expands to form a spherical knob which is the initial of the first-formed spore. Into this initial a daughter nucleus passes. The expansion of the first conidium leads to the rupture of the phialide wall near its tip, and the remnants of the broken phialide wall persist as a cap around the first-formed conidium. Before the rupture of the phialide wall, a layer of wall material is laid down (Fig. 153D). This layer becomes the outer wall layer of the conidium within which an inner wall later develops (Fig. 153H). Nuclear division continues within the phialide, and cytoplasm and a wall are laid down around a daughter nucleus to form a second conidium which is extruded from the broken tip of the phialide (Fig. 153E). The second conidium and all subsequently formed conidia differ from the first in that they are not enveloped by remnants of the broken phialide wall. The cytoplasm of the second conidium is at first continuous with that of the first by means of a cylindrical **isthmus**. The formation of an inner conidial wall layer around the conidia severs the cytoplasmic connection. The surviving empty isthmus is sometimes termed the **connective**. The process of nuclear division and the formation of new daughter conidia continues within the phialide, so that a chain of conidiospores is formed, with the youngest spore at the base of the chain, and the oldest at the tip.

The fine structure of phialides and phialoconidium ontogeny has been studied in *Aspergillus* by Trinci *et al.* (1968), Oliver (1972), Fletcher (1976), Hanlin (1976), and in *Penicillium* by Fletcher (1971).

Time-lapse photographic studies of conidial development in *P. corylophilum* by Cole & Kendrick (1969a) indicate that, under the conditions they used, each conidium took 50–60 minutes to develop. The details of conidium ontogeny are probably the same in both genera. At the fine structural level, interest centres on the origin of the wall layer surrounding the conidium. In *Penicillium*, Fletcher has described an 'apical plug' of material (Fig. 154A,D) lining the neck of the phialide but distinct from the phialide wall itself. The apical plug forms the primary wall of the conidium and, as each conidium is extruded, a septum formed by centripetal ingrowth of wall material pinches off the conidial protoplast from the protoplast of the phialide. Closure of the septal pore results in the complete delimitation of each conidial protoplast. A new wall layer is formed inside the primary wall. The delimiting wall (Fig. 154D) survives as the connective linking adjacent spores in the chain. The outer, primary wall may become folded and corrugated so that the surface of the spore appears rough or spiny. Mature spores are often pigmented; green in *Penicillium*; yellow, green, brown or black in *Aspergillus*.

In *Penicillium* and *Aspergillus*, the spores remain dry and are dispersed by air currents. In many other fungi, the phialospore wall is slimy, and chains of spores do not form. Instead, at the tip of the phialide, masses of sticky spores adhere together, e.g. in *Trichoderma* (p. 343), *Gliocladium* (p. 343), *Verticil-*

lium (p. 348). In such cases, dispersal is brought about most probably by insects, rain-splash or other agencies. Multinucleate phialoconidia are quite common, and in some cases the phialoconidia are multicellular. For example, in *Fusarium* (p. 347) and *Cylindrocarpon* (p. 346), the conidium is divided by transverse septa into several cells. Subramanian (1971) has discussed phialidic development more fully. It should be noted that the genera with sticky phialoconidia described above are not classified in the Eurotiales but in the Hypocreaceae or Deuteromycotina.

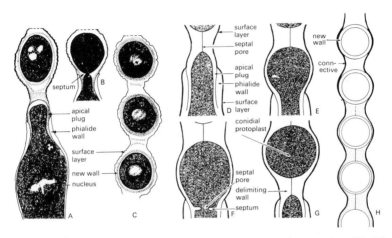

Fig. 154. *Penicillium*: diagrams illustrating phialoconidium ontogeny (modified from Fletcher, 1971). A. *P. clavigerum*: phialide and delimited conidium protoplast. B. *P. clavigerum*: recently delimited protoplast. Note the ingrowth of the septum. C. *P. clavigerum*: distal part of a conidium chain. Note the formation of a new wall around the conidial protoplast. The original wall material between adjacent spores persists as the connective. D–H. Diagrammatic interpretation of mode of conidium formation in *Penicillium*.

The nomenclature of the Eurotiaceae presents problems because perfect and imperfect states of the same fungus may have, in the past, been given separate generic names. For example, the generic name *Eurotium* has been given to the perfect state of some species of *Aspergillus* (e.g. *A. repens*). Another difficulty is that fungi with similar conidial states may have different perfect states. An example of this is *Aspergillus nidulans*, the name for the conidial state of a fungus whose perfect state is *Emericella nidulans*. A further problem is that many species of *Aspergillus* and *Penicillium* have no known perfect state, which would lead to their being classified in the Deuteromycotina or Fungi Imperfecti. A system of classification which separates organisms which are morphologically similar, and are probably related, differing only in the absence of a particular character, is clearly unsatisfactory.

Byssochlamys (Fig. 155)

Byssochlamys is a genus of soil fungi, including one thermotolerant species, *B. verrucosa* (Brown & Smith, 1957; Stolk & Samson, 1972; Samson, 1974; Samson & Tansey, 1975). *Byssochlamys fulva* is of economic interest because it may cause spoilage of canned and bottled fruit. It is fairly common in the soil of orchards, and so may be splashed onto fruit. Its ascospores can survive a temperature of 84–87 °C for 30 minutes and have been known to retain viability up to 98 °C. They are also resistant to high concentrations of sulphur dioxide and alcohol. Moreover, the fungus can grow at low oxygen tension. It can also produce a variety of pectolytic enzymes (Chu & Chang, 1973). These combined properties explain its survival and growth in canned or bottled fruit. *Byssochlamys nivea* has been isolated from soil. In the presence of high sucrose concentrations, the ascospores are resistant to heat damage (Beuchat & Toledo, 1977). In culture, both species reproduce asexually by the formation of chains of conidia derived from tapering open-ended phialides (see Fig. 155A) which occur singly or in groups on the aerial mycelium. Conidial apparatus of this type belongs to the form-genus *Paecilomyces* of the Fungi Imperfecti. Terminal, thick-walled unicellular thalloconidia (sometimes

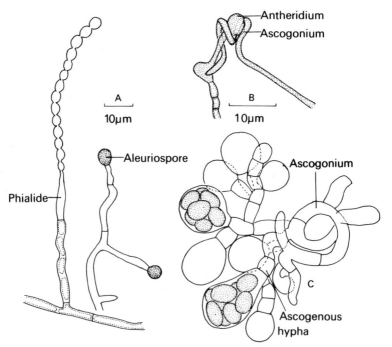

Fig. 155. *Byssochlamys nivea*. A. Phialospores and aleuriospores. B. Coiled ascogonium surrounding an antheridium. C. Ascogonium bearing ascogenous hyphae which in turn bear asci. Note the absence of sterile investing hyphae.

termed aleuriospores) are also found. The asci of *Byssochlamys* develop best in cultures incubated around 30 °C. In *B. nivea*, a club-shaped antheridium becomes encoiled by an ascogonium (Fig. 155B). Later, the coiled ascogonium develops short branches, or ascogenous hyphae, which bear globose, eight-spored asci either terminally or laterally, so that eventually clusters of asci can be found, but there is no sign of sterile hyphae enclosing them (Fig. 155C).

Aspergillus and its perfect states (Figs. 156–158)

The nomenclatural problems which stem from the fact that species of the form-genus *Aspergillus* may either lack perfect states, or possess perfect states which would be classified in different genera of Ascomycetes, have already been referred to. Similar problems arise in the form-genus *Penicillium*. There are two different solutions to the difficulty, which reflect different opinions as to the relationships of the organisms concerned. One solution is to apply the 'conidial' name (*Aspergillus* or *Penicillium*) alike to forms with or without ascocarps (Raper & Thom, 1949; K. B. Raper, 1957; Raper & Fennell, 1965). This is to imply that all the forms with similar conidial states, e.g. species of *Aspergillus*, are sufficiently closely related despite differences in the morphology of the perfect states. The alternative approach is to use the 'perfect' names and recognise that the different ascomycete genera may have somewhat similar conidial states. The similarity may or may not indicate relationship. It might result from convergent evolution. The latter view is gaining support (Benjamin, 1955; Subramanian, 1972). Although logically the exclusively conidial forms should then be classified in the Deuteromycotina (Fungi Imperfecti), they will be discussed here, along with 'perfect' forms from which they may possibly have been derived, presumably by the loss of the capacity to reproduce sexually (p. 307). It will be recalled (see p. 252) that parasexual recombination may provide alternative opportunities for the generation of variation on which natural selection might operate.

Subramanian (1972) has shown that the perfect states of *Aspergillus* can be assigned to nine ascomycete genera. These include the genus *Edyuillia*, the perfect state of *Aspergillus athecius*, in which the asci develop in naked clusters, not surrounded by inverting hyphae. This species is believed to represent either a primitive or reduced form related to *Eurotium* (Raper & Fennell, 1965). Other ascomycete genera are *Eurotium*, which includes the perfect states of the *Aspergillus glaucus* group, and *Emericella*, the perfect states of forms related to *Aspergillus nidulans*.

Eurotium (Figs. 156, 157)

Members of this genus are widely distributed in nature, especially in soil. They are responsible for spoilage of foodstuffs, especially those with high osmotic concentration. A typical example is *Eurotium repens* (Fig. 156) which is

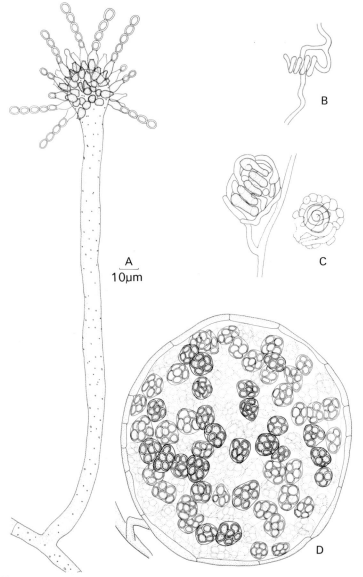

A
10μm

B

C

D

Fig. 156. *Eurotium repens*. A. Conidiophore. B. Ascogonium. C. Ascogonium surrounded by sterile hyphae. D. Cleistocarp showing mature and immature asci.

common on mouldy jam. In culture, on media low in sugar content, only conidia are formed. The segment of the mycelium from which the conidiophore arises persists as a swollen foot cell (Fig. 156A). The tip of the conidiophore swells to form a club-shaped vesicle bearing directly on its surface a cluster of bottle-shaped phialides which give rise to chains of green conidia in basipetal succession. On media of high sugar content (e.g. 20% sucrose), yellow spherical ascocarps develop also. Aerial hyphae develop coiled ascogonia (Fig. 156B) and, although there are reports of associated antheridia (Benjamin, 1955), they are not always visible. The ascogonium becomes invested by sterile hyphae which grow up from the stalk of the ascogonium. The ascogonium becomes septate, and from its segments ascogenous hyphae develop which penetrate and dissolve the surrounding pseudoparenchyma derived from the investing hyphae. Globose asci develop from croziers at the tips of the ascogenous hyphae and, when ripe, the ascocarp consists of clusters of asci surrounded by a single-layered yellow-coloured peridium. The peridium breaks open irregularly. The asci do not discharge violently: but the spores escape as the ascus walls break down. The ascospores are broadly lenticular, and without obvious surface ornamentation.

Eurotium repens, like nearly all ascocarpic forms, is homothallic, and cultures can be transferred by means of conidia, ascospores or hyphal tips. If repeated transfers are made by conidia only, the ability to form ascocarps declines, but can be restored by making subcultures from ascospores (see Fig. 157, Mather & Jinks, 1958). This suggests that the formation of ascocarps and conidiophores is partially controlled by cytoplasmic determinants, and it also suggests a way in which the non-ascocarpic forms might have evolved.

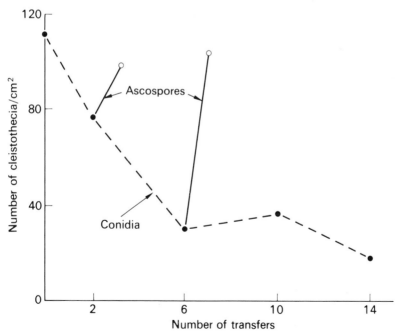

Fig. 157. *Aspergillus glaucus*. Change in density of cleistothecia in subcultures transferred either by means of conidia or ascospores (after Mather & Jinks, 1958).

Emericella (Fig. 158)

Emericella differs from *Eurotium* in a number of respects. Whilst in *Eurotium* the ascocarp is eventually surrounded by a single-layered peridial envelope, that of *Emericella* is enclosed by chains of very thick-walled cells termed Hülle cells (Fig. 158B). Whilst the ascospores of *Eurotium* are colourless and unorna-

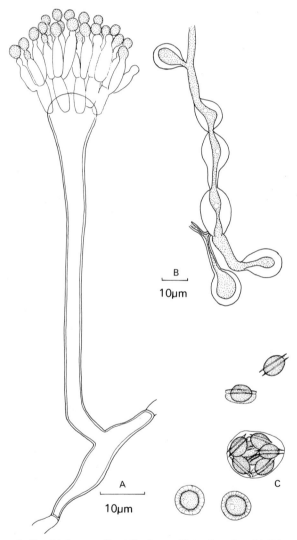

Fig. 158. *Emericella nidulans.* A. Conidiophore. Note that the phialides are not borne directly on the vesicle. B. Hülle cells, thick-walled cells surrounding the ascocarp. C. Ascus and ascospores. Note that the ascospores bear a double flange. A and C to same scale.

mented, those of *Emericella* are red and bear a double equatorial flange, so that the spores resemble pulley-wheels (Fig. 158c). The conidiophores also differ in that the phialides are not borne directly on the vesicle but on a series of cylindrical cells termed metulae.

The best-known species is the soil fungus, *E. nidulans* (*Aspergillus nidulans*), so called because of the nest-like arrangement of ascocarps surrounded by Hülle cells. This species has been widely used in genetical studies on sexual and parasexual recombination (Roper, 1966; Clutterbuck, 1974). A related fungus, *E. heterothallica* (*Aspergillus heterothallicus*), as the name implies, is heterothallic (Kwon & Raper, 1967). Single ascospore isolates fail to form ascocarps. The discovery of heterothallism should prompt further investigation of other species of *Aspergillus* in which ascocarps have not yet been reported, and if they develop ascocarps, it should be possible to assign them to the appropriate 'perfect' genus.

Aspergillus species lacking ascocarps

A large number of non-ascocarpic species of *Aspergillus* are known (Raper & Fennell, 1965). Some of these are of considerable economic importance, for example in industrial fermentations. *A. niger* is used in the production of citric acid, gluconic acid and other products (Smith, 1969). Some strains of the fungus are pathogens of plants, especially in the tropics; for example, crown rot of groundnuts and boll rot of cotton. The fungus is also used in a bioassay of soil for trace elements, e.g. copper. *A. oryzae* is used in fermentations of rice and soya products, and in the industrial production of proteolytic and amylolytic enzymes (Hesseltine, 1965). *A. flavus*, a common soil fungus, may infest products such as groundnut meal and dried foods, and produce a carcinogenic toxin called aflatoxin known to induce liver cancer in man and poultry (Wogan, 1965).

Penicillium and its perfect states (Figs. 159–161)

Penicillium presents the same taxonomic and nomenclatural problems shown by *Aspergillus*. It is a form-genus based on conidial morphology. Some species have perfect states which can be assigned to different genera: *Eupenicillium*, *Talaromyces* and *Carpenteles*. The different kinds of ascocarp represented by these generic names appear to be correlated with the type of conidiophore, especially with the complexity of conidiophore branching. However, most species of *Penicillium* have no known ascocarps.

It is one of the most cosmopolitan genera of fungi, occurring whenever substrata and conditions are suitable for growth, on all kinds of decaying materials, and the spores are almost universally present in air, so that it is a frequent contaminant of cultures. The contamination of a bacterial culture by *P. notatum* resulted in inhibition of bacterial growth, and this observation led

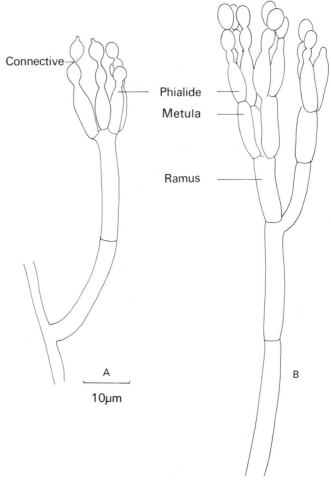

Connective

Phialide

Metula

Ramus

A

10μm

B

Fig. 159. A. *Penicillium spinulosum*. B. *Penicillium expansum*. Comparison of conidio-phore structure.

Fleming to the discovery of the antibiotic penicillin (Fleming, 1944). Although *P. notatum* was first used in antibiotic production, an intensive search for antibiotic production by other species led to its replacement by the related *P. chrysogenum* (K. B. Raper, 1952). Other economically important species are *P. camemberti* and *P. roqueforti* which play a part in cheese fermentation, and *P. griseo-fulvum*, a source of the antibiotic griseo-fulvin which causes distor-tion of fungal hyphae and has proved useful clinically in the treatment of skin and nail infections (Brian, 1960). *Penicillium italicum* and *P. digitatum* cause rotting of citrus fruits, whilst *P. expansum* causes a brown-rot of apples.

The classification of the genus is difficult and over 100 species have been recognised (Raper & Thom, 1949). The characteristic conidial apparatus is a

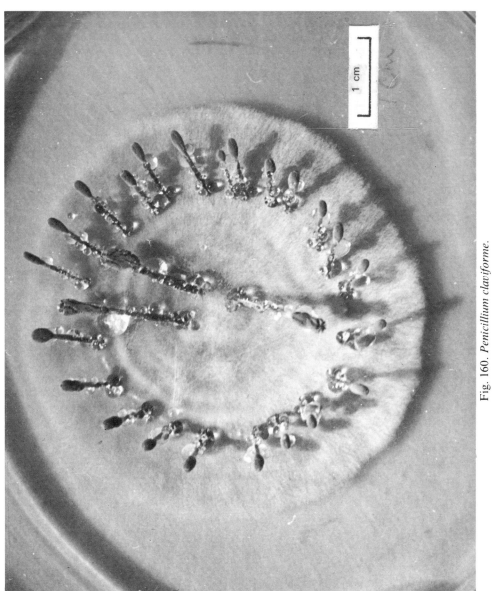

1 cm

Fig. 160. *Penicillium claviforme.*

branched conidiophore, bearing successive whorls of branches terminating in clusters of phialides. In some species (belonging to the section Monoverticillata), e.g. *P. spinulosum* (Fig. 159A), the phialides are borne directly on the conidiophore. More commonly they are borne on a further whorl of branches or **metulae**, and these may in turn arise on a further verticil of branches, the **rami**, e.g. *P. expansum* (Fig. 159B). In some species, e.g. *P. claviforme*, the individual conidiophores may be aggregated together into club-shaped fructifications or **coremia** (see Fig. 160).

Perfect states of *Penicillium*:

Eupenicillium

This includes a group of species with relatively simple conidiophores (belonging to the Monoverticillata stricta aggregate or the Divaricata aggregate of species), which produce either sclerotia or ascocarps with very tough peridia (sclerotioid ascocarps). The ascospores have pulley-wheel-like flanges. It is possible that the two groups are related, and is also possible that some of the sclerotial forms may ultimately prove to develop ascocarps (Scott & Stolk, 1967; Stolk, 1968). Forms possessing sclerotioid ascocarps include *E. javanicum* and *E. brefeldianum* (Monoverticillata stricta). The latter species is homothallic, but in culture it frequently develops sectors which are predominantly conidial. This shows how the sclerotial forms with similar conidia, such as *P. thomii*, may have arisen. The Divaricate group of species with sclerotioid ascocarps were previously classified in the genus *Carpenteles*. They clearly resemble a group of sclerotial species.

Talaromyces (Fig. 161)

A group of species with soft cottony loose ascocarps, such as *T. vermiculatus* (*Penicillium vermiculatum*) and *T. stipitatus* (*Penicillium stipitatum*), is included. The loose ascocarp suggests a resemblance to the Gymnoascaceae. The conidiophores of *Talaromyces* belong to a group of Penicillia called the Biverticillata–Symmetrica with long tapering phialides.

3: PYRENOMYCETES

The group Pyrenomycetes covers several developmental lines within Ascomycetes. They have been defined by Müller & von Arx (1973) as ascomycetes with ascomata (ascocarps) entirely surrounded by a peridial wall, and containing unitunicate asci which are primarily arranged in a hymenial layer. The ascocarps are termed **perithecia**, and in general they are provided with an opening or **ostiole** which is lined by **periphyses**. The perithecia may occur singly or may

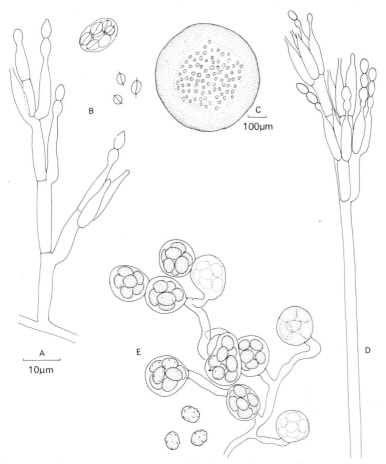

Fig. 161. A. *Talaromyces stipitatus*, conidiophore. B. Ascus and ascospores: note the equatorial frill. C. *Talaromyces vermiculatus*, ascocarp. D. Conidiophore: note the long tapering phialides characteristic of the Biverticillata-Symmetrica. E. Ascogenous hyphae and asci: note that some asci arise in chains.

be clustered together on or within a **stroma**. Although the final form of the perithecium may appear rather uniform, a number of distinct types of development have been distinguished, and these are discussed in relation to each order. Certain groups of Pyrenomycetes include closely related genera which may be ostiolate or astomatous, i.e. lacking an ostiole. Müller & von Arx have classified the majority of Pyrenomycetes into a single order Sphaeriales, whilst some other authorities divide Pyrenomycetes into several orders. Müller & von Arx have included the Erysiphales in the Pyrenomycetes, but this group has been considered earlier in this book (p. 283) as Plectomycetes. The Loculoascomycetes, forms with bitunicate asci, are excluded from the Pyrenomycetes. They may form ascocarps resembling perithecia, but the details of development are different, and the fructifications of Loculoascomycetes should properly be termed **pseudothecia**, although the word perithecium is often loosely applied to both forms of fructification.

The Pyrenomycetes grow on a wide range of habitats, e.g. soil, dung and decaying plant remains. They are particularly common on woody hosts. Some are plant pathogens, e.g. *Claviceps*, *Nectria*, whilst others are fungal symbionts of lichens, e.g. *Verrucaria*. Many have distinctive conidial states.

SPHAERIALES

As defined by Müller & von Arx, this order of Pyrenomycetes includes about fifteen families, some of which would be regarded as orders by other authorities. We will examine representatives from only six families.

Ophiostomataceae

The Ophiostomataceae includes forms with mostly hyaline unicellular ascospores, evanescent asci (i.e. asci whose walls deliquesce within the body of the perithecium) and ascocarps with long necks. The best-known genera are *Sphaeronaemella* and *Ceratocystis*. *S. fimicola* grows on herbivore dung and forms minute, pale-brown perithecia with long necks which accumulate sticky masses of ascospores.

Ceratocystis (Figs. 162–164)

Over eighty species are known (Olchowecki & Reid, 1974; Upadhyay & Kendrick, 1975), some causing serious plant diseases – e.g. *C. ulmi*, the cause of Dutch elm disease; *C. fagacearum*, the cause of oak wilt; *C. fimbriata*, the cause of a rot of sweet potatoes, wilt diseases of rubber and coffee; and *C. adiposa*, the cause of black rot of sugar cane. Other species cause staining of timber, attacking mainly the medullary rays of the sapwood, e.g. *C. piceae* and *C. coerulescens*. The perithecia have a swollen base and a long slender neck often bearing a fringe of hairs at its tip (Fig. 162). The ascus wall breaks down early during the development of the ascospores, so that intact asci are difficult to find. The ascospores are forced along the narrow neck and accumulate as sticky spore drops held in place by the ostiolar fringe (Fig. 162), probably as an adaptation to insect dispersal. The ascospores are small and hyaline, and vary in shape from ellipsoidal to bean-shaped, hat-shaped, quadrangular or needle-shaped. Various kinds of conidial apparatus are known (Figs. 163–164). In *C. ulmi*, conidiophores may be **mononematous** (single – Fig. 164c) or **synnematous** (i.e. clustered together to form a parallel bundle of conidiophores termed a **coremium** or **synnema** – Fig. 164A,B). The individual conidiophores may be colourless or, more usually in synnemata, dark-coloured. The hyaline conidia accumulate in a sticky mass at the tips of the

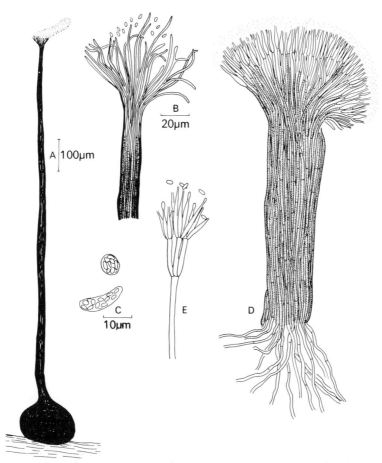

Fig. 162. *Ceratocystis piceae*. A. Perithecium showing spore drop at tip of neck. B. Details of ostiole showing ring of setae. C. Asci. D. Conidial fructification bearing sticky mass of spores. E. Details of apex of a conidiophore. B and D to same scale; C and E to same scale.

synnemata. Details of the development of the conidia from monematous or synnematous conidiophores show that they develop blastically and sympodially, on short denticles. They are cut off from the conidiogenous cell by a septum and, following detachment, a birth scar is visible at the base of the conidium. This conidial state of *C. ulmi* has been classified in the form-genus *Pesotum*, although in older literature it was placed in the form-genus *Graphium* (Crane & Schoknecht, 1973; Harris & Taber, 1973). In addition to this type of conidium, two further conidial forms are known in *C. ulmi*. One is formed by yeast-like budding of conidia, and the other by intra-hyphal budding to form micro-endospores (Fig. 164D,E) (Sansome & Brasier, 1973). In other species of *Ceratocystis*, conidia may be borne on a single stalk with a

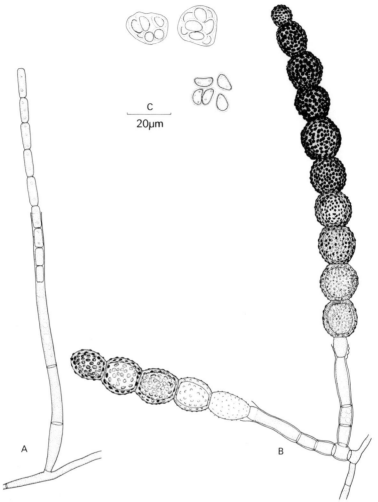

Fig. 163. Conidia and asci of *Ceratocystis* spp. A. *Endoconidiophora* type conidia of *C. coerulescens,* B. Conidia of *C. adiposa.* C. Asci ascospores of *C. adiposa.*

much-branched head, belonging to the form-genus *Leptographium,* or are formed in chains within long tapering phialides, belonging to the form-genus *Endoconidiophora* (see Upadhyay & Kendrick, 1975).

Some species, e.g. *C. ulmi,* are heterothallic; others, e.g. *C. piceae,* are homothallic. The development of the perithecia follows a distinctive course designated by Luttrell (1951) as the *Ophiostoma* type, but the name *Ceratocystis* has priority over *Ophiostoma.*

The ascogonia are free upon the mycelium. Branches from the stalk cell of the ascogonium or from neighbouring vegetative hyphae envelop the ascogonium to form the perithecial initial. The outer layers of hyphae develop into a perithecial wall which

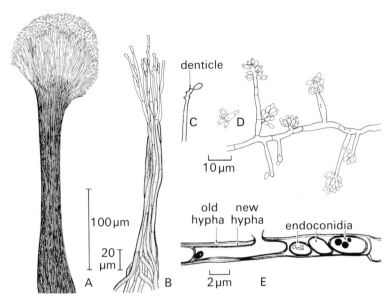

Fig. 164. Asexual reproduction in *Ceratocystis ulmi*. A, B. Synnematous conidiophores consisting of a parallel bundle of dark coloured hyphae branching at the tip to form conidiogenous cells. The conidia are borne in a sticky drop. C. Mononematous conidiophores terminating in conidiogenous cells which are polyblastic and produce conidia sympodially. The conidia are attached by denticles. D. Conidium showing yeast-like budding. E. A new hypha developing with an older hypha and producing endoconidia (after Harris & Taber, 1973).

surrounds a centrum composed of pseudoparenchymatous cells. The asci mature progressively from apex to base of the perithecium along chains of ascogenous cells derived from the ascogonium. Consequently, the asci are produced irregularly throughout the centrum and never form a definite wall layer. As the asci mature the sterile cells of the centrum collapse and disintegrate to form the perithecial cavity. The walls of the asci deliquesce and free the ascospores within the perithecial cavity. Growth of hyphae in the apical region of the perithecial wall produces an elongated neck which is penetrated by a lysigenous ostiole.

The essential feature of this kind of development is that here the asci are produced irregularly throughout the centrum instead of forming a lining to the base of the perithecium.

This kind of development has been found in various species of *Ceratocystis*. Whilst some accounts claim that asci do not arise following crozier formation, later accounts show that typical croziers are formed at the tips of ascogenous hyphae which may arise from a cushion in the base of the perithecium (Gwynne-Vaughan & Broadhead, 1936; Andrus & Harter, 1937; Bakshi, 1951; Rosinski, 1961).

The taxonomic disposition of *Ceratocystis* is not generally agreed. Whilst some authors include it in the Plectomycetes (Nannfeldt, 1932; Hunt, 1956), others have suggested that it might be placed in a separate order, the Ophio-stomatales (Rosinski, 1961) or Microascales (Upadhyay & Kendrick, 1975).

DUTCH ELM DISEASE

Dutch elm disease has recently become epidemic in the southern part of Britain, and is also a serious disease in North America. The disease was first recorded in Britain in 1927, and between 1930 and 1940 about 10% of elm trees in southern England were killed. Later, the disease declined in severity. The disease symptoms are yellowing and browning of foliage, followed by wilting, defoliation and death of branches. The spring wood of infected twigs often shows brown discoloration. The symptoms are the result of occlusion of the vessels, especially those of the current season, by gum and by tyloses derived from the wood parenchyma. Toxins have been isolated from culture filtrates.

The disease is spread by bark-boring beetles, especially *Scolytus multi-striatus* and *Hylurgopinus rufipes*. The fertilised female tunnels into the bark of living trees to lay eggs, and may carry the sticky conidia or ascospores into the tunnel. The larvae tunnel outwards in radiating tunnels and emerge as young beetles in spring and summer. The beetles, often carrying spores from fructifications of the parasitic fungus beneath the bark, fly to young twigs, where they feed on bark before maturation and mating. Infection of the host may thus occur at two stages; during maturation feeding and by the gravid female prior to egg-laying. Spread of the fungus within a tree from inoculation points is confined to the outer, water-conducting sapwood, and may be by mycelial growth or by the movement of conidia within the xylem vessels. Movement by the latter method may be as rapid as 10 cm/day, which is faster than the fungus can grow (Hart & Landis, 1971). Conidia can pass through bordered pits from one vessel segment to another (Pomerlau, 1970).

Between 1965 and 1967, simultaneous outbreaks of a more severe or aggressive form of the disease were noted at several points in southern England adjacent to ports. The disease spread rapidly and a large number of elms died during the following years. Isolations from trees affected with the aggressive form of the disease yielded, in culture, a strain of *C. ulmi* which could be distinguished, by its fluffy appearance and more rapid growth, from the non-aggressive strain which had a waxy appearance and slower growth. Strong circumstantial evidence was obtained, from inspection of cargoes, that the aggressive strain of the fungus was imported from Canada and North America on unbarked logs of Rock-elm *Ulmus thomasii,* destined for boat-building (Gibbs & Brasier, 1973a, b; Brasier & Gibbs, 1978).

The control of Dutch elm disease has, so far, been disappointing, and its ravages have left obvious effects on the landscape. Felling of infected trees has not contained the outbreak. Injections of the wood with the fungicide beno-myl have given some protection, but the treatment needs to be repeated in order to maintain control. The treatment is too costly to be used on a wide scale. Spraying with insecticides to eliminate the vectors, and removal, burning or de-barking of wood to reduce the overwintering larval populations of the bark-beetles, is also possible but difficult. Prohibition of transport of infected wood into unaffected areas is desirable (Burdekin & Gibbs, 1974; Gibbs, 1974). Biological control of the insect vector is also being attempted. In the longer term, the breeding and selection of elm varieties less susceptible to the disease is desirable, but breeding trees is a slow process. *Ulmus pumila* and *U. japonica* hold promise in possessing a degree of natural resistance to the aggressive strain of *C. ulmi* (Gibbs, *et al.*, 1975; Gibbs, 1978).

Sordariaceae

Members of this family form dark-coloured perithecia which are not contained within a stroma. Genera such as *Sordaria* and *Podospora* are typically coprophilous, fruiting on the dung of herbivores, but lignicolous species are also known (Carroll & Munk, 1964). *Neurospora* occurs in nature on burnt soil and vegetation in warmer countries. All three genera have been extensively used in genetical studies. The perithecia usually have an ostiole lined by periphyses, but some genera are **astomous,** i.e. have closed fruit-bodies lacking ostioles. The asci are thin-walled, and the apical apparatus of the ascus is in the form of a thickened annulus or apical plate which does not stain blue with iodine. Free-ended paraphyses are often present but may dissolve at maturity. The ascospores are black and sometimes surrounded by a mucilaginous epispore or have mucilaginous appendages. They are mostly unicellular and germinate by means of a germ pore.

Sordaria (Figs. 165–166)

Perithecia of *Sordaria* are common on the dung of herbivores and occasionally on other substrata (Moreau, 1953; Lundqvist, 1972). The best-known species is *S. fimicola,* which has been used in experiments on nutrition, the physiology of fruiting, spore liberation and genetics. Perithecial development occurs within 9 days on a wide range of media. A longitudinal section of a perithecium (Fig. 165) shows a basal tuft of asci at different stages of development. The asci elongate in turn and only one ascus can occupy the ostiole at a time. The spores are flung out for distances of up to 8 cm. Because each spore is about 13 μm wide and the apical apparatus of the ascus is only about 4 μm wide, the spores are gripped as they leave the ascus, and projectiles which vary in size from one to eight spores may be formed. The larger the number of spores in the projectile, the greater the distance of discharge, doubtless due to the fact that the surface/volume ratio of single spores is greater than that of multiple-spored projectiles, so that the effects of wind resistance are disproportionately high (Ingold & Hadland, 1959a) (Fig. 166).

The ascospores of *S. fimicola* have a distinct mucilage envelope which enables them to adhere to herbage. They can survive for long periods and, on

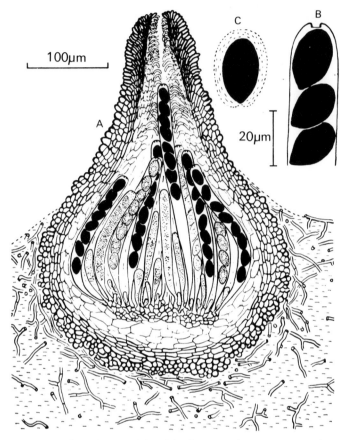

Fig. 165. *Sordaria fimicola*. A. L.S. perithecium growing on agar. B. Ascus apex. C. Ascospore showing mucilaginous epispore (after Ingold, 1965).

drying, gas vacuoles may appear in the spores, but they retain their viability (Ingold, 1956; Milburn, 1970). Spore germination does not readily occur unless the spores are ingested by a herbivore, but chemical treatment with sodium acetate or pancreatin can induce germination. In pure culture, growth and fruiting is stimulated by the addition of biotin (Hawker, 1957). Although fruiting occurs equally well in light or dark, the ascospore discharge is stimulated by light (Ingold & Dring, 1957; Ingold & Hadland, 1959b). The necks of the perithecia are phototropic, and, as in many other coprophilous fungi, this is probably an adaptation to ensure that spores are projected into the air away from the dung substratum.

The development of the perithecium in *Sordaria* may belong to the '*Diaporthe*' type of Luttrell (1951).

The ascogonia are formed within a stroma or free upon the mycelium. Branches from the stalk cell of the ascogonium or from neighbouring vegetative hyphae envelop the

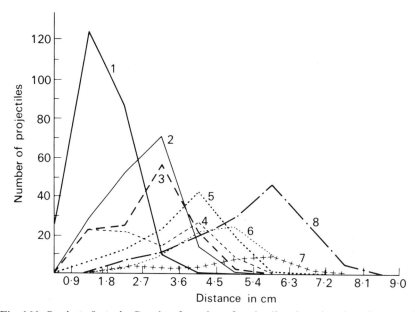

Fig. 166. *Sordaria fimicola*. Graphs of number of projectiles plotted against distance of projection for projectiles containing one to eight ascospores. The figure associated with each graph shows the number of spores per projectile (from Ingold & Hadland, 1959a).

ascogonium to form a spherical mass of tissue, the perithecial initial. The outer layers of this mass become differentiated into a perithecial wall. The central portion develops into a centrum composed of pseudoparenchymatous cells. Expansion and ultimate disintegration of these pseudoparenchyma cells produces the perithecial cavity. The asci expand as a group into the disintegrating centrum pseudoparenchyma and ultimately form a layer lining the base of the perithecial cavity. Growth of hyphae in the apical region of the perithecial wall produces a more or less elongated perithecial neck. The neck is penetrated by a schizogenous periphysate ostiole.

Development of this type has been claimed for *Sordaria fimicola* (Piehl, 1929; Dengler, 1937; Ritchie, 1937), for *S. macrospora* (Parguey-Leduc, 1967b), and for *S. humana* (Uecker, 1976). In *S. humana,* Uecker has shown that paraphyses arise from the floor of the perithecial cavity, and elongate along with the asci, but as the asci mature, the paraphyses dissolve, and only remnants persist. The presence of paraphyses is more characteristic of the '*Xylaria*' type of perithecial development than the '*Diaporthe*' type (for discussion, see Uecker, 1976).

The cytological details of perithecial and ascus development in *S. fimicola* do not present unusual features. The cells of vegetative hyphae may contain over 100 nuclei, and several nuclei enter the young ascogonium. Nuclear fusion occurs in the penultimate cell of typical croziers, and this is followed by four nuclear divisions. The first two are meiotic, and the last two mitotic, so that sixteen nuclei are present in the ascus. Each spore is binucleate (Carr &

Olive, 1958; Heslot, 1958). The fine structure of ascospore development has been described by Furtado & Olive (1970) and Mainwaring (1972). The septa present in the mycelium and most of the cells making up the body of the perithecium are simple septa of the normal Ascomycete type, but the septal pores of the ascogenous hyphae, i.e. those cutting off the cells of the crozier, are unusual in that they are surrounded by a rim or flange, somewhat reminiscent of the dolipore septum found in Basidiomycotina. Similar pores have been found in septa in ascogenous hyphae of other Ascomycetes and, although their function is not understood, their presence appears in some way to be correlated with the stage of development preceding karyogamy and meiosis (Furtado, 1971).

Although *S. fimicola* is homothallic, it has the capacity to hybridise. Wild-type strains have black ascospores, but mutants are known with colourless or pale-coloured spores. If a wild-type strain and a white-spored mutant are inoculated together into a culture, a proportion of hybrid perithecia may develop from heterokaryotic segments of the mycelium. Most of the asci from hybrid perithecia contain four black and four white ascospores, and six arrangements of the two kinds of ascospores are found (Fig. 129). Asci which have four black or four white ascospores at the tip of the ascus are those in which the gene for spore colour segregated at the first meiotic division of the fusion nucleus in the young ascus. In those with two black and two white ascospores at the tip of the ascus, segregation of the gene for spore colour occurred at the second meiotic division. First-division segregation results from the absence of a cross-over between gene and centromere, whilst second-division segregation results from a single cross-over between gene and centromere. Since the likelihood of crossing over depends on the distance between gene and centromere, the frequency of the two kinds of segregation can be used in mapping the position of the gene for spore colour (Catcheside, 1951; Olive, 1956, 1963a). In a low proportion of hybrid asci, spores showing 5:3, 6:2, or very rarely 7:1 segregations are found. These findings have been largely explained in terms of gene conversion (Fincham, 1971; Fincham & Day, 1971). Similar findings have been reported for other Ascomycetes, e.g. *Ascobolus*.

Not all species of *Sordaria* are homothallic. *S. brevicollis* and *S. heterothallis* are heterothallic. Both species form minute unicellular spermatia which are involved in the fertilisation of mycelia of opposite mating type (Olive & Fantini, 1961; Fields & Maniotis, 1963). *S. brevicollis* is a useful fungus for genetical demonstrations (Chen, 1965; Chen & Olive, 1965).

Podospora (Figs. 167–168)

Perithecia of *Podospora* are also to be found on herbivore dung and over sixty species are known (Mirza & Cain, 1969; Lundqvist, 1972). Most species have semi-transparent perithecia within which the outline of the asci can be seen

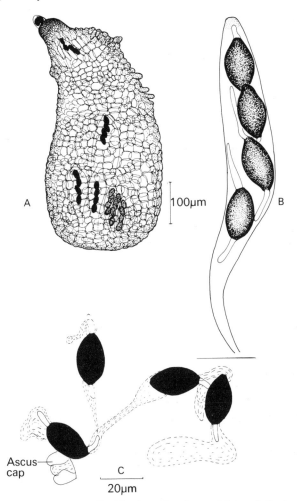

Fig. 167. *Podospora minuta*. A. Perithecium showing asci through the transparent wall. B. Ascus. C. Projectile consisting of four spores attached to the ascus cap and to each other by means of mucilaginous appendages.

(Fig. 167A). The number of spores in the ascus varies from 4 to 512, and although, in the past, spore number has been used as a taxonomic criterion, the species concept has been widened to include forms with a range of spore numbers so that, for example, 8-, 16-, 32- and 64-spored forms of *Podospora decipiens* are recognised (Moreau, 1953; Remacle & Moreau, 1962). The name *Podospora* (podos = foot, spora = seed) refers to the mucilaginous appendage attached to one or both ends of the black ascospore (Fig. 167B). In some of the commonest species, *P. curvula* and *P. minuta,* the spore appendages are attached to the cap of the ascus, and when the ascus explodes, the spores, roped together by their appendages, are shot out as a single projectile (Fig.

167c). As in *Sordaria*, it has been shown that multi-spored projectiles are discharged further than single spores (Walkey & Harvey, 1966a).

Perithecium development in *P. anserina* is initiated from a coiled ascogonium which becomes surrounded by hyphae to form an envelope that will become the perithecium wall. Several layers of thin-walled pseudoparenchymatous cells develop within the perithecium wall. Paraphyses arise from the base of the ascocarp and from the pseudoparenchyma cells, growing inwards and upwards, completely filling the centrum of the perithecium with tightly packed filaments. Ascogenous hyphae develop from the ascogonium at the base of the centrum, and asci develop following the usual nuclear fusion and meiosis (Beckett & Wilson, 1968). At the apex of the ascocarp, an ostiole, lined by periphyses, develops. This type of development seems to be characteristic of the Sordariaceae and, according to Mai (1976), it is distinct from the '*Diaporthe*' type of development to which Luttrell (1951) assigned it.

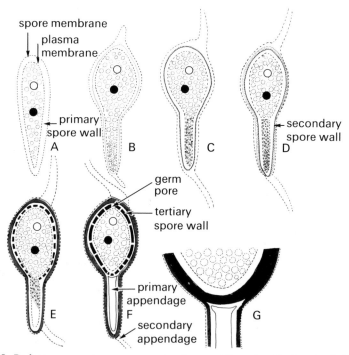

Fig. 168. *Podospora anserina*, ascospore development (based on Beckett *et al.*, 1968). A. Binucleate ascospore initial enclosed by two membranes, between which the primary spore wall develops. B–D. The secondary spore wall develops within the primary wall, and the secondary appendages develop towards each end of the spore by outpushing of the spore membrane. E–G. Development of tertiary, pigmented wall layer. The enucleate tail of the spore is cut off from the body of the spore, and its cytoplasm degenerates, but the tail persists as the primary appendage. Note that the tertiary wall layer does not extend into the primary appendage. At the opposite pole, a thinner area in the tertiary wall marks the position of the germ pore.

The development of the ascospores of *Podospora anserina* has been studied by Beckett *et al.* (1968). The ascospores (four in *P. anserina,* but see below) are delimited by a double membrane system as in other Ascomycetes (Fig. 168). The primary spore wall develops between the two membranes, gradually pushing them apart. The inner membrane continues to function as the plasma membrane of the spore, whilst the outer functions as the spore membrane. As the primary spore wall widens, secondary wall material is laid down towards the outer edge of the primary wall. These primary and secondary walls enclose the whole of the spore, including the spore head and the tail. A tertiary wall layer representing the pigmented layer of the spore head is laid down to the inside of the secondary layer (Fig. 168E–G). The elongated tail of the spore is cut off from the spore head by a septum. The pigmented tertiary wall layer does not extend into the spore tail, which therefore remains colourless. Its contents degenerate. This part of the spore persists as the primary appendage. Secondary appendages develop at the apex of the spore head and at the basal tip of the primary appendage. They arise by outpushings of the spore membrane. A thinner area in the tertiary wall at the end of the spore, opposite the primary appendage, marks the position of the germ pore (Fig. 168F).

Podospora anserina normally has four-spored asci, each ascospore being binucleate. Cultures derived from single binucleate ascospores usually form perithecia readily. Occasionally, however, smaller uninucleate ascospores may occur in some asci and when such spores are grown fruiting does not occur. Instead, perithecia only develop when certain of the uninucleate ascospore cultures are paired together. On each strain derived from a uninucleate ascospore, ascogonia bearing trichogynes, and spermatia develop, but these are self-incompatible: perithecia only develop if trichogynes of one strain are spermatised by spermatia of a genetically distinct strain (Ames, 1934). Thus although the behaviour of the large ascospores suggests that *P. anserina* is homothallic, it is clear that the underlying mechanism controlling perithecial development is a heterothallic one, of the usual one-gene, two-allele type (i.e. with '+' and '−' strains). Whitehouse (1949a) has termed this type of mating behaviour secondary homothallism. The majority of large ascospores (about 97%) contain nuclei of the two distinct mating types (Esser, 1965, 1974). In this fungus a novel kind of incompatibility system has also been discovered. If 'uninucleate' strains of different geographical origin are mated together a phenomenon termed 'barrage' can often be observed as a white zone between the two strains (see upper and lower right-hand sectors in Fig. 169). Instead of the expected reciprocal behaviour of '+' and '−' mating types it is commonly found that deviations, occur, e.g.

1. If reciprocal crosses are made, only one of the two reciprocal crosses is compatible; the other is incompatible; this non-reciprocal incompatibility has been termed semi-incompatibility.

2. Reciprocal incompatibility also occurs: between + and − strains of different races no perithecia are formed.

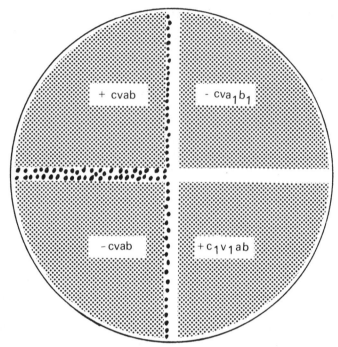

Fig. 169. *Podospora anserina*. Genetic basis for heterogenic incompatibility. A Petri dish has been inoculated with four different mycelia. The presence of perithecia is indicated by heavy stippling. In these crosses the upper and lower left-hand sectors contain fully compatible mycelia homozygous at all loci except the mating type locus (+ and −). The upper sectors show semi-incompatibility because the *a* and *b* loci are heterozygous, whilst the lower sectors are semi-incompatible because the *c* and *v* loci are heterozygous. Semi-incompatibility is demonstrated by the presence of perithecia on only one of the sectors. The two right-hand sectors are fully incompatible because all four loci *a*, *b*, *c*, *v* are heterozygous (after Esser, 1965).

Esser (1965) has explained the behaviour of this fungus on the basis of four unlinked loci, *a, b, c, v*, each with two alleles. Semi-incompatibility occurs in crosses of the type $ab \times a_1b_1$ and $cv \times c_1v_1$, i.e. where two different alleles are involved. Thus, in Fig. 169 the cross $+cvab \times -cvab$ is fully compatible since the two strains differ only in mating type: all the other genes are homozygous. The cross $+cvab \times -cva_1b_1$ is semi-incompatible since only one pair of genes differ (ab vs a_1b_1). The cross $-cvab \times +c_1v_1ab$ is also semi-incompatible but the genes *c* and *v* differ (cv vs c_1v_1). Total incompatibility results from differences between both pairs of genes simultaneously (e.g. cross $-cva_1b_1 \times +c_1v_1ab$). In other words even if two strains differ in mating type they are not fully compatible unless the four genes are identical in each strain. Incompatibility is here brought about by *differences* in genetic make-up, and Esser has coined the term *heterogenic incompatibility* for this kind of behaviour in contrast with the more common type of incompatibility where genetic similarity precludes mating. This type of incompatibility is not confined to *P. anserina* but is also found in *Sordaria* (Olive, 1956) and other fungi.

Kemp (1976) has reasoned that semi-incompatibility in *P. anserina* is the result of unilateral nuclear migration, i.e. by the formation of heterokaryotic hyphae by nuclear migration in one direction, but not in the opposite. Since nuclear migration appears to be independent of cytoplasmic streaming, the failure of nuclear migration might be caused by the failure of microtubules to attach to the nuclear envelope. Kemp believes that it is this process which is under genetical control. He has postulated two types of locus, one producing a specific nuclear migration inhibitor, and the other determining whether or not the nuclear envelope has resistance to the specific migration inhibitor.

Neurospora (Fig. 170)

Species of *Neurospora* have been widely used in genetical and biochemical studies. The best-known are *N. crassa* and *N. sitophila* both of which are eight-spored and heterothallic. *N. tetrasperma* is four-spored and secondarily homothallic. The reasons why *Neurospora* has proved so useful as a tool in biochemical and genetical research are that wild-type strains have simple nutritional requirements (a carbohydrate source; simple mineral salts; and one vitamin, biotin); mutations can be induced readily by irradiation of conidia; growth and sexual reproduction is rapid; and tetrad analysis by means of ascus dissection is relatively easy (Tatum, 1946; Beadle, 1959). The extensive literature has been compiled by Bachmann & Strickland (1965) as a bibliography containing 2,310 references (see also Barratt, 1974). In nature these species of *Neurospora* colonise burnt ground and charred vegetation, and are also found in warm humid environments such as wood-drying kilns and bakeries where they can cause serious trouble because of their rapid growth and sporulation. For this reason *N. sitophila* is sometimes called the red bread mould. The generic name is derived from the characteristically ribbed ascospores (Lowry & Sussman, 1958, 1968; Sussman & Halvorson, 1966; Austin *et al.,* 1974; Hohl & Streit, 1975) (Fig. 170). The ascospores of *N. crassa* are viable for many years and do not germinate readily unless treated chemically (e.g. by chemicals such as furfural) or by heat shock (e.g. 60 °C for 20 minutes). Following such treatment, the spores germinate and produce a coarse, septate, rapidly growing mycelium, each segment of which is multinucleate. In contrast, the conidia are killed by such heat treatment (Sussman, 1966). The ascospores have reserves of lipids and the carbohydrate trehalose.

Within 24 hours, the mycelium can begin asexual reproduction. Upright branches develop from which branched chains of multinucleate pink conidia arise. Further conidia develop by budding of the terminal conidium on a chain and, when the terminal conidium gives rise to two buds, the chain branches (Fig. 170D,E). Conidia of this type belong to the form-genus *Monilia*. The

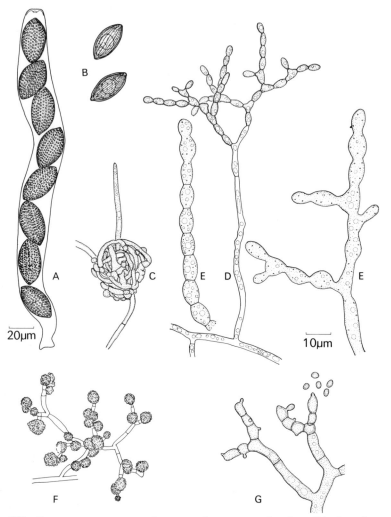

Fig. 170. *Neurospora crassa.* A. Ascus. B. Ascospores showing ribbed surface. C. Protoperithecium showing projecting trichogyne. D. Macroconidia from 1-day-old culture. E. Enlarged view of developing macroconidia. F. Microconidia forming sticky clusters. G. Enlarged view showing origin of microconidia. A, B, C, D, F to same scale; E, G to same scale.

individual segments of the spore chain break apart and are readily dispersed by wind. The spores are formed in vast numbers and, if released into a laboratory, can cause serious contamination of other cultures.

Cultures derived from a single ascospore also develop two other types of reproductive structure. In contrast to the large, dry, wind-dispersed macro-conidia, clumps of smaller oval, sticky microconidia develop laterally (Fig. 170F,G). The conidiogenous cells from which the microconidia develop have been interpreted as reduced phialides (Subramanian, 1971; Lowry *et al.*, 1967; Turian, 1976). Ascogonia, terminated by long tapering trichogynes, and surrounded at the base by hyphae, also develop (Fig. 170c). Such structures are termed protoperithecia or bulbils.

In *N. crassa* and *N. sitophila,* no further development occurs in single ascospore cultures, i.e. each strain is self-incompatible. Incompatibility is controlled by a pair of alleles *A* and *a,* and if two compatible strains are grown together in a culture vessel for a few days, microconidia of one strain can be transferred to the trichogynes of the opposite strain by flooding with sterile water. Transfer of macroconidia of the opposite strain to a trichogyne can also affect fertilisation. Fusion between the trichogyne and the fertilising cell is followed by migration of one or more nuclei from the fertilising cell down the trichogyne into the ascogonium (Backus, 1939). Development of ripe perithecia occurs within 7–10 days, and follows a pattern typical for Ascomycetes generally (Colson, 1934; McClintock, 1945; Singleton, 1953; Hohl & Streit, 1975), although Mitchell (1965) has suggested that there may be some anomalous features. Nelson & Backus (1968) have described perithecial development in *N. terricola* and *N. dodgei,* which are both homothallic. Development resembles that of other Sordariaceae. The centrum of immature perithecia is filled with paraphyses, but they disintegrate as the asci mature.

Mitosis of the nucleus in the ascospore initially results in the spores being binucleate, and this appears to be a general feature of the Sordariaceae.

In *N. tetrasperma* the asci normally contain four binucleate ascospores, and single-spore cultures derived from four-spored asci produce perithecia. As in *Podospora anserina* occasional asci have five or six ascospores, some smaller than the rest and uninucleate. Cultures derived from single small ascospores do not develop perithecia, but, when certain of such cultures are paired, perithecia develop. The genetical basis for incompatibility in uninucleate ascospores is the same as in *N. crassa,* and the reason why single large ascospores give rise to perithecial cultures is that the two nuclei present usually include one of each mating type. The segregation of the two mating-type alleles almost invariably takes place at the first meiotic division in developing asci, i.e. crossing-over occurs very infrequently between the locus for incompatibility and the centromere. During the second and third nuclear divisions in the ascus the spindles overlap and when the four ascospores are cut out they normally contain two nuclei of distinct mating type, or, occasionally, uninucleate ascospores are formed (Fig. 171) (Dodge, 1927).

It is of interest that in *Podospora anserina* which is also secondarily homothallic the same end is achieved by different means. In this fungus segregation of mating-type alleles occurs at the second meiotic division in 98% of the asci. The spindles at the second division lie transverse to the long axis of the ascus, whilst the third-division spindles lie parallel with the ascus. Ascospores are cut out around pairs of genetically different nuclei (Fig. 171) (Franke, 1957).

In *N. crassa* heterokaryons are not normally formed between mycelia of

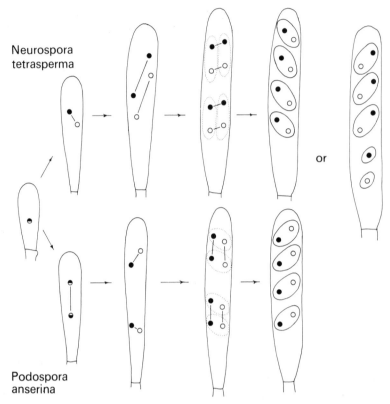

Fig. 171. Two mechanisms resulting in secondary homothallism. Top row, *Neurospora tetrasperma* showing first-division segregation of mating-type alleles and overlapping spindles. Bottom row, *Podospora anserina* showing second-division segregation. At the second meiotic division the spindles lie transversely in the ascus. Third-division spindles are parallel to the ascus (after Fincham & Day, 1965).

opposite mating type (Sansome, 1946) and this implies that plasmogamy normally occurs only between a trichogyne of one strain and a fertilising agent (e.g. a microconidium or macroconidium) of the opposite strain. This condition is termed *restricted heterokaryosis*. Later work has shown that even within one mating type the ability to form heterokaryons is under genetical control. Two genes involved are designated *C* and *D*. Stable heterokaryons are normally only formed between strains which have a like genotype (e.g. *CD + CD, Cd + Cd* or *cd + cd*). When strains of unlike genotype anastomose there is evidence of cytoplasmic incompatibility resulting in vacuolation and disorganisation of cell contents in the region of the anastomosis. Similar cytoplasmic reactions are visible when anastomosis occurs between hyphae of wild-type strains differing in mating type (Garnjobst, 1955; Garnjobst & Wilson, 1956; Wilson *et al.*, 1961). In contrast, heterokaryons are readily formed between different mating-type strains of *N. tetrasperma* (Dodge, 1942), which thus exhibits *unrestricted heterokaryosis*. Whitehouse (1949a) has made the interesting suggestion that whilst restricted heterokaryosis necessitates the intervention of sexual organs in plasmogamy, where unrestricted heterokaryosis occurs, the need for sex organs as a prerequisite of plasmogamy no longer holds, although they may still occur and function. This view is supported by evidence that in many Ascomycetes with unrestricted heterokaryosis there are no differentiated sex organs: plasmogamy occurs between vegetative hyphae.

Melanosporaceae

This is a small family with simple perithecia, often with long projecting necks, but sometimes (as in *Chaetomium*) lacking necks entirely. The asci are club-shaped and thin-walled, and break down within the perithecia so that the ascospores ooze out in tendrils and are not violently discharged. The ascospores are unicellular and black in colour. The only genus we shall consider is *Chaetomium* which some authorities place in a separate family, Chaetomiaceae (Hawksworth, 1971), or order, Chaetomiales (Ames, 1963).

Chaetomium (Fig. 172)

Over 180 species are known (Ames, 1963; Seth, 1972), many of them causing decay of cellulose-rich substrata such as textiles in contact with soil, straw, sacking, dung (Lodha, 1964a,b) and wood, which may undergo a superficial decay known as soft-rot (Savory, 1954). *C. thermophile* is thermophilic (Tansey, 1972).

The perithecia of *Chaetomium* are superficial and barrel-shaped, and they are clothed with dark, stiff hairs. In some species, e.g. *C. elatum,* one of the commonest species, growing on damp rotting straw, the hairs are dichotomously branched. In others, e.g. *C. cochliodes,* the body of the perithecium bears straight or slightly wavy, unbranched hairs, whilst the apex bears a group of spirally coiled hairs. The hairs are roughened or ornamented, and the type of ornamentation is an aid to identification (Hawksworth & Wells, 1973). When the perithecia are ripe, a column-like mass of black ascospores arises from the apex (Fig. 172A). In most species, the spores are lemon-shaped, with a single germ pore. The spore column results from the breakdown of the asci within the body of the perithecium, i.e. the asci do not discharge their spores violently. The young asci are cylindrical to club-shaped, but this stage is very evanescent, and is only found in young perithecia (Fig. 172B), often before the spores have become pigmented, and sometimes before the perithecium has developed an ostiole. The development of perithecia in *Chaetomium* shows some variation between species (Whiteside, 1957, 1961; Corlett, 1966; Berkson, 1966; Ave & Miller, 1967; J. C. Cooke, 1969a,b, 1970). The ascogonia are

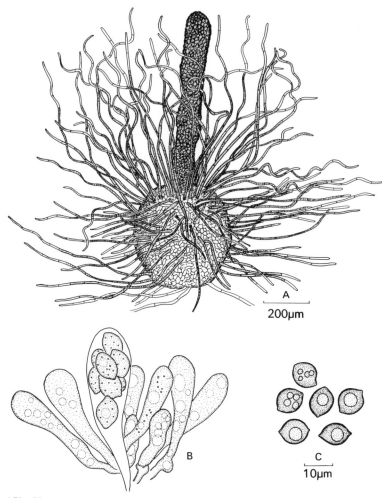

Fig. 172. *Chaetomium globosum*. A. Perithecium showing tendril of ascospores. B. Asci. c. Ascospores.

coiled and lack antheridia. Investing hyphae arising from the ascogonial stalk or from adjacent vegetative cells surround the ascogonium. Perithecial hairs develop early from the external cell layer. The centrum is at first filled with hyaline pseudoparenchyma. At the apex of the perithecium, certain of these cells become meristematic and give rise to the elongate periphyses which line the ostiole. Ascogenous hyphae develop from the ascogonium at the base of the centrum and, at about this time, the surrounding pseudoparenchyma cells of the centrum deliquesce to form a cavity. In some species, croziers have been reported, but in others they may be absent. Paraphyses of two types have been described in some species. In *C. brasiliense*, lateral paraphyses, i.e. paraphyses arising from the pseudoparenchyma cells outside the hymenium, have been reported, whilst hymenial paraphyses have been reported in *C. globosum* (Whiteside, 1961). The paraphyses are very evanescent, and it is possible that they occur in some other species. This type of perithecium development belongs to the '*Xylaria*' type of Luttrell (1951).

Most species of *Chaetomium* are homothallic, but a few, e.g. *C. cochliodes*, are heterothallic (Seth, 1967; Sedlar *et al.*, 1972). *Chaetomium elatum* is reported by some workers to be homothallic, but some isolates appear to be heterothallic (Fox, 1953). Conidial states are rare in *Chaetomium*, but simple phialides and phialospores occur in *C. elatum* and *C. globosum*, whilst *C. piluliferum* forms both phialospores and globose thalloconidia of the *Botryotrichum* type (Daniels, 1961). *Chaetomium trigonosporum* has conidia belonging to the form-genus *Scopulariopsis* (Corlett, 1966).

The fruiting of *Chaetomium* species in culture is often stimulated by the inclusion of cellulose in the form of filter-paper, cloth or jute-fibre. Extracts of jute stimulate sporulation of *C. globosum* in pure culture, and stimulation has also been found when *C. globosum* is grown with *Aspergillus fumigatus*. Attempts to analyse the chemical nature of the stimulus have shown that the effect of jute extract can be partially substituted by the addition of calcium to the medium (Basu, 1951). Sporulation is also stimulated by presence in the medium of sugar phosphates and phospho-glyceric acid, and it has been shown that *A. fumigatus* excretes such compounds into the medium and that sugar phosphates are also present in jute extract (Buston *et al.*, 1953; Buston & Khan, 1956).

The breakdown of cellulose by *Chaetomium* is brought about by a powerful cellulase (Abrams, 1950; Agarwal *et al.*, 1963a,b). When fed with sucrose *C. globosum* breaks down the disaccharide into its two hexose moieties, fructose and glucose, but the glucose moiety is absorbed preferentially. Little sucrose is absorbed directly (Walsh & Harley, 1962). Attack of wood by *C. globosum* is associated with depletion of cellulose and pentosan, but a considerable proportion of the lignin remains (Savory & Pinion, 1958; Levi & Preston, 1965). Some species of *Chaetomium*, e.g. *C. cochliodes*, are antagonistic to soil-borne and seed-borne fungal pathogens, and the possibility of using them to control plant diseases has been investigated (Tveit, 1953, 1956; Tveit & Moore, 1954; Tveit & Wood, 1955). An antibiotic, chaetomin, has been isolated from *C. cochliodes* (Waksman & Bugie, 1944).

Xylariaceae

Typical members of this group have perithecia embedded in stromata. Most members are saprophytic on wood, and occasionally on other substrata such as dung, but a few are parasitic on woody hosts, e.g. species of *Rosellinia*. *Ustulina deusta* causes a butt-rot of beech. General accounts of Xylariaceae have been given by Miller (1930, 1932a,b) and Dennis (1968); see also Martin (1967).

The development of perithecia in the family conforms to what Luttrell (1951) has termed the '*Xylaria*' type. The generic name *Xylosphaera* is sometimes used because, as Dennis (1958a,b) has shown, this name has priority, but the proposal to use the name has not been welcomed by other mycologists, e.g. Petrak (1962), Holm & Müller (1965).

The ascogonia are produced free upon the mycelium or more commonly within a stroma. Branches from the stalk cells of the ascogonium or from neighbouring vegetative hyphae surround the ascogonium and form the perithecial wall. Hyphal branches with free tips (paraphyses) grow upward and inward from the inner surface of the wall over the base and sides of the perithecium. Pressure exerted by the growth of opposed paraphyses expands the perithecium and creates a central cavity. The perithecium becomes pyriform as a result of growth of hyphae in the apical region of the wall to form a neck. The layer of inward-growing hyphae is continuous up the sides and into the perithecial neck. Growth of these hyphae within the neck produces a schizogenous ostiole lined with free hyphal tips (periphyses). The ascogonium produces ascogenous hyphae which typically grow out along the inner wall over the base and sides of the perithecium. Asci derived from the ascogenous hyphae grow among the paraphyses to form a continuous hymenium of asci and more or less persistent paraphyses lining the perithecial cavity. In some forms the paraphyses are evanescent, and the ascogenous hyphae form a plexus in the base of the perithecium. The asci then arise in a single aparaphysate cluster (Luttrell, 1951).

The cytology of ascus development in *Xylaria* (Rogers, 1968, 1969, 1975a; Beckett & Crawford, 1973) and in the related genus *Hypoxylon* (Rogers, 1965, 1975b) follows the usual pattern in most cases. The ascospores may be uninucleate or binucleate. In *X. polymorpha* and *H. serpens*, immature ascospores are divided by a septum near the base, which cuts off a small appendage, but the appendage disappears in mature ascospores, which appear unicellular and with only a truncate base to mark the position of the remnant of the appendage. The wall of the mature ascospore is dark, but there is a hyaline germ slit running the length of the ascospore (Fig. 174A). The ascus tip contains a cylindrical apical apparatus which is amyloid (i.e. it stains bright blue with iodine). The apical apparatus is pierced by a narrow pore and is everted as the ascus explodes (Greenhalgh & Evans, 1967; Beckett & Crawford, 1973).

Daldinia (Figs. 173–174)

Daldinia concentrica grows parasitically on ash *(Fraxinus excelsior)* but can continue fruiting on dead trunks and branches. It is rare on other hosts but grows on wood of birch *(Betula)* or gorse *(Ulex)* which has been burnt. On ash, it forms large (5–10 cm diameter) hemispherical brown stromata annually which contain ripe asci between May and October. In section (Fig. 173), the stromata show a concentric zonation of alternating light and dark bands. The surface of young stromata may be covered with a pale fawn powdery mass of conidia. The conidia are dry and oval in shape, developing successively at the

Fig. 173. A. *Xylaria polymorpha*. Perithecial stromata at base of a stump of *Fagus sylvatica*. B. *Daldinia concentrica*. Perithecial stromata at base of *Fraxinus excelsior*. One stroma has been cut open to show the concentric zonation, and the perithecia in the outer layers.

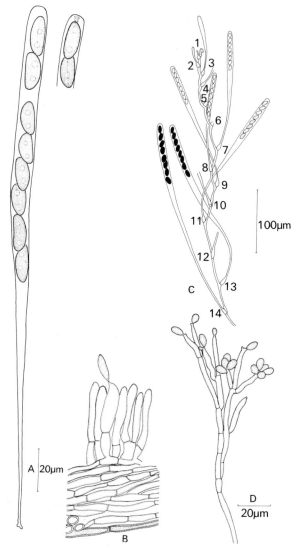

Fig. 174. A. *Xylaria hypoxylon*. Ascus. The ascus tip to the right has been stained with iodine to reveal the apical apparatus. B. *X. hypoxylon* conidiophores. C. *Daldinia concentrica*. Ascogenous hypha. The numbers represent successive asci working backwards from the apex. D. *D. concentrica: Nodulisporium*-type conidia. c after Ingold (1954a).

tips of branched conidiophores by the outgrowth of the wall, and when detached leave a small scar (Fig. 174D). Conidia of this type have been named *Nodulisporium tulasnei*.

Perithecia develop in the outer layers of the stroma, each arising from a coiled archicarp. The perithecial wall is lined by ascogenous hyphae which are unusual in that there is often a considerable distance separating successive asci (Ingold, 1954a) (Fig. 174C). The stroma of *Daldinia* apparently functions as a water reserve and detached stromata will continue to discharge ascospores for about three weeks even if placed in a desiccator (Ingold, 1946a). Spore discharge is nocturnal and the rhythm of spore discharge is maintained for several days if detached stromata are maintained in continuous dark. In continuous light periodic spore discharges ceases after about 3 days but is restored on return to alternating light and dark (Ingold & Cox, 1955). The output of spores from a single stroma of average size is about 10 million a night.

Xylaria (Figs. 173–175)

Stromata of *Xylaria hypoxylon,* the candle-snuff fungus, are common on stumps and fallen branches of deciduous trees. As in most Xylariaceae growing on wood the limits of the mycelium within infected tissues are visible as conspicuous black zone lines. The stromata are branched and cylindrical. At the upper end, the stroma is covered by a white powdery mass of conidia (Figs. 175A, 174B). Perithecia develop later at the base of the stroma and are visible externally as swellings at the surface (Fig. 175B). The apical apparatus of the ascus is visible after staining in iodine as a bright blue cylindrical collar pierced by a narrow pore, even in immature asci (Fig. 174A). *Xylaria polymorpha* (dead men's fingers) grows in late summer and autumn at the base of old tree stumps. The stromata are finger-like and clustered (Fig. 173). The surface is at first covered by an inconspicuous conidial layer, but eventually perithecia develop beneath the surface of the whole stroma, and are not restricted to the basal region as in *X. hypoxylon.* Both species are active wood-rotting fungi inducing decay of the white-rot type.

Hypoxylon (Fig. 176)

This is a large genus of over a hundred species (Miller, 1961; Whalley & Greenhalgh, 1973a,b) forming stromata which are often hemispherical or sometimes flattened on the surface of wood and bark. Different species often show a preference for a particular host. Common species are *H. fragiforme* (= *H. coccineum*) almost confined to freshly dead branches and trunks of *Fagus* (Fig. 176A), *H. multiforme* on *Betula* (Fig. 176B), and *H. rubiginosum* which forms flat stromata on decorticated wood of *Fraxinus*. The young stroma of all these species bears a conidial felt of the *Nodulisporium* type (Chesters & Greenhalgh, 1964). Most species show nocturnal spore discharge (Walkey & Harvey, 1966b).

Hypocreaceae

Members of this family have brightly coloured (white, yellow, orange, red, violet) perithecia which may be single or seated on a stroma. The texture of the

Fig. 175. *Xylaria hypoxylon*. A. Conidial stroma. Conidia are borne on the white tips of the branches. B. Perithecial stroma. Perithecia develop at the base of the old conidial stroma.

Fig. 176. *Hypoxylon fragiforme* and *H. multiforme*. A. *H. fragiforme*: perithecial stromata on *Fagus sylvatica*. One stroma has been broken open to show the perithecia embedded in the outer layers. B. *H. multiforme*: perithecial stromata on *Betula pendula*.

perithecia or stromata is fleshy or leathery. The perithecial ostiole is lined by periphyses. The asci are unitunicate, and contain ascospores which are often two-or-more celled, and may break up inside the ascus to form part-spores. Some authorities include the family in a distinct order, Hypocreales (Rogerson, 1970). Perithecial development conforms to the '*Nectria*' type of Luttrell (1951). The ascogonia, which are formed within a stroma, become surrounded by concentric layers of vegetative hyphae which form a true perithecial wall.

The cells of the inner layer of the wall in the apical region of the young perithecium produce a palisade of inward-growing hyphal branches. These hyphal branches grow downward to form a vertically arranged mass of hyphae with free ends termed apical paraphyses (Luttrell, 1965b). Pressure exerted by the elongation of the apical paraphyses, accompanied by expansion of the wall, creates a central cavity within the perithecium. The free tips of the apical paraphyses ultimately push into the lower portion of the wall so that they become attached at both the top and bottom of the perithecial cavity. Ascogenous hyphae arising from the ascogonium spread out across the floor and sides of the cavity and produce asci by means of croziers. The asci grow upward among the apical paraphyses and form a concave layer lining the inner surface of the wall in the basal region of the perithecium. A schizogenous ostiole lined with periphyses (hyphae with free apices, attached at their bases to the inner wall of the neck) develops in the wall of the perithecium. At maturity, the perithecia may protrude from the stroma and appear to be seated on its surface. Perithecial development of this type has been described for *Nectria* (Strickmann & Chadefaud, 1961; Hanlin, 1961, 1971), and for *Hypocrea* (Doguet, 1957; Hanlin, 1965; Parguey-Leduc, 1967a; Canham, 1969).

Hypocreaceae are saprophytic on plant substrata, in soil, on dung, etc., and some are also plant pathogens. Their conidial states are phialidic and belong to form-genera such as *Fusarium, Cylindrocarpon, Gliocladium, Verticillium* and *Trichoderma*. Cain (1972) has suggested that the Eurotiaceae (which also have phialidic conidia) are derived from this group.

Hypocrea (Figs. 177–179)

Species of *Hypocrea* form brightly coloured fleshy perithecial stromata, with perithecia embedded in the outer layers. The thin-walled asci contain eight two-celled ascospores, and in many species the spores separate into two part-spores before ascus discharge, so that sixteen part-spores are released (Fig. 178c). *H. pulvinata* forms bright yellow stromata on dead fruit-bodies of *Piptoporus betulinus*, the birch polypore. It is possible that this fungus grows parasitically on the polypore. The ascospores are often visible as white tendrils issuing from the ostioles of the perithecia (Fig. 177). In culture, conidia are formed in sticky masses at the tip of single phialides (Fig. 178B). Conidia of this type belong to the form-genus *Cephalosporium* (Rifai & Webster, 1966). Some species of *Hypocrea* have conidia of the *Trichoderma* type, in which whorls of phialides give rise to separate, sticky, green or white spore masses. *Hypocrea rufa* forms conidia referable to *T. viride* (Fig. 179D). *Trichoderma* species are important soil saprophytes, and several are antagonistic to other fungi, including some pathogens (Rifai, 1969). Conidia of the *Gliocladium* type are found in cultures of *H. gelatinosa* (Webster, 1964) (Fig. 179A,B).

Fig. 177. *Hypocrea pulvinata*. Perithecial stromata on fruit-body of *Piptoporus betulinus*. The darker dots are the perithecial ostioles. Tendrils of white ascospores can be seen issuing from some of the ostioles.

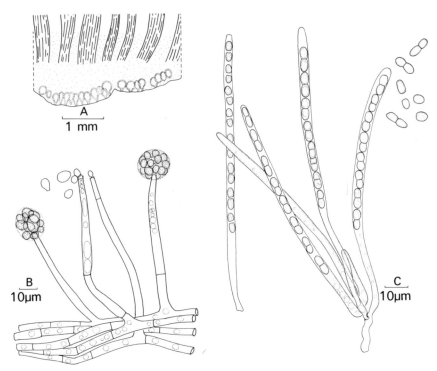

Fig. 178. *Hypocrea pulvinata*. A. L.S. lower portion of fruit-body of *Piptoporus betulinus* showing perithecial stroma of *Hypocrea* in section. B. Conidia produced from upright phialides. C. Asci and ascospores. Note how the two-celled ascospores break up into part-spores.

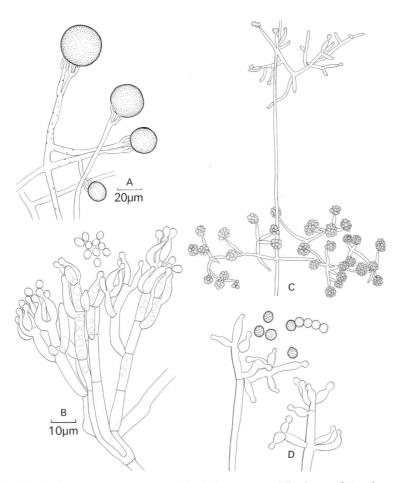

Fig. 179. Conidia of *Hypocrea* spp. A. *Gliocladium*-type conidiophores of *H. gelatinosa*. B. Detail of phialides of *H. gelatinosa*. C. *Trichoderma viride* conidial state of *H. rufa*. Arrangement of conidiophores. D. Detail of phialides of *H. rufa*. A, C to same scale; B, D to same scale.

2 mm

Fig. 180. *Nectria cinnabarina* Conidial and perithecial stromata on a twig of *Acer pseudoplatanus*. The smooth pale structures are conidial pustules and the rough clustered bodies are perithecia which develop around the base of conidial pustules.

Nectria (Figs. 180–184)

Perithecia of *Nectria* are common on twigs and branches of woody hosts. Many are saprophytic but some cause economically important diseases, e.g. *N. galligena* causes apple and pear canker. *Nectria cinnabarina* or coral spot is common on freshly cut twigs, but may occasionally be a wound parasite. The name coral spot refers to the pale pink conidial pustules, about 1–2 mm in diameter, which burst through the bark (Fig. 180). Before the connection with *Nectria* was understood these conidial pustules had been named *Tubercularia vulgaris*. They consist of a column of pseudoparenchyma bearing a dense tuft of conidiophores, long slender hyphae producing phialides at intervals along their length (Fig. 181B,D). Conidial pustules of this type are termed **sporodochia.** The conidia are sticky and form a slimy mass at the surface of the

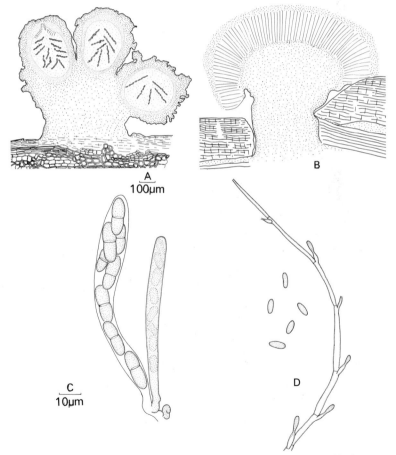

Fig. 181. *Nectria cinnabarina.* A. V.S. perithecial stroma. B. V.S. conidial stroma or sporodochium. C. Asci. D. Conidiophore, phialides and conidia.

sporodochium. They are dispersed very effectively by rain-splash (Gregory *et al.*, 1952). Around the base of the old conidial pustule perithecia arise (Fig. 180), and eventually the pustule may bear perithecia over its entire surface. Perithecial pustules develop in damp conditions in late summer and autumn and are readily distinguished from conidial pustules by their bright red colour and their granular appearance. The pustules are regarded as perithecial stromata, and bear as many as thirty perithecia. Ripe perithecia contain numerous club-shaped asci each with eight two-celled hyaline ascospores (Fig. 181C). They are somewhat unusual in being multinucleate (El-Ani, 1971). There is no obvious apical apparatus to the ascus (Strickmann & Chadefaud, 1961).

Perithecial development begins by the development of ascogonial primordia beneath the surface of the conidial pustule. The details of development have already been described. An important feature is the development of apical paraphyses which grow downwards from the upper part of the perithecial cavity. As the asci develop from ascogenous hyphae lining the base of the perithecial cavity, they grow upwards through the mass of apical paraphyses, which are difficult to find in mature perithecia (Strickmann & Chadefaud, 1961).

Other species of *Nectria* differ from *N. cinnabarina* in a number of ways. In

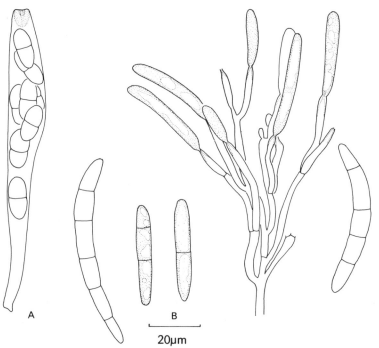

Fig. 182. *Nectria mammoidea.* A. Ascus: note the apical apparatus. B. *Cylindrocarpon*-type conidia.

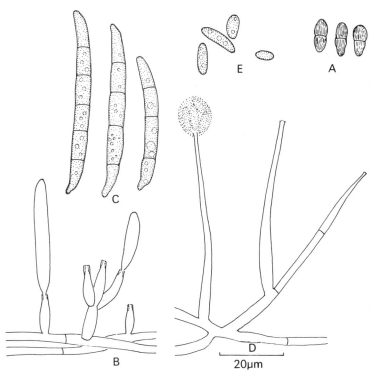

Fig. 183. *Nectria haematococca*. A. Ascospores. B. Phialides bearing macroconidia. C. Macroconidia of the *Fusarium* type. D. Phialides bearing microconidia. E. Microconidia.

many, the perithecia are not grouped together on a stroma, but occur singly (e.g. in *N. galligena*). The asci of some species, e.g. *N. mammoidea*, have a well-defined apical apparatus (Fig. 182A). The conidial states of some *Nectria* species are classified in several form-genera of the Fungi Imperfecti, including *Cephalosporium, Cylindrocarpon, Fusarium* and *Verticillium* (Booth, 1959, 1960, 1966a, 1978) (Figs. 183, 184).

Clavicipitaceae

The fungi in this group have several characteristics. The perithecia are developed on a fleshy stroma; the asci have a well-defined thick apical cap, and the ascospores are long and narrow, often breaking up into short segments. They escape singly and successively through a narrow pore in the ascus cap. Most members are parasitic on grasses (e.g. *Claviceps, Epichloe*) or on insects (*Cordyceps*). Whilst some mycologists accord the group ordinal rank (Nannfeldt, 1932; Gaümann, 1964; Rogerson, 1970), others regard it as a family of the Hypocreales (Bessey, 1950) or Sphaeriales (Miller, 1949; Luttrell, 1951).

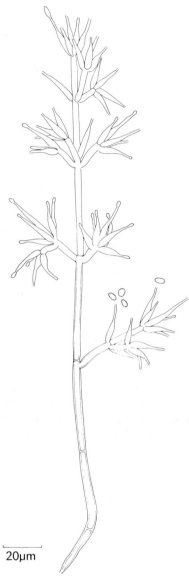

20µm

Fig. 184. *Nectria inventa. Verticillium cinnabarinum* conidial state. The whorled phialides bear globose masses of sticky phialospores.

Claviceps (Figs. 185–186)

Claviceps purpurea, the cause of ergot of grasses and cereals, grows on a wide range of grasses in late summer and autumn. Although common on rye and other cereals in Europe and North America it is not usually troublesome in Britain. In the occasional years in which its incidence is high there is an apparent correlation with high relative humidity and low maximum temperature in June (Marshall, 1960a). This is probably the result of a prolongation of the period during which the cereal hosts are susceptible to infection. A form with small sclerotia on *Molinia, Phragmites* and *Nardus* is regarded by some authorities as a separate species, *C. microcephala*. Grasses and cereals infected with *C. purpurea* develop purple curved sclerotia (ergots) in the place of healthy grain (Fig. 185A). The sclerotia contain a number of toxic alkaloids (McLaughlin *et al.*, 1964) and if they are eaten they can cause severe illness and sometimes death. One effect of the toxins is to constrict the blood vessels, and the impaired circulation may result in gangrene or loss of limbs. Another effect is on the nervous system, resulting in convulsions and hallucinations. In the Middle Ages the symptoms of ergotism were called 'St Antony's fire' and there are numerous records of outbreaks of the disease (see Barger, 1931; Ramsbottom, 1953; Fuller, 1969; Bové, 1970). With improved grain-cleaning techniques the disease is now rare in humans, and the last recorded outbreak in Britain was in the Jewish community in Manchester in 1928 as a result of eating rye-bread. Cattle and sheep which have eaten sclerotia from pasture grasses may also be affected and if pregnant animals are involved there is a risk of abortion (Ainsworth & Austwick, 1973).

The sclerotia of *C. purpurea* are used medicinally to hasten uterine contraction during childbirth. The ergot of commerce is produced by cultivating the fungus on rye *(Secale cereale)*, and crops are produced commercially in Eastern Europe, Spain and Portugal. Attempts are being made to extract the medically important alkaloids from pure cultures of the fungus (Mantle, 1974).

The sclerotia fall to the ground and overwinter near the surface of the soil, and probably need a period of low temperature before they can develop further. The main reserve substance of the sclerotium is lipid, which may account for 50% of the dry weight. It is likely that the chilling period is necessary before enzymes capable of mobilising the lipid reserves develop (Cooke & Mitchell, 1966, 1967, 1969, 1970; Mitchell & Cooke, 1968). The following summer, they develop one or more perithecial stromata about 1–2 cm high, shaped like miniature drum-sticks (Fig. 185C). The perithecial stromata are positively phototropic (Hadley, 1968). The enlarged spherical head contains a number of perithecia, embedded in the stroma, each surrounded by a distinct perithecial wall (Fig. 186A). The cytological details of perithecial development have been studied in *C. purpurea* by Killian (1919) and in *C. microcephala* by Kulkarni (1963). In the outer layers of the head of

Fig. 185. *Claviceps* and *Epichloe*. A. Head of rye (*Secale cereale*) bearing several sclerotia (ergots) of *Claviceps purpurea*. B. Rye inflorescence at anthesis bearing drops (arrowed) of the honeydew or *Sphacelia* conidial stage of *Claviceps purpurea*. C. *Claviceps purpurea*, germinated sclerotium showing the stalked perithecial stromata. D. *Epichloe typhina*, perithecial stroma on leaf sheath of *Agrostis tenuis*.

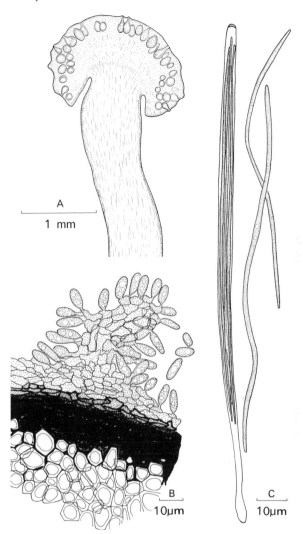

Fig. 186. *Claviceps purpurea.* A. L.S. perithecial stroma. B. T.S. young sclerotium showing the formation of the conidia on the surface. C. Ascus and ascospores. Note the cap of the ascus.

the perithecial stroma, club-shaped multinucleate antheridia and ascogonia undergo plasmogamy. Ascogenous hyphae made up of predominantly binucleate segments develop from the base of the ascogonium, and the tips of the ascogenous hyphae form croziers with binucleate penultimate segments. The penultimate cell elongates to form the ascus, and fusion between the two nuclei occurs. There are numerous asci in each perithecium, each containing a bundle of eight filiform ascospores. The ascus bears at its tip a conspicuous cap (Fig. 186C). Ascospore release coincides approximately with anthesis of the grass or cereal host, and infection of the developing ovary occurs. Successful infection of *Secale* from cultures derived from a single ascospore show that the fungus is homothallic (Esser, 1978). Whether ascospore infection occurs via the stigmata or the meristematic tissue at the base of the ovary is not certain, but within a few days of infection, a crop of conidia are formed, which are also capable of infecting young ovaries (Figs. 185B; 186B). Sometimes grasses which ripen early may develop conidia capable of infecting later-flowering cereals. This has been noted for *Alopecurus myosuroides* which becomes infected from ascospores and forms conidia which infect wheat (Mantle & Shaw, 1976). The conidia are unicellular and are budded off from the surface of the infected ovary. They are enveloped in a sticky, sweet liquid called honeydew, which is attractive to insects. The honeydew contains glucose, fructose, sucrose and other sugars (Mower & Hancock, 1975a). There is considerable variation in conidial dimensions from different grass hosts (Loveless, 1971; Loveless & Peach, 1974). Infection of a grass flower by *Claviceps* results in increased translocation of water and sucrose towards the diseased flower. Within the infected host tissue, conversion of sucrose from the host to mono-, di- and oligo-saccharides by the fungus creates a continuing 'sink' for sucrose translocation, and evaporation at the surface of the diseased grain results in increased osmotic concentration of the sugars, possibly resulting in the increased rate of translocation (Mower & Hancock, 1975b). Cereal grains normally store starch, but the sclerotium which replaces the seed contains none. *Claviceps purpurea* probably produces an inhibitor of starch phosphorylase which inhibits starch formation by its host. The fungus is unable to utilise starch and it has been suggested that the inhibitor prevents the conversion of sugars to starch and preserves the sugars of the host in forms available to the fungus (Campbell, 1959). Insects ingest conidia and honeydew and, on reaching uninfected flowers, may bring about new infections. The conidia do not germinate in the concentrated honeydew but only after it has been diluted. Before it was realised that these conidia belonged to *C. purpurea* they had been described under the name *Sphacelia segetum*. Conidia are formed freely in cultures derived from ascospores or from sclerotial tissue, and can be used experimentally to bring about infection. Although earlier studies had indicated that there was considerable degree of physiological specialisation it now appears that an isolate from one grass host may be capable of infecting a wide range of other cereals and grasses (Campbell, 1957). The period when the host is susceptible to infection varies, and corresponds with the time when the glumes are open (Campbell & Tyner, 1959). The course of infection following experimental inoculation with conidia is from the base of the young ovary, not from the stigmata (Campbell, 1958), and within about 5 days of infection a further crop of conidia may develop. Eventually the whole of the ovary may become converted into a sclerotium which is often considerably larger than the normal grain.

The control of ergot in cereals is difficult. Although several techniques are available, none is completely effective. Use of ergot-free seed would reduce infection, but inoculum may survive from a previous crop, or be provided by infected grasses bordering the field. Systemic fungicides would need to be applied in sufficient amounts to produce an effective concentration at the surface of the ovary. Since insects are vectors of secondary infections, combined use of fungicides and insecticides may be efficacious (Puranik & Mathre, 1971). Since infection occurs at anthesis, selection for self-pollinated cleistogamous strains of cereals may also help to reduce infection. A plant-breeding programme to develop resistant varieties is dependent on a source of natural resistance, and there are indications that certain Spring and Durum wheats do possess a certain degree of resistance (Platford & Bernier, 1970).

Epichloe (Figs. 185 and 187)

Epichloe typhina causes 'choke' of pasture grasses, and is common on grasses such as *Dactylis, Holcus* and *Agrostis*. At flowering time, the uppermost leaf sheath becomes surrounded by a white mass of mycelium 2 cm or more in length, and at the surface small unicellular conidia are produced (Fig. 187B). Later, the conidial stroma becomes thicker and turns orange in colour as perithecia are formed (Fig. 185C). The perithecia contain numerous asci, each with a well-defined apical cap, and eight long narrow ascospores which break up within the ascus to form part-spores (Fig. 187D) (Ingold, 1948). The mycelium is systemic, for the most part intercellular, unbranched and mainly located in the pith, but in the region of the inflorescence primordium, intracellular penetration of the vascular bundles is found. Because of the systemic condition, grasses infected with *Epichloe* can be propagated clonally and cultivated to yield plants, almost all of which are infected. Perithecial stromata are formed only on tillers containing inflorescence primordia, and by manipulation of daylength or auxin treatment in such a way as to stimulate induction of inflorescence primordia, it has been shown that the formation of stromata is correlated directly with the presence of an inflorescence primordium rather than with external conditions (Kirby, 1961).

In most infected grass hosts, although flowering is suppressed, vegetative growth is only slightly reduced, and the number of vegetative tillers may be increased. Thus choke is not a serious disease in pasture crops, but can be serious in crops grown for seed, especially in cocksfoot *(Dactylis)* (Large, 1952). Infection in *Festuca rubra* can be carried over by mycelium in the seeds produced by occasional flowering heads formed on infected plants, but flowering of infected *Dactylis* is rare. Normal inflorescence development from infected *Dactylis* tillers can be induced by application of gibberellic acid, and seed derived from infected plants does not produce infected offspring (Emecz & Jones, 1970). Experimentally, it has not been possible to infect seeds of *Dactylis*, and the only effective method of infection is by application of

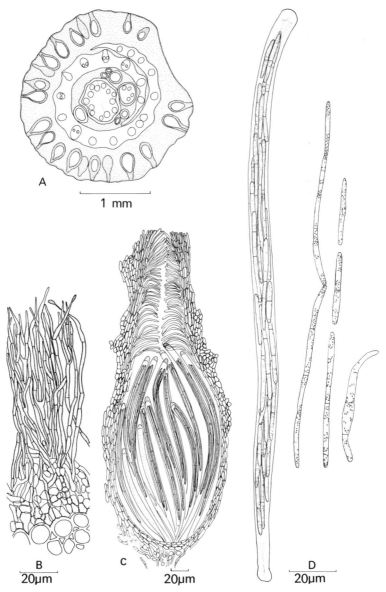

Fig. 187. *Epichloe typhina.* A. T.S. stem and leaf sheath of *Agrostis* surrounded by a perithecial stroma. Note the axillary shoots between the leaf sheath and stem. B. Part of conidial stroma. C. A single perithecium. Note the periphyses lining the ostiole. D. Ascus and ascospores. Note the apical apparatus of the ascus.

ascospores or conidia to the cut ends of stubble (Western & Cavett, 1959). If this is the natural route of infection in other grasses, it may explain the greater incidence of the disease in *Agrostis* in heavily grazed pastures (Bradshaw, 1959). However, throughout its range of distribution, *Epichloe* is attacked by a parasitic fly, *Phorbia phrenione*, which feeds on conidia, conidiophores and hyphae. The conidia can germinate after passage through the fly's gut, and it is possible that the fly may be a vector of the disease (Kohlmeyer & Kohlmeyer, 1974).

Perithecia develop on the surface of the conidial stroma and possess true perithecial walls. Paraphyses arise from the inside of the upper part of the perithecial wall, and as the asci develop, most of these paraphyses disappear, but some may persist to line the ostiole (Fig. 187c). Most authorities interpret the ascus wall as single, but Doguet (1960) has suggested that it might be bitunicate, and that the spores may be released by a mechanism similar to the 'Jack-in-a-box' type found in some other Ascomycetes. The apical apparatus consists of a thickened ring pierced by a narrow canal continuous with the cytoplasm of the ascus (Fig. 187d).

Systemic infection by *Epichloe* in certain hosts, e.g. *Festuca arundinacea*, may be symptomless, i.e. no conidial or perithecial stromata occur. Cattle grazing on infected *Festuca* have been known to suffer from gangrene-type symptoms similar to those caused by *Claviceps* alkaloids (Bacon *et al.*, 1977).

Cordyceps (Fig. 188–189)

About 150 species of *Cordyceps* are known, most of them parasitic on insects. Others are parasitic on spiders or on the subterranean fruit-bodies of *Elaphomyces*. *Cordyceps militaris* forms club-shaped orange-coloured stromata which project above the surface of the ground in autumn from buried Lepidopterous larvae and pupae (Fig. 188a). Several genera of Lepidoptera and some Hymenoptera are susceptible. The stromata bear numerous perithecia. The asci have conspicuous apical caps and contain eight long narrow ascospores which break up after discharge into numerous short segments (Fig. 189b). If the ascospores alight on the integument of a susceptible pupa, germ tubes may penetrate, possibly aided by their ability to hydrolyse chitin (Huber, 1958; McEwen, 1963). After infection, cylindrical hyphal bodies appear in the haemocoel of the pupa. The hyphal bodies increase by budding, and the buds are distributed within the insect's body. After death, mycelial growth follows, and the body of the insect becomes transformed into a sclerotium, from which the perithecial stromata later develop. In pure culture, a conidial state of the *Paecilomyces* type is formed (Fig. 189c). Earlier observations that coremia of the *Isaria* type were also conidia of *Cordyceps* are erroneous (Petch, 1936). When Lepidopteran pupae are inoculated by introducing conidia into the body cavity, death follows within 5 days. Mature fruit-bodies can develop within 45–60 days of infection (Shanor, 1936). The

Fig. 188. A. *Cordyceps militaris*: perithecial stroma arising from lepidopterous pupa. B. Enlargement of tip of stroma to show perithecia. C. *Cordyceps ophioglossoides* attached to ascocarp of *Elaphomyces*.

fungus has been grown and induced to form perithecial stromata in pure culture on rice grain, supplemented with nitrogen in the form of haemoglobin or casein. Light is required for fruiting. The nutritional requirements of the fungus are simple, and it is difficult to explain its restriction in nature to Lepidopterous hosts (Basith & Madelin, 1968).

Development of perithecia has been studied in *C. agariciformia* (probably synonymous with *C. canadensis*) (Jenkins, 1934) and in *C. militaris* (Varitchak, 1931). Coiled septate ascogonia arise in the peripheral layers of the perithecial stroma. The segments of the ascogonium become multinucleate and give rise to ascogenous hyphae from which asci develop in a single cluster at the base of the perithecium. The perithecial wall arises from hyphae which develop from the stalk of the ascogonium or from surrounding hyphae. Paraphysis-like hyphae grow inwards from the perithecial wall, but at maturity these hyphae are dissolved and disappear.

Cordyceps ophioglossoides grows on the subterranean fruit-bodies of *Elaphomyces,* forming bright yellow mycelial strands over the surface. Brown

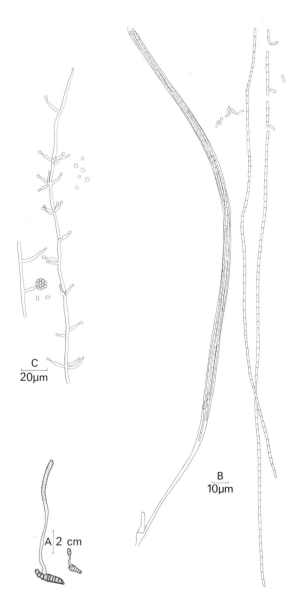

Fig. 189. *Cordyceps militaris.* A. Two perithecial stromata attached to pupae. B. Ascus and ascospores. The ascospore to the right contains eighty-two segments. Note the ascus cap. C. Conidiophores and conidia.

club-shaped perithecial stromata grow above the soil surface in autumn (Fig. 189c).

4: DISCOMYCETES

This is an assemblage of Ascomycetes in which the ascocarp is generally cup-shaped or saucer-shaped, with a hymenium which is usually freely exposed when ripe. Exceptions are the hypogaeous Discomycetes (Tuberales) which form subterranean fruit-bodies with an enclosed hymenium. Korf (1973) and Kimbrough (1970) have reviewed the classification of Discomycetes, and Dennis (1968) has written a useful guide to identification. An important feature in classification is the presence or absence of an operculum at the ascus apex. The inoperculate forms (Helotiales, Phacidiales) often form rather smaller and less conspicuous fruit-bodies than the operculate forms (Pezizales). The Pezizales are mostly terrestrial or coprophilous, whilst the Helotiales usually grow saprophytically or parasitically on plants. The Phacidiales include a number of important leaf pathogens. Many lichens have Discomycetes as their fungal component or mycobiont. The lichen-forming habit has probably evolved many times amongst different groups of Discomycetes. There are two different approaches to the classification of lichenised fungi: to adopt a completely separate classification for them, or to attempt a classification which groups them with their free-living relatives, where known (Poelt, 1973; Cooke & Hawksworth, 1970).

HELOTIALES

We need not study the detailed classification (see Nannfeldt, 1932; Dennis, 1968; Korf, 1973) but will consider a few representatives. Whilst most of the Helotiales are saprophytic, the group includes a number of important plant pathogens such as species of *Sclerotinia* and its segregates, and *Trichoscyphella willkommii,* the cause of larch canker.

Sclerotinia (Figs. 190–193)

The characteristic feature of this genus is the formation of stalked apothecia which grow from stromata developed within the infected host tissue. The apothecia usually develop in spring from over-wintered stromata. The stroma is a food storage organ and is differentiated into two parts, a medulla of hyaline cells and a rind of dark thick-walled cells. Two generalised types of stroma have been distinguished. To quote from Whetzel (1945):

The *sclerotial stroma* (commonly called the *sclerotium*) has a more or less characteristic form and a strictly hyphal structure under the natural conditions of its development. While elements of the substrate may be embedded in its medulla they occur there only incidentally and do not constitute part of the reserve food supply. The *substratal stroma* is of a diffuse or indefinite form, its medulla being composed of a loose hyphal weft or network permeating and preserving as a food supply a portion of the suscept or other substrate (e.g. culture media).

Various types of macroconidia are formed and some authors (e.g. Whetzel, 1945; Dennis, 1956) have separated species with different macroconidia into distinct genera or subgenera. For example *Sclerotinia fuckeliana* (= *Botryotinia fuckeliana*) has *Botrytis cinerea* as its conidial state (Groves & Loveland, 1953) whilst *Sclerotinia fructigena* (= *Monilinia fructigena*) has *Monilia fructigena* as its conidial state. Other species may have no macroconidial state, e.g. *S. curreyana* a parasite of *Juncus effusus* and *S. tuberosa* a parasite of *Anemone nemorosa*. Apothecia of the last two species are common in May. In *S. curreyana* the apothecia arise from black sclerotia in the pith at the base of the *Juncus* stem (see Fig. 190A). Infected stems look paler than healthy stems, and by feeling down to the base of an infected stem the sclerotium can be felt as a swelling, between finger and thumb. The sclerotium has an outer layer of dark cells and a pink interior which includes some of the stellate pith cells of the host (Fig. 191A). One to several apothecia may grow from a single sclerotium. The ascospores are released in late spring and infect the current season's stems. In culture germinated ascospores form a mycelium which develops microconidia, from small phialides (Fig. 191D). Similar clusters of microconidia can be found on infected *Juncus* later in the season and line cavities beneath the epidermis in the upper part of infected culms. Whetzel (1946) has used the term *spermodochidium* for these microconidial fructifications (Fig. 191E). It is probable that the microconidia play a role in fertilisation.

The apothecia of *S. tuberosa* (Fig. 190B) are about 2 cm in diameter and arise from sclerotia within rhizomes of *Anemone nemorosa* (Spaeth, 1957;

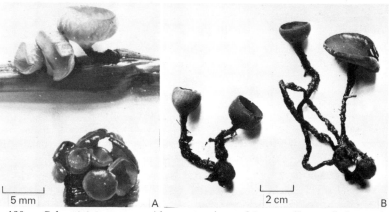

5 mm   A   2 cm   B

Fig. 190. A. *Sclerotinia curreyana*. Above: stem base of *Juncus effusus* splitting to show a black sclerotium bearing three apothecia. Below: sclerotium removed from the host bearing eight apothecia. B. *Sclerotinia tuberosa*. Two groups of apothecia rising from sclerotia formed on rhizomes of *Anemone nemorosa*. The apothecia are formed at the surface of the soil.

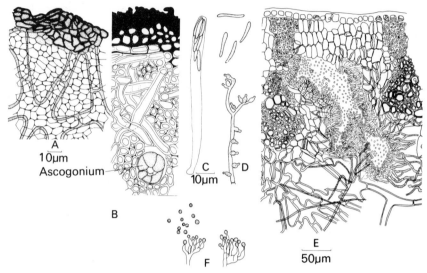

Fig. 191. *Sclerotinia curreyana.* A. T.S. sclerotium. Note the stellate pith cells of the host, *Juncus effusus.* B. T.S. sclerotium showing an ascogonium. C. Ascus and ascospores. D. Microconidia in culture. E. T.S. spermodochidium on *Juncus effusus.* Note the cavity lined by phialides. F. Microconidia from host.

Siegel, 1958). They may also occur on garden *Anemone* associated with black-rot disease. Microconidia are formed in culture. Electron microscope studies of the ascus wall show that it has a two-layered wall, but the two layers do not separate from each other – i.e. the ascus is not functionally bitunicate.

Fig. 192. *Sclerotinia fructigena.* Apple showing brown rot caused by this fungus, and bearing conidial pustules.

The ascus apex contains a thickened dome of wall material with a central canal. As the ascus explodes, the apical apparatus is everted (Schoknecht, 1975).

Sclerotinia fructigena is the cause of brown-rot of apples, pears and some other fruits, and although the apothecial state has not been found in Britain the disease is common and is transmitted by means of conidia. Apples and pears showing brown-rot bear buff-coloured pustules of conidia often in concentric zones (Fig. 192). Sporulation is stimulated by light, and adjacent zones correspond to daily periods of illumination. The conidia are formed in chains which extend in length at their apices by budding of the terminal conidium. Occasionally more than a single bud is formed, and this results in branched chains (see Fig. 193D). Conidial formation of this type is characteristic of the form-genus *Monilia* of the Fungi Imperfecti. Infection of the fruit is commonly through wounds, caused mechanically or by insects such as codling moth, wasps and earwigs (Croxall *et al.*, 1951). Fruit left lying on the ground is the source of infection in the following season. Infected fruit becomes mummified during the winter, but in the following year such fruit may develop conidial pustules. The disease may develop in stored apples, and in some varieties a twig infection (spur canker) may occur. A similar group of diseases

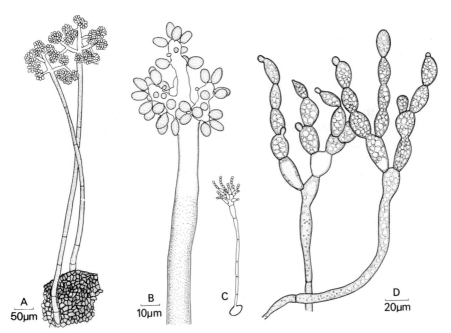

Fig. 193. A–C. *Botrytis cinerea*. A. Conidiophores developing from a sclerotium. B. Apex of conidiophores showing origin of conidia as blastospores. C. Conidium germinating to produce phialides and microconidia (after Brierley, 1918). D. *Sclerotinia fructigena*. Conidia of the *Monilia* type.

of apple and plum is caused by *Sclerotinia laxa* which also has a *Monilia* conidial state. The apothecial state is rarely found (Wormald, 1921, 1954).

Although the apothecia of *Sclerotinia fuckeliana* are not commonly collected, the conidial state, *Botrytis cinera,* is abundant on all kinds of moribund plant material, and the fungus is also associated with a wide range of disease often referred to as grey mould. It is probable that the name *Botrytis cinerea* is a collective name used to describe a number of closely similar, but genetically distinct, species, possibly with distinct apothecial states. For this reason many authors prefer to write of a *Botrytis* of the *cinerea* type. Serious diseases caused by this fungus are grey mould of lettuce, tomato, strawberry and raspberry, die-back of gooseberry and damping-off of conifer seedlings. The macroconidia are formed on infected host tissue as dark-coloured branched conidiophores. The tips of the branches are thin-walled and bud out to form numerous elliptical multinucleate conidia (blastospores) (Hughes, 1953) which are easily detached by wind, or are thrown off as the conidiophores twist hygroscopically (Fig. 193A,B). Microconidia are formed from clusters of phialides (Fig. 193C). The microconidia are capable of germination (Brierley, 1918) but they are also involved in sexual reproduction. Sclerotia are formed at the surface of infected tissue and the fungus overwinters in this form. In spring the sclerotia may develop to give rise to tufts of macroconidia or, much less commonly, to apothecia. *Sclerotinia fuckeliana* is heterothallic with ascospores of two kinds. In a single ascospore culture macroconidia, microconidia and sclerotia develop, but not apothecia. Apothecia will develop if microconidia of one mating type are applied to sclerotia of opposite mating type (Groves & Drayton, 1939; Groves & Loveland, 1953). In Whitehouse's (1949a) terminology *S. fuckeliana* shows physiological heterothallism. Another example of the same kind is *Sclerotinia gladioli* (= *Stromatinia gladioli*) (Drayton, 1934). In this species both sclerotial and substratal stromata are formed. The stromata bear columnar receptive bodies within which are ascogonial coils. When microconidia of opposite mating type are applied to the receptive bodies apothecia develop. A second kind of behaviour is found in *Sclerotinia narcissi* (Drayton & Groves, 1952). Of the eight spores formed in the asci of this fungus four produce mycelia bearing microconidia but no sclerotia, whilst the other four produce mycelia bearing sclerotia and stromata. Apothecia develop on the strains forming stromata if microconidia are transferred to them. Thus the mating behaviour of *S. narcissi* differs from that of *S. gladioli,* and we can say that *S. narcissi* is sexually dimorphic. This type of behaviour is not common in Ascomycetes and it is possible that an incompatibility system of this has been derived from the more usual system exemplified by *Sclerotinia gladioli* by aberrations which prevent the normal sequence of development of sexual organs in basically hermaphroditic forms (Raper, 1959). A third kind of mating behaviour is seen in *S. sclerotiorum* which is homothallic. A single ascospore culture produces microconidia and sclerotia which bear ascogonial coils beneath the rind. Transfer of microconidia to the sclerotia on the same mycelium results in the formation of apothecia (Drayton & Groves, 1952). A similar process of self-fertilisation also occurs in *Sclerotinia porri* (= *Botryotinia porri*) (Elliott, 1964).

Microconidia of some species of *Sclerotinia* fail to germinate when transferred to nutrient media. In the related fungus *Gloeotinia temulenta,* the cause of blind seed disease of ryegrass, it has been shown that the cytoplasm of the microconidia is deficient in RNA. In contrast, the microconidia of *Neurospora crassa* which do germinate, are heavily charged with RNA (Griffiths, 1959). It has therefore been suggested that, since RNA is essential for protein synthesis, the microconidia are incapable of growth.

Sclerotinia sclerotiorum causes a range of diseases (*Sclerotinia* rot, stalk break) in cultivated plants such as carrots, potatoes, tomatoes and hops, and on wild plants. There is no conidial state. Sclerotia formed on decaying crop debris survive in the soil. Sclerotia form readily in culture and have been the subject of extensive investigations into physiology of development (see R. C. Cooke, 1969, 1970, 1971a,b; Trevethick & Cooke, 1973), morphology (Calonge, 1970; Colotelo, 1974; Kosasih & Willetts, 1975; Wong & Willetts, 1975), survival and germination (Coley-Smith & Cooke, 1971).

Trichoglossum (Fig. 194)

Trichoglossum is a representative of the Geoglossaceae, or earth-tongues, which form club-shaped stalked fruit-bodies, growing usually on the ground,

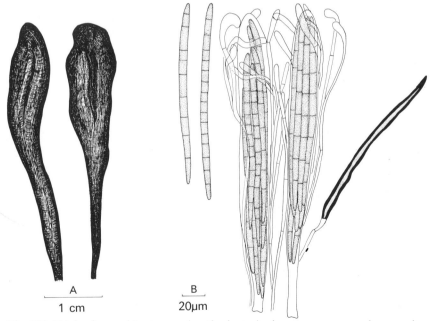

A   B
1 cm   20µm

Fig. 194. *Trichoglossum hirsutum*. A. Apothecia. B. Asci, ascospores, paraphyses and a hymenial seta.

but sometimes on dead leaves or amongst *Sphagnum* (e.g. *Mitrula*). Accounts of the family have been given by Nannfeldt (1942) and Dennis (1968). *Trichoglossum hirsutum* has black, somewhat flattened fruit-bodies up to 8 cm high, and grows in pastures and lawns. The ascospores are long, dark and septate, and the asci are interspersed by black, thick-walled, pointed hymenial setae whose function is not known (Fig. 194B). The presence of hymenial setae separates *Trichoglossum* from *Geoglossum* which grows in similar habitats. The elongate ascospores of *Trichoglossum* and *Geoglossum* are discharged singly through a minute pore at the tip of the ascus. When the ascus is ripe, the pore bursts and one ascospore is squeezed into it, blocking it. The pressure of the ascus sap behind the spore causes the spore to protrude, at first slowly, but when about half the spore is projecting, the spore gathers velocity and is rapidly discharged. Another ascospore immediately takes the place of the first and the process of discharge is continued until all eight ascospores are released (Ingold, 1953).

PHACIDIALES

The best-known example is *Rhytisma acerinum,* the cause of tar-spot disease of sycamore. Some other plant pathogens also belong here, e.g. *Lophodermium pinastri,* the cause of leaf-cast of pine. For details of the group see Dennis (1968), Korf (1973).

Rhytisma (Figs. 195–196)

Rhytisma acerinum is common on leaves of sycamore, *Acer pseudoplatanus*, forming black shining lesions (tarspots) about 1–2 cm wide (Fig. 195). The lesions arise from infection by ascospores released from apothecia on overwintered leaves. Lesions become visible to the naked eye in June or July, some two months after infection, as yellowish spots which eventually turn black. Sections of the leaf at this stage show an extensive mycelium filling the cells of the mesophyll, and especially the cells of the upper epidermis. Within the epidermal cells spermogonia develop. These are flask-shaped cavities which give rise to uninucleate curved club-shaped spermatia measuring about 6×1 μm (Fig. 196A,B). The spermatia exude from the upper surface of the centre of the lesion through ostioles in the spermogonial wall. The spermatia do not germinate, even on sycamore leaves, and it is believed that they play a sexual role (Jones, 1925). Apothecia begin development in the portion previously occupied by spermogonia and the hymenium is roofed over by several layers of dark cells formed within the upper epidermis. The asci complete their development on the fallen leaves and are ripe about May when sycamore leaves of the current season have unfolded (Aragno, 1968). The hymenium is exposed by means of cracks in the surface layer of the fungal stroma (Fig. 196B) and the asci discharge their spores, sometimes by puffing. Although the

Fig. 195. *Rhytisma acerinum*. A. Leaf of sycamore, *Acer pseudoplatanus* with tar spot. B. Tar spot from an overwintered leaf showing the cracking of the surface to reveal the hymenia.

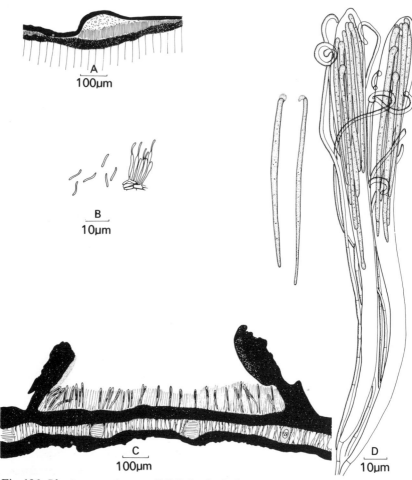

Fig. 196. *Rhytisma acerinum*. A. T.S. living leaf of *Acer pseudoplatanus* in June showing spermogonium. B. Details of cells forming spermatia. C. T.S. overwintered leaf of *Acer* showing the opening of the lips of the epithecium to reveal the hymenium. D. Asci, paraphyses and ascospores. Note the mucilaginous appendage at the upper end of the ascospore.

spores are only discharged to a height of about 1 mm above the surface of the stroma they are carried by air currents to leaves several metres above the ground. The ascospores are needle-shaped and have a mucilaginous epispore especially well developed at the upper end (Fig. 196D), and this probably helps in attaching them to leaves. Infection occurs by penetration of the term tubes through stomata on the lower epidermis.

Rhytisma acerinum is absent from densely populated areas, probably because the germination of ascospores is inhibited by sulphur dioxide. The incidence and frequency of colonisation of sycamore leaves by the tar-spot fungus can be used as accurate visual indices of air pollution (Bevan & Greenhalgh, 1976; Greenhalgh & Bevan, 1978).

LECANORALES

This is a group of inoperculate Discomycetes which live symbiotically with algae in lichen thalli. Other fungi also form lichens, e.g. the Pleosporales, Hysteriales, Sphaeriales, Basidiomycetes and some Fungi Imperfecti (Santesson, 1953; Ciferri & Tomaselli, 1955; Poelt, 1973), but the majority of lichens belong in this group. About a quarter of the known species of fungi are lichenised. Lichens inhabit the surfaces of rocks and trees, and some grow on infertile soil. They may play a rôle in rock weathering and soil formation (Syers & Iskander, 1973). Aquatic lichens are also known attached to rocks in the sea or in fresh-water streams. Although they contain two organisms, lichens are classified into species, genera and families. Over 1,300 British species are known (Smith, 1918, 1921, 1926; Duncan, 1959, 1963; James, 1965; Alvin, 1977). General accounts of lichens have been written by des Abbayes (1951), Ahmadjian (1967a), Hale (1967), Ahmadjian & Hale (1973) and Richardson (1975). Members of the Chlorophyceae (green algae) and Myxophyceae (blue-green algae) may be the algal components of lichens (Ahmadjian, 1967b). It is possible to grow the algal and fungal partners of many lichens separately in pure culture, and physiological studies of the separate components and of intact lichen thalli have been made (Quispel, 1959; Smith, 1962, 1963; Ahmadjian, 1965). Attempts have also been made to synthesise algal thalli *in vitro* from cultures of the two components, but typical lichen thalli have rarely been formed (Ahmadjian, 1973).

In most lichen thalli, the algae are confined to a special region, the algal zone, interspersed by fungal hyphae (see Figs 199 and 200). Above the algal zone, there is often a cortex made up entirely of closely packed fungal cells, and below the algal zone, there is the medulla made up of loosely woven, thick-walled hyphae. Although algal cells are occasionally penetrated by fungal haustoria (Peveling, 1973), it is probable that digestion of the algal cells is not a normal feature of lichen symbiosis. When $^{14}CO_2$ is supplied to intact lichen thalli in the light, organic compounds containing ^{14}C accumulate first in the algal zone, and later in the medulla (Smith, 1961). In lichens with blue-green algae as the phycobiont, e.g. *Peltigera polydactyla*, the form in which carbohydrate moves from alga to fungus is glucose, but in those with green algae, it moves in the form of polyhydric alcohols, e.g. ribitol, erythritol and sorbitol. In experiments in which disks of lichen tissue are allowed to absorb ^{14}C-labelled sugars, much of the labelling can later be detected in the fungal tissue in the form of sugar alcohols such as mannitol and arabitol (Smith *et al.*, 1969; Richardson, 1973). There is also evidence that in lichens which contain blue-green algae such as *Nostoc*, nitrogen is fixed by the alga, and the products of fixation are passed on to the fungus (G. D. Scott, 1956; Millbank &

Kershaw, 1973). No such fixation has been demonstrated in lichens with green algae, although in some of these, pockets of blue-green algae within the thallus, termed cephalodia, may be fixing nitrogen. Experiments with pure cultures of lichen algae and lichen fungi show that growth of either may be stimulated with culture filtrates of the other component (Quispel, 1959).

Although lichens are often regarded as a classical example of mutualistic symbiosis, i.e. an association in which both partners obtain mutual benefit, it is doubtful what advantage the phycobiont gains from the association, and there is no evidence of movement of nutrients or other substances from the fungus to the alga (D. C. Smith, 1973, 1978).

The reproduction of these lichen fungi is by means of ascospores which are violently discharged from the apothecia. If the spores germinate close to a suitable algal partner, a new lichen thallus may be initiated. Many species also reproduce by means of detachable fragments of the thallus which contain both fungal and algal cells, and sometimes the propagules form a powdery mass of granules (soredia) on the surface of special upright branches of the thallus, e.g. in *Cladonia* (Fig. 198).

Lichens are particularly sensitive to aerial pollutants, and especially to sulphur dioxide, so that they tend to disappear towards the centres of urban and industrial areas. Because different lichens show a differential sensitivity to pollution, their presence can be used as an accurate index of the level of SO_2 pollution. One of the least sensitive lichens is *Lecanora conizaeoides* which, in the absence of competitors, may dominate suitable surfaces in city centres (Gilbert, 1973; Hawksworth & Rose, 1976). Because of their capacity to absorb minerals, lichens take up and retain radio-active nuclides formed as a result of nuclear explosions. These nuclides may be incorporated in a food chain, lichen→reindeer→man, leading to their accumulation in human tissues (Tuominen & Jaakkola, 1973).

Xanthoria (Figs. 197A, 199)

The most abundant species is *X. aureola* (= *X. parietina* var. *aureola*) which forms bright yellow crusts on the surface of rocks, roofs, trees and farm buildings, especially near the sea (Fig. 197A). It is particularly common in places enriched by manure, e.g. dust from cattle yards, or from birds. The thallus is lobed and attached to the substratum by rhizoids. The algal component is the green alga *Trebouxia*, which forms single globose cells. The apothecia are saucer-shaped, and about 2–3 mm in diameter, on the upper surface of the thallus, and the algal zone extends into the apothecial margin (Fig. 199). The ascospores are at first one-celled, but ingrowth from the wall of the ascospore eventually divides the contents of the spore into two. The yellow colour of the thallus is due to the presence of the quinone parietin.

Fig. 197. A. *Xanthoria aureola*, thallus with apothecia. B. *Peltigera polydactyla*. Note the rhizinae growing from the lower side of the thallus and the folded marginal apothecia.

Fig. 198. *Cladonia pyxidata.* Primary foliose thallus bearing funnel-shaped podetia growing at the base of a Millstone grit wall. Note the granular soredia outside and inside the podetia.

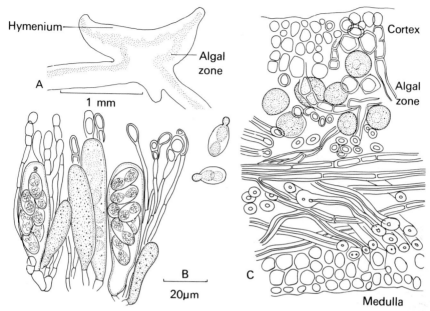

Fig. 199. *Xanthoria aureola.* A. V.S. thallus and apothecium showing the extension of the algal zone into the apothecium. B. Asci, paraphyses and two germinating ascospores. C. V.S. thallus.

Peltigera (Figs. 197B, 200)

Species of *Peltigera* form large lobed leaf-like thalli attached to the ground or to rocks by groups of white rhizinae. The commonest species are *P. polydactyla* (Fig. 197B) and *P. canina* which grow amongst grass on heaths, and on sand dunes and on rocks amongst moss. The algal component is the blue-green alga *Nostoc.* The apothecia are reddish-brown, folded extensions of the thallus which do not contain algal cells. The red colour of the apothecia is due to pigments in the tips of the paraphyses (Fig. 200B).

Cladonia (Figs. 198, 201)

There are numerous species of *Cladonia,* some of them extremely common, growing on heaths and moors, on rocks and walls. There are two kinds of thallus. The primary thallus is prostrate and lobed, and the secondary thallus is upright and cylindrical, often consisting of a hollow podetium which may open out into a cup, around the margins of which apothecia develop as in *C. pyxidata* (Fig. 198). The podetia frequently bear the granular soredia, containing algal and fungal cells (Fig. 201). In wind-tunnel experiments, using *C. pyxidata,* Brodie & Gregory (1953) showed that soredia were removed from the funnel-shaped podetia at winds of 1·5–2·0 m/second although they were

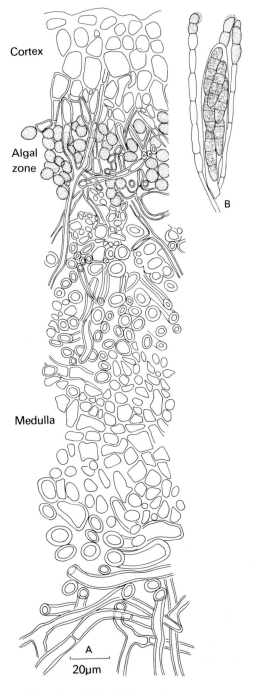

Cortex

Algal
zone

Medulla

A

20μm

B

Fig. 200. *Peltigera polydactyla*. A. V.S. thallus. B. Ascus, ascospores and paraphyses.

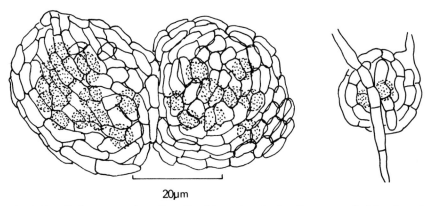

Fig. 201. *Cladonia pyxidata.* Soredia. Note the algal cells surrounded by fungal hyphae.

not removed from horizontal glass slides at the same wind-speeds. They suggested that funnel-shaped structures generate eddy currents when placed in a wind-stream and that the eddy currents effectively remove soredia.

In some species of *Cladonia,* e.g. *C. sylvatica,* the primary basal thallus quickly disappears and a secondary thallus, made up of much-branched cylindrical axes, persists. *Cladonia rangiferina,* reindeer moss, is common in the arctic tundra. It grows slowly, at less than 1 cm/year. It has been estimated that, in the arctic, lichen thalli may be up to 4,000 years old.

PEZIZALES

In the Pezizales, the asci are operculate, opening by a lid or operculum, and in this respect the group differs from the Helotiales in which the asci are inoperculate. The detailed classification will not be discussed (see Dennis, 1968; Korf, 1973), but a number of common examples illustrating some of the major groups will be described.

Pyronema (Figs. 202, 203)

Pyronema is found on bonfire sites and on sterilised soil. There are two common species, *P. omphalodes* (= *P. confluens*) and *P. domesticum.* In *P. omphalodes* the apothecia are often confluent and lack marginal hairs, whilst in *P. domesticum* the apothecia are discrete, and surrounded by tapering hairs (see Fig. 202A). *Pyronema domesticum* forms sclerotia in culture whilst *P. omphalodes* does not (Moore & Korf, 1963). In earlier studies the distinction between the two species was often not appreciated and some of the work on '*P. confluens*' may well have been done on *P. domesticum.* Both species are homothallic and grow well in agar culture or on sterilised soil and form their pink apothecia, 1–2 mm in diameter, within 4–5 days. There have been

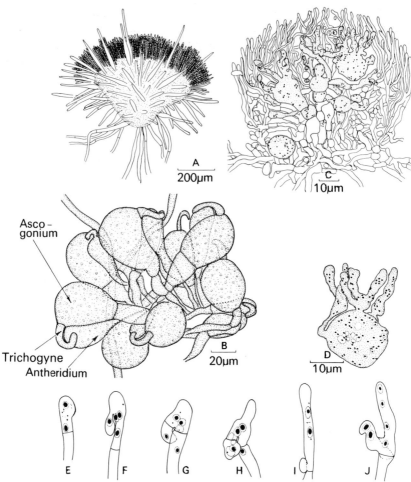

Fig. 202. *Pyronema domesticum.* A. Apothecium showing hymenium and excipular hairs. B. Group of ascogonia and antheridia. C. V.S. through developing apothecium showing several ascogonia producing ascogenous hyphae and the development of paraphyses and excipulum from the ascogonial stalks. D. Enlarged view of the asco-gonium and developing ascogenous hyphae. E–J. Stages in development of asci. E. Binucleate tip of ascogenous hyphae beginning to form a crozier. F. Quadrinucleate stage. G. Septation of crozier to form a binucleate penultimate cell. H. Development of ascus from binucleate cell. I. First meiotic division complete. J. Second meiotic division complete. Note the proliferation of a new ascogenous hypha from the stalk cell.

numerous accounts of the cytology of apothecial development (for references see Moore, 1963), but earlier claims for a double nuclear fusion and reduction (brachymeiosis) are no longer accepted (Hirsch, 1950; I. M. Wilson, 1952; McIntosh, 1954). In *P. domesticum* the apothecia arise from clusters of paired ascogonia and antheridia formed by repeated dichotomy of a single hypha.

The ascogonia are fatter than the antheridia and each ascogonium is surmounted by a tubular recurved trichogyne which makes contact with the tip of
the antheridium (Fig. 202B). The ascogonia and antheridia are multinucleate
and, following fusion of the antheridium with the trichogyne, numerous
antheridial nuclei enter the ascogonium. Contrary to earlier reports there is no

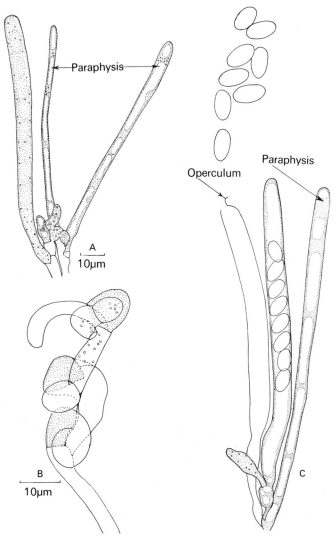

Fig. 203. *Pyronema domesticum*. A. An immature ascus (left) showing the ascogenous
hyphae from which it developed continuing to proliferate. B. More magnified view of
tip of an ascogenous hypha showing repeated proliferation. The three stippled cells
represent penultimate cells of croziers probably destined to develop into asci. C.
Mature asci, one discharged and showing an operculum. A paraphysis is also shown
apparently arising from the ascogenous hypha.

nuclear fusion at this stage. Multinucleate ascogenous hyphae develop from the ascogonium (Fig. 202D). They are septate and often branched and their tips are recurved to form croziers. The tip of the crozier is at first binucleate (Fig. 202E) and each nucleus divides mitotically. Septa develop between the nuclei to cut off a uninucleate terminal cell, a binucleate penultimate cell containing two non-sister nuclei, and a uninucleate antepenultimate cell or stalk cell (Fig. 202 F,G). The binucleate penultimate cell usually develops into an ascus. The two nuclei fuse and the diploid fusion nucleus undergoes meiosis, followed by mitosis so that eight haploid nuclei are formed around which the ascospores are cleaved. The fine structure of developing asci of *Pyronema* has been studied by Reeves (1967). Occasionally, instead of forming an ascus, the binucleate cell may grow out to form a new crozier. Proliferation of the original crozier can also occur by fusion of the uninucleate terminal cell with the stalk cell, which then grows out to form a new crozier. In this way a complex branched system bearing several asci may arise from the apex of a single ascogenous hypha (Fig. 203B).

The remaining tissues of the apothecium develop from the hyphae bearing the ascogonia. The paraphyses develop after the differentiation of the sex organs and are mostly fully developed before the ascogenous hyphae appear, so that the asci develop between the paraphyses. Paraphyses apparently also develop from the ascogenous hyphae (Fig. 203C). Pseudoparenchymatous cells develop around the base of the ascogonia and fill space not occupied by paraphyses or ascogenous hyphae. To the outside of the apothecium, they form a specialised layer, or excipulum, which in *P. domesticum* bears the tapering excipular hairs (Fig. 202A).

In some isolates of *P. domesticum*, ascogenous hyphae have been reported to develop from ascogonia in which the trichogynes have failed to fuse with antheridia – i.e. plasmogamy has been by-passed. Despite this, the further development of the apothecia followed the normal pattern (Moore-Landecker, 1975).

Pyronema is one of a group of Discomycetes which are associated with burnt ground. The reasons for their preference for this habitat are not fully understood, but it is known that *Pyronema* has a very rapid growth rate and appears to be relatively intolerant of fungal competitors (El-Abyad & Webster, 1968a,b).

Ascobolus (Figs. 204, 205)

Most species of *Ascobolus* are coprophilous, growing on the dung of herbivorous animals, but *A. carbonarius* grows on old bonfire sites (van Brummelen, 1967). Common coprophilous species are *A. furfuraceus* (= *A. stercorarius*) which is to be found on old cow dung, often along with *A. immersus* (Fig. 205). Whilst these species are heterothallic, others, e.g. *A. crenulatus* (= *A. viridulus*), are homothallic. Characteristic features of all species are the

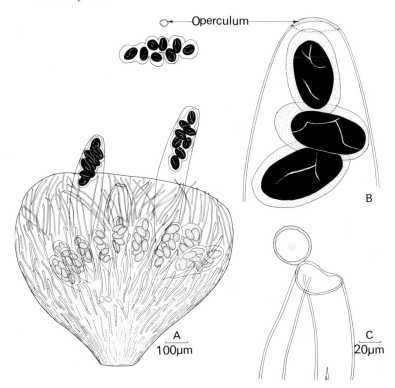

Fig. 204. *Ascobolus immersus*. A. Apothecium showing two projecting asci. Immature asci can be seen below the general level of the surface. A single projectile consisting of eight adherent ascospores is shown above the apothecium. Note the operculum which has also been projected. B. Tip of ripe ascus showing operculum. C. Tip of discharged ascus. In this case the operculum has remained attached to the ascus tip.

purple colour of the ascospore and the operculate phototropic asci. *A. furfuraceus* forms yellowish saucer-shaped apothecia up to 5 mm in diameter, and when it is mature the surface of the apothecium is studded with purple dots which mark the ripe asci. As the asci mature they elongate above the general level of the hymenium. The ascus tips are phototropic and this ensures that when they explode the spores are thrown away from the dung. The ascospores have a mucilaginous epispore which aids attachment. In *A. immersus,* which has very large ascospores (about 70 μm $\times$ 30 μm). the epispores cause all the eight ascospores to adhere to form a single projectile about 250 μm long which is capable of being discharged for distances up to 30 cm horizontally. The spores attach themselves to herbage and when eaten by a herbivore germinate in the faeces. It is likely that digestion stimulates spore germination for most spores fail to germinate on nutrient media, but can be stimulated to do so by treatment with a hydroxide or by bile salts (Yu, 1954). The purple pigment in

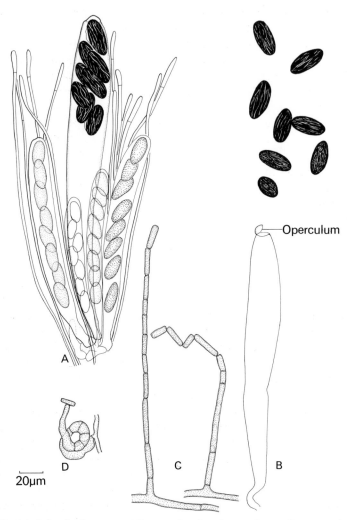

Operculum

20μm

Fig. 205. *Ascobolus furfuraceus*. A. Group of asci and paraphyses. One ascus is mature and contains purple-pigmented ascospores. B. The same ascus as shown in A after discharge. The ascus decreases in size after discharge. Note the operculum. C. Arthrospores (or oidia) developed in 5-day-old culture. D. Coiled ascogonium formed in a single ascospore culture within 48 hours of adding oidia of the opposite mating type. The trichogyne of the ascogonium has grown towards the oidium and has fused with it.

the spore wall develops late, and immature spores are colourless. The spore wall frequently bears longitudinal colourless striations.

A single ascospore culture of *A. scatigenus* (=*A. magnificus*) does not produce apothecia. Sex organs, coiled ascogonia, and antheridia are formed only when mycelia of different mating type are grown together. Ascogonia and antheridia develop on both strains, i.e. each strain is hermaphrodite, but it is self-incompatible: the antheridia of one strain do not fertilise the ascogonia borne on the same mycelium (Gwynne-Vaughan & Williamson, 1932). The ascospores of this fungus are of two types, 'A' and 'a', and fertilisation can only occur between an 'A' ascogonium and an 'a' antheridium and reciprocally. There is thus a gene for mating type represented in two allelic states 'A' and 'a', and incompatibility is controlled by this gene irrespective of the presence of both types of sex organ on each strain. There is no morphological difference between the two compatible mating types and Whitehouse (1949a) has used the term 'two-allelomorph physiological heterothallism' to describe this situation.

A similar situation occurs in *A. furfuraceus* but here each strain first produces chains of arthrospores or oidia (see Fig. 205C). The oidia can germinate to form a fresh mycelium, and they also play a part in sexual reproduction. It has been shown that mites and flies may transport oidia of one strain to the mycelium of the alternate strain, and following this apothecia develop (Dowding, 1931). The process of fertilisation has been studied in more detail by Bistis (1956, 1957) and by Bistis & Raper (1963). If an 'A' oidium is transferred to an 'a' mycelium, the oidium fails to germinate and within 10 hours an ascogonial primordium appears on the 'a' mycelium (Fig. 205D). The ascogonium consists of a broad coiled base and a narrow apical trichogyne which shows chemotropic growth towards the oidium and eventually fuses with it. There is evidence that this sequence of events is under hormonal control, and it has been suggested that a fresh 'A' oidium is not immediately capable of inducing development of ascogonial primordia, but must itself be first sexually activated by a secretion from the 'a' mycelium. Following activation, the oidium can induce ascogonial development. By substitution experiments, it has been shown that an 'A' ascogonium can be induced to fuse with an 'A' oidium, i.e. an oidium of the same mating type, but apothecia fail to develop from such fusions. In compatible crosses, fertile apothecia develop within about 10 days of fertilisation, each ascus producing four 'A' and four 'a' spores. The development of apothecia of *A. furfuraceus* has been studied by Gamundi & Ranalli (1963), Gremmen (1955), Corner (1929), Wells (1970, 1972) and O'Donnell *et al.* (1974). The ascogonium becomes surrounded by sheath hyphae which develop from the ascogonial stalk, and the paraphyses and excipular tissues develop from the sheath hyphae. The ascogonium gives rise to numerous ascogenous hyphae. Van Brummelen (1967) has distinguished two kinds of ascocarp development in *Ascobolus*. In *gymnohymenial* forms, the hymenium is exposed from the first until the maturation of the asci. In *cleistohymenial* forms the hymenium is enclosed, at least during its early development. *A. furfuraceus* and *A. immersus* are examples of cleistohymenial development.

A. immersus has proved a useful tool in interpreting the fine structure of the gene. Although the wild-type strains have purple ascospores, several series of mutants with pale spores have been found. When crosses are made between

380

Fig. 206. *Aleuria aurantia*. A mature apothecium and two developing apothecia.

certain pairs of such mutants, wild-type recombinants can result from two types of event: (a) crossing over, giving reciprocal recombinants and (b) conversion, yielding non-reciprocal recombinants corresponding to only one of the four products of meiosis.

A conversion is detected by the presence of coloured and colourless spores in the ratio 6:2, whilst crossing over results in a 4:4 ratio. In certain series of mutants, coversions are distinctly more frequent than crossing over, and this has led to the postulation of a genetic unit termed a *polaron,* a linear structure

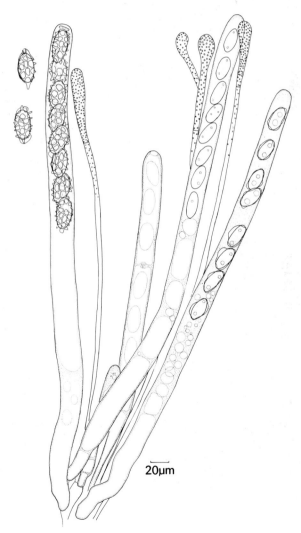

20µm

Fig. 207. *Aleuria aurantia.* Asci, ascospores and paraphyses. The tips of the paraphyses are filled with orange granules.

Fig. 208. A. *Morchella esculenta.* B. *Morchella elata.*

Fig. 209. *Helvella crispa.*

on which mutant sites are located and within which only non-reciprocal changes (conversions) can occur. Conversions have been interpreted as implying a double replication of one part of a chromatid whilst the corresponding part of the other is not replicated (Lissouba *et al.*, 1962; Fincham & Day, 1971; Decaris *et al.*, 1974).

OTHER MEMBERS OF THE PEZIZALES

There are numerous other common representatives of this group. *Coprobia granulata* is abundant on cow dung, forming orange apothecia about 3 mm in diameter. It is heterothallic, but sex organs are lacking; plasmogamy is brought about by fusion of mycelia of opposite mating type (Gwynne-Vaughan & Williamson, 1930). Species of *Peziza* form saucer-shaped apothecia up to 5 cm or more in diameter, growing on the ground, on rotting wood or dung. The ascus tip of *Peziza* stains blue with iodine and this character distinguishes it from *Aleuria*. *Aleuria aurantia,* sometimes called the orange-peel *Peziza,* forms bright orange saucer-shaped apothecia in the autumn up to 10 cm in diameter (Fig. 206). The ascospores have a coarsely reticulated surface (Bellemère & Melendez-Howell, 1976) (Fig. 207). The orange colour is caused by orange granules in the club-shaped tips of the paraphyses. *Morchella esculenta* (Fig. 208A) is edible and forms stalked apothecia about 10–15 cm high. It is found in woods or grassland on limestone in April and May. The hymenium lines the shallow depressions on the upper part of the fruit-body, whilst the ridges between lack asci. The ascus tips of *Morchella* are curved so that the spores are projected outwards, and do not impinge on the opposite side of the depressions in which the asci are formed. *Helvella crispa* forms white-stalked apothecia about 10 cm high in autumn. The hymenium is borne on a saddle-shaped head borne on a sterile stipe (Fig. 209).

Hypogaeous Discomycetes

TUBERALES

These are truffles, ascomycetes which form subterranean fruit-bodies, in which the hymenium is not open to the exterior. The asci do not discharge their spores violently, and lack a specialised apical apparatus or operculum. Many of the fruit-bodies have a strong smell and flavour, and are excavated and eaten by animals such as squirrels and rabbits, and it is therefore possible that dispersal is brought about in this way. The truffles of commerce are *Tuber melanosporum* (the Périgord truffle) and *T. magnatum* (the white truffle of Piedmont), and a number of other species. Certain species form sheathing mycorrhizas with trees; for example, *T. albidum* and *T. maculatum* with *Pinus strobus,* and *T. aestivum, T. brumale* and *T. melanosporum* with hazel, *Corylus avellana* (Fontana & Palenzona, 1969; Fassi & Fontana, 1969; Palenzona, 1969; Delmas, 1976).

The Périgord truffle is associated with the roots of *Quercus* spp. in France, and truffles are cultivated there by growing the appropriate species of oak (Malençon, 1938; Singer, 1961). Crops of truffles develop after about seven years, and are sometimes collected with the aid of trained pigs or dogs who can detect the fruit-bodies by their smell. Skilled truffle collectors can also detect the position of truffles, guided by truffle flies, which also seek out the fruit-bodies (Ramsbottom, 1953; Delmas, 1976).

There are several genera of Tuberales (Hawker, 1954, 1974). Whilst some authors include the genus *Elaphomyces* here, others place it in the Eurotiales.

Tuber (Figs. 210, 211)

Fruit-bodies of *Tuber* spp. can be collected by raking the soil under trees. Scrapings of rabbits and squirrels are often a useful indication of likely sites. Common species are *T. rufum, T. puberulum* and *T. excavatum*. The fruit-bodies are globose and up to about 3 cm in diameter. In section they consist of an outer peridium, often of thick-walled cells, and a gleba, or central fertile part, traversed by darker 'veins' which represent the hymenium. In some species the veins communicate with the exterior by one or more pores. For example in *T. excavatum* there is a basal cavity from which extensions protrude into the gleba. The asci are globose and often contain fewer than eight spores, often only two to four. The ascospores are thick-walled and ornamented by spines (e.g. *T. rufum*, Fig. 210) or by reticulate foldings of the outer wall of the spore (e.g. *T. puberulum*, Fig. 211D).

The development of fruit-bodies of *T. excavatum* have been studied by Bucholtz (1897) and Hawker (1954). The young fruit-body is a disk-like mass of hyphae which becomes corrugated on the lower side, and by arching of the upper surface becomes globose with a large basal opening leading to a hollow chamber. The convoluted lower surface forms a system of complex branched channels, called the 'venae externae'. A palisade-like layer of paraphyses develops over the surface of the venae externae and also forms within the flesh of a fruit-body lining a system of internal cavities, the 'venae internae'. Beneath the palisade, binucleate cells appear and give rise to ascogenous hyphae bearing the globose asci. The paraphyses grow out to form a loose weft of hyphae filling the venae externae. The cytology of ascus development has been followed in *T. brumale* and *T. aestivum* (Greis, 1938, 1939). The binucleate condition is established by fusion of two uninucleate cells, and from the fusion cell ascogenous hyphae develop containing numerous pairs of nuclei. Crozier formation occurs as in many other Ascomycetes.

It has been suggested (Fischer, 1938) that the Tuberales may have evolved from Ascomycetes with exposed hymenia as found in the existing Pezizales. Included in the Tuberales are genera such as *Genea* (see Fig. 211A,B) in which the fruit-body resembles an apothecium with a pore opening on the upper

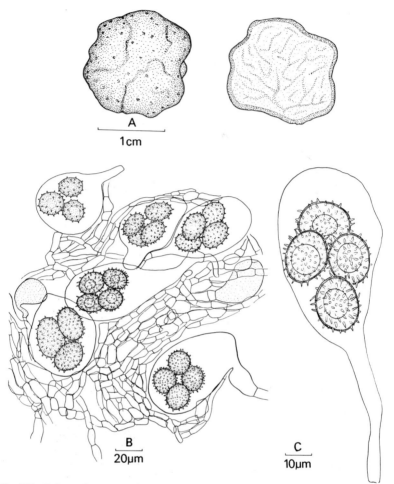

Fig. 210. *Tuber rufum.* A. Fruit-body in surface view and in section showing the veins. B. Portion of hymenium. C. Ascus.

side. The asci of *Genea* are cylindrical and eight-spored, and in this respect resemble more closely the asci of epigaeous Discomycetes.

Elaphomyces (Figs. 212, 213)

Elaphomyces, the hart's truffle, is probably the most common British hypogaeous fungus, and fruit-bodies of *E. granulatus* (Figs. 212, 213) and *E. muricatus* can be collected throughout the year beneath the litter layer under various trees, but especially beech. *Elaphomyces muricatus* is often parasitised by *Cordyceps ophioglossoides* forming yellow mycelium around the subterranean fruit-bodies, and a club-shaped perithecial stroma above ground. The

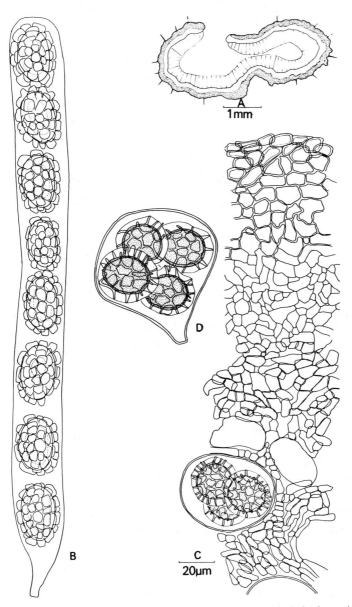

Fig. 211. A, B. *Genea hispidula.* C, D. *Tuber puberulum.* A. V.S. fruit-body. B. Ascus. C. V.S. fruit-body. D. Ascus.

Fig. 212. *Elaphomyces granulatus.* Two ascocarps, one cut open to show contents.

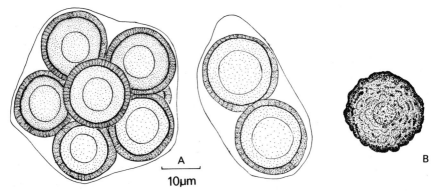

Fig. 213. *Elaphomyces granulatus.* A. Asci with two and seven ascospores. B. Mature ascospore.

Elaphomyces fruit-bodies vary in size from about 1–4 cm. When cut open an outer rind (or cortex) can be distinguished from a central mass containing the globose asci, traversed by lighter sterile 'veins'. The asci in *E. granulatus* usually contain six spores and in *E. muricatus* two to four. The spores are dark brown and thick-walled when mature and the conditions necessary for their germination are not known.

5: LOCULOASCOMYCETES

The characteristic feature of this group is that the ascus is bitunicate; it has two separable walls. The outer wall does not stretch readily, but ruptures laterally or at its apex to allow the stretching of a thinner inner layer. The fruit-body within which the asci develop is regarded as an ascostroma. This has been defined as an aggregation of vegetative hyphae not resulting from a sexual stimulus (Wehmeyer, 1926), but Holm (1959) has questioned the accuracy of this definition, since examples are now known where the ascocarps do develop following a sexual stimulus (Shoemaker, 1955). Within the developing asco-carp, one or more locules are formed by the down-growth of pseudopara-physes (see below) and the development of asci. One or more ostioles develop by the breakdown of a pre-formed mass of tissue. Where a single locule develops, a structure resembling a perithecium results and, although this term is commonly used, these structures should strictly be termed **pseudothecia**. The name given to the group was coined by Luttrell (1955) and corresponds to the Ascoloculares of Nannfeldt (1932). Several large orders are included (Luttrell, 1973; von Arx & Müller, 1975), but we shall consider only one, the Pleospor-ales.

PLEOSPORALES

This is a large group of ascomycetes, including some economically important genera of plant pathogens such as *Cochliobolus* and *Pyrenophora*, parasitic on grasses and cereals; *Ophiobolus, Pleospora* and *Leptosphaeria*, common sapro-phytes or weak parasites of herbaceous plants; and *Sporormia*, a saprophyte of dung.

The development of pseudothecia in these forms conforms to the *Pleospora* type of Luttrell (1951). Ascogonia arise within a stroma and, in the region of the ascogonia, a group of vertically arranged septate hyphae appears, each one arising as an outgrowth from a stroma cell. These hyphae are capable of elongating by intercalary growth and are termed **pseudoparaphyses** (Luttrell, 1965b). Pseudoparaphyses arise near the upper end of the cavity and grow downwards. Their tips soon intertwine and push between the other cells of the stroma so that free ends are seldom found. They may thus be distinguished from **true paraphyses** formed in other fungi from hyphae attached at the base of the cavity, extending upwards and free at their upper ends. They may also be distinguished from **apical paraphyses** which are attached above, arising from a clearly defined meristem near the apex of a perithecium, forming a well-defined palisade of hyphae free at their lower ends (see the *Nectria* type of development, p. 339). In the *Pleospora* type of development, asci arise amongst the pseudoparaphyses at the base of the cavity and grow upwards between them. The ostiole develops lysigenously, i.e. by breakdown or separation of pre-existing cells. Development of this general type has been described in *Pleospora herbarum* (Wehmeyer, 1955; Corlett, 1973), *Leptosphaeria* (Dodge, 1937), *Sporormia* (Arnold, 1928; Morisset, 1963) and other fungi (see Luttrell, 1951, 1973).

Leptosphaeria (Fig. 214)

Leptosphaeria species grow on moribund leaves and stems of herbaceous plants. There are probably some 200 species, many growing on a wide range of hosts, but some confined to one host plant. Although most are saprophytic or only weakly parasitic, some are troublesome parasites, e.g. *L. avenaria*, the cause of speckled blotch of oats, and *L. coniothyrium*, the cause of cane blight of raspberry. *L. acuta* grows in abundance at the base of overwintered stems of nettles *(Urtica dioica)*. The black shining pseudothecia are somewhat conical and flattened at the base. The bitunicate asci elongate within a pre-formed group of branching pseudoparaphyses, and close examination of the direction of growth and branching suggests that the pseudoparaphyses may be both ascending and descending. The ostiole of the perithecium is formed lysigenously by breakdown of a pre-existing mass of thin-walled cells (Fig. 214A).

The bitunicate structure of mature asci is difficult to discern because, as the ascus expands, the inner wall protrudes through a thin area in the outer wall at the ascus tip (Fig. 214E) and then the inner wall extends. Thus the tip of the ascus in expanded asci is single-walled. The ascospores are discharged successively in the space of about 5 seconds, and the ascus shows a slight decrease in length as each spore escapes (Hodgetts, 1917).

Associated with the thick-walled conical pseudothecia on the nettle stems are thinner-walled, slightly smaller globose pycnidia with cylindrical necks (Fig. 214B). The cavity of the pycnidia is lined by small spherical cells which give rise to numerous rod-shaped pycnospores which are capable of germination (Fig. 214C). Such pycnidia have been named *Phoma acuta*, and culture studies show that this stage is the conidial state of *L. acuta* (Müller & Tomasevic, 1957). Various different kinds of conidial apparatus have been found in other species of *Leptosphaeria* (Lucas & Webster, 1967; Lacoste, 1965).

Pleospora (Figs. 215, 216)

Wehmeyer (1961) recognises over 100 species, but this is probably an underestimate. Most species form fruit-bodies on moribund herbaceous stems,

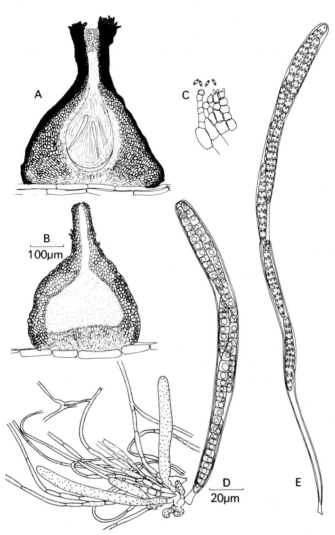

Fig. 214. *Leptosphaeria acuta*. A. L.S. immature pseudothecium. Note the young asci (stippled) elongating between a pre-formed mass of pseudoparaphyses, and the thin-walled cells at this stage blocking the ostiole, later dissolved. The centrum subsequently enlarges, dissolving the pseudoparenchyma surrounding it. B. L.S. pycnidium. C. High-power drawing of cells lining the pycnidium showing the origin of the conidia. D. Cluster of developing asci from a young pseudothecium. Note the branching of the pseudoparaphyses. E. Stretched bitunicate ascus showing rupture of the outer wall at its apex.

apparently as saprophytes, but it is likely that some are weak parasites. Of these, *P. bjoerlingii* (= *P. betae*) is the cause of black-leg of sugar beet. *P. herbarum* attacks a wide range of cultivated hosts, causing such diseases as net-blotch of broad bean and leaf spot of clover, lucerne and other hosts. It may be seed-borne. This fungus is also abundant on overwintered herbaceous

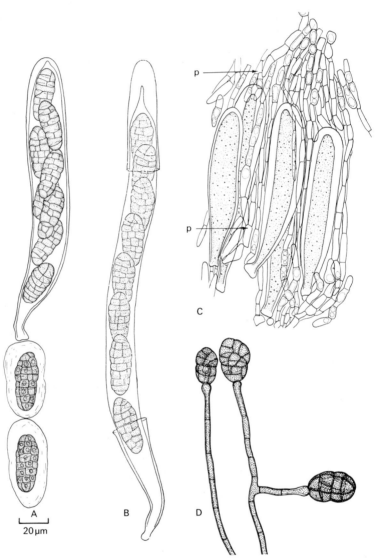

Fig. 215. *Pleospora herbarum*. A. Ascus and ascospores showing mucilaginous epispore. B. Stretched bitunicate ascus showing rupture of outer wall. C. Developing asci and pseudoparaphyses. The arrows (p) indicate points of branching of ascending and descending pseudoparaphyses. D. Conidia of *Stemphylium* type.

stems of numerous maritime plants. The large black, somewhat flattened pseudothecia contain broad sac-like bitunicate asci, with eight yellowish-brown, slipper-shaped ascospores with transverse and longitudinal septa (Fig. 215A). The conidial state, *Stemphylium*, is often associated with the pseudothecia. The *P. herbarum* complex includes a number of similar species forming conidia which are critically distinct from each other in morphology and dimensions (Simmons, 1969). The conidia develop singly from the tips of conidiophores swollen at their apices, as blown-out ends. A narrow neck of cytoplasm connects the developing spore to its conidiophore through a pore, and Hughes (1953) has termed conidia of this type porospores (Fig. 215D), but the term poroconidium is also used (Ellis, 1971a). Electron microscope studies (Carroll & Carroll, 1971) show that conidial development is blastic, involving the whole of the wall at the apex of the conidiogenous cell. The cytoplasmic connection between the conidiogenous cell and the conidium is narrow, and is surrounded by two layers of thickened wall material. The conidia of *P. herbarum* are formed more readily in cultures illuminated in near-ultra-violet

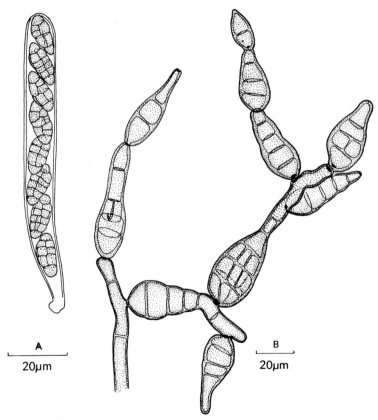

A
20μm

B
20μm

Fig. 216. *Pleospora infectoria*. A. Ascus. B. *Alternaria* conidial state.

light (Leach, 1968). Light and low temperature also stimulate pseudothecial development (Leach, 1971). The fungus is homothallic. According to Meredith (1965a), the conidia are violently jolted from the tip of the conidiophore.

Pleospora infectoria, which forms pseudothecia on grass and cereal culms in spring, has smaller ascospores. In culture, this fungus forms branching chains of beaked spores, and new spores are formed at the tip of the chain (Fig. 216). Conidia of this type belong to the form genus *Alternaria*, and are also poroconidia. Spores of *Alternaria* are abundant in the air in late summer and autumn, and may be a cause of inhalant allergy (Hyde & Williams, 1946).

Sporormia (Fig. 217)

Pseudothecia of *Sporormia* are common on dung of herbivores, and the fungus is also occasionally isolated from soil. *S. intermedia* is one of the most common species, and has thin transparent perithecial walls through which the asci can be seen (Fig. 217). The ascus has a well-defined double wall. The outer wall does not stretch readily and as the ascus becomes turgid the outer wall ruptures and the thin elastic inner wall stretches considerably. The tip of the ascus projects through the ostiole and the ascus then explodes. The ascospores of *S. intermedia* are four-celled and surrounded by a mucilaginous envelope. The spore may break up into individual cells, each capable of germination. Spore discharge is nocturnal (Walkey & Harvey, 1966b).

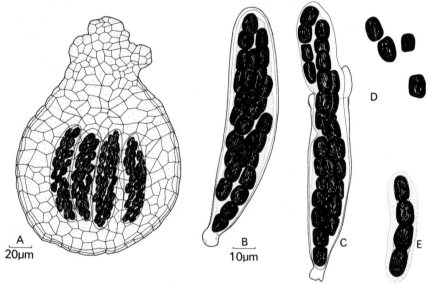

Fig. 217. *Sporormia intermedia*. A. Pseudothecium with asci visible through the transparent wall. B. Ripe unextended ascus showing the double wall. C. Elongating ascus showing rupture of the outer wall and extension of the inner. D. Ascospore separated into its four component cells. E. Intact ascospore.

4

Basidiomycotina
(Basidiomycetes)

Many of the familiar larger fleshy fungi are members of this group, which includes the toadstools, bracket fungi, fairy clubs, puff-balls, stinkhorns, earth-stars, bird's nest fungi and jelly fungi. Most of these are saprophytes, causing decay of litter, wood or dung, and some are serious agents of wood decay such as *Serpula lacrymans (Merulius lacrymans)* the dry-rot fungus (Cartwright & Findlay, 1958). Some of the toadstools which are associated with trees form mycorrhiza, a symbiotic association (Harley, 1969), but some are severe parasites, e.g. *Armillariella (Armillaria) mellea,* the honey agaric, which destroys a wide range of woody and herbaceous plants. Whilst the fleshy fungi enjoy a notorious reputation for being poisonous, the majority of toadstools are harmless, and several species besides the field mushrooms are good to eat (Ramsbottom, 1953). Two important groups of plant pathogens, the rusts (Uredinales) and smuts (Ustilaginales), are usually also classified in the Basidiomycetes. In nature, these organisms are confined to living host plants.

The characteristic spore-bearing structure is the basidium. In contrast with the endogenous spores of the ascus, basidia bear spores exogenously, usually on projections termed sterigmata. The number of spores per basidium is typically four, but two-spored basidia are quite common. In *Phallus impudicus,* the stinkhorn, there may be as many as nine spores per basidium. Basidia vary considerably in structure, and the form of the basidium is an important criterion in classification. In the toadstools and their allies the basidium is a single cylindrical cell, undivided by septa, typically bearing four basidiospores at its apex (Fig. 218). Such basidia are termed **holobasidia**. In the Uredinales and Ustilaginales the basidium develops from a thick-walled cell (teliospore or chlamydospore) and is usually divided into four cells by three transverse septa. Transversely segmented basidia are also found in the Auriculariaceae, but here the basidia do not arise from resting cells, In the Tremellaceae the basidia are longitudinally divided into four cells, whilst in the Dacrymycetaceae the basidium is unsegmented but forked into two long arms, to form the so-called tuning fork type of basidium. Segmented basidia are sometimes termed **phragmo-basidia** (or **heterobasidia**). Some of these different kinds of basidia are illustrated in Fig. 219.

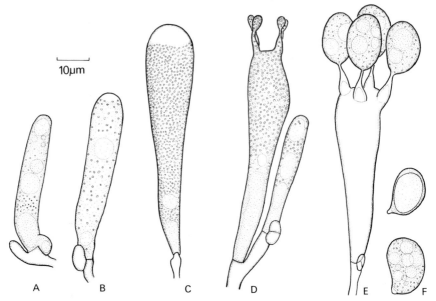

Fig. 218. *Oudemansiella radicata*. Stages in development of basidia. A. Young basidium with numerous vacuoles. Note the clamp connection at the base and the formation of a further basidial initial. B. Later stages showing the development of a clear apical cap. C. Localisation of vacuoles towards the base of the basidium. D. Development of sterigmata and spore initials. Note the enlargement of the basal vacuole. E. Fully developed basidium. The spores are full of cytoplasm, whilst the body of the basidium contains only a thin lining of cytoplasm, surrounding an enlarged vacuole. F. Discharged basidiospores.

DEVELOPMENT OF BASIDIA

The development of a basidium can be illustrated by reference to a gill-bearing fungus such as *Oudemansiella radicata* (= *Collybia radicata*) which grows on old stumps of deciduous trees (Figs. 218, 237A). The basidium arises as a terminal cell of a hypha making up the gill tissue on the underside of the cap of the fruit-body. The basidia are packed together to form a fertile layer or hymenium. A basidium is at first densely packed with cytoplasm, but soon several small vacuoles appear. Later, a single large vacuole develops at the base of the basidium and, by the enlargement of this vacuole, cytoplasm is pushed towards the end of the basidium. A clear cap is visible at the tip, and it is here that the sterigmata develop. Corner (1948) has postulated that there must be four elastic areas of the wall from which the sterigmata extend, and he has suggested that the force for the development of the basidia and the spores comes from the enlargement of the basal vacuole which acts like a piston ramming the cytoplasm into the spores. Ripe basidia thus contain very little cytoplasm, but contain merely a large vacuole (Fig. 218E).

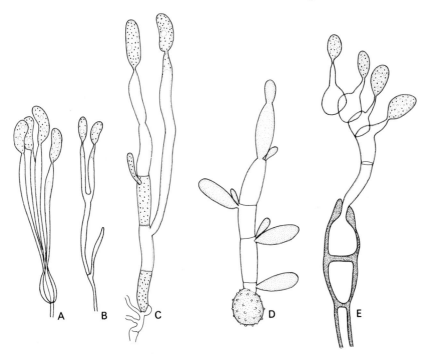

Fig. 219. Some different kinds of basidia. A. Longitudinally divided basidium of *Exidia glandulosa* (Tremellaceae). B. Tuning-fork type basidium of *Calocera viscosa* (Dacry-mycetaceae). C. Transversely divided basidium of *Auricularia auricula* (Auricularia-ceae). D. Germinating chlamydospore of *Ustilago avenae* (Ustilaginales). E. Germinating teliospore of *Puccinia graminis* (Uredinales).

Young basidia are binucleate, and nuclear fusion occurs here. The resulting fusion nucleus undergoes meiosis immediately, so that four haploid daughter nuclei result, and one is distributed to each basidiospore. In some basidia, meiosis is followed by a mitotic division, so that some basidiospores are binucleate. The plane of spindle-formation during meiosis may be transverse to the long axis of the basidium, and such fungi are said to be **chiastobasidial**. The contrasting condition, in which the spindles are orientated parallel to the long axis of the basidium, is termed **stichobasidial**. The division planes may have some taxonomic significance. For a fuller discussion of the cytology of basidia, see Raper (1966a), Olive (1953, 1965), Kühner (1977), and Wells (1977).

The fine structure of basidial development has been studied by several workers. Nuclear fusion triggers the development of endoplasmic reticulum vesicles which are scarce in basidia before nuclear fusion has occurred (Clemençon, 1969). During meiosis, the nuclear membrane may break down (e.g. in *Coprinus radiatus;* Lerbs, 1971) or may persist (e.g. in *Agaricus bisporus;*

Thielke, 1976). During meiosis in *Coprinus radiatus,* there is an accumulation of endoplasmic reticulum within the plasmalemma close to the basidial wall, and its position there has been compared with that of the ascus vesicle in developing asci (Lerbs, 1971). During meiotic prophase, there is extensive synthesis of membrane complexes which accumulate in the upper part of the basidium. The growth of the sterigmata has been likened to tip growth of vegetative hyphae (McLaughlin, 1973). Numerous vesicles have been reported within sterigmata which may contain precursors for the synthesis of wall materials in developing basidiospores. The movement of nuclei through the sterigmata into the basidiospores is associated with the development of cytoplasmic and intranuclear microtubules in *Lentinus edodes* (Nakai & Ushiyama, 1978). The nuclei moving into the basidiospores become elongated. There may be a discontinuity in the nuclear envelope, with the intranuclear microtubules entering through the discontinuity in the envelope. The cytoplasmic microtubules develop an association with a spindle pole body at the leading edge of the nucleus as it moves into the basidiospore.

The variation in structure of basidia has raised problems of nomenclature of the different component parts, and Talbot (1973a) has proposed definitions which are adopted here in simplified form.

Basidium: a fungal cell or organ which bears spores terminally and singly in extensions of its wall after karyogamy and meiosis . . . have occurred within the cell or organ. The term includes probasidium, metabasidium (and pariobasidium if present) and sterigmata as parts of the whole mature basidium.

Probasidium: the morphological part or developmental stage of the basidium in which karyogamy, or enlargement of a single haplo-parthenogenetic nucleus, occurs; i.e. the primary basidial cell.

Metabasidium: the morphological part or developmental stage of the basidium in which meiosis . . . occurs. In most basidia it replaces and obliterates the probasidium; in some basidia remnants of the probasidium are evident at the base of the metabasidium in the form of a sac, cyst or stalk cell, which may or may not be delimited by a septum. Subordinated to the metabasidium is the *pariobasidium:* the distal and functional portion of a metabasidium in which effete probasidial remnants are apparent when the basidium is mature, usually as a result of probasidial development in two phases.

Sterigma: an extension of the metabasidium wall, through which one or more nuclei are transferred from the metabasidium to the developing terminal basidiospore. The sterigma is composed of a basal filamentous or inflated part, the *protosterigma,* and an apical spore-bearing point, the *spiculum.*

Holobasidium: a basidium whose metabasidium is not divided by primary septa (i.e. septa formed in direct association with nuclear division and separating the daughter nuclei), but may sometimes become adventitiously septate (i.e. may develop septa formed independently of nuclear division and especially in association with change in the local concentration of cytoplasm).

Phragmobasidium: a basidium whose metabasidium is divided by primary septa, usually cruciate or transverse.

FINE STRUCTURE OF BASIDIOSPORES

The structure of developing and mature basidiospores has been studied by transmission and scanning electron microscopy (Wells, 1965; Perreau-Bertrand, 1967; Clemençon, 1970; Pegler & Young, 1971; Kühner, 1973; Nakai & Ushiyama, 1974b; Nakai, 1975; see also p. 85 and Fig. 38).

The walls of the sterigma and spore initial are continuous, i.e. the spore initial is formed by inflation of the apex of the sterigma. The contents of the enlarging basidiospore initial are in direct cytoplasmic continuity with the lumen of the sterigma. As the spore develops, successive wall layers are deposited, derived from two membranes known as the internal basidial layer

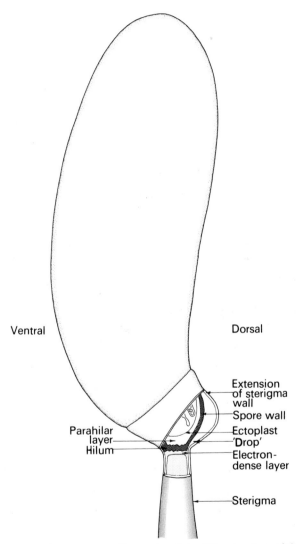

Fig. 220. *Schizophyllum commune*. Reconstruction of the structure of the sterigma and basidiospore based on electron micrographs of thin sections (after Wells, 1965).

and the external basidial pellicle. In some species, remnants of these membranes may persist in mature spores (Fig. 38, and see below). The first wall layer to appear is the **episporium**, which is the thickest and most obvious layer formed on the inside of the spore adjacent to the plasmalemma. The episporium may, in certain forms, give rise to a layer outside it, the **exosporium**. At the pore opposite the point of attachment, the episporium and exosporium may be thinner to form the germ pore. The final wall layer to appear is the **endosporium**, the innermost layer. In the region of the hilum (see Figs. 38 and 220), the endoplasmic reticulum of the basidiospore retracts, and the endosporium disrupts the continuity between the sterigma and spore. The shape of the mature spore is the result of the turgor pressure of its contents, and differential hardening of the wall layers. Globose basidiospores, such as those of *Oudemansiella mucida* or *Amanita vaginata*, result from the stretching of wall layers which are equally elastic. More usually, the spore is asymmetrical, with the adaxial or dorsal face less strongly curved than the abaxial or ventral face, presumably due to the more elastic nature of the curved wall during development (Figs. 38, 220). In many Basidiomycetes, the spore wall is not smooth, but is ornamented by spines or folds usually derived from the exosporium. The spore may be colourless or pigmented. The pigment, where present, may be in the cytoplasm or in the episporium (or in both). Spore shape, size, ornamentation and pigmentation are important taxonomic characters at the generic or family level. In some basidiospores, the spore is surrounded by one or two further layers, the **perisporium** and the **ectosporium**. The perisporium, probably derived from the internal basidial layer, often disappears early in spore development, but in *Coprinus* it persists as a loosely attached, colourless layer which envelops the spore as the perisporial sac. The ectosporium is a fine membrane, scarcely visible by light microscopy, which probably originates from the external basidial pellicle.

THE MECHANISM OF BASIDIOSPORE DISCHARGE

Observations on ripe basidia show that the individual spores are in turn projected violently from the tip of the sterigma. The term **ballistospore** has been applied to violently projected spores (Derx, 1948). Whilst most basidiospores are ballistospores, some, e.g. those of the Gasteromycetes, are not. These non-violently projected spores have been termed **statismospores**. Shortly before discharge occurs a swelling appears on the adaxial side where the spore and the sterigma join. Although observed earlier by other workers, Buller (1909, 1922) drew especial attention to it, and it is often referred to as 'Buller's drop'. Buller assumed that the drop consisted largely of water. Prince (1943) showed that the drop could form under water, whilst Corner (1948) showed that the drop could persist in material preserved in alcohol–formalin. He also found that a drop may appear, increase in size, and then diminish and disappear before discharge. These observations suggest that the drop is surrounded by a membrane, and may consist of a ballooning outwards of the

sterigma wall. Electron micrographs of thin longitudinal sections of sterigmata with attached spores (Wells, 1965) support this view (Fig. 220). Opinions
vary on the content of the drop. Corner (1948), Müller (1954) and Wells (1965)
conclude that the membrane contains liquid, whilst Olive (1964) believes that
it contains gas. Evidence that the drop contains liquid is shown by the fact that
a drop is often carried with the spore (Buller, 1922) and that spores immediately after discharge may be surrounded by liquid (Müller, 1954). Spore
discharge may occasionally take place without the appearance of the drop,
and in some species, e.g. in the rust-fungus *Cronartium ribicola,* the drop has
not been observed (Bega & Scott, 1966). It is therefore possible that the drop is
not an essential part of the spore discharge mechanism.

A number of possible mechanisms have been suggested for basidiospore
discharge. In considering them, we cannot dismiss the possibility that several
distinct mechanisms may operate.

1. *Rounding-off of turgid cells:* in his studies of spore development in the rust
fungus *Gymnosporangium nidus-avis,* Prince (1943) illustrated the presence of
a flat septum across the neck of the sterigma. Following spore release, the
hilum of the spore and the end of the sterigma were convex (Fig. 221). It was
therefore concluded that the basidiospore bounced off the sterigma by rounding-off of secondary walls laid down on either side of the original septum.
Wells (1965) has illustrated the development of a flat septum formed by
inward growth of wall material in electron micrographs of developing basidia
of *Schizophyllum commune* (see Fig. 220). After discharge the apex of the
sterigma is rounded, although the hilum of the spore remains flat, and Wells
concludes that turgor pressure within the basidium contributes to the force of
basidiospore discharge. The rounded papilla-like hilum of the spore after
discharge in the rust-fungus *Cronartium ribicola* has led Bega & Scott (1966)
to suggest that a rounding-off mechanism may be involved in this fungus.
This mechanism is known to occur in other fungi, e.g. the conidia of some
Entomophthorales, the aeciospores of rust fungi and the conidia of some

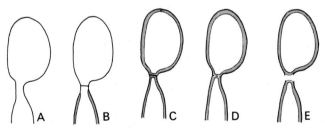

Fig. 221. Basidiospore development and discharge in *Gymnosporangium nidus-avis*
(after Prince, 1943). A. Basidiospore which has reached full size. B. Formation of
primary cross-wall across the neck of the sterigma. C. Laying down of secondary walls
on either side of the primary wall. D. Dissolution of primary wall. E. Spore bounces off
sterigma due to rounding off of both turgid secondary walls.

Fungi Imperfecti. Buller (1909), who first proposed it for basidiospores, termed it *jerking discharge*, to contrast it with *squirting discharge* as found for example in Ascomycetes.

2. *Jet propulsion:* Brefeld (1877) claimed that basidiospores were propelled by tiny jets of liquid from the sterigmata, and Corner (1948) has supported the idea. Objections to the hypothesis are:

1. There is no appreciable reduction in basidial volume as spore discharge proceeds (Buller, 1922).
2. Since spores are discharged successively, for the continued maintenance of basidial turgor it is also necessary to postulate that the sterigma tip be sealed quickly after spore projection. Electron micrographs of sections through sterigmata after spore release show no evidence of a pore (Wells, 1965; Bega & Scott, 1966).

3. *Surface tension:* Buller (1922) raised the possibility that the surface tension of the drop might contain sufficient energy for spore projection. Ingold (1939) calculated the surface energy of the drop, assuming it to be water, and found it to be more than adequate to provide energy for spore release. It was not shown, however, how this energy, which is *potential energy,* could be transformed into the necessary *kinetic energy* to perform the work of spore projection, and the idea has received no support.

4. *Explosive discharge:* Olive (1964) studied spore release in the mirror-yeast *Sporobolomyces* (probably a Basidiomycete) and inferred that the bubble which appears at the junction of the sterigma and spore contains gas, possibly carbon dioxide. This conflicts with Müller's (1954) observation on *S. salmonicolor* that the bubble contains liquid. Olive believes that the explosion of the gas-containing bubble causes the spore to be jolted from the sterigma. Support for the gas-explosion hypothesis comes from experiments by Ingold & Dann (1968) on the effect of external gas pressure on spore discharge in *Schizophyllum*. The number of spores discharged against a pressure of 50 bars was negligible as compared with the numbers discharged at atmospheric pressure. In contrast spore discharge in *Entomophthora coronata*, which occurs by a rounding-off mechanism, is little affected by external pressure. Further support for the idea that the bubble may contain gas (and possibly also liquid) comes from experiments by van Niel *et al.* (1972) on *Sporobolomyces holsaticus*, a yeast-like fungus which is probably a Basidiomycete. These workers noted that the spore sagged on its sterigma shortly before discharge, and this was interpreted as due to a weakening of the attachment of the spore to the sterigma, probably by enzymatic hydrolysis of the outer wall layer. Spores which show the characteristic sagging can be removed from the tip of the sterigma by micromanipulator needles. Spores so detached were

observed to be jolted from their position on the needle by the explosion of a bubble. Sections of discharged spores in the region of the hilum show a ruptured wall layer in the position previously occupied by the bubble. A circular rupture was also observed above the hilar appendix in scanning electron micrographs of basidiospores of *Lentinus edodes* (Nakai & Ushiyama, 1974b). Calculations of the forces required to project a *Sporobolomyces* spore over a horizontal distance of 100 μm show that a bubble of radius 2 μm containing gas at a pressure of 5 bars (over 500 kN/m^2) would be needed to power the gas jet. Presumably, for larger basidiospores, still larger pressures would be needed. Since it is unlikely that the spore wall is impermeable to gases, van Niel *et al.* have postulated that gas might be evolved suddenly, following the mixing of two non-gaseous substances which, up to the time of spore discharge, had remained separate in the spore.

5. *Electrostatic repulsion:* If electrostatically charged plates are placed under basidiomycete fruit-bodies it is found that the majority of spores collect on one or other of the plates, indicating that the spores themselves carry an electrostatic charge (Buller, 1909; Gregory, 1957). It has been suggested that electrostatic repulsion may be partly responsible for spore projection. Savile (1965) believes that basidiospores are released by the gas-explosion proposed by Olive, and that electrostatic forces are also involved in directing accurately the path of the spore. Two objections can be raised against this hypothesis:

(a) The magnitude of the charge is probably inadequate to displace the spore laterally for a sufficient distance. Swinbank *et al.* (1964) determined the charge on the basidiospores of *Serpula lacrymans,* the dry-rot fungus, and concluded that it is unlikely that charges play any important part in the escape of spores from fruit-bodies.

(b) The charge on the spore is probably acquired in the process of separation from the sterigma. Before separation the spore and sterigma would have the same potential, and if the spore acquired its charge on separation, the sterigma would simultaneously acquire a charge of the same magnitude, but of opposite sign. Thus since spore and sterigma have charges of opposite sign they would tend to attract, rather than repel, each other.

There is still no general agreement on the mechanism of basidiospore discharge (see Ingold, 1971). In the face of observations that discharged spores are surrounded by a drop of liquid, more convincing evidence for the existence of a gas bubble will be needed before the gas-explosion hypothesis is generally accepted. Studies of the fine structure of the sterigmata of further Basidiomycetes may clarify the issue. In many Basidiomycetes the hymenia are almost vertical, and the basidia lie horizontally. The spores on release are shot forward horizontally for a short distance, usually less than 1 mm, and then turn abruptly through a right angle and fall at a characteristic slow terminal velocity. The peculiar right-angled trajectory (Fig. 222) is a consequence of the

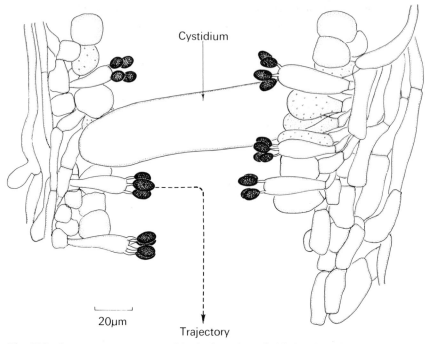

Cystidium

20μm

Trajectory

Fig. 222. *Coprinus atramentarius.* Vertical section of gill showing ripe basidia and a cystidium extending from one gill to another. The arrow indicates the possible trajectory of the basidiospores.

large surface-to-volume ratio of basidiospores, which results in a disproportionately high wind resistance. This trajectory has been termed a sporabola by Buller (1909). The form of the trajectory has been discussed more fully by Ingold (1965).

NUMBERS OF BASIDIOSPORES

The number of basidiospores produced by fruit-bodies can be extremely high. It has been calculated (Buller, 1909) that the detached cap of a mushroom *Agaricus campestris* produced 1.8×10^9 spores in 2 days, at an average rate of 40 million per hour. Estimates for some other fungi are given in table 5.

The mycelium of all these fungi is perennial, and a single mycelium can bear numerous fruit-bodies, so it is clear that any single basidiospore has an infinitesimal chance of successfully establishing a fruiting mycelium.

THE STRUCTURE OF THE MYCELIUM

On germination many basidiospores produce a mycelium (the primary mycelium) which may at first be multinucleate, but later septa are laid down cutting the mycelium into uninucleate segments. The septa on the primary mycelium are simple cross-walls (Fig. 223A). The nuclei in a given haploid mycelium are

Table 5. *Number of spores produced by certain Basidiomycetes*

	Total number of spores	Spore fall period	Number of spores discharged per day
Calvatia gigantea	7×10^{12}	—	—
Ganoderma applanatum	$5{\cdot}46 \times 10^{12}$	6 months	3×10^{10}
Polyporus squamosus	5×10^{10}	14 days	$3{\cdot}5 \times 10^9$
Agaricus campestris	$1{\cdot}6 \times 10^{10}$	6 days	$2{\cdot}6 \times 10^9$
Coprinus comatus	$5{\cdot}2 \times 10^9$	2 days	$2{\cdot}6 \times 10^9$

After Buller (1922).

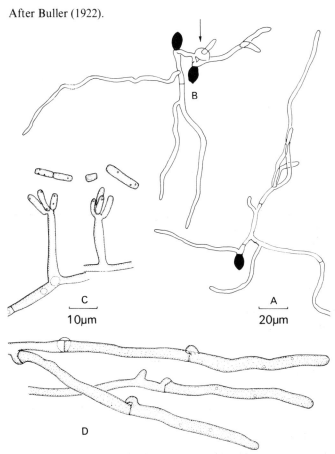

Fig. 223. *Coprinus cinereus.* A. Basidiospore 24 hours after germination. Note the simple septa. B. Two basidiospores showing fusion of germ tubes (arrowed). C. Oidia formed on the monokaryotic mycelium. D. Dikaryotic mycelium showing stages in the development of clamp connections. A and B to same scale; C and D to same scale.

derived from the original single nucleus which enters the basidiospore so that all are usually identical, or **homokaryotic**. Since each segment of the mycelium is uninucleate it can also be described as **monokaryotic** (Jinks & Simchen, 1966). In the majority of Basidiomycetes homokaryotic mycelia do not fruit, and before fruiting can occur fusion is necessary between two homokaryons of different mating type (see below). When two compatible homokaryotic mycelia come into contact breakdown of the walls separating them occurs to achieve cytoplasmic continuity. Nuclear migration follows. Although plasmogamy occurs at this point, nuclear fusion or karyogamy is delayed, and a mycelium develops with binucleate segments, each segment containing two genetically distinct (i.e. compatible) nuclei. This mycelium is termed the secondary mycelium. Such mycelia are **dikaryotic** (strictly **heterokaryotic dikaryons**). The two nuclei in each cell of a dikaryon usually divide simultaneously, and nuclear division is described as **conjugate**. In many cases, but by no means all, dikaryotic mycelia bear at each septum a characteristic lateral bulge termed a **clamp connection**, or clamp (Figs. 223, 224). Whilst it is reasonable to infer that mycelia which bear clamps are usually dikaryotic, the converse is not true: there are numerous fungi in which the dikaryotic myce-

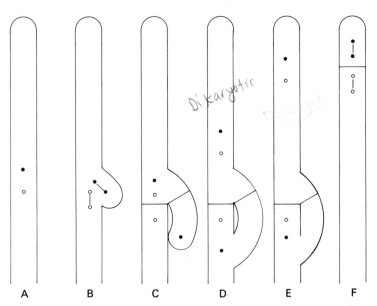

Fig. 224. Diagrammatic representation of clamp formation. A. A dikaryotic hyphal tip. B. Simultaneous nuclear division and formation of a backwardly directed lateral branch into which one of the daughter nuclei passes. C. Formation of two cross-walls cutting off a terminal cell which contains two compatible nuclei; and the lateral branch with a single nucleus. D. Fusion of lateral branch with subterminal cell which is now dikaryotic. E. Later stage. F. Hypothetical nuclear division and segmentation in a dikaryotic hypha without clamp formation. Note that the hyphal tip would become homokaryotic.

lium does not bear clamps. This is often true of the hyphae making up a fruit-body (Furtado, 1966). The clamp connection is a device which ensures that when a dikaryotic mycelium segments, each segment contains two genetically distinct nuclei. In the absence of clamps or some other mechanism for re-arrangement of nuclei there would be a tendency for dikaryotic mycelia to segment at their apices into homokaryotic segments (see Fig. 224). Some authors regard the clamp connection of Basidiomycetes as homologous to the crozier at the tip of ascogenous hyphae in Ascomycetes, and have argued that the Basidiomycetes evolved from Ascomycete ancestors (Teixeira, 1962; Gaümann, 1964; Olive, 1965; Raper & Flexer, 1971). Evidence from fossil material shows that clamp connections were present in the Middle Pennsylvania era (Carboniferous) (Dennis, 1970).

The fine structure of Basidiomycete septa is much more complex than that of an Ascomycete. Close to the limits of resolution of the light microscope, it is possible to discern that, in both monokaryotic and dikaryotic mycelia, each septum is pierced by a narrow septal pore, 0·1–0·2 μm wide, surrounded by a barrel-shaped thickening, the septal swelling (see Fig. 225, after Bracker & Butler, 1963). This kind of septum has been found in numerous Basidiomycetes (Auriculariaceae, Tremellaceae, Aphyllophorales, Agaricales) but not,

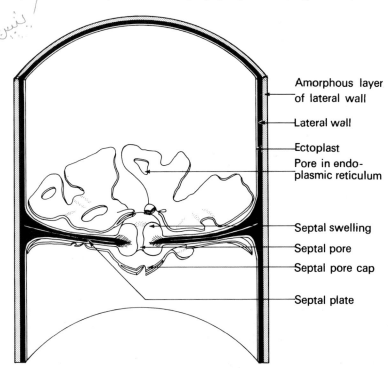

Amorphous layer of lateral wall

Lateral wall

Ectoplast
Pore in endoplasmic reticulum

Septal swelling
Septal pore
Septal pore cap

Septal plate

Fig. 225. Fine structure of a basidiomycete septum (*Rhizoctonia solani*, after Bracker & Butler, 1963).

apparently, in the Ustilaginales or Uredinales. It is known as the **dolipore septum** (Moore & McAlear, 1962; Moore, 1965, 1975; Moore & Marchant, 1972; Thielke, 1972; Ellis *et al.*, 1972). The septal pore is overarched with a perforated cap which is an extension of the endoplasmic reticulum. This cap is termed the **parenthesome** or pore cap. Despite these apparent barriers, there is good cytoplasmic continuity between adjacent cells, and organelles such as mitochondria have been observed within the septal pore of *Rhizoctonia* (Bracker & Butler, 1964). However, in *Polyporus rugulosus,* the perforations of the pore cap are too small to allow the passage of mitochondria (Wilsenach & Kessel, 1965). There is some variation in the structure of the dolipore septum in different parts of a mycelium or fruit-body. For example, in the central canal of the dolipore separating the basidium from the cell behind it in *Agaricus bisporus,* there are two outer ridges, and also a median membrane which prevents direct cytoplasmic contact between the two cells (Thielke, 1972). In the subhymenium cells (i.e. the cells immediately underlying the basidial layer) of *Agrocybe praecox,* the parenthesome is capped by a hemispherical cap of fibrillar material on the outer side only – i.e. on the side of the septum facing away from the basidium. Possibly this outer cap serves a function in separating the basidium from the remaining cells of the fruit-body (Gull, 1976). Migration of nuclei from one cell to another is accompanied by a breakdown of the complex dolipore apparatus (Giesy & Day, 1965; Janszen & Wessels, 1970).

Cell walls of most Basidiomycetes that have been examined are composed of microfibrils of chitin and also glucans with $1 \rightarrow 3$ linked and $1 \rightarrow 6$ linked β-D glucosyl units (Aronson, 1965; Scurfield, 1967; Bartnicki-Garcia, 1968, 1973; Hunsley & Burnett, 1970; Wessels *et al.*, 1972).

MATING SYSTEMS IN BASIDIOMYCETES

About 10% of Basidiomycetes which have been tested are homothallic (Raper, 1966a). Three types of homothallic behaviour may be distinguished:

1. *Primary homothallism:* in *Coprinus sterquilinus* a single basidiospore germinates to form a mycelium which soon becomes organised into binucleate segments bearing clamp connections at the septa. There is no genetical distinction between the two nuclei in each cell, and this mycelium is capable of forming fruit-bodies.

2. *Secondary homothallism:* in *Coprinus ephemerus* f. *bisporus* the basidia bear only two spores, but the spores are heterokaryotic. After meiosis two nuclei enter each spore and a mitotic division may follow. On germination, a single spore germinates to form a dikaryotic mycelium capable of fruiting. Occasional spores give rise to non-clamped mycelia which individually do not fruit, but if the non-clamped mycelia are paired in certain combinations fruiting occurs, showing that the fungus is basically heterothallic. This situation is

closely parallel to that found in certain four-spored Ascomycetes, such as *Neurospora tetrasperma*, and occurs in a number of other two-spored Basidiomycetes (Raper, 1966a). The cultivated mushroom, *Agaricus bisporus*, has a mating system of this type. Although most basidia bear two spores, four-spored basidia do occur, and monosporous cultures derived from four-spored basidia when crossed, produce fruiting mycelia in certain combinations. It has been suggested that a simple bipolar system (see below) is operating (Miller, 1971; Raper *et al.*, 1972).

3. *Unclassified homothallism:* the four-spored wild mushroom, *Agaricus campestris*, is homothallic in the sense that a mycelium derived from a single spore is capable of fruiting. There is nuclear fusion in the basidium, followed by two nuclear divisions, presumably meiotic. However, paired nuclei, conjugate nuclear divisions and clamp connections have not been observed. Another species whose life cycle is difficult to interpret is *Armillariella mellea* (= *Armillaria mellea*). Most of the cells of the mycelium are monokaryotic, and there is no evidence of clamp connections in the mycelium or the rhizomorphs (see p. 65). Fruit-body primordia arise from the monokaryotic rhizomorphs, and the cells of the young primordia are also monokaryotic. However, cells making up the gill tissue are dikaryotic, and these dikaryotic hyphae are associated with clamp connections, whilst the monokaryotic cells formed, for example, in the remaining cells of the stem and cap, have no clamps. Estimations of nuclear volume in monokaryotic and dikaryotic cells suggest that the nuclei of monokaryotic cells are diploid, whilst those of dikaryotic cells are haploid. It is presumed that, during the formation of gill initials, the diploid nuclei undergo haploidisation, by an unknown mechanism. Within the basidia, nuclear fusion and meiosis occur, and a single meiotic product enters each basidiospore. In the spore, the nucleus divides mitotically, and one daughter nucleus from each spore migrates back into the body of the basidium and degenerates (Korhonen & Hintikka, 1974; Tommerup & Broadbent, 1975; Ullrich & Anderson, 1978). Migration of nuclei back into the basidium, following a post-meiotic nuclear division, has been reported in other Hymenomycetes such as *Boletus* (Duncan & Galbraith, 1972).

Amongst the remaining 90% of Basidiomycetes reported to be heterothallic, we can distinguish three conditions.

1. *Bipolar:* in species such as *Coprinus comatus* (the shaggy ink-cap) and *Piptoporus betulinus* (the birch polypore), when mycelia obtained from single spores from any one fruit-body are mated together, dikaryons are formed in half the crosses. This can be explained on the basis of a single gene (or factor) with two alleles. Because only a single factor is involved, the genetical basis for the bipolar condition is described as **unifactorial**. Segregation of the two alleles at meiosis ensures that a single spore carries only one allele. Dikaryons are

only formed between monokaryons carrying different alleles at the locus for incompatibility. In fact, it is known that, in a population of fruit-bodies collected over a wide area, there may be numerous alleles at the incompatibility locus (see below). About 25% of Basidiomycetes examined have been shown to be bipolar. The Uredinales and most Ustilaginales have mating systems of this type.

2. *Tetrapolar:* in species such as *Coprinus cinereus* (often referred to as *C. lagopus*) and *Schizophyllum commune*, when monosporous mycelia derived from a single fruit-body are intercrossed, fertile dikaryons result in only one-quarter of the matings. The genetical basis proposed for this situation is that incompatibility is controlled by two genes, with two alleles at each locus. Because two separate factors are involved, the genetic basis is said to be **bifactorial**. Thus we can denote the two genes as A and B and their two alleles as A_1, A_2 and B_1, B_2 respectively. Consider the cross of a monokaryon bearing A_1B_1 with another bearing A_2B_2. This would result in a fertile dikaryon which we can write as $(A_1B_1 + A_2B_2)$. Such a dikaryon would form spores following meiosis and the spores would be of four kinds: A_1B_1, A_2B_2 (parentals) and A_2B_1 and A_1B_2 (recombinants). In most cases studied the proportions of the four kinds of spore are equal, showing that the A and B loci are not linked, but borne on different chromosomes. The result of crossing the four different types of monokaryotic mycelia is shown in table 6.

It can be seen that fertile dikaryons are only formed when the alleles present at each locus in the opposing monokaryons differ – i.e. in crosses of the type $A_1B_1 \times A_2B_2$ or $A_2B_1 \times A_1B_2$. Where there is a common allele at one or both loci the cross is unsuccessful. Thus the success of inbreeding within the spores of any one fruit-body is only 25% in tetrapolar as compared with 50% in bipolar forms.

In both *Schizophyllum* and *C. cinereus* it was soon discovered that although a single spore from one fruit-body was compatible with only one-quarter of its fellow spores, if crosses were made instead by crossing spores from different

Table 6. *Crossing experiment with all four kinds of monokaryotic mycelium in a tetrapolar Basidiomycete. The + sign denotes the formation of a fertile dikaryon*

	A_1B_1	A_2B_2	A_2B_1	A_1B_2
A_1B_1	—	+	—	—
A_2B_2	+	—	—	—
A_2B_1	—	—	—	+
A_1B_2	—	—	+	—

fruit-bodies, successful matings would often occur in every cross, i.e. a spore from one fruit-body could mate successfully with 100% of the spores from a different fruit-body. The explanation of this phenomenon is that, instead of the single pair of alleles at each locus necessarily present in any one dikaryotic mycelium, in a population representing the species as a whole a large number of alleles is present. Suppose that a second fruit-body had the composition $(A_3B_3 + A_4B_4)$, then all the four kinds of spore it produced, A_3B_3, A_3B_4, A_4B_3 and A_4B_4 would be compatible with all the spores of the original fruit-body, on the assumption that the essential requirement for fertility is that in any cross both alleles should differ at both loci. On this basis although inbreeding would be only 25% successful, outbreeding would be much more successful, approaching 100%. This 100% successful outbreeding would imply the existence of an infinite number of alleles. Analysis of the number of A and B alleles in a world-wide sampling of 57 collections of *Schizophyllum commune* yielded 96 A factors and 56 B factors (Raper *et al.*, 1958). Extrapolation from these results led to the conclusion that the total numbers of incompatibility factors in the world population of *S. commune* are 339 (217 to 562) A factors, and 64 (53 to 79) B factors (Raper, 1966a). A later estimate gave 450 A factors and 90 B factors (Raper & Flexer, 1971). For *C. cinereus*, 164 A factors and 79 B factors have been estimated (Day, 1963), and similar numbers have been estimated for other tetrapolar Hymenomycetes. Much lower numbers of alleles have been estimated for the Gasteromycetes (e.g. *Cyathus striatus*, 4 A and 5 B; *Crucibulum vulgare*, 3 A and about 16 B), and it has been suggested that the small number of incompatibility factors may be related to the specialised method of dispersal of the basidiospores within peridiola (White-house, 1949b).

The large number of alleles in *Schizophyllum* raises questions about the structure and function of the incompatibility genes. It has been shown that when certain strains are crossed pairs of 'new' alleles may appear in the progeny, and when the 'new' strains are inter-crossed, parental alleles may re-appear. From this evidence it was concluded that the 'A' locus in *Schizophyllum* consists of two sub-loci, α and β, with a number of alleles at each sub-locus. On the basis of this model we can suppose that A_1 has the structure of $\alpha_1-\beta_1$, and A_2 has the structure $\alpha_2-\beta_2$. Recombination within the A 'locus' would give $\alpha_1-\beta_2$ and $\alpha_2-\beta_1$, which behave effectively as 'new' mating-type factors. In order to explain the large number of A factors it is necessary to postulate nine different alleles at the A_α locus and about fifty different alleles at the A_β locus (Raper, 1966b; Raper & Flexer, 1971). The 'B' locus also includes two sub-loci, B_α and B_β, each with an estimated number of nine alleles (Koltin *et al.*, 1967; Parag & Koltin, 1971; Stamberg & Koltin, 1973). Separate functions have been ascribed to the A and B loci. Working with *Cyathus stercoreus* Fulton (1950) ascribed to the A locus the formation of clamp connections, since only when this locus is heterozygous are clamp connections formed. The B locus controls nuclear migration, which only takes place if the mated mycelia have different *B* alleles. The same distinctive functions of the *A* and *B* loci have been claimed for other tetrapolar Basidiomycetes, including *S. commune* and *C. cinereus* (Raper, 1966a; Fincham & Day, 1971), so that it is no longer necessary to allocate the symbols in an arbitrary way.

As yet there is no clear understanding of the reasons for the high degree of specificity of action of the large number of alleles at the two loci, but for a discussion of some hypotheses see Raper (1966a), Dick (1965) and Parag (1965).

3. *Octopolar:* in *Psathyrella coprobia*, a **trifactorial** mating system has been reported (Jurand & Kemp, 1973). The three factors *A, B* and *C* are thought to be inherited independently, i.e. they are unlinked. The function of the *A* factor is probably identical to that of bifactorial species, whilst the *B* factor is concerned with fruit-body initiation. All three factors must be heterozygous for the occurrence of nuclear migration and the formation of mature fruit-bodies. This is the first report of trifactorial control of mating in fungi and, until confirmed, should be treated with reserve.

OIDIA

Some Hymenomycetes reproduce by means of oidia, and the oidia may have a sexual rôle (Brodie, 1936). They are uncommon in homothallic and two-spored forms (Kemp, 1975b). Oidia are of two main types, wet and dry. *Coprinus cinereus* has oidia of the wet type, produced only on monokaryons on specialised erect oidiophores (Fig. 223). Several oidia are formed at the tip of the oidiophore and coalesce in a sticky globule. The oidia are dispersed by insects (Brodie, 1931, 1932). They are cylindrical, uninucleate and smooth-walled (Heintz & Niederpruem, 1970). *Psathyrella coprophila* also forms wet oidia. The surface of the oidium is extended into numerous filamentous appendages (Fig. 226) enclosed in a capsule probably composed of mucopoly-saccharide (Jurand & Kemp, 1972). Dry oidia develop usually as chains of cylindrical arthroconidia (Fig. 226), and oidia of this type have been described on monokaryons of *Coprinus micaceus* and *Clitocybe truncicola*. Dry oidia are formed on monokaryons and dikaryons in *Flammulina velutipes* and on dikaryons of *Peniophora gigantea* (Fig. 226). The occurrence of oidia on dikaryons suggests that here they may play an asexual role.

Oidia formed on monokaryons, whether wet or dry oidia, normally function as spermatising agents, and usually germinate poorly. If an oidium is placed a little distance ahead of a growing monokaryotic hypha, the growing hypha may change direction as if stimulated chemotropically towards the oidium (Fig. 226). The chemotropic response has been detected over distances of the order of 75 μm. This is remarkable in view of the fact that the diameter of the approaching monokaryon is about 2·5 μm, and the growth zone of the hyphal tip about 0·5 μm (Kemp, 1975a). The chemotropic growth response has been termed the 'homing reaction', and it appears to be elicited not only by

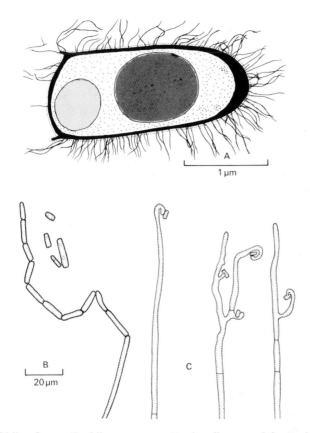

Fig. 226. Oidia of some Basidiomycetes. A. *Psathyrella coprophila*: tracing from an electron micrograph of a longitudinal section of an oidium. The larger dark body within the oidium is the nucleus. Filaments extend from the wall (after Jurand & Kemp, 1972). B. *Peniophora gigantea*: formation of chains of oidia by disarticulation of an aerial hypha. C. *Coprinus cinereus*: homing reaction of hyphal apices or lateral branches of monokaryotic mycelia towards oidia placed near them 4 hours earlier.

compatible oidium–hypha combinations but between incompatible combinations. It may also be elicited by different species. Where the oidium and approaching hypha are compatible, plasmogamy – i.e. fusion between the hyphal tip and the oidium – takes place, and this is followed by nuclear migration and the eventual establishment of a dikaryon (Bistis, 1970). Plasmogamy may also occur between an oidium and an unrelated approaching hypha, i.e. one belonging to a different species. In this case, the introduction of a nucleus from the oidium into a cell of the hypha results in a lethal response, the death of the hybrid cell, and possibly of some adjoining cells. Kemp (1975a,b) has argued that the lethal response is important in maintaining interspecific barriers.

NUCLEAR MIGRATION

Dikaryotisation of a monokaryon may take place by plasmogamy with a compatible oidium, another compatible monokaryon, or a dikaryon which may donate a compatible nucleus (see below). The introduction of a compatible nucleus is followed by its division and migration so that eventually the majority of the cells of the original monokaryon become dikaryotic. The rate of migration of the introduced nuclei is often considerably greater than the growth rate of a dikaryotic hypha. Nuclear migration rates for a number of fungi are: *Coprinus cinereus* 0·5–1·0 mm/hour; *C. congregatus* 4 cm/hour; *Schizophyllum commune* 1·5–5·4 mm/hour (Snider, 1965, 1968; Ross, 1976). Two alternative explanations of rapid nuclear migration have been proposed: either the nuclei are actively motile or are moved passively in streaming cytoplasm. At present, the balance of evidence is in favour of passive movement (Snider, 1968), but it is interesting that microtubules, which are known to be associated with the movements of *dividing* nuclei and other forms of cell motility, are associated with migrating nuclei in *Schizophyllum commune* (Raudaskoski, 1972a) and *Coriolus versicolor* (Girbardt, 1968).

The structure of the dolipore septa would seem to present a formidable barrier to nuclear migration, but electron micrographs of hyphae within which migration is taking place show that the pore is degraded, and that nuclei can pass through the broken-down septum (Giesy & Day, 1965; Jersild *et al.*, 1967; Koltin & Flexer, 1969; Janszen & Wessels, 1970). It has been shown, in matings of *Schizophyllum* with different B alleles, i.e. matings in which nuclear migration can occur, that the *R*-glucanase enzyme activity is higher than in matings with common B alleles, and it has been suggested that *R*-glucanase is involved in septal dissolution (Wessels & Koltin, 1972; Raudaskoski, 1972b).

The Buller Phenomenon: Buller (1931) discovered that, if a homokaryon of *C. cinereus* was opposed to a dikaryon, it was possible for the homokaryon to be converted to the dikaryotic state. The same phenomenon has been reported in other tetrapolar and in bipolar forms. Conversion or dikaryotisation is brought about by nuclear migration from the dikaryon into the monokaryon. A number of different kinds of combination are possible (Papazian, 1950; Raper, 1966a).

I. Legitimate
    A. Compatible: homokaryon compatible with both components of the dikaryon:
        e.g. bipolar $- (A_1 + A_2) \times A_3$
            tetrapolar $- (A_1B_1 + A_2B_2) \times A_3B_3$
    B. Hemicompatible: homokaryon compatible with only *one* of the components of the dikaryon:

e.g. bipolar – $(A_1 + A_2) \times A_2$
tetrapolar – $(A_1B_1 + A_2B_2) \times A_1B_1$.

II. Illegitimate (noncompatible): homokaryon compatible with neither component of the dikaryon:

tetrapolar – $(A_1B_1 + A_2B_2) \times A_1B_2$ or A_2B_1.

A number of surprising features were discovered in such pairings.

In compatible pairings using *Schizophyllum* it was found that the selection of a compatible nucleus from the dikaryon was not a matter of chance. Consider the fully compatible 'di–mon' mating $(A_1B_1 + A_2B_2) \times A_3B_3$. If a conversion of the A_3B_3 homokaryon by one of the compatible nuclei of the dikaryon were entirely random, dikaryons of the type $(A_1B_1 + A_3B_3)$ and $(A_2B_2 + A_3B_3)$ would be equally frequent. In fact there is evidence of preferential selection of one mating type, but the reasons for the selection are obscure (Ellingboe & Raper, 1962; Raper 1966a). A second unexpected feature is the discovery that dikaryotisation can occur in incompatible pairings. One reason for this phenomenon is that somatic recombination between the nuclei of the original dikaryon can occur to give rise to a nucleus compatible with the monokaryon (Raper, 1966a).

EVOLUTIONARY CONSIDERATION OF MATING SYSTEMS IN BASIDIOMYCETES

Consideration has been given to the evolutionary relationships of the various kinds of mating system. Within a single genus such as *Coprinus* there are homothallic, bipolar and tetrapolar representatives. A close degree of similarity has been found in the details of incompatibility control in unrelated tetrapolar forms such as the Gasteromycetes, Polyporales and Agaricales. The genetical complexity of the tetrapolar condition makes it highly improbable that such a condition could have evolved independently on several occasions. For these and other reasons Raper (1966a) has argued that the primitive condition is the tetrapolar one and that the bipolar and homothallic states are secondary. However, Stamberg & Koltin (1973), from an analysis of the effects of inbreeding and outbreeding potential of a unifactorial or bifactorial mating system with either simple or complex loci, and with one or more alleles at the sub-loci, have argued the reverse, i.e. that, in *Schizophyllum*, the incompatibility system evolved from a one-locus unifactorial system.

CLASSIFICATION OF BASIDIOMYCOTINA

There are many opinions on the classification of Basidiomycetes (Shaffer, 1975). We shall follow the scheme proposed by Ainsworth (1973).

Key to classes of Basidiomycotina

1 Basidiocarp lacking and replaced by teliospores or chlamydospores (encysted probasidia) grouped in sori or scattered within the host tissue; parasitic on vascular plants   **Teliomycetes (p. 483)**
Basidiocarp usually well developed; basidia typically organised as a hymenium; saprobic or rarely parasitic   2

2 Basidiocarp typically gymnocarpous or semiangiocarpous; basidia phragmobasidia (Phragmobasidiomycetidae) or holobasidia (Holobasidiomycetidae); basidiospores ballistospores
Hymenomycetes (p. 416)
Basidiocarp typically angiocarpous; basidia holobasidia; basidiospores not ballistospores   **Gasteromycetes (p. 469)**

It is not possible in an abbreviated key to present all the distinctions between the various groups. This classification is by no means a natural one and its origin can be traced to Elias Fries, a nineteenth-century Swedish mycologist who was largely dependent on macroscopic criteria. For example, it is well known that a group such as the Gasteromycetes is not made up of closely related forms. It has been suggested that some Gasteromycetes are closely related to the agarics *Russula* and *Lactarius,* whilst the Gasteromycete *Rhizopogon* is closely related to *Boletus* (see Heim, 1948, 1971, and Corner, 1954 for discussion). Much research will be necessary to delimit the truly natural groupings amongst the Basidiomycetes, and for the present a Friesian system is being adopted as a framework. Corner (1966) has summed up the position admirably:

It is now agreed that the Friesian classification is artificial and, for the world flora, unworkable. It was based on gross features which microscopic study has shown to be the result of parallel evolution as well as of common heritage. It mixed artificial grades and natural series. In its place, more fundamental methods are being developed by microscopic anatomy and microchemistry. Probably, however, because of the labour which these methods involve, not one quarter of the world flora has been studied and confirmed in the necessary detail. Therefore, there is no alternative comprehensive classification.

1. HYMENOMYCETES

This is the largest group of Basidiomycetes, and includes many of the well-known toadstools, bracket fungi, fairy clubs, jelly fungi and the like. The basidia are often arranged in a palisade-like fashion to form a hymenium which is fully exposed at maturity, in contrast with the Gasteromycetes where it is enclosed.

The main subclasses are separated on basidial structure. The Phragmobasidiomycetidae form phragmobasidia, whilst the Holobasidiomycetidae have holobasidia (see p. 395).

Holobasidiomycetidae

McNabb & Talbot (1973) have classified the Holobasidiomycetidae into six orders, of which we shall study three: Agaricales, Aphyllophorales and Dacrymycetales. The first two have holobasidia of the conventional type, whilst the Dacrymycetales have unicellular forked basidia of the 'tuning-fork' type (Fig. 176).

An important distinction between the Agaricales and Aphyllophorales is that the fruit-bodies of the Agaricales are fleshy, being usually composed of thin-walled hyphae which inflate. Such construction is termed **monomitic**. In contrast, the fruit-bodies of many Aphyllophorales are often more complex, being composed of thin-walled **generative hyphae**, which may be accompanied by either thick-walled unbranched **skeletal hyphae**, or thick-walled much-branched **binding hyphae**, or both. Their construction may thus be monomitic, **dimitic** or **trimitic**..

In many Agaricales, the developing hymenophore is surrounded by one or more veils, but these are not present in the Aphyllophorales. A further distinction is that in the Agaricales the nuclear spindles in the basidia are transverse (chiastobasidial), whilst in the Aphyllophorales forms with longitudinally orientated spindles (stichobasidial) occur.

AGARICALES

Singer (1975) and A. H. Smith (1973) have classified the group into some sixteen to eighteen families. Singer's concept has been followed broadly, except that *Schizophyllum* has been considered along with the Aphyllophorales. I have also not followed Singer in placing the Polyporaceae in the Agaricales. An outline classification of Agaricales, listing the characters of the families, and with keys to families and genera, has been provided by A. H. Smith (1973). Traditionally, all the gill-bearing Hymenomycetes were placed in a single family, the Agaricaceae, but modern taxonomic treatments have resulted in subdivision into more homogeneous families. Agarics are mostly saprophytic, and play an important rôle in the decay of woodland and grassland litter, wood, dung and composts. A few are parasitic, e.g. *Armillariella mellea* (= *Armillaria mellea*), a serious pathogen of woody hosts. Many agarics and boleti form mycorrhiza with forest trees (Harley, 1969). Most of the fleshy forms are edible, and some are specially cultivated for food, notably *Agaricus bisporus*, the cultivated white mushroom in Europe and North America, *Volvariella volvacea,* the Padi straw mushroom in the tropics, and *Lentinus edodes,* the Shiitake of East Asia (Singer, 1961; Atkins, 1966). Some are poisonous, especially *Amanita phalloides* and *A. verna* (Ramsbottom, 1953), whilst others have hallucinogenic properties, notably *Psilocybe mexicana* and related species (Heim & Wasson, 1958; Heim, 1963; Singer, 1975).

Fruit-bodies of agarics usually arise from a dikaryotic mycelium which may

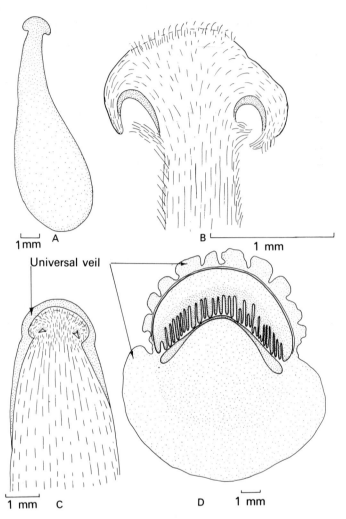

Fig. 227. Sporophore development in some Agaricales, illustrated by longitudinal sections (after Reijnders, 1963). A. *Clitocybe clavipes*. Gymnocarpic development. B. *Lentinus tigrinus*. Secondary angiocarpy resulting from extension of hyphae from pileus margin and stipe to enclose the previously differentiated hymenophore. C. *Stropharia semiglobata*. Primary angiocarpy. Note the universal veil enclosing the upper part of the primordium. In mature fruit-bodies this becomes gelatinous. The hymenophore is also enclosed by a partial veil. D. *Amanita rubescens*. Tangential section. Note the break-up of the universal veil to form scales on the surface of the pileus. The gill chamber is enclosed by a partial veil.

be short-lived or perennial. In some cases the individual hyphae may be aggregated into mycelial strands or rhizomorphs. The general form of an agaric fruit-body is umbrella-shaped, with a central stalk or stipe, supporting a cap or pileus with numerous radially arranged gills of lamellae on the lower side of the cap. In some species, especially those growing on wood, the stipe may be excentric (lateral) or absent. In the related Boletaceae the lower side of the cap bears a series of tubes opening to the exterior by individual pores. The hymenium covers the face of the gills or lines the tubes. The young fruit-body may be enclosed in a mass of tissue called the **universal veil** which is broken as the pileus expands leaving a basal cup-like **volva**, and sometimes scales on the pileus as in *Amanita* spp. In some agarics the hymenium may be protected by a **partial veil** stretching from the edge of the cap to the stem. Where the partial veil is thin and diaphanous as in *Cortinarius* spp. it is termed the **cortina**, but in some genera, e.g. *Agaricus, Armillariella* and *Amanita*, the partial veil is composed of a firmer tissue which remains as a distinct ring or **annulus** attached to the stem (see Fig. 227).

The arrangement of the veils reflects the development of the fruit-body. Reijnders (1963) has given a comprehensive survey of development. Several kinds of development occur:

1. *Gymnocarpic:* the hymenophore is naked from the time of its first appearance and is never enclosed by tissue. The pileus develops at the tip of the stipe and the hymenophore differentiates on the lower side. Gymnocarpic development is found in several unrelated genera, e.g. *Cantharellus cibarius, Xerocomus subtomentosus (Boletus subtomentosus), Russula emetica, Lactarius rufus* and *Clitocybe clavipes* (Fig. 227A).

2. *Angiocarpic:* here the hymenophore, or the part of the primordium from which the hymenophore will be formed, is enclosed by tissue during part of its development. Two kinds of angiocarpic development have been distinguished:

(a) *Primary angiocarpy:* the pileus margin, the hymenophore, and sometimes the pileus and stipe differentiate beneath the surface of the primary tissue of the primordium (protenchyma). *Stropharia semiglobata* and *Amanita rubescens* are primarily angiocarpic (Fig. 227C,D).

(b) *Secondary angiocarpy:* the hyphae from an already differentiated surface grow out towards the exterior to enclose the primordium or part of it. The hyphae may extend from the margin of the pileus towards the stipe, or from the stipe to the pileus, or both.

In *Lentinus tigrinus* hyphae from both the pileus margin and the stipe extend to enclose the developing gills (Fig. 227B).

Reijnders (1963, 1975) has classified the different kinds of development further and has provided further terms to describe them. He has argued that gymnocarpic development is more primitive than angiocarpic.

FRUIT-BODY ANATOMY

The tissue of the sporophore in the Agaricales is pseudoparenchymatous, consisting of an aggregation of hyphae. The hyphae are usually thin-walled and dikaryotic (generative hyphae), and may or may not bear clamp connections. The fruit-body expands due to inflation of the cells. Although no differentiation into skeletal or binding hyphae occurs, specialised tissues or cells may arise. In a study of the fine structure of the sporophore of *Agaricus campestris*, Manocha (1965) has shown that the stipe contains two kinds of cells: wide inflated cells and narrower thread-like cells. A similar differentiation is found in *Coprinus cinereus*. According to Borriss (1934) when portions of stipe tissue are plated on to suitable media it is only the thinner hyphae which give rise to vegetative growth. A further example of differentiation is seen in *Lactarius* which has a system of laticiferous hyphae containing latex which exudes if the flesh is broken (Lentz, 1954). Characteristic swollen spherical cells (sphaerocysts) give the flesh of genera such as *Lactarius* and *Russula* a spongy texture (Fig. 228B). The surface layers of the cap often show considerable modification.

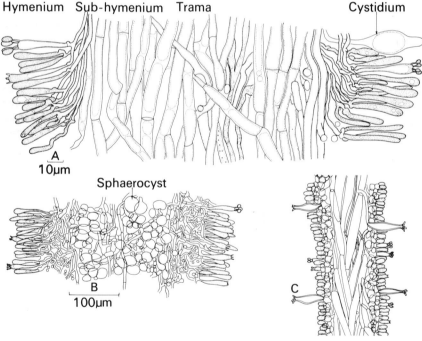

Fig. 228. L.S. gills of various agarics of the aequi-hymenial type. A. *Flammulina velutipes.* B. *Russula cyanoxantha.* Note the globose sphaerocysts. C. *Pluteus cervinus.* Note the characteristic **V** arrangement of the cells of the trama (inverted trama). Note also the hooked cystidia. B and C to same scale.

Two main types of gill structure have been distinguished:

1. *Aequi-hymeniiferous* (aequi-hymenial): most agaric genera have gills of this type. In longitudinal section the gills are wedge-shaped. The term aequi-hymeniferous refers to the fact that the hymenium develops in an equal manner all over the surface of the gill – i.e. basidial development is not localised at any one point on the gill. The wedge-shaped section may be an adaptation to minimise wastage of spores should the fruit-body be tilted from the vertical. Buller (1909) calculated that for the field mushroom *Agaricus campestris*, a displacement of 2°30′ from the vertical would still allow all the spores to escape. Slight adjustments in the orientation of the stipe and of the gills themselves may further help to minimise wastage (Fig. 230).

2. *Inaequi-hymeniferous* (inaequi-hymenial): this type of gill is characteristic of the genus *Coprinus* (popularly known as ink-caps) where the gills are not wedge-shaped in section, but parallel-sided. The term inaequi-hymeniiferous refers to the fact that the hymenium develops in an *unequal* manner, with basidia ripening in *zones*. In *Coprinus* a wave of gill maturation begins at the base and passes slowly upwards. After the basidia at the lower edge of the gill have discharged their spores the gill tissue undergoes digestion (deliquescence) into an inky black liquid which drips away from the cap. The geotropic *gill curvature* characteristic of aequi-hymenial types is absent, but the stipe curves to bring the gills into an approximately vertical position.

The structure of an aequi-hymeniiferous gill as seen in longitudinal section is shown in Fig. 228. There is a central group of longitudinally running threads termed the **trama (hymenophoral trama)**. In some cases the trama may consist of more than one layer (Singer, 1975; Douwes & von Arx, 1965). Then, the layer immediately outside the central layer of the trama is termed the hymeno-podium, and the trama is said to be bilateral (see *Amanita rubescens*, Fig. 229). In forms with a simple trama the ends of the tramal hyphae turn outwards to

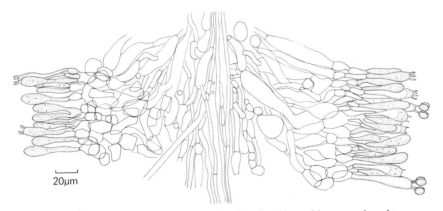

20μm

Fig. 229. V.S. gill of *Amanita rubescens*, showing the bilateral hymenophoral trama.

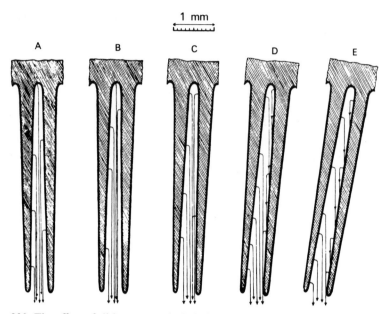

Fig. 230. The effect of tilting an agaric fruit-body with wedge-shaped gills. Note that appreciable tilting of the fruit-body can take place before the trajectories of most spores are interrupted (after Buller, 1909).

form a distinct layer of shorter cells, the **subhymenium**, which lies immediately beneath the **hymenium**, consisting of a palisade-like layer of ripe basidia, developing basidia (basidioles) and sometimes other structures such as cystidioles and cystidia. All these represent the terminal cells of hyphae making up the fruit-body. Possibly they are homologous with basidia. Fusion of paired nuclei within cystidia has been reported (Lentz, 1954). Cystidioles are thin-walled, sterile elements of the hymenium, about the same diameter as the basidia, and usually protruding only slightly from the hymenial surface. Cystidia are more varied. They are often enlarged conical or cylindrical cells which may arise in the hymenium along with the basidia (hymenial cystidia) or sometimes deeper, for example in the trama (tramal cystidia). In many species of *Coprinus* (inaequi-hymenial) the cystidia may stretch across the space between the gills (Fig. 222) and probably serve to space the gills apart. In *Pluteus* the tramal cystidia bear hook-like tips (Fig. 228c) whose function is not understood. The suggestion that they might deter animals such as slugs from eating the gill tissue is not supported by feeding experiments (Buller, 1924). Various terms have been used to describe cystidia. Those on the gill face are termed pleurocystidia: those on the gill margin cheilocystidia. But cystidia are not confined to the hymenium. Similar structures have been found on the surface of the pileus (pileocystidia) and the stipe (caulocystidia). We are still ignorant of the function of these structures, although some are concerned with secretion. For a fuller discussion see Romagnesi (1944), Lentz (1954) and Singer (1975).

THE PHYSIOLOGY OF FRUIT-BODY DEVELOPMENT

Much remains to be learnt about the control of morphogenesis in Hymeno-mycete fruit-bodies. Some of the problems will be discussed in relation to three agarics and one polypore which have been intensively investigated. The agarics are *Agaricus bisporus* (the cultivated white mushroom), *Coprinus cinereus* (= *C. lagopus*) a common coprophilous fungus, and *Flammulina velutipes* (= *Collybia velutipes*) a lignicolous toadstool which fruits through-out the winter. The polypore is *Polyporellus brumalis* (= *Polyporus brumalis*) also lignicolous, fruiting in winter and spring, and somewhat unusual in having a centrally stalked fruit-body (Fig. 231). With the exception of *A. bisporus* which fruits well only on compost, the other species fruit readily, even on synthetic liquid media, and this enables the chemical composition of the medium to be varied. By changing the level of CO_2 and humidity of the atmosphere and the light intensity it is possible to control the differentiation of the fruit-body. The effects of nutrition will not be considered here.

1. *Light:* Light intensity, duration and wavelength are important components of any treatment. Light may have an effect on induction of sporophore initials, a tropic effect, or an effect on fruit-body form, e.g. ratio of stipe length to pileus diameter. In *C. cinereus* fruiting may occur both in light and in darkness, but takes place earlier in the light (after 10 days in continuous light and 15 days in continuous darkness according to Madelin, 1956). Sensitivity of fruiting to the effect of light occurs after about 7 days' growth, and exposure to white light as brief as 1 second at 2,500 lux or 5 seconds at 1 lux was effective. Further increase in duration had no further effect on the numbers of sporophores produced. This behaviour contrasts with that of *F. velutipes* where light also stimulates sporophore production. Here the yield of sporo-phores increases with increasing duration of light (Plunkett, 1953). Curiously, the influence of light on sporophore development in *C. cinereus* can be replaced by mechanical agitation (Stiegel, 1952).

It has long been known that low light intensity, or absence of light, may result in sporophores of curious shape, often with elongated stipes and poor pileus development. Fruit-bodies of this kind, have been found on timber in mine workings and caves. Light is necessary for pileus formation in *F. velutipes* (Plunkett, 1953, 1956).

Light may also stimulate directional growth of the stipes of many lignico-lous and coprophilous hymenomycetes, e.g. *C. cinereus, P. brumalis*, whilst in terrestrial agarics such as *Agaricus campestris* sporophores at all stages of development appear to be non-phototropic (Buller, 1909). For fungi develop-ing in crevices in dung or wood the phototropic stimulus for stipe growth

Fig. 231. Fruit-bodies of some Hymenomycetes, not to same scale. A. *Flammulina velutipes*. B. *Coprinus cinereus*. C. *Polyporellus brumalis*. D. *Agaricus bisporus.*

followed by a light-operated stimulus to cap development or expansion would probably be of survival value.

2. *The aeration complex*

(a) *CO_2 concentration:* the CO_2 content of the atmosphere may have a profound influence on sporophore development. In *A. bisporus* the concentration of CO_2 in the substratum of commercial compost beds rarely falls below 0.3%, about ten times the normal atmospheric concentration, and may rise to 20% or higher during the growth of the mycelium. Concentrations above 1.5% stimulated stipe elongation, but prevented cap expansion. Normal sporophore development occurs at about 0.2% CO_2 (Tschierpe, 1959; Tschierpe & Sinden, 1964). It has been suggested that CO_2 levels within the compost would stimulate stipe elongation which would carry the cap up into regions where the CO_2 concentration was sufficiently low to allow cap expansion and spore release. Expansion of the cap and release of spores within the compost would clearly be disadvantageous. In *F. velutipes* increasing CO_2 concentration above normal concentrations up to 5% causes a reduction in pileus diameter, but also results in shorter stipes. The presence of light and low CO_2 levels are necessary for normal sporophore development (Plunkett, 1956).

(b) *Humidity and evaporation:* in controlled conditions the rate of evaporation from developing sporophores may affect sporophore shape. In *P. brumalis* cap diameter is reduced as transpirational water loss is decreased. Increased transpiration results in more rapid translocation rates into the fruit-body as indicated by the movement of dyes (Plunkett, 1956, 1958). In comparable experiments with *F. velutipes* quite low rates of transpiration inhibited sporophore expansion.

3. *Gravity:* the hymenium of most agarics and polypores must be arranged more or less vertically in order to ensure maximum efficiency of spore release (Buller, 1909). Many fruit-bodies can adapt to changes in position either by growth of new tubes in the case of polypore fruit-bodies on wood, or by adjustments in the curvature of the stipe. In some cases, e.g. *P. brumalis*, the fruit-body may respond to a phototropic stimulus during the early part of its development, and apparently acquire a capacity for geotropic adjustment relatively late. It is of interest to enquire how the changing response of the sporophore to these two different stimuli is achieved. It is likely that both stimuli are operative throughout development. Perception and response to both stimuli occur in the growing apical part of the stipe. Once the pileus initial has formed it shades the stipe so that the response to gravity becomes relatively more important during the later stages of fruit body development (Plunkett, 1961). By arranging to illuminate cultures from below Plunkett has induced pileus formation to occur upside down, i.e. with the pores facing upwards. In such fruit-bodies a normal hymenium developed.

4. *Temperature:* there is usually a range of temperature between which sporophore development occurs. Increase of temperature within the lower part of this range hastens development. For *C. cinereus* the optimum temperature for mycelial growth is about 37 °C, but fruit-bodies are not formed above about 30 °C. In *F. velutipes* the optimum temperature for growth is 25 °C, but sporulation does not occur at this temperature. Continuous incubation in the

light at 20 °C induces fruit-body formation. A short period of incubation (12 hours at 15 °C) can stimulate fruit-body initiation and following such treatment fruit-body development may occur at 25 °C (Kinugawa & Furukawa, 1965).

5. *Biological stimulation: Agaricus bisporus* can be readily grown in pure culture, but it does not fruit on agar media. Fruiting can be induced in the laboratory if mushroom spawn (mycelium grown on cereal grain) is covered by a layer of non-sterile casing soil (San Antonio, 1971). The addition of charcoal to sterilised casing soil will permit fruiting under aseptic conditions (Long & Jacobs, 1974). During commercial cultivation of mushrooms, spawn is inoculated into compost, but sporophores only form after the compost has been covered with a layer, about 2·5 cm, of casing soil. If the casing soil is sterilised, fruiting is inhibited, and it was therefore suspected that the casing soil might contain micro-organisms capable of stimulating fruiting (Eger, 1963). Several species of bacteria, belonging to the genera *Arthrobacter*, *Bacillus* and *Rhizobium,* have been isolated, capable of stimulating fruiting (Park & Agnihotri, 1969). Another organism active in stimulating fruiting is *Pseudomonas putida*, which is itself stimulated to develop by volatile metabolites from *Agaricus* mycelium (Hayes *et al.,* 1969; Hayes, 1972). This is an example of synergistic interaction between soil organisms, and it is likely that there are many interactions of this kind in the natural environment.

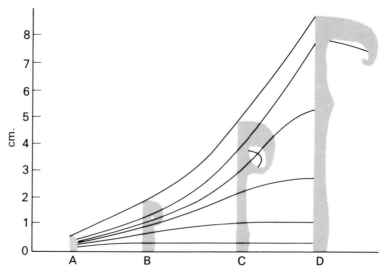

Fig. 232. Diagrammatic growth curve of *Agaricus bisporus* showing four different stages. The lines are drawn through homologous points. The horizontal axis represents a rough approximation of time (after Bonner *et al.*, 1956).

FRUIT-BODY EXPANSION

The rapid expansion of mushroom fruit-bodies is a well-known phenomenon. The force of expansion can be considerable. *Coprinus atramentarius* is capable of cracking asphalt paving, and Buller (1931) showed that *C. sterquilinus* can lift a weight over 200 grams, many times its own weight. Studies of the growth of various agarics, e.g. *Agaricus bisporus* (Bonner *et al.*, 1956), *C. cinereus* (Borriss, 1934; Gooday, 1974b), have shown that the final rapid stage of expansion is due almost entirely, if not entirely, to cell extension, i.e. of a number of cells originally laid down in the young primordium. The most rapid zone of extension is in the zone of the stipe immediately beneath the cap (Figs. 232, 233). In *F. velutipes,* the expanding fruit-body primordium is dependent on its subtending mycelium for a supply of nutrients for about two-thirds of its development. In the final stages, it appears to require only water for expansion to proceed at the normal rate (Gruen & Wu, 1972a,b). Glycogen is present in the cells of the young stipe of *C. cinereus* and tends to disappear as it matures (Borriss, 1934). The osmotic concentration of the cell sap of the stipe cells actually shows a decrease as they mature. There is an increase in elasticity of the stipe cells, but this seems hardly adequate to explain the fifty-fold increase in growth rate measured during the final stages of stipe extension. Possibly there is an increase of the permeability of the stipe cells to water. Elongation of stipe cells is accompanied by an increase in chitin content of the stipes, and N-acetylglucosamine is incorporated into the chitin of the cell walls (Gooday *et al.*, 1976). In *C. cinereus,* removal of the cap at a late stage does not affect the elongation of the stipe (Gooday, 1974b). In this respect, *Agaricus bisporus* and

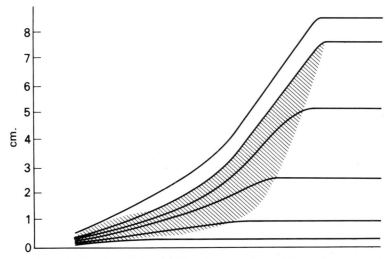

Fig. 233. Growth curve of *Agaricus bisporus.* Those regions of the stalk which are in the process of elongation are stippled (after Bonner *et al.*, 1956).

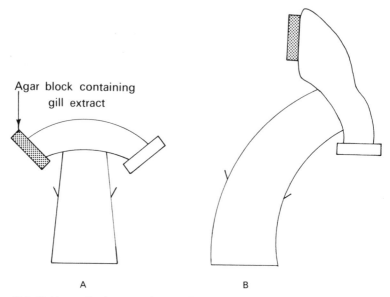

Agar block containing
gill extract

A   B

Fig. 234. Evidence for hormonal control of stipe and pileus expansion in *Agaricus bisporus*. A fruit-body is prepared (A) by removal of all the pileus tissue except for a segment across the diameter. All the gill tissue is also removed. An agar block containing an alcoholic extract of gill tissue is attached to one side of the prepared pileus, and a similar agar block lacking the extract is placed opposite. Expansion of the stipe and pileus tissue on the side to which the test substance is applied is shown in B (after Hagimoto & Konishi, 1960).

Flammulina velutipes differ, because in these species there is evidence that extension of the stipe is controlled by substances produced by the cap, and that the maturing gills produce a hormone which induces cell extension (Gruen, 1969). In *A. bisporus* fruit-bodies, in which all but a narrow strip of pileus tissue, and all the attached gill, are cut away, the unilateral attachment of agar blocks containing an alcoholic extract of gill tissue causes positive curvature of the stipe (Fig. 234). Similar excision experiments, in which only part of the pileus tissue of *Flammulina velutipes* is left attached to the stipe, failed to result in stipe curvature (Gruen, 1969). The nature of the hormone remains to be determined. It is unlikely to be one of the naturally occurring higher plant hormones, since the addition of these compounds does not yield comparable results (Gruen, 1959, 1963; Hagimoto & Konishi, 1959, 1960; Hagimoto, 1963).

FRUIT-BODY FORM AND FUNCTION

It is a common observation that the larger agaric sporophores tend to have stipes which seem disproportionately wide when compared with those of smaller fruit-bodies. The mass of the cap of an agaric is proportional to its volume, i.e. is proportional to the *cube* of its radius, and this mass must be

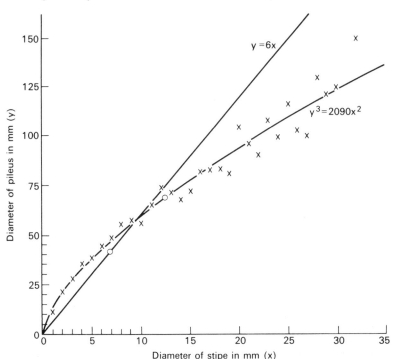

Fig. 235. Diameters of pileus averaged against diameters of stipes of agarics; 1,166 determinations. The straight line is a plot connecting the origin to the arithmetic mean value, giving a line whose equation is $y = ax$. The curve is of the form $y^3 = ax^2$, which is in accord with the Principle of Similitude (after Bond, 1952).

supported by the cross-sectional area of the stipe which is in proportion to the *square* of its radius. This is in accord with the principle of similitude (D'Arcy Thompson, 1961). If this principle holds good for agaric sporophores a plot of pileus diameter (y) against stipe diameter (x) should not be linear but should have the form $y^3 = ax^2$. Ingold (1946b; 1965) and Bond (1952) have shown that the relationship holds good (Fig. 235).

SOME COMMON AGARICS AND BOLETI

A detailed survey of the Agaricales would be out of place here, and those interested are referred to more specialised texts such as Singer (1975), Kühner & Romagnesi (1953), Lange (1935–40), Lange & Hora (1963), Moser (1967), Henderson *et al.* (1969) and Watling (1970). A check list giving modern nomenclature of British agarics is available (Dennis *et al.*, 1960). A list of some important families is given below, together with the names of some common genera. Criteria used in classification are the presence or absence of annulus and volva, spore colour, method of attachment of gills to stem, the structure of the trama, presence or absence of sphaerocysts, cystidia and other specialised structures, and the chemical reactions of the flesh and spores.

Hygrophoraceae

Species of *Hygrophorus* are often brightly coloured and viscid, with a waxy texture. They are especially common in pastures.

Tricholomataceae

Common genera: *Laccaria, Clitocybe, Tricholoma, Armillariella (Armillaria), Collybia, Oudemansiella, Marasmius* and *Mycena*.

Armillariella mellea, the honey agaric, has already been mentioned as a parasite of woody hosts. The fruit-bodies are edible. They are remarkably variable in form, and it is possible that one name is being applied to several taxa (Singer, 1970). The fungus is a root-parasite and can spread by basidiospores, root-to-root contact, or the extension of bootlace-like rhizomorphs from food-bases in infected stumps (Fig. 30). Similar, but flattened, rhizomorphs form beneath the bark, showing where the cambium has been destroyed. Much is known about the distribution of rhizomorphs which, in ideal conditions, are capable of growing about 1 m per year. Their distribution is related to soil type. On moist sites, they are concentrated in the top 10 cm, but on dry sites they are found deeper in the profile. In experiments in which rhizomorphs are allowed to develop on woody inocula buried at different depths, those at a depth of 30–60 cm tended to grow towards the surface along gradients of increasing oxygen and decreasing carbon dioxide concentrations (Morrison, 1976). Soil disturbance, caused by timber extraction and soil preparation for a new tree crop, provide a stimulus to fresh rhizomorph development which places the later crop at considerable risk (Redfern, 1973). Rhizomorph development does not take place at high temperatures, and this may explain their absence from tropical soils (Swift, 1968). Rhizomorph formation occurs readily in culture and is stimulated by ethanol, other low molecular weight alcohols, oils, fatty acids, indole-3-acetic acid, and *o*- and *p*-aminobenzoic acid. The common soil fungus *Aureobasidium pullulans* also produces a volatile stimulus to rhizomorph initiation, and this has been identified as ethanol (Garraway & Weinhold, 1968; Pentland, 1968; Garraway, 1970; Sortkjaer & Allermann, 1972).

A number of antibacterial and antifungal antibiotics are produced in culture by *Armillariella mellea*. The antibiotics are effective against Grampositive bacteria, and also some fungi including the common soil fungus *Trichoderma* (Richard, 1971; Oduro *et al.*, 1976).

Attack of a fresh host is often accompanied by the development of white sheets of mycelium between the wood and the bark. Attack does not always result in death, but in a localised root rot, although extensive killing, e.g.

group dying of conifers, is common where conditions are suitable. In decayed wood, the fungus may persist for many years and may form pseudosclerotial plates ('zone lines') of dark bladder-like cells (Lopez-Real, 1975; Lopez-Real & Swift, 1975).

The control of disease caused by *Armillariella mellea* is rendered difficult by its prolonged survival in decaying stumps. Removal of infected stumps, although laborious, may be worth while in gardens, nurseries and orchards because it reduces the food base from which the fungus can spread. Biological control by infecting stumps with competing fungi may be effective (Redfern, 1968). Replacement of *Armillariella* by competing fungi can be hastened by chemical treatment of stumps with substances such as ammonium sulphamate, or the herbicide 2,4,5-trichlorophenoxyacetic acid (2,4,5-T) which controls regrowth and prolonged viability of hardwood stumps. Prompt colonisation of competing fungi can be ensured by inoculating treated cut stumps with suspensions of basidiospore or mycelial fragments. Promising competitors to *A. mellea* are *Coriolus versicolor* and *Phlebia merismoides* (Rishbeth, 1976). Other types of chemical treatment, e.g. by the use of fumigants such as carbon disulphide or methyl bromide are aimed at killing or weakening the *Armillariella* within the host tissues. Such damaged mycelium may then be attacked by other soil fungi such as *Trichoderma*. It is of interest that treatment with methyl bromide reduces antibiotic production by *Armillariella* (Munnecke *et al.*, 1970; Ohr & Munnecke, 1974).

Curiously, this destructive parasite is a mycorrhizal associate of the colourless orchid *Gastrodia elata*, within the tuber of which the fungal hyphae undergo digestion.

Oudemansiella radicata (= *Collybia radicata*) grows on stumps and may form a long tapering root (pseudorhiza) which extends downwards through the soil to buried wood (Buller, 1934). The basidia and spores are large and convenient for the study of development (Figs. 218 and 237).

Marasmius oreades is called the fairy ring agaric because this and other species often fruit in circles. Even when fruit bodies are not apparent, dark green circles of grass may indicate the presence of the fungus. The circular form results from radial outgrowth from a central point, with the mycelium towards the centre dying off. Measurements of the annual rate of radial advance and of the diameter of fairy rings show that some are probably several centuries old (Shantz & Piemeisel, 1917; Bayliss-Elliott, 1926; Ramsbottom, 1953; Parker-Rhodes, 1955). Analysis of the mating types found in isolations from sporophores all collected from a common ring show that they are identical, which tends to confirm the origin of the ring from a single point (Burnett & Evans, 1966). The appearance of the ring varies throughout the year, but at certain times there may be two dark green rings separated by a brown ring where the grass is dying. The dead zone might be due to harmful effects of the mycelium on the grass roots, or to the lower water content of the soil in this region. It is of interest that HCN, a substance commonly found in basidiomycete sporophores, has been obtained from mycelia of *M. oreades* in pure culture and from affected turf (Lebeau & Hawn, 1963; Ward & Thorn, 1965; Filer, 1966). Studies of the soil fungi associated with the mycelial zone of *M. oreades*, which may be confined to the top few inches of soil, show that fewer species of fungi can be isolated from this zone than from the soil beyond it (Warcup, 1951). The fruit-bodies of *M. oreades* are edible and can be dried.

Mycena species have delicate brittle sporophores which are common in woodland litter growing amongst grass or on decaying leaves, twigs or stumps *(M. galericulata)*. A characteristic feature of the genus is that the gill edge bears cystidia (cheilocystidia). In some species, e.g. *M. galopus,* the broken stipe exudes latex.

Amanitaceae

Amanita (Figs. 229, 236)

Several deadly poisonous toadstools belong to this genus, notably *A. phalloides* (the death cap), *A. verna* and *A. virosa*. The poisonous principles, termed amanita-toxins, are complex cyclopeptides, and include α, β- and γ-amanitin, and phalloidin (Tyler, 1971). Some other species, whilst definitely poisonous, are less deadly. They include *A. muscaria,* the fly agaric, which has hallucinatory properties. The hallucinogenic substances are isoxazole derivatives, ibotenic acid and the related decarboxylation product muscimol (Tyler, 1971). Some other species, such as *A. rubescens, A. vaginata* and *A. fulva* are good to eat. The last two species were earlier classified in a separate genus *(Amanitopsis)* because of the absence of an annulus, but the distinction is no longer maintained. Many species of *Amanita* are mycorrhizal. Macroscopically, members of the genus can be recognised by their white gills and spores, the presence of an annulus (ring) on the stem, and a volva, which is visible as a cup at the base of the stem. Remnants of the volva may persist as volva scales on the cap. A characteristic microscopic feature is the bilateral hymenophoral trama (Fig. 229).

Agaricaceae

Agaricus (Fig. 231)

The best-known representatives are *A. campestris*, the field mushroom, and *A. bisporus*, the cultivated mushroom. The cultivated species was derived from naturally growing mycelium which spawn-collectors were able to distinguish from other kinds (Ramsbottom, 1953; Singer, 1961; Bohus, 1962). The pink coloration of the young gills is due to cytoplasmic pigment in the spores. Later, the gills turn a purplish-brown, due to the deposition of dark pigments in the spore wall. Whilst most of the larger specimens of *Agaricus* are edible, *A.*

xanthodermus is slightly poisonous. The generic name *Psalliota* has earlier been used for these species. For an account of the biology and cultivation of edible mushrooms, see Chang & Hayes (1978).

Lepiota (Fig. 236)

L. procera and *L. rhacodes* are two handsome scaly toadstools (parasol mushrooms) which are edible. A feature of both is that the ring is free to move up and down the stipe.

Coprinaceae

Coprinus (Fig. 231)

Some aspects of the biology of the ink-caps have already been studied. The characteristic feature of the genus is the parallel-sided (inaequi-hymenial) gill which, in most species, undergoes deliquescence from the base upwards immediately following spore discharge. In *C. curtus* and some other species,

Fig. 236. Fruit-bodies of some common agarics. A. *Amanita excelsa*. Note the cracking of the volva on the cap of the youngest specimen, and the annulus on the stem. B. *Amanita fulva* (the tawny grisette). Note the cup-like volva at the base of the fruit-body. Here there is no annulus. C. *Armillariella mellea* (the honey agaric). Cluster of fruit-bodies near the base of a living birch tree. Note the annulus. D. *Lepiota procera* (the parasol mushroom). In mature specimens the ring on the stem is movable.

the cap expands by widening of downwardly directed grooves which divide the gills into Y-shaped forms. Autodigestion of the basal part of the vertical limb also occurs (Buller, 1931). In *C. plicatilis*, in which the cap expands in this way, there is no deliquescence (Buller, 1909). Deliquescence in *C. cinereus* is accompanied by chitinase activity which develops about 2 hours before spore release. Exposure of living hymenial cells of *Coprinus* to the fluid which exudes from autolysing fruit-body tissue results in the release of binucleate protoplasts, and it is therefore concluded that the chitinase is partially responsible for the breakdown of cell walls. The chitinase is localised into vacuoles or lysosomes. Other hydrolytic enzymes are also present (Iten, 1969–1970; Iten & Matile, 1970). Many species, e.g. *C. cinereus,* are coprophilous, but there are forms which grow on wood, e.g. *C. micaceus* above ground, or from buried wood, e.g. *C. atramentarius. Coprinus comatus* is a large terrestrial species (the shaggy ink-cap or lawyer's wig) which is edible. In this species, large cystidia line the inner free edge of the gill. Buller (1924) has shown that, in many species of *Coprinus*, the basidia are dimorphic, either long or short. The short forms are not merely immature basidia, but it appears that basidiospores ripen at two separate levels on the hymenium. This arrangement makes it possible to crowd a large number of basidia in a given area without interference.

C. cinereus has been the subject of many genetical investigations (for references, see Guerdoux, 1974). Monokaryotic cultures may form sclerotia on aerial or submerged mycelium. The sclerotium develops from a single cell. Repeated branching of hyphae results at first in an undifferentiated mass of cells. Later, the central cells accumulate glycogen, whilst the outer cells, which do not accumulate glycogen, become thick-walled to form a protective rind. The aerial sclerotia may function as propagules and be dispersed (Waters *et al.,* 1975a,b; Moore & Jirjis, 1976). The development of fruit-bodies can similarly be traced to a single cell. Branching results in a complex lattice of hyphae which differentiate to form the specialised tissues of the mature fruit-body (Matthews & Niederpruem, 1972, 1973; Cox & Niederpruem, 1975).

Panaeolus

Most species grow on dung or on rich soil. The fruit-bodies resemble those of *Coprinus*, but the gills are aequi-hymenial and do not deliquesce. The surface of the gill appears mottled because the basidia in a given area tend to ripen together (Buller, 1922). Common species on dung are *P. campanulatus, P. sphinctrinus* and *P. semi-ovatus* (sometimes placed in a separate genus *Anellaria*).

Strophariaceae

Stropharia

The best-known species in *S. semi-globata* which is a very common coprophilous agaric with a yellow hemispherical viscid cap. *Stropharia aeruginosa* (the verdigris-cap) is an attractive bluish-green agaric with white scales, found amongst grass in pastures and woods.

Hypholoma

H. fasciculare, the sulphur tuft forms clumps of yellow fruit-bodies at the base of dead deciduous tree stumps. It is distinguishable from *H. sublateritium*, which grows in the same situations, because the latter has a larger brick-red cap. The generic name *Naematoloma* is sometimes used.

Psilocybe

Mention of *Psilocybe* has already been made in connection with the hallucino-genic properties of certain species. *Psilocybe semilanceata* is widespread in lawns and pastures.

Pholiota

Pholiota squarrosa grows parasitically, and forms clumps of large yellowish-brown scaly fruit-bodies at the base of trees such as ash and elm. It is somewhat reminiscent of *Armillariella mellea* from which it is distinguished by its rusty brown spores.

Cortinariaceae

Cortinarius

Over 200 species are known in Britain alone. Almost all species are terrestrial and many are mycorrhizal associates of trees. Although determination of species is difficult, the genus is easy to recognise by the clay-coloured gills and remnants of the cortina on the stem.

Boletaceae

Boletaceae resemble some gill-bearing fungi in having fleshy fruit-bodies, similar development, and a hymenium which may or may not be protected during development by a partial veil. The hymenium, however, lines tubes which open to the exterior by pores on the lower side of the cap. The Paxillaceae (see below) share many features common to the Boletaceae. Modern classification of the Boletaceae divides them into about thirteen genera, most of which were previously included in the genus *Boletus* (Watling, 1970; Smith, 1973; Becker, 1975). Important features in classification at the generic level include spore colour (yellow, pink, brown, purple-black, olive), spore shape (ellipsoid or subfusiform), presence or absence of cystidia, texture

Fig. 237. Fruit-bodies of some common agarics and boleti. A. *Oudemansiella radicata*. This fungus grows attached to tree stumps and has a long tapering rooting base to the stem. B. *Paxillus involutus.* Note the decurrent gills and inrolled cap margin. C. *Suillus grevillei*. Note the ring on the stem and the pores on the underside of the cap. This fungus is a mycorrhizal associate of *Larix*. D. *Boletus chrysenteron*. Here there is no ring on the stem.

of the cap (glutinous, dry or scaly) and markings on the stem (punctate, veined-reticulate, scaly). The flesh of several Boletaceae turns blue when exposed to the air or when bruised, and this is due to the oxidation of an anthraquinone pigment to a blue colour (boletoquinone) by the enzyme laccase. Most Boletaceae are mycorrhizal associates of coniferous or deciduous trees.

Boletus sensu stricto (Fig. 237D)

Common species of *Boletus* include:

B. chrysenteron which occurs in mixed woodland associated with broad-leaved trees. It is commonly parasitised by an Ascomycete *Apiocrea chrysosperma* (conidial state = *Sepedonium chrysospermum*) which also grows on *Paxillus involutus*.

The *Boletus edulis* complex, which occur in coniferous or deciduous woodland. It is now recognised that *B. edulis* represents a 'species complex' which can be discriminated into a number of distinctive species and varieties. Members of the *B. edulis* complex are considered to be amongst the best of the edible toadstools and are often sold in markets in continental Europe.

B. satanus, characteristically associated with beech or oak on calcareous soils, has red pores, and a red stem bearing a red network of veins. It is poisonous.

B. parasiticus is unusual in that it grows parasitically, forming its own sporophores on those of *Scleroderma*.

Leccinum

Species of *Leccinum* have scaly stems. *L. scabrum (Boletus scaber)* and *L. versipelle (Boletus versipellis)* are common mycorrhizal associates of birch *(Betula)*.

Suillus (Fig. 237C)

Species of Suillus have glutinous caps, and some have a ring on the stem which is the remains of the partial veil. They are all mycorrhizal with conifers, e.g. *S. luteus, S. bovinus* with pine *(Pinus)* and *S. grevillei (Botetus elegans)* with larch *(Larix)*.

Paxillus

Some authorities classify the gill-bearing genus *Paxillus* in the Boletaceae, whilst others place it in a separate family, the Paxillaceae. Apart from the close structural similarity, it is interesting that both *Paxillus* and *Boletus* are parasitised by a mould, *Apiocrea chrysosperma*, which does not attack other agarics. *Paxillus involutus* is a common mycorrhizal associate of birch.

Russulaceae

Russula and *Lactarius*

These two genera are very closely related to each other. This is shown by the close similarity of their spores, the presence of sphaerocysts, and of latici-ferous hyphae in the flesh (Fig. 228B). When the flesh of a *Lactarius* sporo-phore is broken latex exudes which may be colourless or brightly-coloured. In some cases the latex changes colour, e.g. in *L. deliciosus,* the edible saffron milk-cap, where the latex is at first bright orange, but turns green. Latex does not exude from *Russula* sporophores. Numerous species of both genera are abundant in woodland where many of them are mycorrhizal associates of trees. *Russula ochroleuca* is common in mixed deciduous woodland, whilst *R. fellea,* distinguished from the former by its bitter taste and its honey-coloured gills, is confined to beech woods. *Lactarius turpis* is a mycorrhizal associate of birch, whilst *L. rufus,* distinguished from other reddish-brown species by its peppery taste, is associated with conifers.

APHYLLOPHORALES

Donk's (1964) concept of this group has been adopted. He has defined it as

an artificial order of holobasidious Hymenomycetes, opposed to the Agaricales, forming distinct fruit bodies. Fruit body developing centrifugally with one-sided hymenophores, or clavarioid with amphigenous hymenium, not developing within a universal veil, the hymenium not covered by a partial veil and exposed during the maturation of the basidiospores. Hymenophore smooth (hymenium may be folded), toothed or tubulate, exceptionally, and then mostly imperfectly lamellate.

In place of the traditional Friesian system of classification based on gross macroscopic features of the fruit-body and details of hymenial arrangement, modern systems of classification lay stress on microscopic features, especially the details of anatomy, colour and staining reactions of hyphae and spores. Critical examination of Basidiomycete fruit-bodies in this way demonstrates close relationships between fungi hitherto classified in quite separate groups under the Friesian system. The emergence of more natural groupings of species and genera is reflected in the arrangement of families and creation of new genera. Donk (1964) and Talbot (1973b) have recognised over twenty families within the Aphyllophorales, but the arrangement is still tentative and as a result of further research it may well prove desirable to create further families. We shall study examples from only a few of these families, viz.: Auriscalpiaceae, Clavariaceae, Ganodermataceae, Hydnaceae, Polypora-ceae, Schizophyllaceae and Stereaceae.

Polyporaceae

In older systems of classification Hymenomycetes in which the hymenium lined tubes opening to the exterior by means of pores, were all classified here. It was later recognised that there is only a limited number of ways in which a hymenium can be arranged so that basidiospores can be discharged successfully, and it became clear that the tubular arrangement of the hymenium had probably evolved independently several times. Thus *Boletus* is now classified in the Agaricales. A striking demonstration of convergent evolution was the discovery that *Aporpium caryae* which for over 120 years had been treated as a polypore *(Polyporus caryae)* had cruciate–septate basidia characteristic of the Tremellaceae.

Microscopic analysis and studies of fruit-body development have led to splitting of the old heterogeneous Polyporaceae into a number of smaller, more homogeneous, families. For example, *Fistulina hepatica,* the beef-steak fungus parasitic chiefly on *Quercus* and *Castanea,* is now placed in a separate family, the Fistulinaceae. *Ganoderma* has been removed to a separate family, the Ganodermataceae. The residual Polyporaceae is undoubtedly still heterogeneous and we may expect further separations as relationships become more clearly defined. It is in this residual sense that the Polyporaceae are here considered.

The Polyporaceae or bracket fungi are important economically because they include a number of serious pathogens of coniferous trees, e.g. *Heterobasidion annosum* (= *Fomes annosus*) the cause of heart-rot or butt-rot of conifers and *Piptoporus betulinus* (= *Polyporus betulinus*) the cause of heart rot of birch. Many are also responsible for decay of timber (Cartwright & Findlay, 1958; Fergus, 1960). Two types of decay have been recognised. In fungi capable of cellulose decay, the brown lignin content of the wood is left undestroyed, and the condition is described as a brown rot. Where lignin destruction takes place the rotted wood has a white appearance so that the term white rot is applied. The ability to degrade lignin by a fungus appears to be correlated with the production of an extracellular oxidase in pure culture, and this can be detected by the oxidation of gallic or tannic acid to a brown colour or by the rapid oxidation of an alcoholic solution of guiacum to a blue colour (Nobles, 1958a,b, 1965). Analysis of sound and decayed wood shows that the cellulose content is reduced in both brown and white rots, and the lignin content remains practically constant in the brown rots but is decreased in the white rots (Findlay, 1940). Isolation of fungi from decaying timber often results in monokaryotic growth in culture, suggesting that the monokaryotic mycelium may have a prolonged independent existence. All polypores which have been studied are heterothallic, about 44% bipolar and 56% tetrapolar. Multiple alleles have been demonstrated in both bipolar and tetrapolar forms (Whitehouse, 1949b; Nobles, 1958a,b; Raper, 1966a). The dikaryotic mycelium which may or may not bear clamp connections is often perennial in large tree trunks and may give rise to a fresh crop of sporophores annually. In some cases, e.g. species of *Fomes,* the fruit-body itself may be perennial. Typically the fruit-bodies develop as fan-shaped brackets lacking stipes, but there are forms with lateral, i.e. excentric stipes, e.g. *Polyporus squamosus* and a few with centrally stalked fruit-bodies, e.g. *Polyporellus brumalis* (= *Polyporus brumalis*) and *Coltricia perennis.* The last is unusual in being terrestrial, not

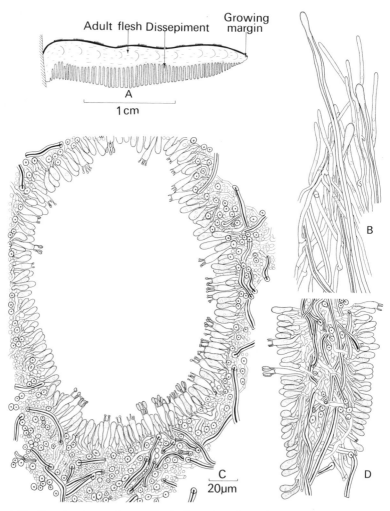

Fig. 238. *Coriolus versicolor,* A. Vertical section through sporophore. B. Group of hyphae teased from the growing margin. Only generative and skeletal hyphae are found here. In the adult flesh binding hyphae are also present (see Fig. 239). C. Transverse section across a pore showing the hymenium. D. Longitudinal section of part of a dissepiment. The tissue contains only generative and skeletal hyphae. B, C, D to same scale.

lignicolous. When fruit-bodies of polypores and other wood-rotting fungi develop on the underside of logs they may be appressed to the surface of the wood and are then described as resupinate.

In order to understand the structure of a polypore fruit-body it is necessary to tease out a thick section of the flesh with fine needles under a dissecting microscope. This technique of hyphal analysis was developed by Corner (1932, 1953) and has also been described by Teixeira (1962). It may be necessary to study hyphae from various parts of the fruit-body; the growing margin, the adult flesh, the context immediately above the tubes and the dissepiments, i.e. the tissues separating the tubes (Fig. 238). Corner has described three distinct kinds of hyphae, although not all may be present in any species:

1. **Generative hyphae:** thin-walled near the growing margin, often thicker-walled behind, with or without clamps, usually with distinct cytoplasmic contents. This kind of hypha is universally present in all polypore fruit-bodies at some stage of development. The generative hyphae give rise to basidia and also to two other kinds of hyphae.

2. **Skeletal hyphae:** unbranched thick-walled hyphae with a narrow lumen which arise as lateral branches of the generative hyphae. The skeletal hyphae form a rigid framework.

3. **Binding hyphae:** much-branched, narrow, thick-walled hyphae of limited growth. These hyphae tend to weave themselves between the other hyphae of the flesh.

Several other kinds of hyphae have been described, some as intermediates between these three principal systems. For example, the terms 'aciculiform', 'arboriform' and 'vermiculiform' have been applied to skeletal-like hyphae, whilst the term 'gloiopherous hypha' has been used for cells with dense oily contents (Lentz, 1973).

The three basic kinds of hyphae are illustrated in Fig. 239. Where all three kinds are present together, the fruit-body is said to be trimitic (Greek μτοζ, a thread of the warp). *Coriolus versicolor (Polyporus, Polystictus, Trametes versicolor)* is a good example of a trimitic polypore (Fig. 240A). The growing margin and the tissue of the dissepiments are dimitic, consisting of generative and skeletal hyphae only. Binding hyphae are only found in the adult flesh some distance behind the growing margin.

There are two types of dimitic construction in polypores:

(a) Dimitic with binding hyphae (i.e. generative and binding hyphae). This kind of construction is found in *Laetiporus sulphureus* (= *Polyporus sulphureus*) (Fig. 240B). The dissepiments are, however, monomitic.

(b) Dimitic with skeletal hyphae. The fruit-bodies of *Heterobasidion annosum* are of this type (Fig. 240C).

Where generative hyphae only are present, the construction is described as monomitic. Fruit-bodies of *Bjerkandera adusta* (= *Polyporus adustus*) are monomitic. Here the walls of the generative hyphae may thicken with age.

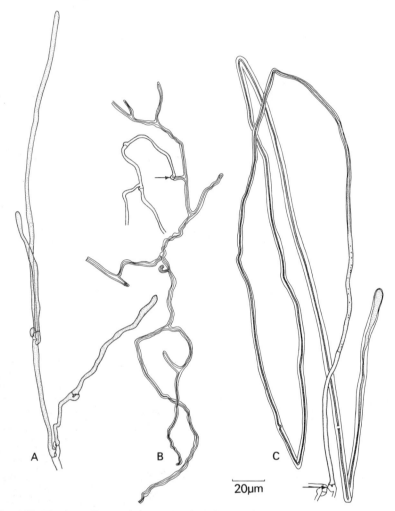

Fig. 239. Hyphae dissected from the fruit-body of a trimitic polypore (*Coriolus versicolor*). A. Generative hyphae, characterised by thin walls, dense cytoplasmic contents and clamp connections. B. Binding hypha, branched and thick-walled. The arrow shows the origin from a generative hypha. C. A skeletal hypha, unbranched and thick-walled. The arrow shows the origin from a generative hypha.

The distinction between the different kinds of construction is best appreciated by attempting to tear the fruit-bodies of these fungi apart. *Coriolus versicolor* tears with difficulty in contrast to the cheese-like consistency of *B. adusta*. Various modifications to the different hyphal systems may occur with age. For example in *L. sulphureus* the generative hyphae may become inflated. In *Polyporus squamosus* the binding hyphae arise relatively late following inflation of the generative hyphae, converting the sappy flesh of the fully grown fruit-body to a drier firmer texture. In *Piptoporus betulinus* also,

Fig. 240. Fruit-bodies of some Aphyllophorales. A. *Coriolus versicolor* as seen from above. B. *Laetiporus sulphureus*. C. *Heterobasidion annosum* at the base of living pine tree. D. *Piptoporus betulinus* on dead birch trunk.

binding hyphae arise very late, entirely replacing the generative hyphae. The dissepiments show a different construction, being dimitic with skeletal hyphae.

The Polyporaceae is a large group, probably containing over 1,000 species. Over forty genera occur in Britain. Pegler (1966, 1967, 1973) has given keys. Notes on interesting features of some common polypores are given below.

Coriolus versicolor (Fig. 240A) is a common saprophyte on hardwood stumps and logs, causing a white rot. Both the mycelium and the fruit-bodies are resistant to desiccation. The annual trimitic fruit-bodies have a zoned velvety upper surface which readily absorbs rain. There is often considerable variation in the morphology of the fruit-bodies. Variation in fruit-body form on a single log can be traced to distinct columns of decayed wood when the log is sectioned on a band saw. Dark-brown zone lines separate the columns. All isolations from within a column yield an identical dikaryon, and the zone lines mark intraspecific antagonism between distinct dikaryons, showing that, within the log, the fungus does not behave as a single 'unit mycelium' but as a series of discrete populations (Rayner & Todd, 1977, 1978). Monokaryotic mycelia form arthroconidia. This species is tetrapolar with multiple alleles. Extensive studies of its structure have been made by both light and electron microscopy (Girbardt, 1958, 1961).

Heterobasidion annosum (sometimes called *Fomes annosus*) (Fig. 240C) is the cause of heart-rot or butt-rot of conifers, and occasionally of deciduous trees such as birch. The disease is especially common on alkaline soils and is an important cause of heart-rot in coniferous woods (Peace, 1962). The perennial fruit-bodies are formed close to the ground at the base of stumps and are readily identified by their orange-brown colour with a white margin. They are dimitic with skeletals. Spores are produced throughout the year. The conidial state (Fig. 241) has been called *Oedocephalum lineatum*. The focus of the disease in plantations is often from stumps which have become infected from spores. Colonisation of the roots of the stump is followed by spread to adjacent healthy roots by root-to-root contact: the fungus does not grow freely in the soil. Modern methods of control are aimed at preventing stump infection by treatment with creosote, sodium nitrite or other disinfectants. An interesting alternative treatment is to inoculate stumps with spores of saprophytic basidiomycetes such as *Peniophora gigantea* which competes with the parasite in pine, rotting the stumps and preventing successful colonisation (Anon, 1967a; Rishbeth, 1952, 1961).

Piptoporus betulinus is a common sight on dead and dying birch trees (Fig. 240D). It is probably a wound parasite, basidiospores entering where branches have broken off. Infected trees show a brown rot of the heart wood at first with cubical cracking, later powdery. Although the brown rot indicates cellulose decay, there is also evidence of oxidase activity which may also suggest lignin breakdown (Macdonald, 1937). The fungus is bipolar with about thirty alleles at the mating-type locus. Studies of the distribution of alleles in successive

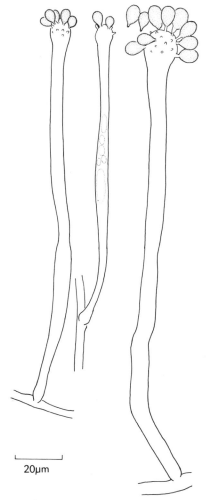

Fig. 241. *Heterobasidion annosum.* Conidiophores and conidia.

annual crops of sporophores on the same birch trunk show that several dikaryotic mycelia may exist side by side in the same trunk, and that each mycelium does not always produce a fruit-body in any one year (Burnett & Partington, 1957; Burnett, 1965).

Polyporus squamosus is a wound parasite of deciduous trees such as elm and sycamore. The mycelium may persist on dead stumps and logs, and form successive annual crops of sporophores during the early summer. The fruit-bodies are distinctly fleshy, and Singer (1975) includes this fungus in the Agaricales.

Ganodermataceae

This family is usually included in the Polyporaceae. The distinguishing feature is that the spore is double-walled, with a dark-coloured inner layer bearing an ornamentation which pierces the hyaline outer one, so that the spore appears to have a spiny surface (Fig. 242) (Heim, 1962; Furtado, 1962; Donk, 1964).

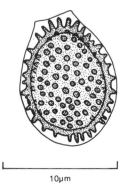

10μm

Fig. 242. *Ganoderma applanatum.* Basidiospore. The truncate portion is the apex of the spore.

The hyphal structure of the fruit-body is trimitic. A characteristic feature is that the skeletal hyphae are of two types: (a) *arboriform*, showing an un-branched basal part with a branched tapering end, and (b) *aciculiform*, unbranched and usually with a sharp tip (Hansen, 1958; Teixiera & Furtado, 1963; Furtado, 1965). The best-known example is *Ganoderma applanatum* (sometimes called *Elfingia applanata*), a wound parasite which causes active heart-rot of beech and other trees (Fig. 243). Both cellulose and lignin are attacked. The sporophores are perennial, brown, woody, fan-shaped brackets, often 50 cm and occasionally almost 1 m across (Herrick, 1953). A new layer of hymenial tubes is formed each year beneath the layer of the previous year. The tubes may be up to 2 cm in length and about 0·1 mm in diameter. Thus the tube may be 200 times as long as broad, and the fall of the spore down this tube raises problems. The hard rigid construction of the sporophore minimises lateral disturbance from the vertical alignment of the tubes. Gregory (1957) has shown that the majority of the spores carry a positive electrostatic charge, but whether this charge has any relevance to the positioning of the spores during their fall seems doubtful. It has been calcu-lated that a large specimen may release as many as 20 million spores per minute during the five or six months from May to September (Buller, 1922). Spore discharge can continue even during periods of drought, doubtless associated with uptake of water from the tree host (Ingold, 1954b, 1957). Spores plated on to media suitable for germination may take 6–12 months to

Fig. 243. *Ganoderma applanatum*, A. Two sporophores attached to a beech tree. B. Detached sporophore split vertically to show two layers of hymenial tubes.

develop germ tubes. Pairings between monosporous mycelia show that the fungus is tetrapolar, with multiple alleles (Aoshima, 1953). The large output of spores may be an adaptation to the low probability of a spore infecting a tree trunk (van der Plank, 1975).

Auriscalpiaceae

Hymenomycetes in which the hymenium covers teeth or spines were classified by Fries in the Hydnaceae. It has since become clear that this gross arrangement occurs in a number of unrelated forms (Harrison, 1971, 1973). This is well shown by *Pseudohydnum gelatinosum* (Fig. 258) whose basidia are longitudinally divided (cruciate), and which therefore has affinities with the Tremellaceae. Some genera formerly placed in the Hydnaceae are now placed in families such as the Auriscalpiaceae and Hericiaceae (Donk, 1964), so that the Hydnaceae now contains only *Hydnum* and a few residual genera which may eventually be accommodated elsewhere. The Auriscalpiaceae includes not only forms with toothed hymenium such as *Auriscalpium*, but also *Lentinellus* which has gills (Maas-Geesteranus, 1963a; 1975). *Auriscalpium vulgare* (sometimes called *Hydnum auriscalpium*) grows on buried pine cones, forming

Fig. 244. *Auriscalpium vulgare*. Two fruit-bodies growing from a pine cone.

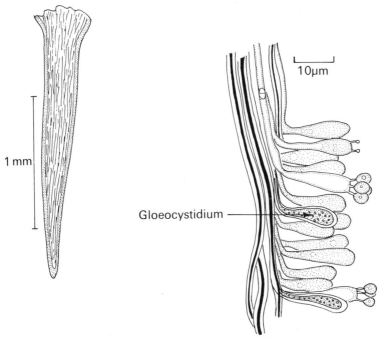

Gloeocystidium

Fig. 245. *Auriscalpium vulgare*. A. Spine from underside of cap, bearing hymenium. B. Portion of hymenium showing gloeocystidia.

stalked, one-sided, brown, hairy fruit-bodies at the surface of the ground during autumn and winter (Fig. 244). The hyphal construction is dimitic with skeletals (Ragab, 1953). The hymenium is formed on vertical, finger-like down-growths from the underside of the pileus. Interspersed amongst the basidia are irregularly enlarged, thin-walled hyphal tips with highly refractile contents termed gloeocystidia (Fig. 245) (Lentz, 1954). The sporophores are capable of proliferation, and show rapid readjustment to the vertical position if displaced laterally (Harvey, 1958).

Hydnaceae

Fruit-bodies of *Hydnum repandum* (Fig. 246A) and *H. rufescens* grow in woodland. They are more or less mushroom-shaped, with a stalked cap which may be central or lateral. The construction is monomitic, with generative hyphae which become inflated, giving the fruit-body a fleshy texture (Maas-Geesteranus, 1963b; 1975). The hymenium covers the tapering spines which develop from the lower side of the cap. In contrast with *Auriscalpium*, gloeocystidia are absent.

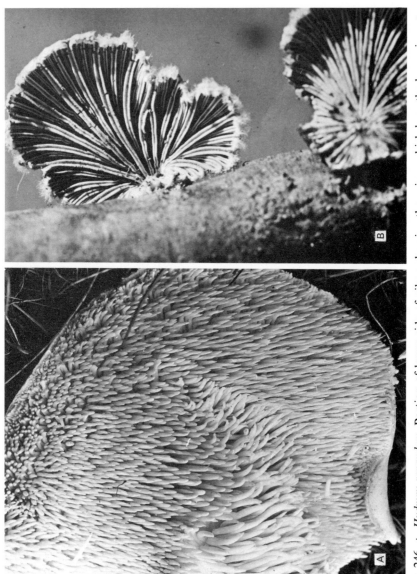

Fig. 246. A. *Hydnum repandum*. Portion of lower side of pileus showing the spines which bear the hymenium. B. *Schizophyllum commune*. Lower side of fruit-body showing the split 'gills'.

Schizophyllaceae

Traditionally *Schizophyllum* has been classified in the Agaricales but details of fruit-body development (Essig, 1922; Wessels, 1965) and the development of the 'gills' are 'so peculiar that homology with the agaric gill is out of the question' (Donk, 1964). *Schizophyllum commune* grows saprophytically or

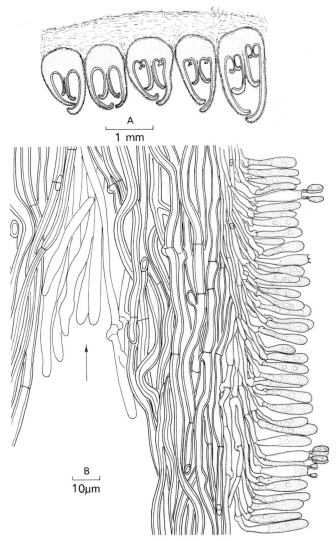

Fig. 247. *Schizophyllum commune*. A. V.S. portion of sporophore in the dried state showing the divided inrolled 'gills'. B. High-power drawing of part of a 'gill' in the region of the split (arrowed). Note the thin-walled hyphae in this region, contrasting with the thicker-walled hyphae making up the rest of the flesh.

parasitically on woody substrata forming fan-shaped, laterally attached fruit-bodies with a furry upper surface. The name *Schizophyllum* refers to the longitudinally split gills which may be a xeromorphic adaptation (Fig. 246B). In dry weather the gills curve inwards so that the hymenial surface is protected by a series of adjoining folds. The curvature is believed to be due to the shrinkage of the hymenial layers on drying. Since the remaining tissue of the gill is composed of thick-walled clamped hyphae which do not shrink so readily, inward curvature follows. In the region of the split the cells are thinner-walled (Fig. 247). Fruit-bodies which have been kept dry for 2 years rapidly take up water through the hairy upper surface and within 2–3 hours the gills straighten out as the hymenium expands. Spore discharge commences after 3–4 hours (Buller, 1909). Material which Buller subjected to freeze-drying in 1910 and 1912 has been re-wetted, and after fifty-two years some of the sporophores have revived and produced spores (Ainsworth, 1965).

Schizophyllum has been studied intensively by workers interested in its mating system (Raper, 1966a), genetics (Raper & Hoffmann, 1974), in nuclear migration (Snider, 1965), morphogenesis (Niederpruem & Wessels, 1969) and taxonomy (Linder, 1933; Cooke, 1961). Dikaryotic mycelia fruit readily in culture.

The fruit-body develops as an inverted cup with a hymenium developing over the entire lower surface and, by more rapid growth on one side, the fruit-body may become fan-shaped. The split gills arise by marginal proliferation, and their number is increased by downgrowths from the flesh of the fruit-body. The cytology of basidial development shows no unusual features. Mature basidiospores are binucleate, following mitosis of the nucleus which enters each spore (Ehrlich & McDonough, 1949).

Coniophoraceae

Serpula lacrymans causes dry rot, and is one of the most serious agents of timber decay in buildings (Cartwright & Findlay, 1958). Both hardwoods and softwoods are attacked. Only wood with a moisture content above about 20–25% of the oven dry weight is liable to attack by the fungus. Properly dried and seasoned timber has a moisture content of 15–18% and in a properly ventilated house this soon falls to 12–14% or lower (Findlay & Savory, 1960). If woodwork becomes wet through contact with damp masonry, through faulty construction or poor ventilation then infection from air-borne basidiospores is likely to follow. The mycelium within the wood develops chiefly at the expense of the cellulose, and water produced by the breakdown of cellulose (sometimes termed the water of metabolism) may render the timber on which it is growing moister, which in turn stimulates further growth. The epithet '*lacrymans*' (weeping) refers to the beads of moisture sometimes found on decaying timber. Well-rotted timber is shrunken with transverse cracks and has a dry crumbly texture. Sheets of mycelium may extend over the timber and

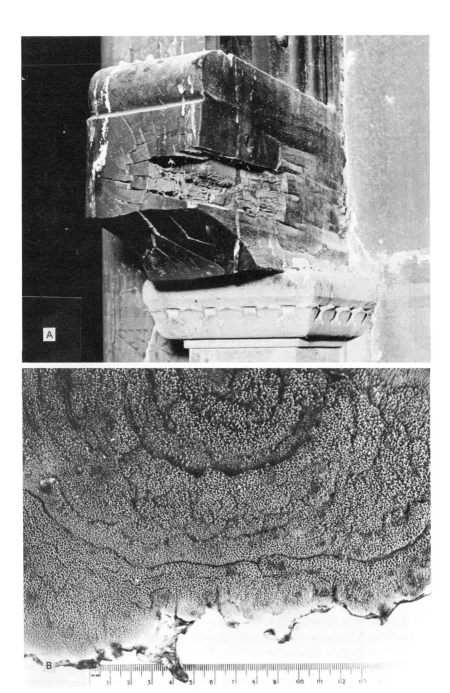

Fig. 248. *Serpula lacrymans*, the dry-rot fungus. A. Beam supporting a roof arch of a church showing the typical cracking transverse to the grain of the wood. The wood also shows shrinkage. B. Fruit-body seen from the lower side showing the shallow pores.

adjacent brickwork, and the fungus is also capable of spreading over long distances (several metres) by means of mycelial strands up to 5 mm in diameter. The strands can penetrate mortar and spread from room to room of houses. The internal hyphae are modified for rapid conduction, enabling water to be transported, and enabling colonisation of relatively dry timber. Even if infected timber is removed the strands can initiate fresh infections.

Fruit-bodies develop as flat fleshy surface growths. On the lower side a brown honeycomb-like arrangement of shallow pores supports the hymenium (Fig. 248). The construction is monomitic. Immense numbers of rusty-red basidiospores can be produced which may form deposits visible to the naked eye.

In the control of dry rot it is important to strip out all infected timber and to sterilise adjoining brickwork. New timber can be treated with disinfectants. The most important measure, however, is to ensure, by proper construction, that the moisture level of the timber remains below the point at which infection is likely.

Serpula lacrymans is essentially a fungus of buildings and is rarely found away from human habitation. It has, however, been collected on spruce logs in the Himalayas, at an altitude of 8,000–10,000 ft (Bagchee, 1954). It is widely distributed throughout northern Europe, but is confined to the cool temperate zone. Its optimum temperature (23 °C) is rather low, and its maximum temperature is about 26 °C. These temperature characteristics may explain its distribution and why it is not found on exposed timber (Cartwright & Findlay, 1958).

Clavariaceae

Hymenomycetes with branched or unbranched cylindrical or clavate fructifications were previously aggregated into this family, but again, microscopical analysis shows that such an arrangement groups together unrelated forms, and it is clear that the Clavarioid type of fructification has evolved independently in several unrelated basidiomycete groups. Corner (1950) has monographed the Clavarioid fungi and has attempted to sort them into natural (i.e. related) groups, and Petersen (1973) has given keys to genera. Microscopic structure shows that relatives of certain Clavarioid forms lie with Hydnoid, Polyporoid and Agaricoid genera.

Clavaria and its allies (Figs. 249, 250)

This is a large genus of pasture and woodland fungi with cylindrical or club-shaped branched or unbranched fructifications. The flesh is made up of thin-walled hyphae which lack clamp connections, and may become inflated and develop secondary septa (Fig. 249). The hymenium which covers the whole surface of the fruit-body usually consists of four-spored basidia with or

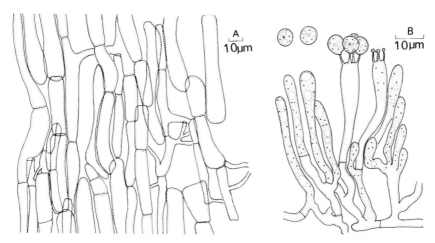

Fig. 249. *Clavaria vermicularis.* A. Hyphae from the flesh. Note the absence of clamp connections. B. Part of the hymenium. There are no cystidia. Drawings prepared from var. *sphaerospora.*

without basal clamps, bearing colourless spores. A typical species is *C. vermicularis* (Fig. 250A) which grows in grassland forming tufts of whitish simple fruit-bodies. *Clavaria argillacea* forms yellow club-shaped fructifications on moors and heaths and peat bogs. Here the basidia bear a wide loop-like clamp at the base. There are numerous common representatives of the Clavariaceae (Lange & Hora, 1963). *Clavariadelphus pistillaris* which forms exceptionally large club-shaped fruit-bodies (7–30 × 2–6 cm) grows in deciduous woods. The construction is monomitic, with clamps at the septa. As the fruit-body matures the hymenium becomes thicker by the development of further layers of basidia. *Clavulinopsis corniculata* with branched yellow fruit-bodies, and *Clavulinopsis fusiformis* with simple yellow fruit-bodies are frequently found in acid grassland. The flesh is composed of clamped hyphae without secondary septa. The hymenium becomes thickened with age, and the basidia are typically four-spored.

Clavulina cristata is a very variable fungus with highly branched fructifications (Fig. 250B). In microscopical details its construction is very similar to *Clavulina rugosa* which is also variable but generally forms whitish, less richly branched fruit-bodies. A characteristic feature of the genus is that the basidia are two-spored, narrowly cylindrical, and often septate after spore discharge. The hymenium thickens with age. The construction is monomitic, with clamped inflated hyphae (Fig. 251). Donk (1964) places *Clavulina* in a separate family, the Clavulinaceae.

Some of the more richly branched fairy clubs are placed in the genus *Ramaria*, distinguished by the tougher flesh and yellow- to brown-coloured basidiospores, which are often rough. *Ramaria stricta* is unusual in growing

Fig. 250. A. *Clavaria vermicularis* fruit-bodies. B. *Clavulina cristata* fruit-bodies.

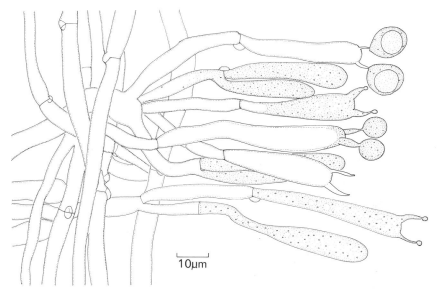

Fig. 251. *Clavulina rugosa*. Portion of the flesh and hymenium. Note the clamped hyphae and the narrow two-spored basidia.

on rotten wood. The hyphae of the flesh have thick walls. The mycelial hyphae are dimitic with skeletals.

Stereaceae

In this family the fruit-body is flattened, appressed or resupinate, sessile or stalked, with the hymenium smooth (unfolded) and on one side of the fructification. The construction is usually dimitic with skeletals, but monomitic and trimitic forms are known. There are about ten genera (Talbot, 1973b). Most species are lignicolous and saprophytic, but some are important parasites.

Stereum (Figs. 252, 253)

Species of *Stereum* are common on decaying stumps and branches. *Stereum hirsutum* forms clusters of yellowish, fan-shaped, leathery brackets (Fig. 252A) on various woody hosts, and is important as a cause of decay of sapwood of oak logs after felling. *Stereum gausapatum* is another common fungus on oak, which can grow parasitically on living trees and cause 'pipe rot' of the heartwood. When bruised, the fruit-bodies 'bleed', i.e. exude a red latex. This phenomenon is also found in *S. rugosum* on broad-leaved trees and *S. sanguinolentum* on conifers. In all these cases, there are specialised laticiferous hyphae in the flesh which may extend through to the hymenium (see Fig. 253 of *S. rugosum*). These hyphae are interpreted as modified skeletal hyphae.

A fungus which superficially resembles *Stereum* is *Chondrostereum purpur-*

Fig. 252. A *Stereum hirsutum*. Sporophores on beech stump. The hymenial surface is on the lower face of the fruit-body. B. *Thelephora terrestris*. Fruit-bodies seen from above. The hymenium is on the lower side.

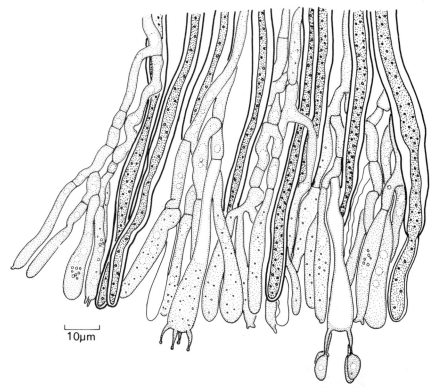

10μm

Fig. 253. *Stereum rugosum.* Section of hymenium. The thick-walled hyphae with dense contents are laticiferous hyphae which exude a red latex when damaged causing the fruit body to 'bleed'.

eum (= *Stereum purpureum*), a member of the Corticiaceae (Talbot, 1973b). *C. purpureum* is an important parasite of Rosaceae, including fruit trees such as plum and cherry, causing 'silver leaf' disease. The silver sheen on the leaves is caused by the separation of the epidermis from the palisade mesophyll due to toxic secretions from the mycelium of the fungus, not in the leaves themselves, but in the branches beneath. Infection is through wounds. Control measures include the protection of pruning cuts, and the removal of all wood bearing sporophores from orchards (Anon., 1967b). The fungus is not confined to Rosaceae, but is common on stumps of birch soon after felling.

Thelephoraceae

Here the fruit-body is generally monomitic, in contrast with the prevailing dimitic construction of the Stereaceae, and cystidia are usually absent. Most of the genera are terrestrial, but some are lignicolous.

Thelephora (Figs. 252, 254)

Thelephora terrestris grows on light acid soil, sometimes running over twigs and other debris. It has been shown to form mycorrhiza with *Arbutus menziesii*, a member of the Ericaceae (Zak, 1976). The fan-shaped fructification superficially resembles a *Stereum*, but is made up of clamped generative hyphae only. The basidiospores are brown and warty.

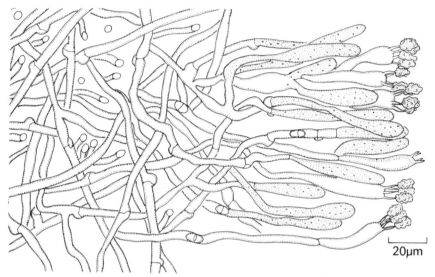

Fig. 254. *Thelephora terrestris.* Section of fruit-body and hymenium. Note the monomitic construction.

DACRYMYCETALES

This order is characterised by the forked basidium (Figs. 219ʙ, 256ʙ). Traditionally, the group has been classified with other jelly-fungi in the Phragmobasidiomycetidae, but there is a growing body of opinion favouring the transfer of the group to the Holobasidiomycetidae (McNabb & Talbot, 1973). All members are saprophytic on decaying wood, forming gelatinous or waxy fruit-bodies of varied shape, often coloured yellow to orange. There is a single family, Dacrymycetaceae, with about nine genera (McNabb & Talbot, 1973; Reid, 1974).

Dacrymyces (Figs. 255, 256)

The orange gelatinous cushions, about 1–5 mm in diameter, so common on damp rotting wood, are the fructifications of *D. stillatus.* Close inspection with a hand lens reveals that the cushions are of two kinds: soft, bright orange,

Fig. 255. A. *Dacrymces stillatus* basidial cushions. B. *Calocera viscosa* growing from a buried conifer stump. C. *Calocera cornea*.

hemispherical cushions, and firmer, pale yellow, flatter cushions. The bright orange cushions are conidial pustules which consist of hyphae whose tips are branched and fragment into numerous multicellular conidia or *arthrospores* (Fig. 256C). The cells are packed with oil globules containing carotene. Such conidia are readily dispersed by rain splash and are obviously similar in function to the splash-dispersed conidia of *Nectria cinnabarina*. The basidial cushions are attached centrally to the woody substratum. The surface layers are composed of clusters of forked basidia (Fig. 256B) which bear pairs of

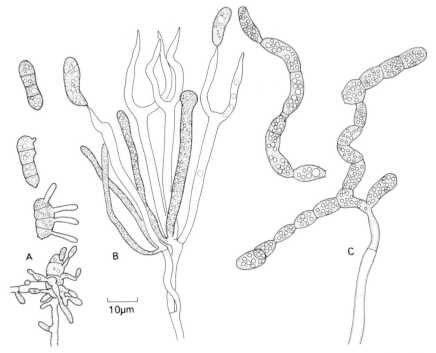

Fig. 256. *Dacrymyces stillatus.* A. Basidiospores showing germination. The basidiospore at the bottom is producing blastospores. B. Basidia. Note that the attached basidiospores are unicellular. They become three-septate on germination. C. Arthrospores from a conidial pustule.

spores. At the time of discharge, the basidiospores are unicellular, but before germination they usually develop three transverse septa. Germination may be by means of germ tubes from any of the cells, or small conidia may develop on short conidiophores formed on germination (Fig. 256A). Similar conidia may arise on older hyphae.

Calocera (Figs. 219, 255, 257)

At first sight, the cylindrical orange outgrowths of *C. viscosa* from coniferous logs, or the smaller *C. cornea* from hardwood logs, could be mistaken for species of *Clavaria* (Figs. 255B,C). The gelatinous consistency and the characteristic forked basidia (Fig. 257) place them in the Dacrymycetaceae. At maturity, the basidiospores of both species are one-septate. They germinate by the formation of germ tubes or globose conidia (McNabb, 1965).

Phragmobasidiomycetidae

In contrast to the Holobasidiomycetidae, the members of this subclass have phragmobasidia, i.e. basidia in which the metabasidium is divided by septa. A

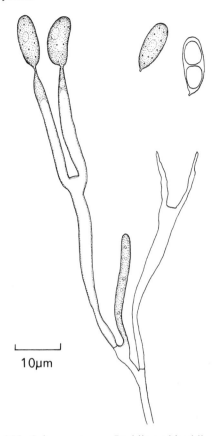

10μm

Fig. 257. *Calocera viscosa*. Basidia and basidiospores.

characteristic feature is that the basidiospores may germinate by repetition – i.e. by producing secondary spores. The fruit-body is often gelatinous or waxy, and for this reason the group is referred to as jelly fungi. The group is sometimes included in the Heterobasidiomycetes along with the Uredinales and Septobasidiales (Donk, 1972a,b). McNabb (1973) recognises three orders: Tremellales, Auriculariales and Septobasidiales. The last order is symbiotic or parasitic on scale insects.

TREMELLALES

The Tremellales are distinguished by their longitudinally divided basidia (Figs. 217, 218). Most species are saprophytic on wood, or are associated with other wood-rotting fungi such as species of *Stereum*. They form fruit-bodies which are gelatinous and often brightly coloured when wet, drying to a cartilaginous texture. The hymenium may occur on one or on both sides of the fructification. In *Pseudohydnum gelatinosum* (sometimes called *Tremellodon gelatinosum*), the fruit-body has a short excentric stalk and a pileus on the lower side of which conical teeth project, resembling those of a *Hydnum* (Fig.

Fig. 258. A. *Tremella frondosa*: fruit-body on oak stump. The hymenium is on both faces. B. *Exidia glandulosa*: fruit-body on lime. The hymenial surface bears black warts and is on one face of the fruit-body. C. *Pseudohydnum gelatinosum*: fruit-bodies seen from below and above. The hymenium is borne on the spines.

258). Polyporoid fructifications with Tremellaceous basidia are found in *Aporpium caryae*. Common representatives are *Exidia* and *Tremella*.

Exidia (Figs. 258, 259)

Exidia glandulosa, sometimes called Witches' butter, forms black rubbery fructifications on decaying branches of various woody hosts, especially lime and oak (Figs. 258B, 259). The hymenium is borne on the lower side of the fructification and in this species is studded with small black warty outgrowths. Wells (1964a,b) has described the fine structure of *E. nucleata*.

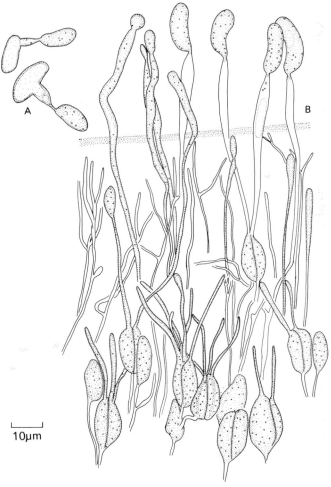

10μm

Fig. 259. *Exidia glandulosa*. A. Basidiospores showing germination by repetition. B. Section of hymenium. Note the longitudinally divided basidia with long epibasidia penetrating to the surface.

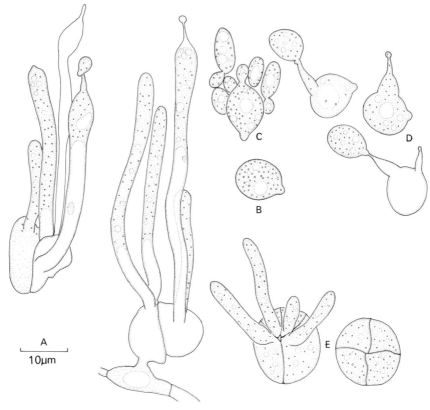

Fig. 260. *Tremella frondosa*. A. Basidia showing epibasidia with various stages of spore development. B. Freshly discharged basidiospore. C. Basidiospore germinating on malt extract agar to form yeast-like cells which also undergo budding. D. Basidiospores germinating in water by repetition, i.e. by producing ballistospores. E. Immature basidia seen from above, showing the division of the basidium into four cells by longitudinal septa.

Tremella (Figs. 258, 260)

Here the fructification consists of flattened contorted folds, with the hymenium on both faces. *Tremella frondosa* grows on oak and beech stumps, forming flesh-coloured to pale-brown fruit-bodies. The basidiospores may germinate by repetition, by yeast-like budding (Fig. 260C,D) or by germ tube. *T. mesenterica* forms yellow to orange lobed fructification on various woody hosts such as oak, willow and beech, usually in association with *Stereum hirsutum*. Heterothallism in *T. mesenterica* is tetrapolar, and there is evidence of a hormone-like substance which stimulates directional growth of compatible germ tubes (Bandoni, 1963, 1965). In culture, this fungus grows in a yeast-like manner. Budding occurs most frequently at one pole of the cell (Bandoni & Bisalputra, 1971).

Fig. 261. *Auricularia auricula*. A. Fruit-body on branch of elder, as seen from above. B. Fruit-body seen from the lower, hymenial surface.

AURICULARIALES

In this order, the basidium (strictly the metabasidium) is divided by transverse septa. Members of the group may be saprotrophic or parasitic on plants or other fungi, forming gelatinous fruit-bodies of varied form. *Helicobasidium brebisonii* is the cause of violet root rot of several cultivated plants, and may also cause tuber rot in potato. Its basidium is curved or coiled. It has been claimed that the Auriculariales may be related to the stock from which the Uredinales evolved. NcNabb (1973) distinguishes eighteen genera.

Auricularia (Figs. 261, 262)

The Jew's ear fungus *A. auricula* (sometimes known as *Hirneola auricula-judae*) forms rubbery, ear-shaped fruits on elder branches (Fig. 261), and is a weak parasite. A wide range of other hosts has been reported. Duncan &

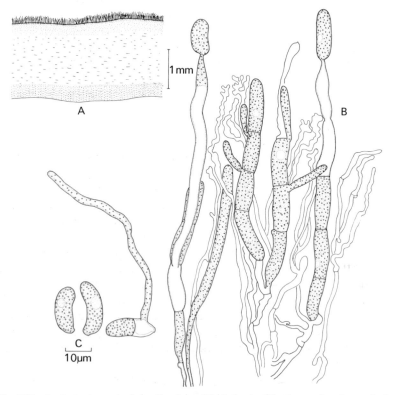

Fig. 262. *Auricularia auricula*. A. Section of fruit-body. The hymenium is on the lower side. B. Squash preparation showing basidia. Note the transverse segmentation and the long epibasidia. The basidia are associated with branched hyphae. C. Basidiospores, one germinating. Note the formation of a septum.

Macdonald (1967) divide *A. auricula* into three 'units'; a European deciduous unit, a North American deciduous unit, and a North American coniferous unit. Interfertility studies between these populations show that the European populations are intersterile with the American deciduous unit, and virtually intersterile with the American coniferous unit. Crosses between the two American units show that they are only partially interfertile. A section through the flesh shows a hairy upper surface, a central gelatinous layer containing narrow clamped hyphae and a broad hymenium on the lower side. Details of the anatomy of the fruit-body are useful in classification (Lowy, 1951, 1952). The fruit-body can dry to a hard brittle mass, but on wetting it quickly absorbs moisture and discharges spores within a few hours. The basidia are cylindrical and become divided by three transverse septa into four cells (Fig. 262). Each cell of the basidium develops a long cylindrical *epibasidium* which extends to the surface of the hymenium and terminates in a conical sterigma bearing a basidiospore, which is projected (ballistospore). On germination a second ballistospore may be formed: alternatively sickle-shaped conidia or a germ tube may develop. The fungus is heterothallic, and there are indications of multiple alleles. Barnett (1937) claimed that it was bipolar, but Banerjee (1956) states that it is tetrapolar. Another common species is *A. mesenterica* which forms thicker hairy fan-shaped fruit-bodies on old stumps and logs of elm and other trees. It causes active wood decay, and may occasionally be parasitic.

2. GASTEROMYCETES

This is an unnatural assemblage of basidiomycetes which share the common negative character that their basidiospores are not discharged violently from their basidia. Instead of the asymmetrically poised basidiospores found in the Hymenomycetes, Gasteromycete basidiospores are usually symmetrically poised on their sterigmata or are sessile. Dring (1973) has termed the basidiospores of Gasteromycetes **statismospores**. Commonly, the basidia open into cavities within a fruit-body, and the basidiospores are released into these cavities as the tissue between them breaks down or dries out. Sometimes, as in *Lycoperdon*, the fruit-body opens by a pore through which the spores escape, but in forms with subterranean fruit-bodies there is no special opening, and it is possible that the spores are dispersed by rodents or other soil animals. In *Phallus* and its allies, the spores are exposed in a sticky mass attractive to insects, whilst in the Nidulariaceae the spores are enclosed in separate glebal masses or peridioles which are dispersed as units.

All members of the group are saprophytic and grow on such substrata as soil, rotting wood and other vegetation, and on dung. *Rhizopogon*, which forms subterranean fruit-bodies, and *Scleroderma* can form mycorrhiza with forest trees. There are two aquatic Gasteromycetes. *Nia vibrissa* grows on driftwood in the sea, forming globose, yellowish basidiocarps a few millimetres in diameter. Its basidiospores bear four or five radiating appendages (Doguet, 1967, 1968, 1969). *Limnoperdon* forms small, floating fruit-bodies in marshes (Escobar *et al.*, 1976).

Evidence from the microscopic structure of the fruit-bodies and spores has led to the conclusion that certain genera of Gasteromycetes are related to agaric genera. It is believed that the Gasteromycete forms may have evolved from Hymenomycete ancestors, possibly as an adaptation to xerophytic conditions. Ingold (1971) regards the Gasteromycetes as a biological group which, having lost the explosive spore discharge mechanism of their Hymenomycete ancestors, have attempted a remarkable series of experiments in spore liberation. For a discussion of the problems of Gasteromycete phylogeny see Heim (1948, 1971), Singer & Smith (1960), Smith (1973) and Dring (1973).

Dring (1973) recognises nine orders, of which we shall study examples from only four. The Hymenogastrales, which are not considered further, are made up almost entirely of subterranean forms, sometimes referred to as false truffles (Fig. 265). Their fruit-bodies are dug up and eaten by rodents. An account of British species is provided by Hawker (1954).

LYCOPERDALES

Common examples of this group are the puff-balls *Lycoperdon*, *Calvatia* and *Bovista*, and the earth-star *Geastrum*. Their fruit-bodies may begin development beneath the surface of the soil, but mature above ground.

Lycoperdon (Figs. 263A,B, 264)

The fruit-bodies are pear-shaped or top-shaped. Most species grow on the ground, but *L. pyriforme* occurs on old stumps, rotting wood and sawdust heaps. Fruit-bodies commonly arise on mycelial cords. The individual cells or the mycelium usually contain paired nuclei, but clamp connections are absent (Dowding & Bulmer, 1964). A longitudinal section of a young fruit-body of *L. pyriforme* (Fig. 264A,B) shows that it is surrounded by a two-layered rind or peridium, but as the fruit-body expands the outer pseudoparenchymatous exoperidium may slough off or crack into numerous scales or warts whilst the tougher endoperidium made up of both thick-walled and thin-walled hyphae remains unbroken, apart from a pore at the apex of the fruit-body. The tissue within the peridium is termed the **gleba**. It is differentiated into a non-sporing region or **sub-gleba** at the base which extends as a columella into the sporing region, or fertile part of the gleba in the upper part of the fructification. The glebal tissue is sponge-like, containing numerous small cavities, and in the upper fertile part the cavities are lined by the hymenium. The tissue separating the hymenial chambers is made up of thick- and thin-walled hyphae. The thin-walled hyphae break down as the fruit-body ripens, but the thick-walled hyphae persist to form the **capillitium** threads between which the spores are contained. The basidia lining the cavities of the gleba are rounded and bear

Fig. 263. A. *Lycoperdon pyriforme*. Fruit-bodies growing from a stump buried beneath leaves. B. *Lycoperdon perlatum*. Fruit-bodies amongst grass. C. *Calvatia gigantea*. Fruit-bodies growing with *Urtica dioica*. The coin is 2·5 cm in diameter. D. *Scleroderma aurantium*.

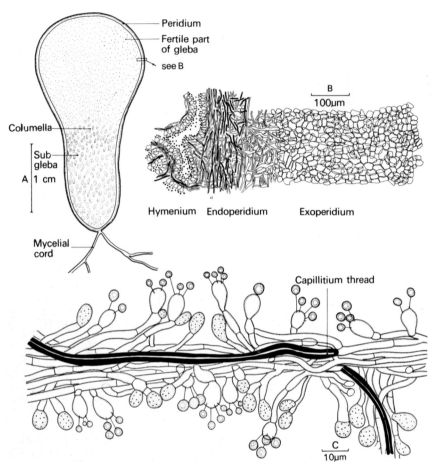

Fig. 264. *Lycoperdon pyriforme.* A. L.S. fruit-body. B. Portion of peridium and gleba. Note the pseudoparenchymatous exoperidium and the fibrous endoperidium. C. Portion of gleba showing basidia, thin-walled hyphae and capillitium threads.

one to four basidiospores symmetrically arranged on sterigmata of varying length (Fig. 264C). Young basidia are binucleate, and nuclear fusion and meiosis occur in the usual way (Ritchie, 1948; Dowding & Bulmer, 1964). The spores are not violently projected from the sterigmata. As the glebal tissue breaks down and dries the spores are left as a dusty mass inside the fruit-body. The thin upper layer of the endoperidium is elastic and acts as a bellows. When rain-drops impinge on this layer small clouds of spores are puffed out (Gregory, 1949). Little is known of the mating behaviour of *Lycoperdon.*

Calvatia (Fig. 263C)

C. gigantea forms fruit-bodies about the size of a rugby football. There is no definite pore; the peridium breaks away to leave a brown spore mass. Buller

(1909) has estimated that the output of a specimen measuring $40 \times 28 \times 20$ cm was 7×10^{12} spores. Even larger specimens, 120–150×60 cm, have been recorded (Kreisel, 1961). The spores are spherical with scattered warts (Gregory & Henden, 1976).

When attempts are made to germinate the spores of this and other puff-balls in the laboratory the percentage germination is extremely low; often fewer than 1 spore in 1,000 germinates. Germination may take several weeks, and is stimulated by the growth of yeasts (Bulmer, 1964; Wilson & Beneke, 1966). Interest in this fungus has been aroused by claims that an extract of it, calvacin, can inhibit the growth of certain types of tumour (Beneke, 1963). Because meiosis in the entire fruit-body is almost simultaneous it has been possible to follow enzymic changes in tissue extracts during and after meiosis (Bulmer & Li, 1966).

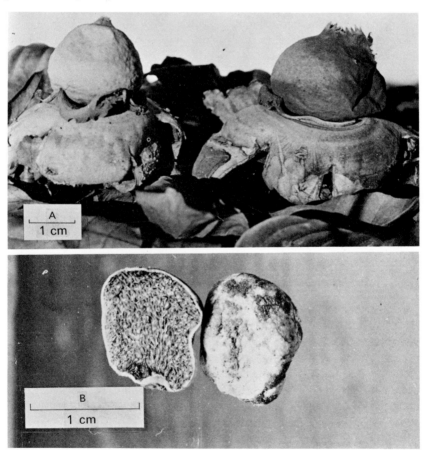

Fig. 265. A. *Geastrum triplex*. Note the recurved exoperidium. B. *Hymenogaster tener*. The fruit-bodies are subterranean. One has been cut open and shows the hymenial chambers.

Geastrum (Fig. 265A)

Species of *Geastrum*, earth-star (Fig. 265A), produce fruit-bodies that begin development beneath soil or litter, or at the soil surface. *G. triplex* grows amongst the leaves of beech, sycamore and pine. The young fruit-body is onion-shaped. The exoperidium is more complex than that of *Lycoperdon*, consisting of a brown outer layer made of narrow hyphae mostly running longitudinally and a paler pseudoparenchymatous inner layer. As the fruit-body ripens the whole of the exoperidium splits open from the tip in a stellate fashion and, due to swelling of the pseudoparenchyma cells of the exoperidium, the triangular flaps curve outwards and make contact with the soil, lifting the inner part of the fruit-body into the air (Fricke & Handke, 1962). The endoperidium is thin and papery and opens by an apical pore. Spores are puffed out when rain-drops strike it (Ingold, 1971). The gleba contains a columella (sometimes termed a pseudocolumella) and capillitium much as in *Lycoperdon*. Basidial development can only be observed in young unexpanded fruit-bodies. The basidia are pear-shaped, with four to six (or possibly eight) spores borne on a knob-like extension of the pointed end (Palmer, 1955).

NIDULARIALES

Here the fruit-bodies are globose or funnel-shaped, and the gleba is differentiated into one or more spherical to lens-shaped **peridioles** or glebal masses which contain the basidiospores. Common examples are *Cyathus*, *Crucibulum* and *Sphaerobolus*. The first two genera are sometimes termed bird's nest fungi. A detailed account has been given by Brodie (1975).

Cyathus (Figs. 266A, 267)

The funnel-shaped fruit-bodies of *C. olla* can be found in autumn growing amongst cereal stubble. *C. striatus*, recognised by the furrowed inner wall of its cup, grows on old stumps and twigs, whilst *C. stercoreus* grows on old dung patches. This last species can be made to fruit readily on chaff cased in soil (Warcup & Talbot, 1962), and since the peridioles retain viability for many years, it provides a convenient example to study. The fungus can also be induced to fruit on liquid and on agar media (Lu, 1973). Light is essential for fruit-body formation. The first sign of development is the appearance of brown mycelial strands at the soil surface, on which knots of hyphae differentiate. In young fruit-bodies, the mouth of the funnel is closed over by a thin papery epiphragm (Fig. 267A), which ruptures as the fruit-body expands. Within the funnel, the peridioles develop. They are lens-shaped, slate-blue in colour and attached to the peridium by a complex **funiculus**. In earlier stages of development in this and other species of *Cyathus*, the peridioles are separated by thin-walled hyphae which disappear at maturity (Walker, 1920). The

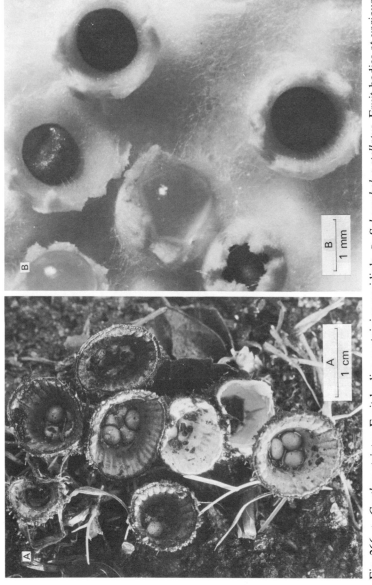

Fig. 266. A. *Cyathus striatus*. Fruit-bodies containing peridiola. B. *Sphaerobolus stellatus*. Fruit-bodies at various stages of development. Four still contain glebal masses, and two (left) have discharged.

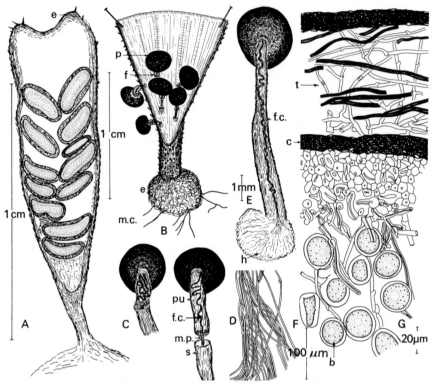

Fig. 267. *Cyathus stercoreus*. A. L.S. immature fruit-body showing peridioles in section. B. Fruit-body cut open and pinned back to show the attachment of the peridioles. C–E. Details of structure of funiculus. C. Condition of funiculus before stretching. D. Stretched funiculus. Note the funicular cord coiled within the purse. E. Funicular cord extended after rupture of the purse. The base of the funicular cord is frayed out to form the hapteron. F. Portion of funicular cord. Note the spirally coiled hyphae. The thickenings are modified clamp connections. G. Detail of peridiole wall and contents. (b, basidiospore; c, cortex; e, epiphragm; em, emplacement; f, funiculus; f.c., funicular cord; h, hapteron; m.c., mycelial cords; m.p., middle-piece; s, sheath; p, peridiole; pu, purse; t, tunica.)

peridioles are surrounded by a **tunica**, made up of loosely interwoven hyphae, and a dark thick-walled **cortex**, in turn lined by a mass of very thick-walled hyaline cells (Fig. 267G). The inner part of the peridiolum is made up of thin-walled hyphae between which basidia develop. The basidia form four to eight basidiospores. The basidia disappear soon after the formation of spores, but the spores continue to enlarge and become thick-walled (Fig. 267G). Most of the thin-walled hyphae also break down, possibly providing nutrients for the enlarging spores. Martin (1927) calls the thin-walled cells 'nurse hyphae'. Nuclear fusion and meiosis occur in the basidia (Lu & Brodie, 1964). It is reported that *C. stercoreus* is probably tetraploid (Lu, 1964).

The fruit-bodies of *Cyathus* and *Crucibulum* have been aptly termed splash-cups. The peridioles are splashed out by the action of rain-drops to distances of over 1 m. The key to understanding the mechanism of discharge lies in the structure of the funiculus. In *Cyathus* (Fig. 267C,D,E), the funiculus is made up of the following structures (Brodie, 1951, 1956, 1963).

(a) **Sheath:** a tubular network of hyphae attached to the inner wall of the peridium.

(b) **Middle piece:** the innermost hyphae of the sheath unite to form a short cord termed the middle piece.

(c) **Purse:** the middle piece flares out at its top where its hyphae are attached to a cylindrical sac, the purse, which is firmly attached to the peridiole at a small depression.

(d) **Funicular cord:** folded up within the purse is a long strand of spirally coiled hyphae (Fig. 267F).

(e) **Hapteron:** the free end of the funicular cord is composed of a tangled mass of adhesive hyphae.

Rain-drops, which may be as much as 4 mm in diameter and have a terminal velocity of about 8 m/second, fall into the cup. Drops of this size are most likely to drip from woodland canopy (Savile & Hayhoe, 1978). The force creates a strong upward thrust which tears open the purse. The spirally wound funicular cord swells explosively, stretching to a length of about 2–3 mm in *C. stercoreus*, whilst in *C. striatus* it may be as long as 4–12 cm. The sticky hapteron at the base of the funicular cord helps to attach the peridiole to surrounding vegetation and the momentum of the peridiole may cause the funicular cord to wrap around objects. The peridioles of *C. stercoreus* are presumably eaten by herbivorous animals and it is known that the basidiospores on release from the peridiole are stimulated to germinate by incubation at about body temperature, but whether animals play a significant rôle in the dispersal of peridioles of other bird's-nest fungi is uncertain. The funiculus of *Crucibulum* is different from that of *Cyathus*, with a longer middle piece, a very short purse and a funicular cord which is composed of relatively few hyphae only slightly coiled.

Both *Cyathus* and *Crucibulum* show tetrapolar heterothallism with relatively few alleles (generally not more than fifteen) at each locus and this is the most usual condition found within the Nidulariales (Burnett & Boulter, 1963).

Sphaerobolus (Figs. 266B, 268)

S. stellatus forms globose orange fruit bodies about 2 mm in diameter attached to rotten wood, rotting herbaceous stems, sacking and old dung of herbivores such as the cow and the sheep. Ripe fruit-bodies open to form a star-like arrangement of two cups fitting inside each other attached only by the triangular tips of their teeth (Fig. 266B). Within the inner cup is a single brown peridiole or glebal mass about 1 mm in diameter. By sudden eversion of the inner cup the glebal mass is projected for a considerable distance. Buller (1933) has given a detailed account of glebal discharge. He showed that the glebal mass could be projected vertically for more than 2 m, and horizontally for over 4 m (see also Ingold, 1971, 1972). The fungus can readily be cultivated if a glebal mass is placed in a plate of oat agar and incubated for about three weeks in the light. A section through an almost mature, but unopened, fruit-body is shown in Fig. 268A. The glebal mass is surrounded by a peridium in which six distinct layers can be distinguished. Three of these layers form the structure of the outer cup. The three layers making up the inner cup consist of an outer layer of tangentially-arranged interwoven hyphae, a layer of radially elongated cells forming a kind of palisade, and a thin layer of pseudoparenchyma whose cells undergo deliquescence before glebal discharge to form a liquid which bathes the gleba and lies in the bottom of the inner cup. Before the

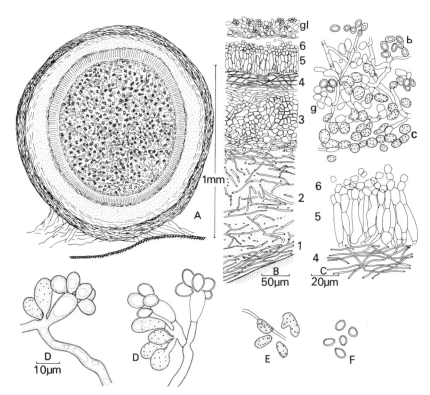

Fig. 268. *Sphaerobolus stellatus*. A. V.S. nearly ripe fruit-body showing the central glebal mass surrounded by a six-layered peridium. B. Details of peridial layers: 1, outermost layer composed of interwoven hyphae; 2, layer in which the hyphae are separated by extensive mucilage; 3, pseudoparenchymatous layer; 4, fibrous layer; 5, palisade layer; 6, layer of lubricating cells; gl, outer layers of glebal mass. C. Enlarged portion of layers 4, 5 and 6 and portion of glebal mass: c, cystidia; g, gemmae; b, basidiospores. D. Clusters of basidia from unripe fruit bodies. There are usually 4–6 basidiospores. E. Gemmae. F. Basidiospores. C, E and F to same scale.

fruit-body opens the cells of the palisade layer are rich in glycogen, but this disappears during ripening and is converted to glucose (Engel & Schneider, 1963), which causes the osmotic concentration of the cells to rise, so that they absorb water and become more turgid. The swelling of the palisade layer is restrained by the tangentially arranged hyphae, and this sets up strains within the tissues of the inner cup which are only released by its turning inside out. Light is necessary for development, and the opening of the fruit-body is phototropic, ensuring that the glebal mass is projected towards the light (Alasoadura, 1963). Peridiole discharge follows a diurnal rhythm, discharge occurring during the light. In continuous light rhythmic discharge ceases, but in continuous darkness a culture which has previously been exposed to alternate periods of 12 hours of light and darkness continues to discharge peridioles rhythmically at times corresponding to the previous light periods, indicating an endogenous rhythm (Engel & Friederichsen, 1964).

The spherical glebal mass is surrounded by a dark-brown sticky coat derived from the breakdown of the cells of the innermost peridial layer. Immediately within the brown outer coat are layers of rounded cells some-times termed cystidia (Fig. 268c). Apparently these cells are incapable of germination and their function is not known. The rest of the glebal mass consists of thick-walled oval basidiospores and thinner walled gemmae. Four to eight basidiospores develop on the basidia about 2 days before discharge (Fig. 268d), but as the glebal mass ripens the basidia disappear. Gemmae arise either terminally or in an intercalary position on hyphae within the glebal mass. Fat-containing cells are also present. The sticky glebal mass adheres readily to objects on which it is impacted and after drying it is very difficult to dislodge, even by a jet of water. Glebal masses are viable for several years. Projectiles adhere to herbage and may be eaten by animals. This may explain the presence of fruit-bodies on dung.

On germination the glebal masses give rise to clamped hyphae which usually arise directly from the gemmae and not from the basidiospores. The mating system of *Sphaerobolus* is not quite clear. Clamps have been reported on mycelia derived from cells which had the appearance of spores (Walker, 1927) and occasionally monosporous mycelia will fruit (Fries, 1948). How-ever, most basidiospores germinate to give mycelia with simple septa. Pairings of monosporous mycelia have indicated that the fungus is usually heterothal-lic, but the results have been interpreted as indicating a bipolar condition (Lorenz, 1933) and a tetrapolar condition (Fries, 1948). It would be of interest to see if multiple alleles exist.

SCLERODERMATALES

Scleroderma (Figs. 263d; 269)

Two common species of earth-ball are *S. aurantium* and *S. verrucosum*. Earlier, before the distinction between the two species was understood, the

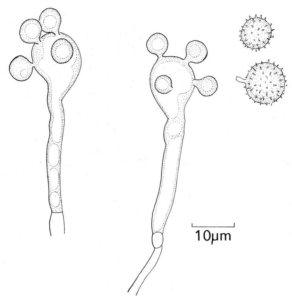

Fig. 269. *Scleroderma verrucosum*. Basidia and basidiospores. Note that the spores are almost sessile.

name *S. vulgare* was used to include both. Earth-balls are found in autumn in acid woodland and heaths growing with such trees as *Pinus, Betula, Quercus* and *Fagus* with which they may form mycorrhiza (Fries, 1942; Harley, 1969). In mature fruit-bodies, the peridium is apparently a single fairly thick layer. Although the glebal mass may be traversed by a system of sterile veins, there is no columella and no capillitium. The basidiospores are sessile (Fig. 269). When the fruit-body is ripe, it cracks open irregularly and the dry spores escape. There is no well-developed bellows mechanism as in *Lycoperdon* or *Geastrum*.

PHALLALES

Common examples of this order are *Phallus impudicus*, the stinkhorn, and *Mutinus caninus* the dog's stinkhorn.

Phallus (Figs. 270B, 271)

In late summer and autumn stinkhorns can be detected readily by their smell. The fruit-bodies arise from 'eggs' about 5 cm in diameter which in turn develop on extensive white mycelial cords which can usually be traced underground to a buried stump (Grainger, 1962). A longitudinal section of an 'egg' (Fig. 271A) shows a thin papery outer and inner peridium and a wider mass of jelly making up the middle peridium. The central part of the fruit-body is

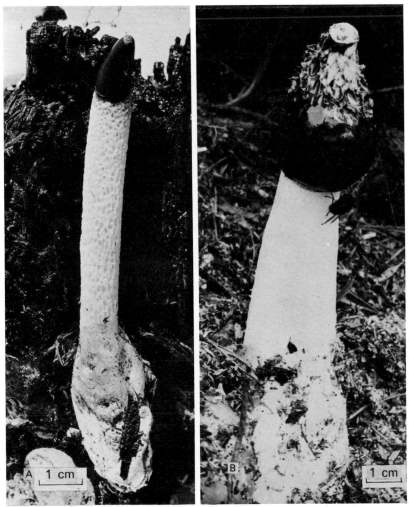

Fig. 270. A. *Mutinus caninus*. Expanded fruit body attached to rotting wood. B. *Phallus impudicus*. Ripe fruit body with blue-bottle feeding on the gluten.

differentiated into a cylindrical hollow stipe and a folded honeycomb-like receptacle which bears the fertile part of the gleba. Within the young gleba are cavities lined by basidia, bearing up to nine spores (Fig. 271B), but as the glebal mass ripens the basidia disintegrate. Fruit-bodies expand very rapidly: within a few hours the stipe may elongate from about 5 cm to a length of about 15 cm. This sudden expansion is probably at the expense of water stored within the jelly in the middle peridium. The mean weight of expanded stipes is more than twice that of unexpanded ones (Ingold, 1959). Expansion of the stipe is accompanied by breakdown of glycogen and its conversion to sugar

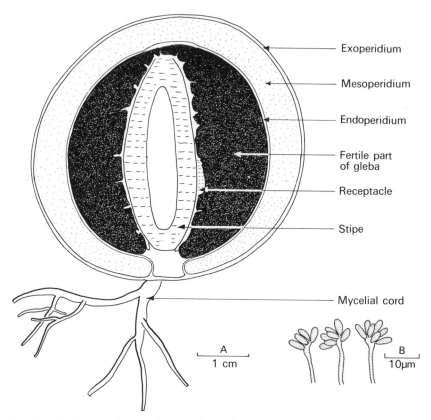

Fig. 271. *Phallus impudicus*. A. L.S. 'egg' showing the unexpanded stipe. B. Basidia.

(Buller, 1933). A similar conversion has been reported in the related fungus *Dictyophora indusiata*. In the unexpanded stipe the cells are folded, but expand to almost twelve times their original volume as the stipe elongates (Kinugawa, 1965). About the same time as the stipe of *Phallus* is elongating the fertile glebal mass begins to secrete a strong-smelling substance together with sugar, both of which are attractive to flies, especially blue-bottles and other insects which normally feed on carrion and on dung (Schremmer, 1963). The smell has been identified as a mixture of the following substances: methylmercaptan, hydrogen sulphide, phenylacetaldehyde, α-phenylcrotonaldehyde, dihydrochalcone, phenylacetic acid, acetaldehyde, acetic acid, proprionic acid, formaldehyde (List & Freund, 1968). Within a few hours, the green spore mass is removed. The spores are defaecated, apparently unharmed, on to surrounding vegetation and the soil, often within a short time of ingestion, but the details of their germination are unknown. How the mycelium succeeds in colonising tree stumps is not understood. There have been few studies of the nutrition and physiology of *Phallus*, but *P. ravenelii* has been shown to make good vegetative growth on a wide range of carbohydrates, and to require thiamine (Howard & Bigelow, 1969).

The general form of *Mutinus* is similar to *Phallus*, but the fruit-bodies are smaller (Fig. 265A). The upper part of the stipe is orange in colour, and the smell is less obvious. The receptacle bearing the glebal mass is not reticulate.

3: TELIOMYCETES

Two important groups of plant pathogens are included here: the smut fungi (Ustilaginales) and the rust fungi (Uredinales). It is very doubtful if the two groups are related. They differ from the Basidiomycotina already studied in several respects. Although their mycelia are septate, the septa are of the simple type, lacking dolipores and parenthesomes. The structures regarded as the equivalent of basidia consist of a thick-walled teliospore (within which nuclear fusion takes place) and a promycelium (metabasidium) which usually gives rise to four or more sporidia. A distinction between the Ustilaginales and the Uredinales is that, in the Uredinales, there are typically only four sporidia formed on the promycelium but, in the Ustilaginales, the number is large and indefinite. The Ustilaginales grow readily in liquid or on agar, and some can be induced to complete their life cycle in pure culture. In contrast, most Uredinales only grow with difficulty in pure culture. Ecologically, they are biotrophic.

As plant parasites, the Ustilaginales are confined to Angiosperms, but the Uredinales parasitise ferns, Gymnosperms and Angiosperms, and in many cases have complex life cycles which may be completed on two unrelated hosts. The Uredinales have a long fossil history and were probably parasitic on ferns in Carboniferous times.

USTILAGINALES

The Ustilaginales or smut fungi are parasitic on Angiosperms, where they often cause diseases of economic significance. Over 1,000 species are known, parasitic on over seventy-five families of Angiosperms (Durán, 1973). Certain families of flowering plants seem particularly favourable hosts for smut fungi, notably the Cyperaceae and Gramineae (Fisher & Holton, 1957), and it is as parasites of cereals that these fungi are especially important. The term smut refers to the mass of dark powdery spores formed in sori on the leaves, stems or in the flowers or fruits of the host plant (see Fig. 272). Various terms have been used for such spores: teliospores, chlamydospores, brand spores, melanospores or ustospores (Moore, 1972). Young teliospores are dikaryotic and arise on a dikaryotic mycelium which is often systemic. Usually there are no specialised haustoria, but short branches of the mycelium may enter host cells. The intracellular hyphae resemble haustoria of other plant pathogens in

Fig. 272. Symptoms of smut diseases. A. *Ustilago violacea* anther smut of Caryophyllaceae on *Silene alba*. B. *Ustilago hypodytes*, stem smut of grasses on *Elymus arenarius*. C. *Ustilago nuda*, loose smut of barley. D. *Ustilago avenae*, loose smut of oats on *Arrhenatherum elatius*. E. *Tilletia caries*, bunt or stinking smut of wheat. F. *Urocystis anemones*, anemone smut on *Ranunculus repens*.

that they are associated with an invagination of the host plasmalemma, and are surrounded by an encapsulation which forms the interface between the host and the parasite. Moreover, they are often partially or sometimes completely enclosed by a sheath probably derived from the host cell wall (Fullerton, 1970). During sorus development, the dikaryotic hyphae proliferate and mass together in intercellular spaces, often destroying the softer internal host tissues, but remaining enclosed in the host epidermis. Virtually all the hyphal segments may become converted to spores. The sporogenous hyphae are composed of binucleate cells. After nuclear fusion, the walls thicken and gelatinise. Meanwhile, the cells enlarge to become globose. The spore initials are at first enclosed in a gelatinous matrix, which disappears at maturity. The ripe, uninucleate, diploid teliospores have thick, usually dark walls, which may be almost smooth or ornamented by spines or reticulations (Langdon & Fullerton, 1975; Hess & Weber, 1976). In some genera, e.g. *Urocystis*, there is a central fertile cell surrounded by a group of non-fertile cells (Fig. 279). The teliospores are commonly dispersed by wind and germinate by a variety of methods (Holton *et al.*, 1968; Zambettakis, 1970, 1973). Those of *Ustilago avenae*, the cause of loose smut of oats and false-oat grass (*Arrhenatherum elatius*) produce a germ tube or promycelium into which the diploid nucleus passes and undergoes meiosis. Thereafter, the promycelium develops three

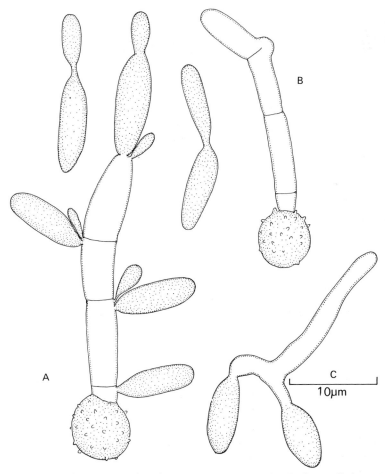

Fig. 273. *Ustilago avenae*. A. Germinating teliospore showing the four-celled promycelium, each cell of which is producing sporidia. Budding of detached sporidia is also shown. B. Germinating teliospore showing fusion of the two terminal cells to initiate a dikaryon. C. Fusion of germ tubes from two basidiospores to initiate a dikaryon.

transverse septa to cut off four cells each containing a single haploid nucleus. Within each cell, the nucleus divides, and a daughter nucleus passes into a sporidium budded off from the cell. Nuclear division may be repeated and each cell may form numerous sporidia. Detached sporidia may also form further spores by budding (Fig. 273A). In this yeast-like phase, many smuts are capable of prolonged saprophytic growth, but parasitic growth by the haploid phase does not usually occur (Hüttig, 1931; Holton, 1936). The four-celled promycelium is regarded as a metabasidium, and the sporidia as basidiospores. *Ustilago avenae*, like most other smuts, is heterothallic, and is bipolar, i.e. the sporidia are of two mating types (Whitehouse, 1951). Mating is controlled by a single gene with two alleles, a_1 and a_2. Basidiospores of opposite mating type fuse by means of a short conjugation tube (Fig. 273C) to initiate a dikaryotic mycelium, which is capable of infecting a new host. Haploid basidiospores or mycelia are incapable of infection. In some promycelia, the dikaryon may be initiated by fusion of adjacent cells of the promycelium (Fig. 273B).

The teliospores of *U. violacea*, the cause of anther smut of Caryophyllaceae, germinate differently (Figs. 272A, 274). The promycelium is here often only three-celled, and is readily detached from the teliospore and may continue to develop sporidia after separation. A succession of promycelia may develop from one teliospore (Wang, 1932). The bipolar nature of incompatibility in smuts was first demonstrated in this fungus (Kniep, 1919).

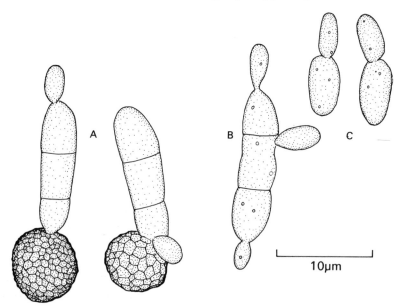

Fig. 274. *Ustilago violacea*. A. Germinating teliospores showing the three-celled promycelium. B. Detached promycelium producing sporidia. C. Detached sporidia-producing buds.

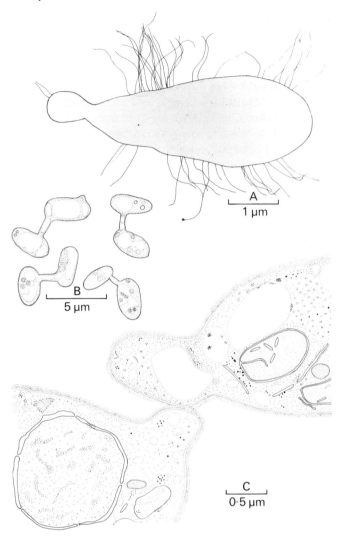

Fig. 275. *Ustilago violacea*. A. A yeast-like cell bearing numerous fimbriae (tracing from electron micrograph by Day & Poon, 1975). B. Conjugation between sporidia (based on Poon *et al.*, 1974). C. Development of conjugation pegs from both partners, involving fusion of the surface coatings (glycocalyces). The plasma membranes are not yet affected (based on Poon *et al.*, 1974).

The process of sporidial conjugation has been intensively studied in *Ustilago violacea*. Within 2 hours after sporidia of opposite mating type are mixed together on a suitable medium, a series of events occurs (Poon *et al.*, 1974) (see Fig. 275):

    1. Intimate pairing of cells of opposite mating type.
    2. Initiation of pegs from the cell walls at the point of contact.

3. Elongation of the pegs.
4. Dissolution of opposing cell walls and plasma membranes.
5. Elongation of the resulting conjugation tube.

The pairing of sporidia of opposite mating type in *U. violacea* and in *U. maydis* is accompanied by the development of fimbriae extending from the walls of the sporidia. Sporidia bearing over 200 long and innumerable short fimbriae have been seen. The fimbriae are 7–10 nm in diameter, mostly 0·5–5 μm in length, although some are over 10 μm long. They appear to be composed largely of protein. Such fimbriae are strikingly reminiscent of the fimbriae of certain bacterial cells and it is likely that they play an essential role in conjugation (Poon & Day, 1975; Day & Poon, 1975). It has been suggested that, after cell pairing, probably mediated by the secretion of an adhesive 'globular material', the fimbriae connect the cells and act as conduits for a 'mutual and reciprocal exchange of information between the two mating types before the assembly of the conjugation tube' (Day & Poon, 1975).

In *Ustilago maydis*, maize smut, sporidial fusion and the formation of a dikaryon usually occur as in *U. avenae*. It has also been shown that mycelia

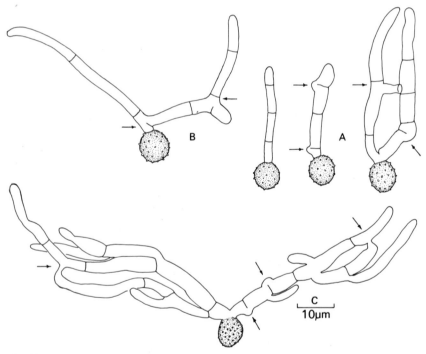

Fig. 276. *Ustilago nuda*. Teliospore germination. A. Three teliospores 20 hours after germination, showing various stages of development of the promycelium. The arrows indicate points where cell fusion has occurred. B. A later stage showing the extension of mycelium from the fusion cells. C. Two-day-old germinating teliospore showing repeated cell fusions.

derived from single sporidia can bring about infection. Such infections give rise to teliospores which form sporidia of two mating types, and it is therefore possible that the sporidia from which these so-called solo-pathogenic lines developed were not homokaryotic: possibly they were dikaryotic or diploid (Christensen, 1932; Whitehouse, 1951; Fischer & Holton, 1957). A more complicated type of heterothallism occurs in this species. Incompatibility may be controlled by two loci, and there is also evidence suggesting the existence of multiple alleles at one of the loci. It is not at present possible to determine whether the second locus is concerned with sexual compatibility or with pathogenicity of the dikaryon (Holliday, 1961a; Christensen, 1963; Halisky, 1965; Holton *et al.*, 1968). A further feature of interest in this fungus is the demonstration of parasexual recombination (Holliday, 1961b). *U. maydis* has been the subject of many genetical investigations (Holliday, 1974).

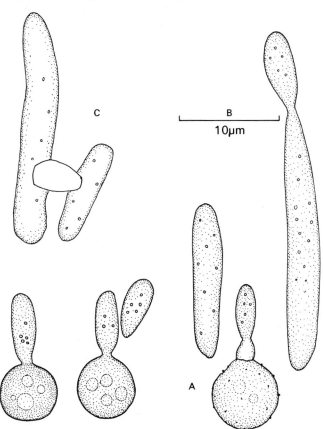

Fig. 277. *Ustilago longissima*. A. Teliospore germination by successive development of sporidia. Note the absence of an extended promycelium. The first-formed sporidia are short. B. Sporidium showing budding. C. Sporidia conjugating. A dikaryotic mycelium arises following conjugation.

In *Ustilago nuda*, the cause of loose smut of wheat and barley (Fig. 272c), there are no sporidia. A promycelium develops on germination and becomes septate, but the dikaryophase is established by fusion of germ tubes derived from the individual uninucleate cells of the promycelium (Fig. 276) (Malik, 1974). Possibly this kind of development is connected with the biology of infection (see below).

Ustilago longissima, the cause of leaf stripe of *Glyceria* spp., has teliospores which, on germination, do not produce an obvious promycelium, but merely a short tube from which sporidia are budded off successively (Paravicini, 1917; Wang, 1934) (Fig. 277).

Spore germination in *Tilletia caries*, the cause of bunt or stinking smut of wheat (Fig. 272E), follows yet another pattern. The young spores are again binucleate, and the two nuclei fuse to form the single diploid nucleus of the mature spore. The fusion nucleus divides meiotically and one or more mitotic divisions follow so that eight to sixteen nuclei are formed. The promycelium is often, but not invariably, non-septate, and from its tip arise narrow curved

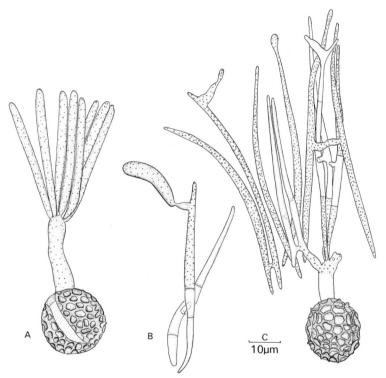

Fig. 278. *Tilletia caries.* A. Germinating teliospore showing a non-septate promycelium and a crown of primary sporidia. B. Two detached primary sporidia showing conjugation. A secondary sporidium has developed from one of the primary sporidia. C. Primary sporidia attached to the promycelium showing conjugation.

uninucleate primary sporidia, corresponding in number with the number of nuclei in the young promycelium (Fig. 278). The primary sporidia conjugate in pairs by means of short conjugation tubes, often whilst still attached to the tip of the promycelium (Fig. 278C). Detached primary sporidia may also conjugate. During conjugation a nucleus from one primary sporidium passes into the other sporidium which therefore becomes binucleate. Each H-shaped pair of primary sporidia develops a single lateral sterigma on which a curved binucleate spore develops (Fig. 278B). This spore is projected violently from the sterigma, and a water-drop has been found associated with the spore shortly before discharge. The violently projected spore is sometimes referred to as the secondary sporidium or secondary conidium. Buller & Vanterpool (1933) have argued that because of its characteristic method of discharge this spore is to be interpreted as a basidiospore, and they use the term primary sterigma for the primary sporidia, and secondary sterigma for the structure

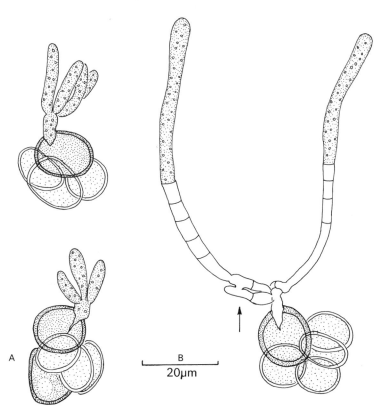

Fig. 279. *Urocystis anemones.* A. Spore balls showing germination. A promycelium develops from a fertile cell, bearing a crown of three or four sporidia. 24 hours. B. Later stage of development. The sporidia have fused (arrow) and a septate mycelium develops from the fusion cell. 48 hours.

subtending the secondary sporidium. It is these spores which bring about infection of the host.

Whilst the teliospores of *Ustilago* and *Tilletia* are unicellular those of *Urocystis* consist of one or more central fertile cells surrounded by a group of thin-walled sterile cells (Fig. 279). In *Urocystis agropyri*, the cause of a leaf-stripe smut of wheat and *Agropyron*, the fertile cell develops a non-septate promycelium into which the fusion nucleus passes, and then undergoes meiosis. A whorl of four sporidia develops at the tip of the promycelium, and these sporidia conjugate in pairs whilst still attached, as in *Tilletia*. After conjugation the two nuclei migrate into an infection hypha which develops from the conjugation tube (Thirumalachar & Dickson, 1949). In *U. anemones* which causes smut of *Anemone* and *Ranunculus* leaves (Fig. 272F) three or four sporidia develop at the tip of a non-septate promycelium. Fusion of the sporidia in pairs precedes the development of branches from their tips (Fig. 279B).

INFECTION

Infection of the host is by means of the dikaryotic mycelium, initiated by sporidial fusion or by fusion of branches of the promycelium. The actual route of infection varies as selected examples will show.

Loose smuts of cereals. The smutted heads of wheat or barley arise on plants infected systemically with *Ustilago nuda*. The embryos of the seeds contain mycelium of the fungus and infection occurs at flowering time. It was earlier thought that the point of entry of the mycelium was the stigma of healthy flowers, but it has been shown that the normal entry point is the young tissue at the base of the ovary (Batts, 1955). The teliospores of *U. nuda* are short-lived, rarely surviving more than a few days under normal conditions. *Ustilago avenae*, the cause of loose smut of oats, can infect at flowering time, but in this case if spores are dusted on to healthy seed, infection of the seedling may occur as it germinates. Spores of this fungus have been shown to be viable after thirteen years.

Covered smuts of cereals. The teliospores of *Tilletia caries* are surrounded by the pericarp of the wheat grain and are not released until the grain is threshed, when they are dusted over the surface of the grain. The teliospores are viable for up to 15 years and, when grain bearing them is sowed, the spores germinate along with the grain, and the coleoptiles of the seedlings are infected following penetration of the germ tube of the secondary sporidium. Infection is systemic, the mycelium growing in the tissues of the shoot. By suitable techniques, it is possible to isolate the parasitic, dikaryotic mycelium from infected host tissues (Trione, 1972). Although infected plants may grow less vigorously than uninfected ones, they show no obvious outward sign of infection until the ears are almost ripe. The glumes of infected heads are pushed wider apart than healthy glumes by the rounded bunted grains which have a bluish appearance (Fig. 272E). Crushed bunt balls have a fishy smell caused by the presence of trimethylamine. For this reason, the disease is sometimes termed 'stinking smut'.

EFFECTS OF INFECTION ON THE HOST

In the case of the loose and covered smuts of cereals, it is obvious that the replacement of the flower or grain by teliospores of the parasite will reduce the yield of the crop. Other less obvious effects also occur. Comparisons of healthy wheat plants with those infected by *Ustilago nuda* have shown that the growth rate of the shoots and roots of infected plants is reduced. The mycelium of the fungus is practically confined to the nodes and the ears, and does not extend into the leaf lamina. The volume of mycelium expressed as a proportion of the total volume of the plant is of the order of 0.01–0.1%, and its effects on the host are surprisingly large. When radioactive $^{14}CO_2$ was fed to leaf laminae in the light, evidence was obtained of translocation of labelled assimilates to the 'sink' formed by the fungus in the ears. Translocation of assimilates was principally in the form of sucrose, but infected areas contained trehalose and the polyols mannitol and erythritol (Gaunt & Manners, 1971a,b,c). Another effect that has been noted is that infection of developing barley grains by loose smut can reduce their absolute weight and quality, but it does not significantly affect their germination.

Infection of meristematic shoot tissues of maize by dikaryotic mycelia of *U. maydis* causes hypertrophy and neoplastic growth of infected tissue, especially young leaves and developing grain. The host nuclei may become polyploid (Callow & Ling, 1973; Callow, 1975). There is no evidence of systemic infection. It is probable that the teliospores survive for several years in the soil (Christensen, 1963).

CONTROL OF SOME CEREAL SMUT DISEASES

Control of loose and covered smuts presents very different problems. Whilst the surface of grain is merely *contaminated* with the spores of covered smuts, in the case of loose smuts the ripe grain is already *infected* by a mycelium within the embryo. Control of covered smuts by means of fungicidal dusts is therefore simple, and it is standard practice for seed grain to be treated by seed merchants in this way. In most countries with well-developed agriculture, bunt of wheat is now a rare disease. The incidence of bunt balls in seed samples sent to the Official Seed Testing Station at Cambridge fell from 12–33% in the period 1921–1925 to 0.2–0.3% in 1955–1957 (Marshall, 1960b). Control of loose smuts at first depended on the discovery by Jensen that the mycelium of *Ustilago nuda* could be killed by hot water treatment which the grain itself could survive. Barley or wheat grain was soaked in cold water for 5 hours, followed by a dip in hot water; 10 minutes at 54 °C for wheat and 15 minutes at 52 °C for barley (Fischer & Holton, 1957). The method, although efficacious, is obviously risky, and has the further disadvantage that the grain must be dried or sowed immediately afterwards. Modern methods of control are chemical, based on the application of systemic fungicides (i.e. fungicides absorbed by the plant and translocated), often in the form of a seed-dusting. Effective compounds are oxathiin derivatives (e.g. carboxin, pyracarbolid, vitavax and plantvax) and benomyl (Brooks, 1972b; Tarr, 1972). The oxathiin derivatives are not phytotoxic except at very high concentration. Their selective action in killing the mycelium of the parasite appears to be associated with inhibition of the enzyme succinate dehydrogenase (Ben-Yephet *et al.*, 1975; White & Thorn, 1975).

Since infection of next season's grain occurs at flowering, one obvious method of control is to inspect crops grown for seed at flowering time and assess the incidence of smutted heads. Only crops which contain fewer than one smutted ear in 10,000 ears are approved for use as seed stocks (Doling, 1966). It is also possible to detect the presence of loose smut mycelium within the embryos by microscopic examination following embryo extraction, and in this way the health of seed samples can be checked (Popp, 1958; Morton, 1961; Laidlaw, 1961). Good correlation has been discovered between estimates of seed-borne infection and incidence of disease in the field (Rennie & Seaton, 1975). A further method of control is to select for cleistogamous flowers. If the flowers do not open, the possibility of embryo infection is reduced, since teliospores cannot penetrate the cereal flower (Macer, 1959). As with rust fungi, the control of smuts by breeding resistant host varieties is complicated by the existence of several physiological races of the pathogen (Johnson, 1960; Halisky, 1965).

ANTHER SMUT OF CARYOPHYLLACEAE.

One of the most curious smut diseases is the anther smut of Caryophyllaceae, in which diseased plants have pollen replaced by purple teliospores. Examination of the fine structure of the teliospore surface from collections on different host species suggests that more than one fungus species is involved (Durrieu & Zambettakis, 1973; Brandenburger & Schwinn, 1974).

When female plants of the dioecious red or white campion (*Silene dioica* and *S. alba*) are infected, the flowers are stimulated to produce anthers, which are absent from healthy female plants. The anthers of the infected females become filled with teliospores, and the ovaries may be poorly developed. Male plants of these hosts are also infected. This phenomenon of induced hermaphroditism has aroused much interest (for references see Fischer & Holton, 1957). It has been suggested that sex-expression depends upon local concentrations of auxin or auxin-inhibitors (Heslop-Harrison, 1957; Baker, 1947), but no evidence has yet been presented to show that hormone or hormone-oxidase levels of flower tissues are significantly affected by infection (Garay & Sagi, 1960). Attempts to extract sex hormones from culture filtrates of the fungus have likewise proved negative (Erlenmeyer & Geiger-Huber, 1935).

Infection of sea campion (*Silene vulgaris* subspecies *maritima*) results in dwarfing, and comparison of extracts from healthy and diseased plants suggests that the dwarfing may be due to a low level of gibberellin in infected plants. In pure culture, the fungus produces indole aceto-nitrile (IAN), and this substance is also present in diseased plants. Apparently the diseased host is unable to convert IAN to indole acetic acid (Evans & Wilson, 1971).

Although the teliospores replace the pollen in infected plants, there is no evidence that the disease is seed-borne. Teliospores are carried by flower-visiting insects. When teliospores are transferred to healthy flowers, mycelium extends beyond the flower into other host tissues and may become systemic. Later-formed flowers are often smutted. Infection of young seedlings, underground shoots and axillary buds can also result in a systemic perennial infection (Hassan & Macdonald, 1971; Evans & Wilson, 1971) (for references see Baker, 1947; Ainsworth & Sampson, 1950).

RELATIONSHIPS OF THE USTILAGINALES

As indicated on p. 483, there are no very good reasons for thinking that the Ustilaginales are related to the Uredinales. Indeed, it is doubted by some whether they are even related to the main stock of Basidiomycotina. The absence of dolipore septa and parenthesomes is one piece of evidence for this view. Moore (1972) has proposed that they should be placed in a new category equivalent in rank to the Basidiomycotina, termed the Ustomycota. It is possible that some of the yeast-like fungi described in the next section are related to Ustilaginales.

BASIDIOMYCETOUS YEASTS

There are some yeast-like organisms which may be related to the Basidiomycotina. Since, for some of them, a perfect state has not been discovered, it may be more appropriate to classify them with the Deuteromycotina (Fungi Imperfecti). The evidence that such organisms are related to Basidiomycotina is of several kinds:

1. Production of ballistospores in some forms – i.e. spores which are violently projected from a sterigma. Kreger-van Rij (1969) has used the term Ballistosporogenous Yeasts for this group.

2. The presence in some species of clamp connections.

3. Life cycles resembling those of other Basidiomycetes, sometimes with dikaryotic states.

4. DNA ratios. The nucleotide composition expressed as the percentage of guanine plus cytosine residues of total DNA (GC content) appears to have taxonomic significance. Storck (1966) found that the GC ratios for different groups of fungi were characteristic:

Ascomycotina	38–54%
Hemiascomycetes	39–45%
Zygomycotina	38–48%
Basidiomycotina	44–63%

Two ballistosporogenous yeasts, *Sporobolomyces roseus* and *S. salmonicolor*, had GC ratios of 50% and 63% respectively (Storck, 1966). Later studies (Storck *et al.*, 1969) gave values ranging from 51·5–65·0% for six species of *Sporobolomyces*.

5. Wall structure. The fine structure of the wall of yeasts with Basidiomycetous affinities is lamellate and, in this respect, different from Ascomycetous yeasts. There are also differences in detail of bud formation (Kreger-van Rij & Veenhuis, 1971).

Sporobolomycetaceae

The ballistosporogenous yeasts are classified into a single family, Sporobolomycetaceae. Phaff (1970) includes three genera, *Bullera*, *Sporidiobolus* and *Sporobolomyces*. The group is sometimes known as the mirror yeasts, because the discharge of ballistospores from the surface of a colony creates a reflection of the outline of the colony. *Sporobolomyces* and *Bullera* usually grow in a yeast-like manner and are distinguished largely on the basis of colour; *Sporobolomyces* forms pink colonies, whilst *Bullera* is white. *Sporidiobolus* may exist in a yeast-like or mycelial phase, and the hyphae may bear clamp-connections and chlamydospores.

Sporobolomyces (Figs. 280–282)

Sporobolomyces roseus is abundant on moribund vegetation, and can be isolated readily by placing senescent plant material (e.g. a grass leaf) over nutrient agar. Scanning electron microscope studies show the fungus to be a common member of the flora of the leaf surface (phylloplane flora). When present in high concentration, the yeast-like cells may adhere to the leaf surface by sheets of mucilage. There is no evidence that the growth of the yeasts causes corrosion of the leaf cuticle (Bashi & Fokkema, 1976). The growth of *S. roseus* on wheat leaves is stimulated by high humidity, and by exogenous nutrients such as pollen grains (Bashi & Fokkema, 1977). There is interest in the suggestion that *Sporobolomyces* on leaf surfaces may compete with foliar pathogens, and that biological control of the pathogens might be possible. Ballistospores projected from the surface of the leaf colonise the agar, and yeast-like colonies develop by budding. Within a few days, further ballistospores are formed. Spores of Sporobolomycetaceae are frequent in the air, especially on warm nights in the summer, and concentrations of hyaline ballistospores probably belonging to *Sporobolomyces* and *Tilletiopsis* may reach a concentration of 10^5–$10^6/m^3$ (Last, 1955; Pady, 1974). These high concentrations of *Sporobolomyces* ballistospores are a cause for concern because it has been shown that the fungus is a respiratory allergen (Evans, 1965). Many members of the Sporobolomycetaceae appear to be associated with lesions caused by other plant parasites. *Sporidiobolus johnsoni* has been isolated from rusted leaves of raspberry. Increased numbers of *Sporobolomyces* colonies are also associated with damage of foliage from nematodes and mites (Last & Deighton, 1965).

The budding phase of *Sporobolomyces roseus* is uninucleate, and nuclear division occurs at about the time of bud formation (Buller, 1933). When spores are formed, a conical sterigma develops vertically and bears a sausage-shaped spore into which a daughter nucleus passes (Fig. 280). The spore is thrown for a distance of about 0·1 mm, apparently by the same mechanism as in many Basidiomycetes (Müller, 1954). Electron micrographs of detached spores show a cylindrical projection with a torn rim (Ingold, 1971). A single sterigma may form a second or even a third spore, and occasionally two sterigmata may arise from one cell. There are reports of nuclear fusion and reduction division in *Sporobolomyces* (see Laffin & Cutter, 1959). In *S. roseus* and *S. salmonicolor*, a true mycelium has been reported in addition to the yeast-like phase.

Further evidence that *Sporobolomyces* is a Basidiomycete is based on the

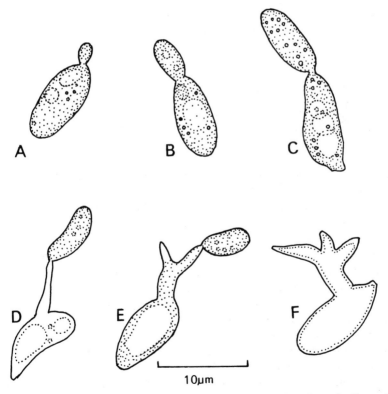

Fig. 280. *Sporobolomyces roseus*. A–C. Various stages in the budding of cells. D. Cell bearing a sterigma and a ballistospore. E. Cell showing two sterigmata. F. Cell with three sterigmata.

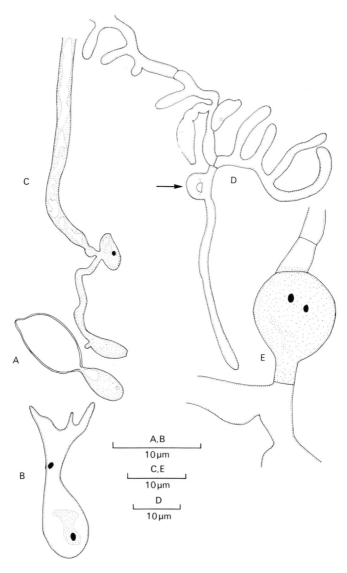

Fig. 281. *Sporobolomyces odorus*. A. Budding cell. B. Cell with sterigmata. C. Conjugation of haploid cells and initiation of dikaryotic hyphae. D. Young dikaryotic hyphae, the original conjugants obscured. The arrow indicates the first clamp connection. E. Late stage of chlamydospore development; the paired nuclei are visible. (After Bandoni *et al.*, 1971.)

mating behaviour of certain isolates. Isolates of *S. odorus,* when paired, showed bipolar heterothallism. Following conjugation between compatible cells, a dikaryotic mycelium bearing clamp connections developed, and, on the dikaryon, globose binucleate chlamydospores (Fig. 281, Bandoni *et al.,* 1971). This type of life cycle is very similar to that of *Sporidiobolus.* According to van der Walt & Pitout (1969) and van der Walt (1970b), *Sporobolomyces salmonicolor* cells may be haploid or diploid. The diploid cells are larger. Both types of cell are capable of budding, and of forming ballistospores. In diploid, but not in haploid, cultures, thick-walled, lipid-rich chlamydospores develop. When such cells were washed and transferred to non-nutrient agar they germinated, usually to produce a short promycelium which reproduced by budding. On the basis of these observations, van der Walt (1970b) has proposed that *Sporobolomyces salmonicolor* has a life cycle of the type shown in Fig. 282. Since the genus *Sporobolomyces* is an 'imperfect' genus – i.e. based on a non-sexual state – he has transferred *S. salmonicolor* to the new genus *Aessosporon.* He has suggested that the fungus is a member of the Tilletiaceae (Ustilaginales, see p. 483). He has termed the chlamydospores formed in diploid cultures teliospores. Whether all members of the Sporobolomycetaceae are related to Ustilaginales is an open question. Others have suggested relationships with different groups of Basidiomycotina – e.g. Dacrymycetaceae (Bulat, 1953), Tremellaceae (Olive, 1952; Martin, 1952).

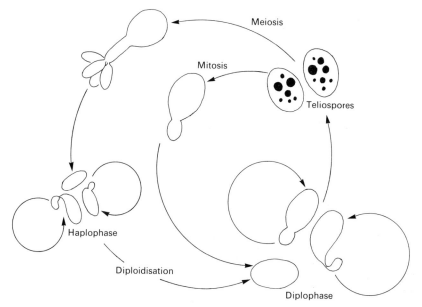

Fig. 282. The postulated life cycle of *Aessosporon salmonicolor* (after van der Walt, 1970b).

Non-ballistosporogenous yeasts

There are other non-ballistosporogenous yeasts with Basidiomycete affinities, possibly with Ustilaginales. They include *Rhodosporidium*, a genus known from terrestrial and marine habitats, which is the perfect state of several species of *Rhodotorula* (Banno, 1967; Fell *et al.*, 1973; Fell, 1976), and *Leucosporidium*, which is the perfect state of several species of *Candida* (Fell *et al.*, 1969). Another yeast-like Basidiomycete is *Cryptococcus neoformans* which causes a serious disease in man and animals (European blastomycosis, cryptococcosis). The fungus may exist saprophytically in pigeon droppings or in bird nesting areas. The perfect state of this fungus is *Filobasidiella*, which is unusual in producing basidiospores in chains (Kwon-Chung, 1975, 1976). The genus has been classified in the Ustilaginales, but has the unusual character for that order in possessing dolipore septa.

UREDINALES

The popular name for Uredinales, rust fungi, refers to the reddish-brown colour of some of the spores. Ecologically, rusts are biotrophs, i.e. haustorial parasites of Angiosperms, Gymnosperms and Pteridophytes. Until recently, only limited success has resulted from attempts to grow them in pure culture or in tissue cultures of their hosts (for references, see Scott & MacLean, 1969; Wolf, 1974; Coffey, 1975). It was therefore a surprise when it was shown that *Puccinia graminis*, a well-known rust fungus causing black stem rust of cereals and grasses, could be grown on Czapek-Dox, yeast extract agar (Williams *et al.*, 1966). By 1973, nine species of rust belonging to four different genera had been grown in artificial culture, and it is presumably only a matter of time until more are grown.

Rusts are common on wild and cultivated plants, often causing severe diseases. About 4,000 species are known, classified into 100 genera (Cummins, 1959; Gäumann, 1959; Wilson & Henderson, 1966; Laundon, 1973). Many have complex life cycles involving five distinct types of spore, although some show modifications in which one or more of these spore stages may be absent (Petersen, 1974). Forms with all five spore stages are described as **macrocyclic**, and it is generally believed that the macrocyclic condition is the primitive one, from which forms with abbreviated life cycles have evolved.

The different types of spore found in the rust life cycle have been given various names. They occur in pustules or sori which have also received various names. I have followed the nomenclature of Hiratsuka (1973, 1975) for the names of spore types and sori. These are defined in tables 7 and 8 based on Hiratsuka (1975). It is customary to assign a Roman numeral 0, I, II, III or IV to the distinct spore types, and these numbers provide a convenient shorthand for describing the range of spores found in a given rust.

Another distinctive feature of many rusts is that they complete their life

Table 7. *Generally accepted terminology of rust fungi*

Spore state	0	I	II	III	IV
Sorus	spermogonium pycnium	aecium aecidiosorus aecidium	uredinium uredosorus uredium	telium teleutosorus	basidium[a] probasidium promycelium
Spore	spermatium pycniospore	aeciospore aecidiospore	urediniospore uredospore urediospore	teliospore teleutospore	basidiospore sporidium

[a] What Hiratsuka terms a basidium is, strictly, a metabasidium (see p. 398).

Table 8. *Definition of spore states of rust fungi*

TELIOSPORES: basidia[a]-producing spores.

BASIDIOSPORES: monokaryotic spores produced on basidia[a] usually as the result of meiosis.

SPERMATIA: monokaryotic gametes.

AECIOSPORES: nonrepeating vegetative spores produced usually from the result of dikaryotization, which germinate to initiate dikaryotic mycelium, thus usually associated with spermogonia.

UREDINIOSPORES: repeating vegetative spores produced usually on dikaryotic mycelium.

[a] What Hiratsuka terms a basidium is, strictly, a metabasidium (see p. 398).

cycle on two different host plants which are usually quite unrelated to each other. For example, *Puccinia graminis* attacks grasses and cereals, but also grows on *Berberis* spp. Rusts with two different host plants are **heteroecious**. Others complete the entire life cycle on a single host, i.e. *P. menthae*, mint rust, and are described as **autoecious**.

The detailed classification of rusts into families and genera need not concern us, but is briefly considered on p. 513.

The classical example of a macrocyclic heteroecious rust is *Puccinia graminis*.

Puccinia (Figs. 283–187)

Puccinia graminis is the cause of black stem rust of cereals and grasses. In Britain it is not a severe parasite, possibly because the summer temperatures are not high. Occasional epidemics of the disease on cereals in the south and west of Britain probably originate from spores which have blown across the

Fig. 283. *Puccinia graminis*. A. Wheat straw showing telia as black raised pustules. B. Wheat leaf showing uredinia which appear as reddish-brown powdery masses.

sea from Spain and Portugal, and there is no strong evidence that the fungus overwinters on cereals in Britain (Ogilvie & Thorpe, 1966). The most common cereal host is wheat; occasionally oats, barley and rye. Grass hosts include *Agropyron, Agrostis* and *Dactylis*. A symptom of infection on wheat leaves (Fig. 283B) is the appearance of brick-red pustules, uredinia, between the veins. In *P. graminis*, the uredinia are mainly distributed on the internodes and leaf sheaths. The uredinia contain stalked, one-celled spores or urediniospores

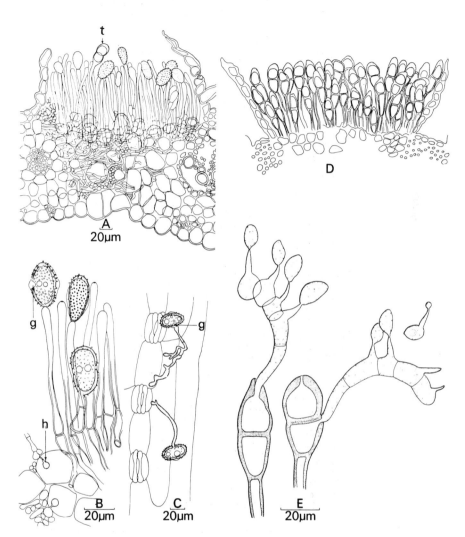

Fig. 284. *Puccinia graminis*. A. T.S. wheat leaf through a uredinium. The stalked unicellular urediniospores are protruding through the ruptured host epidermis. A bicellular teliospore (t) is also present. B. Higher-power detail of urediniospores. Note the germ pores (g) and the haustoria (h) in the host cells. C. Germination of two urediniospores on wheat leaf. Note the directional growth of the germ tubes towards a stoma. D. T.S. leaf sheath through a telium. The stalked teliospores are projecting through the ruptured epidermis. Drawing to same scale as A. E. Germination of teliospores to form metabasidia bearing sterigmata and basidiospores. One basidiospore is giving rise to a secondary spore.

which burst through the host epidermis (Fig. 284A,B). The urediniospores are dikaryotic and arise from a dikaryotic mycelium which is intercellular, forming spherical intracellular haustoria. Electron micrographs of haustoria (Ehrlich & Ehrlich, 1963) show that they are not surrounded by the host plasma membrane, nor do they lie freely in the host cytoplasm. They are surrounded by a distinct encapsulation, possibly consisting of metabolic products either of fungal or of host origin. Small vesicles are also found in the host cytoplasm surrounding haustoria. The haustoria are often closely appressed to the host cell nucleus, which may break down (Manocha & Shaw, 1966). Experiments have been done in which urediniospores were radioactively labelled by growing the parasite on wheat leaves exposed to $^{14}CO_2$. When such labelled urediniospores were used to infect fresh leaves, the radioactivity accumulated in organelles of the fresh host such as the chloroplasts. There was no heavy accumulation of label around the haustoria. These observations indicate a two-way interchange of materials between fungus and host (Ehrlich & Ehrlich, 1970).

In contrast with many Basidiomycetes, the septa in the intercellular mycelium contain simple pores (Ehrlich *et al.*, 1968). The urediniospores have a spiny wall. Near the middle of the spore, the wall has four thinner areas or germ pores. The urediniospores are detached by wind and blown to fresh wheat leaves upon which they germinate by extruding a germ tube from one of the germ pores (Fig. 284C). The germ tube usually penetrates the leaf through a stoma, and beneath the stoma the tip of the germ tube expands to form a sub-stomatal vesicle from which branches develop to give rise to intercellular mycelium and haustoria. Within about 7–21 days of infection, a new crop of urediniospores is formed and this can cause a rapid build-up of infection within a crop. A single uredinium may contain from 50,000 to 400,000 spores, and there may be four to five generations of urediniospores in the growing period of a wheat crop. Later in the season, a second kind of spore may be visible along with the urediniospores. These spores are thicker-walled and two-celled (Fig. 284A,D). They are called teliospores. Eventually, pustules are formed containing teliospores only. These pustules are termed telia, and appear as black raised streaks along leaf-sheaths and stems of infected plants (Fig. 283A). The individual cells of the teliospores each contain a pair of nuclei and, during the later stages of development, these two nuclei fuse to form a diploid nucleus. The teliospores represent the overwintering stage and only develop further after a period of maturation corresponding to winter dormancy. They survive the winter on stubble, and the following spring (April or May) they germinate. Each cell of the teliospore emits a curved, four-celled metabasidium, and each cell of the metabasidium bears a single basidiospore on a sterigma (Fig. 284E). Prior to the development of the basidium, the diploid nuclei of the teliospore divide meiotically to give four haploid nuclei, one of which enters each basidiospore. The basidiospores are projected from the basidium, but are incapable of infecting wheat. Instead, they infect young

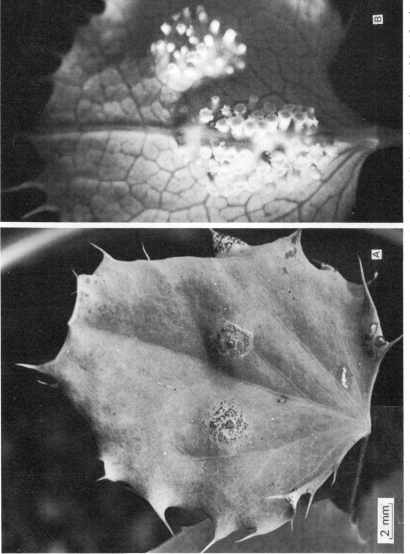

Fig. 285. *Puccinia graminis*. A. Pycnial pustules on the upper surface of a leaf of *Berberis vulgaris*. Note the drops of nectar. B. Aecia on the lower side of a *Berberis* leaf. The outer frilly layer is the white peridium, within which is a mass of orange-coloured aeciospores.

leaves of the alternate host, barberry, *Berberis vulgaris*. Infection of a *Berberis* leaf by a single basidiospore results in the formation of a haploid mycelium which appears as a yellowish, circular pustule (Fig. 285A). On the upper surface of the leaf, penetrating the epidermis, several yellowish, flask-shaped structures termed pycnia develop. The mouth of the flask is lined by a bunch of unbranched, tapering, pointed, orange-coloured hairs, the periphyses, and among the periphyses are thinner-walled, branched hyphae, the flexuous hyphae. Lining the body of the pycnium are tapering cells which give rise to minute uninucleate pycniospores exuded through the mouth of the pycnium and held in a sticky, sweet-smelling drop of liquid by the periphyses (Figs. 285A; 286A). According to Hughes (1970), the tapering cells from which pycniospores arise in rusts are phialides, but in *Puccinia sorghi*, Rijkenberg & Truter (1974) have shown that the detachment of successive pycniospores gives rise to a series of annellations in the neck of the sporogenous cell, and they therefore regard this cell as an annellophore (annellide). Within the mesophyll of the berberry leaf, the haploid mycelium gives rise to several spherical structures or proto-aecia. The proto-aecia are mostly made up of large-celled pseudoparenchyma, but along the upper wall is a crescent-shaped mass of smaller, denser cells.

Single haploid pustules are incapable of further development, unless cross-fertilisation occurs. The sweet-smelling nectar containing pycniospores attracts insects which feed on the nectar. Insects visiting several pustules

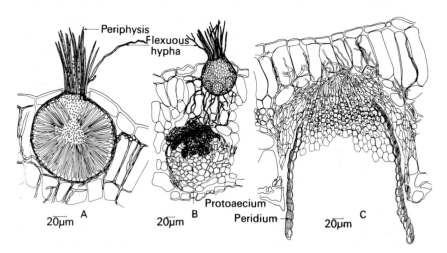

Fig. 286. *Puccinia graminis*. A. T.S. pycnium on leaf of *Berberis vulgaris*. The pycnium is penetrating the upper epidermis. The wall of the pycnium is lined by tapering cells which give rise to pycniospores. B. T.S. leaf of *Berberis* showing a pycnium and a proto-aecium which has become dikaryotised. This is shown by the presence of binucleate cells. C. T.S. leaf of *Berberis* showing an aecium in section. The aecium has burst through the lower epidermis of the host leaf. Note the chains of cells consisting of alternate large and small cells. The large cells are the aeciospores.

transfer pycniospores from one to another. The haploid pustules are of two mating types (+ and −), and if a + pycniospore is brought close to a − flexuous hypha, a short germ tube is formed which anastomoses with the flexuous hypha (Craigie, 1927a,b; Buller, 1950). Nuclear transfer from the pycniospore to the flexuous hyphae then occurs, and migration and multiplication of the introduced nucleus follows (Craigie & Green, 1962). This results in the dikaryotisation of the original haploid pustule, and about 3 days after transfer of pycniospores, binucleate cells become clearly visible in the deep staining tissue which forms the roof of the proto-aecium (Fig. 286B). The binucleate cells now give rise to chains of cells which are also binucleate, but composed of alternately long and short cells. The longer cells enlarge and become aeciospores, but the short cells become crushed and flattened as the spore chain develops (Fig. 286C). During the development of the spore chain, the large pseudoparenchymatous cells of the proto-aecium are also crushed and pushed aside. Surrounding the chains of spores is a specially differentiated layer of cells homologous with the spore chain, whose outer walls are thick and fibrous. This layer of cells forms a clearly defined border or peridium surrounding the spores. Eventually the peridium and spore chain burst through the lower epidermis. The spores are visible as orange-coloured cells enclosed by the white cup-like peridium (Fig. 285B). The cup-like sori are termed aecia and several of them are usually clustered together beneath a pustule, so that this stage is popularly known as the cluster-cup stage. In a section through the centre of a group of young aecia, it is usually possible to find the pycnia penetrating the upper epidermis and the aecia penetrating the lower. The aeciospores are violently projected from the end of the spore chain by rounding-off of the flattened interface between adjacent spores (Dodge, 1924). The dikaryotic aeciospores are incapable of infecting barberry, but can germinate on wheat leaves and penetrate them, giving rise to infections on which urediniospores shortly develop.

The life cycle of *P. graminis* thus occurs on two hosts. On *Berberis*, the haploid infection, following spermatisation, gives rise to dikaryotic aeciospores, which can only infect Gramineae. On the grass or cereal host, the dikaryotic mycelium gives rise to two kinds of spore: urediniospores, which can be thought of as dikaryotic conidia; and teliospores, which are not dispersed but are the organs within which nuclear fusion and meiosis occur (Fig. 287).

The effects of infection on cereal hosts are increased transpiration and respiration, and decreased photosynthesis, resulting in a marked deterioration in quality and quantity of grain (Eversmeyer & Browder, 1974; Buchenau, 1975).

CONTROL OF STEM RUST

The control of black stem rust is possible by several means:

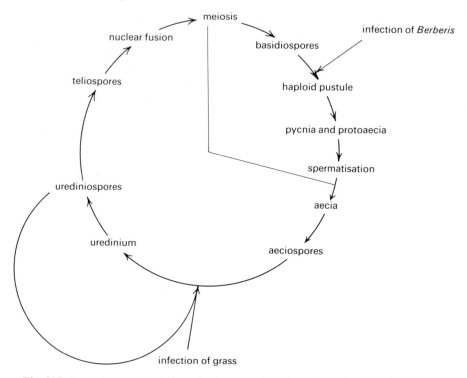

Fig. 287. *Puccinia graminis*. Life cycle diagram. The life cycle can be divided into two parts, a haploid stage on *Berberis*, and a dikaryotic stage initiated by the formation of aecia on *Berberis* and continued on the grass host. The haploid part of the life cycle is included in the smaller sector, and the dikaryotic part is included in the larger sector of the circle.

1. Fungicide application

The application of fungicides to cereal crops can result in control of the disease but, until systemic fungicides were discovered, the effects of the earlier available fungicides were too short-lived or they were too costly to be economically practicable (Rowell, 1968). The development of systemic fungicides such as oxycarboxin holds promise for future control (Rowell, 1973, 1976). In order for the timing of fungicide applications to be most effective, it is necessary to understand the conditions under which severe epidemics of the disease develop. Considerable advances in the understanding of the epidemiology of rust diseases have been made, and it is now possible to forecast the likely incidence of rust epidemics (van der Plank, 1963, 1975; Eversmeyer *et al.*, 1973).

2. Eradication of the alternate host

Before the connection between *Berberis* and wheat rust was established (by de

Bary, 1861–1865), the incidence of the disease in wheat growing close to barberry had been noted and legislation had been introduced (in Rouen in 1660, and in Massachusetts in 1755 according to Large, 1958) for the eradication of barberry. This method is only efficacious in areas where the urediniospores cannot survive the winter, but in areas with a mild winter, urediniospores may survive and bring about infection the following spring.

Another difficulty which renders barberry eradication ineffective is the long-distance transport of urediniospores already mentioned. In the U.S.A., clouds of urediniospores have been tracked from the Mississippi delta into the northern states and Canada, travelling distances of up to 2,000 miles (Craigie, 1945; Stakman & Christensen, 1946; Stakman & Harrar, 1957). Infection of wheat in Canada is thus from inoculum which has developed earlier, to the south, in the U.S.A. Towards the end of the season there may be reversal of the flow of air, so that wheat in Texas and Mexico may be infected from urediniospores derived from northern areas (Pady & Johnston, 1955). Similar long-distance transport of urediniospores has been reported from Europe and India (Johnson *et al.*, 1967). It has even been suggested that the urediniospores of the coffee leaf rust fungus *Hemileia vastatrix* may have been carried by wind from Africa to Brazil (Bowden *et al.*, 1971). Urediniospores can retain their viability for several months given suitable conditions of temperature and humidity.

3. Breeding for resistance

Different grasses and cereals react in different ways to a given strain of *Puccinia graminis* and, on the basis of host susceptibility, it is possible to classify *Puccinia graminis* into six host-specific forms (or *formae speciales*) (Eriksson & Henning, 1894):

1. *P. g.* forma specialis *tritici* on wheat.
2. *P. g.* f. sp. *avenae* on oats and certain grasses.
3. *P. g.* f. sp. *secalis* on rye and certain grasses.
4. *P. g.* f. sp. *agrostidis* on *Agrostis*.
5. *P. g.* f. sp. *poae* on *Poa*.
6. *P. g.* f. sp. *airae* on *Deschampsia (Aira) caespitosa*.

It is possible to produce hybrids between some of these *formae speciales*, e.g. between *P. graminis tritici* and *P. graminis secalis* (Green, 1971b). The hybrid aeciospores do not result in very virulent host reactions. Green has argued that the hybrids resemble a more primitive form of *Puccinia graminis* with low virulence and wide host range, and that evolution in stem rust of cereals is progressing from low virulence and wide host range to high virulence and reduced host range.

Slight but significant differences in dimensions of urediniospores and teliospores from the different hosts occur. For example, the urediniospores of f. sp.

tritici are about 30×18 μm, whilst those of f. sp. *poae* are about 21×14 μm (Batts, 1951). Within a single variety of the rust fungus, further specialisation is found. If the spores of a strain of *P. graminis tritici* are inoculated onto a range of wheat varieties, these hosts will differ in their response. Some may prove to be resistant, others highly susceptible, whilst yet others may be intermediate in their reaction. Spores from a second source may give an entirely different pattern of response. Using the reactions of differential wheat varieties, it has proved possible to classify strains of *P. graminis tritici* into

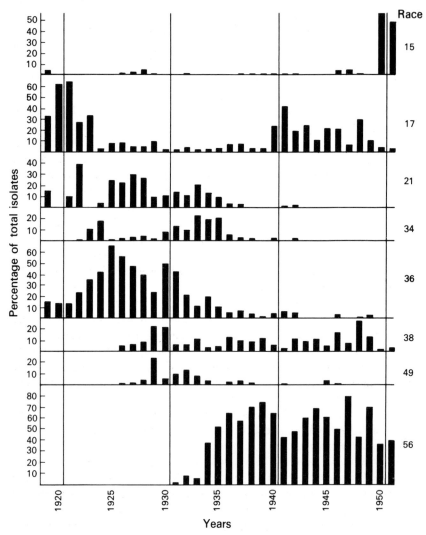

Fig. 288. Diagrammatic representation of the distribution, in Canada, of eight physiological races of wheat-stem rust during the period 1919–51 (after Johnson, 1953).

over 300 physiological races. The existence of this large range of rust races complicates the task of the plant breeder, but fortunately not all these races are prevalent in an area at any one time, so that the task of breeding resistant varieties is practicable and profitable (Johnson, 1953, 1961). The frequency of different rust races varies over the years, reflecting changes in the wheat varieties planted (see Fig. 288). A new race may suddenly build up in frequency, as shown by the appearance of Race 15 (actually a sub-race or biotype referred to as Race 15B) in 1950 in Canada. The appearance of new races presents problems to plant breeders, and it is clear that the task of breeding resistant varieties is a continuing one.

The original scheme for using host differentials was devised by Stakman & Levine (1922), who used twelve hosts ('standard differentials') to discriminate between the physiologic races ('standard races'). The standard differentials frequently failed to reveal important changes in the virulence of the rust population, and this has led to the use of supplementary differentials, which makes possible the discrimination of further physiologic races. This, in turn, has created difficult problems in numbering the rust races, so that quite cumbersome code numbers, e.g. 15B-1LX (Can.), came to be used (Green, 1971a). A more rational way of identifying the physiologic races is to base their classification on the resistance genes of the host plant. The gene-for-gene concept (see Flor, 1971) states that 'for each gene that conditions reaction in the host, there is a corresponding gene in the parasite that conditions pathogenicity'. Using the wheat variety Marquis, a number of 'single gene lines' carrying resistance genes have been identified. The resistance genes have been given the symbols $Sr1$, $Sr2$ . . . etc. Some twenty-five Sr genes in wheat for resistance to stem rust have been identified (van der Plank, 1975; Day, 1974). The origins of the large number of rust races are:

(a) *Recombination during sexual reproduction.* New races of rust are often found adjacent to barberry bushes; and, following inoculation with a single race, a number of variants may be found among the aeciospore progeny.

(b) *A mechanical mixing of nuclei of different races* may occur if urediniospores from the different strains germinate sufficiently close together on a susceptible host for anastomosis of germ tubes or hyphae to occur. Nuclei of different origin may then come together in a common cytoplasm (i.e. to form a distinct heterokaryon), and may be associated in new urediniospores. It has been found that races with distinctive pathogenicity may arise in this way (for references, see Ellingboe, 1961).

(c) *Parasexual recombination.* When two different races were inoculated simultaneously onto a susceptible wheat variety, at least fifteen different races were identified in the urediniospore progeny, which is more than would be expected from simple nuclear re-assortment (Bridgmon, 1959). Similar conclusions have been reached by Watson & Luig (1958), who used the term somatic hybridisation to describe the phenomenon. They have also shown that somatic hybridisation occurs between the two *formae speciales tritici* and *secalis* (Watson & Luig,

1959). One possible explanation is that parasexual recombination has occurred, but the possibilities of gene suppression in the parental strains or of cytoplasmic effects cannot be excluded (Watson, 1957). Another explanation involves the exchange of whole chromosomes between nuclei in the dikaryotic mycelium during synchronous nuclear division (Hartley & Williams, 1971). In *Puccinia graminis*, the haploid chromosome number $n = 6$, and, assuming that the rust is heterozygous for at least one locus on each chromosome, a given dikaryon with a large number of nuclei could carry as many as thirty-two different genotypes. If a new dikaryon were initiated by mixed inoculation of spores from different physiologic races, any of the thirty-two '+' nuclei of one strain could be associated with any one of the thirty-two '−' nuclei of the other strain, creating 2,048 possible new combinations.

(d) *Mutation.* Several examples are known where mutation has occurred giving rise to progeny with increased virulence (Stakman *et al.*, 1930; Day, 1974).

When the urediniospores of *P. graminis* germinate on an unsuitable (i.e. incompatible) host, a variety of reactions occur (Hooker, 1967). In some cases, the host cells die immediately after infection, and the only externally visible signs of infection are small flecks, which are islands of dead tissue. Because the fungus can only grow on living cells, this reaction limits the extension of the infection. The phenomenon is termed *hypersensitivity*. It is not confined to rusts but occurs with other biotrophic pathogens. Although there is evidence that host cell death may result from the production of diffusible toxins in *P. graminis tritici* and other rusts, it is likely that cell death may have more than one cause (Littlefield, 1973; M. C. Heath, 1976).

As already stated, *P. graminis* is not normally a troublesome pathogen in Britain. The most harmful rust on wheat in Britain is *P. striiformis* (= *P. glumarum*) commonly known as yellow rust because of its bright yellow uredinia. In America it is known as the stripe rust. No aecial host has yet been discovered. The disease can appear very early in the season, and it is likely that the fungus can overwinter as mycelium on winter cereals infected in the previous season. *Puccinia recondita* f. sp. *tritici*, brown rust or leaf rust, is also common on wheat, but is relatively harmless in Britain. In America it is of considerable significance (Chester, 1946). The aecia which grow on *Thalictrum* and *Isopyrum* in Europe and Asia have not been found in Britain. Other heteroecious rusts with life cycles resembling *Puccinia graminis* are *P. caricina* with uredinia and telia on *Carex* and pycnia and aecia on *Urtica*; and *P. poarum* with uredinia and telia on *Poa* and pycnia and aecia on *Tussilago*. This species is unusual in having two aecial generations in one year (Wilson & Henderson, 1966). *Puccinia menthae* also has all five spore stages, but is autoecious (i.e. has no alternate host) and is systemic in the aecial state, but the uredinia and telia are associated with discrete mycelia.

The prefix 'eu' is sometimes attached to the generic name of macrocyclic rusts which have a complete life cycle, as a descriptive epithet, i.e. *P. graminis* is a *eu-Puccinia*. Many variants from this type of life cycle are known. One of the best-known is shown by the thistle rust *P. punctiformis* (syn. *P. suaveolens*, *P. obtegens*). This is a systemic autoecious rust attacking *Cirsium arvense*. In spring infected plants are clearly distinguishable by their yellowish appearance and appressed leaves, and by the strong sweet smell associated with numerous pycnia which develop all over the infected shoots. The infections develop from a mycelium perennating in the rootstock. Although the overwintering mycelium is dikaryotic, the mycelium from which the pycnia develop is haploid, and segregation of the two nuclei of the dikaryon presumably occurs as the new shoots are infected. Transfer of pycniospores to compatible flexuous hyphae results in dikaryotisation, but the resulting structures which develop do not resemble normal aecia, but resemble uredinia, and form chocolate-brown masses on the leaves. These sori are best regarded as aecia, and have been termed uredinoid aecia: the term primary uredinium is sometimes used, but it is best avoided (Wilson & Henderson, 1966). The spores from the aecia can infect healthy thistles and normal uredinia develop on these infections. Later, teliospores develop (Buller, 1950; Menzies, 1953). Rusts with uredinoid aecia are brachycyclic (i.e. *P. punctiformis* is a *brachy-Puccinia*). Infected shoots of thistles have a higher endogenous gibberellin content than healthy plants (Bailiss & Wilson, 1967).

Another modification of the life cycle involves the absence of urediniospores. This kind of life cycle is found in heteroecious forms in *Gymnosporangium*, and amongst autoecious forms in *Xenodocus carbonarius*. Rusts showing this kind of life cycle are said to be *demicyclic*.

Some rusts have telia, with or without pycnia, and are said to be *microcyclic*. A common example is *Puccinia malvacearum*, hollyhock rust, which attacks Malvaceae, especially *Althaea* and *Malva*. In this rust only teliospores and basidiospores are known. The teliospores germinate readily, and a single basidiospore can form an infection on which telia arise, i.e. this species is homothallic. The basidiospores give rise to a haploid mycelium with uninucleate segments, and the dikaryotic condition arises shortly before the development of teliospores, either by cell fusions or by cell division not accompanied by the formation of septa (Ashworth, 1931; Buller, 1950). Buller (1950) has argued that all rusts lacking pycnia should prove to be homothallic.

There are other variations in the life cycles of rusts. It is generally believed that the heteroecious macrocyclic forms represent the primitive condition, and that the autoecious condition arose later in evolution. It is also believed that forms with shorter life cycles arose during the coarse of evolution from macrocyclic ancestors. For a discussion of the evidence for these beliefs see Buller (1950) and Jackson (1931). See also Petersen (1974) and Saville (1976).

OTHER RUST FUNGI

Wilson & Henderson (1966) have classified British rusts into three families.

Pucciniaceae

In this family the teliospores are stalked. In *Puccinia* the teliospores are two-celled, whilst in *Uromyces* they are one-celled (Fig. 289A). The separation of these two genera is, however, probably unjustified, and there is evidence that they are very closely related. Common species of *Uromyces* are *U. ficariae*, a microcyclic species forming brown telia on *Ranunculus ficaria*, *U. muscari* on *Endymion non-scriptus*, and *U. dactylidis* with teliospores and urediniospores on *Dactylis*, *Festuca* and *Poa*, and aecia on *Ranunculus* spp. *Triphragmium* has three-celled stalked teliospores (Fig. 289B). The commonest species is *T. ulmariae*, a brachy-form with large bright orange uredinoid aecia on *Filipendula ulmaria*.

In *Phragmidium* the teliospore has several cells (Fig. 289C). *Phragmidium violaceum* is easily recognisable on leaves of *Rubus fruticosus* agg. by the violet leaf-spots it causes. All species of *Phragmidium* are autoecious and confined to Rosaceae.

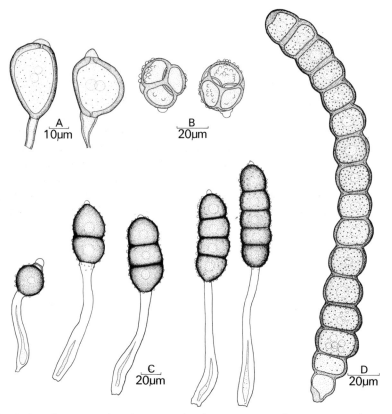

Fig. 289. Teliospores of various rust fungi. A. *Uromyces ficariae*. B. *Triphragmium ulmariae*. C. *Phragmidium violaceum*. D. *Xenodocus carbonarius*.

Xenodocus has teliospores consisting of long chains of up to about twenty dark cells (Fig. 289D). No uredinia occur. The only British species is *X. carbonarius* which forms bright orange-red aecia on *Sanguisorba officinalis. Gymnosporangium* forms telia on *Juniperus.* On this host, the mycelium is perennial and is associated with swelling of the branches. In spring, the swollen shoots produce *Clavaria*-like projections which contain numerous thin-walled, two-celled teliospores. In *G. clavariiforme,* the basidiospores infect *Crataegus* which then bears cylindrical aecia (roestelia).

Coleosporiaceae

Here the teliospores are sessile and form a layer of cells often united laterally. The spores have thin side walls. Before germination, the whole spore becomes divided into four cells arranged in a row, each of which produces a long sterigma bearing a single basidiospore. *Coleosporium tussilaginis* grows on a range of hosts, forming teliospores on *Tussilago, Senecio* and *Melampyrum,* and aecia on pine needles.

Melampsoraceae

Here the unicellular teliospores are also sessile and often form a subepidermal crust. Germination is by an external basidium of the usual type. The aecia lack peridia so that they are diffuse instead of cup-shaped. Such diffuse aecia are termed caeomata (sing. caeoma). Many members of the family grow on ferns and conifers. *Melampsora lini* is an autoecious rust common on *Linum cathar-ticum.* A variety *liniperda* is parasitic on cultivated flax, *Linum usitatissimum.* Some other species of *Melampsora* are parasitic on *Salix,* whilst *M. populnea* forms urediniospores and teliospores on *Populus,* and pycnia and aecia on *Mercurialis perennis. Melampsoridium betulinum* is abundant on leaves of *Betula* where it forms urediniospores and teliospores. The alternate host is *Larix.*

5

Deuteromycotina (Fungi Imperfecti)

Very many fungi are known only in the asexual or mycelial state: their perfect states are either unknown, or may possibly be entirely lacking. Although the lack of sexual reproduction would appear to preclude the normal processes of recombination associated with a sexual cycle, an alternative in the form of parasexual reproduction may provide the mechanism whereby recombination can take place (see p. 252). Thus it is that many very successful plant pathogens, such as *Fusarium* and *Verticillium*, can rapidly respond to selection when faced with host cultivars which have been bred for resistance, or that many of the world's most ubiquitous moulds, such as species of *Penicillium* or *Aspergillus*, may lack perfect states.

The classification and nomenclature of Deuteromycotina present several difficulties:

1. Some Fungi Imperfecti may produce several distinct types of conidial apparatus, and this poses problems of which stage to use in giving the fungus a name. Whilst it is possible to use a separate name for each recognisable state of the fungus, many mycologists would regard it as unsatisfactory to have several names for a single fungus.

2. Unrelated fungi may produce conidial states which are morphologically very similar. We have already seen an example of this kind in the similarity between the conidial states of *Sclerotinia fructigena* (Helotiales) and *Neurospora crassa* (Sphaeriales). In both species, the conidia are hyaline and borne in branched chains formed by apical budding. Presented only with conidia of these two fungi, the morphological similarity is such that it seems reasonable to classify them together. In fact, the conidial states have been assigned to the same genus, *Monilia*. It is clear that the name *Monilia* does not have the same status as some other generic names, because it includes unrelated taxa. The term **'form-genus'** is used for assemblages of this sort, and it is very important to remember that the form-genera of the Fungi Imperfecti may well be made up of species which are unrelated.

3. Although attempts have been made to produce a 'natural' classification of Fungi Imperfecti, the information on which to base such a system is fragmentary and incomplete. A discussion of these problems will be found in Kendrick (1979) and Booth (1978).

TYPES OF CONIDIAL DEVELOPMENT

An enormous range of conidium morphology is found. Variations in shape represent adaptations to dispersal. One important distinction which is probably related to dispersal is whether the conidia are dry or sticky. Dry-spored conidia are generally dispersed by wind, but where the conidia have slimy coats, they cohere together to form aggregations or sometimes tendril-like ribbons, and the spores may be dispersed by insects or by rain-splash (Ingold, 1971; Gregory, 1973). Insect-dispersed conidia are often surrounded by sugary secretions and attractive odours. Despite the great variation in conidium shape, it has proved possible to classify conidial development into a relatively small number of basic methods. The studies of Hughes (1953) have provided a foundation for much of the recent work, and current investigations of conidial development have made use of time-lapse cinematography and scanning and transmission electron microscopy (Kendrick, 1971a; Cole & Samson, 1979).

The main types of conidial development have been outlined on p. 90. Ellis (1971b), in his book *Dematiaceous Hyphomycetes*, distinguishes the following types of development:

(a) thallic: 'the term thallic is used to describe conidium development where there is no enlargement of the recognisable conidium initial or, when such development does occur, it takes place after the initial has been delimited by a septum or septa'. *Endomyces geotrichum* (conidial state *Geotrichum candidum*) has conidia of this type (see p. 269, Fig. 133, and Cole (1975)).

(b) blastic: 'marked enlargement of the recognisable conidium initial takes place before it is delimited by a septum'. Two types of blastic development have been distinguished:

(i) *Holoblastic:* 'where both the outer and inner walls of a blastic conidiogenous cell contribute towards the formation of the conidia'. The *Stemphylium* type conidia of *Pleospora herbarum* (Fig. 215D) develop holoblastically (Carroll & Carroll, 1971). Another good example is *Cladosporium*. This shows apical budding of the conidial chain and, as shown in Fig. 290, a conidium may form two buds at its apex, so that, above that point, the conidial chain is branched. When a holoblastic conidiogenous cell blows out at a single point, it is *monoblastic*, or if at several points, *polyblastic*. *Aureobasidium pullulans* (Fig. 291) forms conidia polyblastically. Thick-walled conidia which develop blastically and are cut off by a break across the conidiophore wall are sometimes termed *aleuriospores*.

(ii) *enteroblastic:* 'where only the inner wall of the conidiogenous cells or neither wall contributes towards the formation of conidia'. When the conidium develops by the protrusion of the inner wall through a channel in the outer wall, development is *tretic*. *Helminthosporium velutinum,* a common fungus on dead wood and bark, shows development of this type (Fig. 292). A

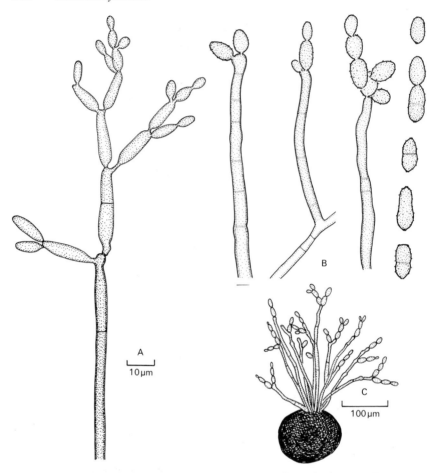

Fig. 290. A. *Cladosporium herbarum*: conidiophore with branching chain of conidia. B. *Cladosporium macrocarpum*: conidiophores and conidia. C. *Cladosporium macrocarpum*: conidiophores arising from a sclerotium.

more common type of enteroblastic development is *phialidic*. Phialidic development has already been described in *Aspergillus* and *Penicillium* (Figs. 153, 154), and is widespread in members of the Eurotiaceae and Hypocreaceae. The essential feature of phialidic development is that the conidia develop in basipetal succession from a specialised conidiogenous cell termed a phialide. The wall of the phialide does not contribute to the wall surrounding the phialoconidium. The conidial chains may be dry, as in *Aspergillus* and *Penicillium,* or, if the conidial wall is slimy, the phialoconidia cohere in mucilaginous blobs, as in *Trichoderma* and *Gliocladium* (Fig. 179). For a fuller discussion of phialidic development, see Subramanian (1971) and p. 301.

Most conidiophores grow in length only at their tips, but often, at the formation of the first conidium, growth ceases. In other cases, after the first

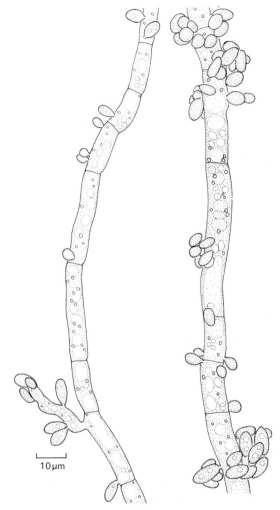

Fig. 291. *Aureobasidium pullulans.* Blastoconidia developing from undifferentiated mycelium. Because several conidia can develop from a single point, development is polyblastic.

conidium is formed and has been detached, the conidiophore continues growth through the point of attachment. Such development is described as *percurrent.* A small collar or *annellation* remains to indicate the new wall growth of the conidiogenous cell before a second conidium develops. Holoblastic conidiogenous cells with closely arranged, percurrent annellations are termed **annellides.** Superficially, they resemble phialides, and care is needed to distinguish the annellations using the light microscope. It may be necessary to use fluorescence microscopy and special staining techniques to visualise them.

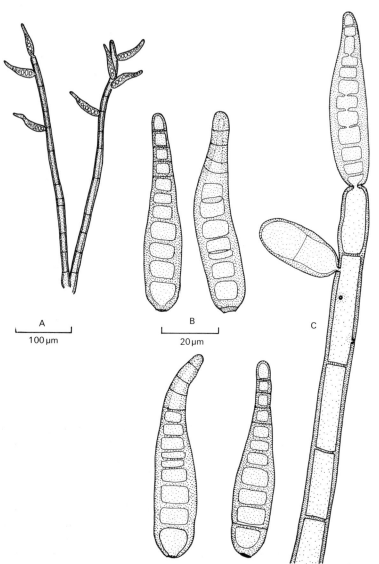

Fig. 292. *Helminthosporium velutinum*. A. Conidiophores and conidia. B. Detached conidia. C. Details of conidial development. Note the narrow channels in the wall through which cytoplasm passes to the developing conidia. This type of development is tretic. B and C to same scale.

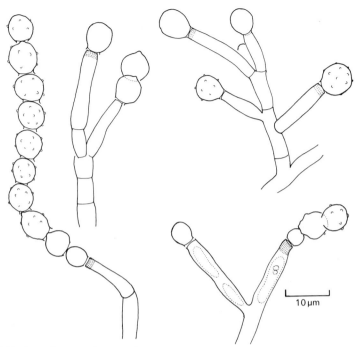

Fig. 293. *Scopulariopsis brevicaulis.* Conidiophores terminating in conidiogenous cells (annellides) from which chains of conidia develop. The stippled areas at the tips of the annellides indicate the region of growth associated with the development of successive conidia.

The condiogenous cell bearing a succession of annellations is sometimes termed an *annellophore.*

Hughes (1971) has given a general account of annellidic development. Cole & Kendrick (1969b) have studied conidial development in *Scopulariopsis brevicaulis* (Figs. 293 and 294) by time-lapse cinephotomicrography, and Hamill (1971) has followed development in the same fungus by transmission electron microscopy. This fungus, which has been isolated from soil, from insects and from various other mouldy substrata, was originally identified as a species of *Penicillium,* but it is now clear that, instead of being phialides, the conidiogenous cells are annellophores. Figure 294 gives an interpretation of conidial ontogeny. The conidiogenous cell is skittle-shaped, and it swells at its apex to form a rounded protrusion which is the first conidium. Cole & Kendrick (1969b) list the following steps in development:

1. The entire tip of the sporogenous cell is converted into the first conidium.

2. The base of this conidium is delimited by a double septum – a newly laid down inner or second wall.

3. The lower part of this double septum covers the new growing apex, and thus becomes the principal integument of the second conidium.

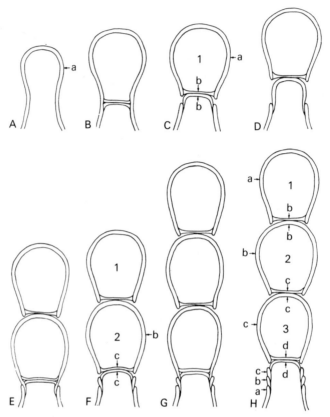

Fig. 294. *Scopulariopsis brevicaulis*. Diagrammatic interpretation of conidium formation. Successive wall layers are labelled a, b, c and d, and conidia are labelled 1, 2 and 3 to indicate their order of formation. (From Cole & Kendrick, 1969b, reproduced by permission of the National Research Council of Canada from the *Canadian Journal of Botany,* **47,** 925–9, 1969.)

4. As the new apex swells, the first conidium secedes in a circumscissile manner, leaving a narrow annular scar derived from the original outer (first) wall of the sporogenous cell.

5. The inner (second) wall, which delimited the base of the first conidium, forms the integument of the second conidium. When this conidium is mature, its base is delimited by a new (third) wall from which is also subsequently derived the outer wall of the third conidium.

6. The second conidium secedes, leaving a second 'annellation' just above the first on the sporogenous cell . . .

7. The formation of each spore entails a slight excess of growth over conversion, and the annellophore gradually lengthens by an accumulation of such small but significant increments as it produces a basipetal sequence of conidia.

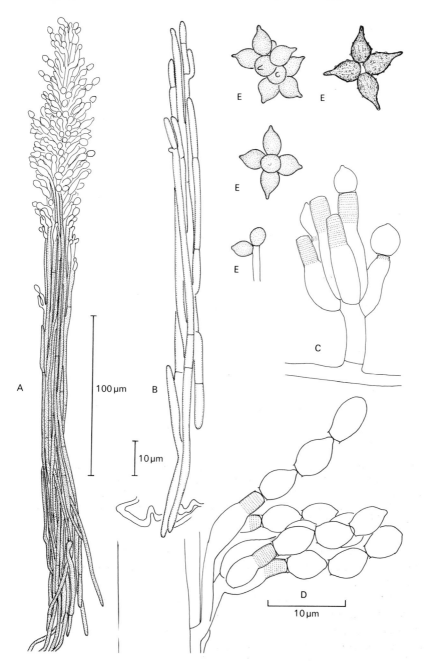

Fig. 295. *Doratomyces stemonitis*. A. Synnema or macronematous conidiophore consisting of a parallel bundle of dark hyphae branching at their tips to form conidiogenous cells and chains of conidia. B. Developing synnema. C. Micronematous conidiophore bearing six annellides. The stippled region is the cylinder of annellide growth. D. Conidiogenous cells (annellides) from a synnema showing chains of annelloconidia. The stippled zone represents the growth cylinder of the annellide. E. *Echinobotryum* conidial state. B and E to same scale. C and D to same scale.

Electron micrographs of thin vertical sections through developing annello-phores (Hammill, 1971; Beckett *et al.*, 1974) show that the septum formation which cuts off each successive conidium is by a centripetal ingrowth of wall material, leaving eventually a circular pore which is plugged by a spherical plug which remains attached to the base of the conidium after secession.

Similar observations have been made on conidial development in *Dorato-myces stemonitis* (Hughes, 1971) and *D. nanus* (Hammill, 1972). *D. stemonitis* has, in addition to its annelloconidia, a distinct *Echinobotryum* conidial state (see Fig. 295). It is of interest that perfect states of both *Scopulariopsis* and *Doratomyces,* where known, belong to the Ascomycete genus *Microascus* (Morton & Smith, 1963).

Although annellophores and phialides appear to represent distinctive types of conidiogenous cell, the number of examples which have been studied in detail is few, and only further study will show whether intermediates exist, i.e. whether these two types of structure represent extremes in a continuum.

VARIATIONS IN CONIDIAL SHAPE AND COLOUR

Although conidial ontogeny can be reduced to a comparatively small number of distinct patterns, there is an enormous range in the form of the fully developed conidium. Conidia may be unicellular, bicellular or multicellular, and multicellular conidia may be divided by septa in one to three planes. The shape of the conidium may be very varied, e.g. globose, elliptical, ovoid, cylindrical, branched, spirally coiled. The colour of the conidia (and the mycelium and conidiophores) may be hyaline, i.e. colourless, brightly col-oured (e.g. pink, green) or dark. The dark pigments are probably melanins. The colour of the conidiophores and conidia are important features used in classification.

Figure 296 illustrates some of the variation in conidium shape, and some of the terms used to describe these variants. An artificial system of classification of conidial fungi has been devised by Saccardo, making use of conidial colour and form to group together similar form-genera. Such a system has its uses as an aid to cataloguing the large number of conidial fungi, but studies of development, and information about the perfect states of some of these conidial fungi, show that the groupings based only on conidial form and colour are very artificial, i.e. they include unrelated elements.

TYPES OF CONIDIAL FRUCTIFICATION

Conidial fructifications (conidiomata) vary in complexity. Where the conidio-phores are little differentiated from vegetative hyphae or are inconspicuous, they are described as *micronematous,* but well-differentiated conidiophores are *macronematous.* When the conidiophores occur singly, they are *monone-matous,* but where they are aggregated to form parallel fascicles of closely appressed hyphae, they are *synnematous.* The fructification can then be termed a *synnema* or *coremium.*

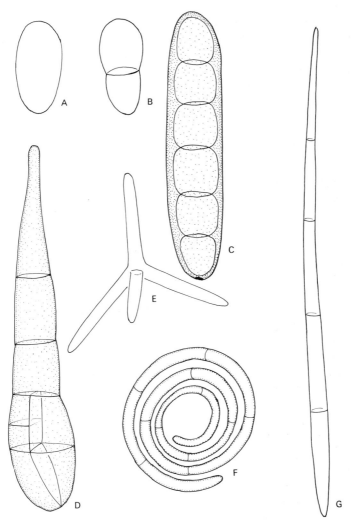

Fig. 296. Variations in conidial form in Deuteromycotina. A. amerospore, i.e. a unicellular spore. B. didymospore, i.e. a bicellular spore. C. phragmospore, i.e. a multicellular spore with transverse walls only. D. dictyospore, i.e. a multicellular spore with transverse and longitudinal septa. E. staurospore, i.e. a spore of stellate shape, with radiating arms. F. helicospore, i.e. a helically coiled spore. G. scolecospore, i.e. a narrowly cylindrical spore. Spore colour is also used in classification. Spores with colourless contents are given the prefix hyalo-, whilst those with dark colours are given the prefix phaeo-. Thus the spore illustrated in B is a hyalodidymospore, whilst C is a phaeophragmospore.

Doratomyces stemonitis has macronematous, synnematous conidiophores (Fig. 295). Shorter aggregations of synnematous conidiophores are termed *sporodochia*. The conidial state of *Nectria cinnabarina* (which has been called *Tubercularia vulgaris*) is a good example of a sporodochium, bearing sticky phialoconidia (Fig. 181). Two other common types of fructification are the *acervulus* and the *pycnidium*. An acervulus is a pseudoparenchymatous aggregation of hyphae which develops beneath the surface of a plant host, and eventually forms a superficial layer of conidiophores. *Colletotrichum graminicola* (Fig. 328) forms acervuli on seeds, shoots and roots of grasses. Here the acervulus is surrounded by stiff, black setae. Pycnidia are flask-shaped or globose fructifications lined by conidiogenous cells (of various types) and opening to the exterior usually by a circular ostiole. *Phoma betae, Ascochyta pisi* and *Septoria nodorum* (see Figs. 329, 330, 331) all possess pycnidia. Pycnospores often ooze out in cirrhi, and are believed to be generally dispersed by rain-splash. Pycnidia may be immersed in the tissues of the plant host, or may be superficial. In some forms, especially those growing beneath the bark of woody hosts, the pycnidia may be compound, i.e. several pycnidial cavities may be enclosed in a single fructification, the pycnidial stroma.

The type of conidial fructification is used in classifying Fungi Imperfecti. Mononematous or synnematous conidiophores are characteristic of the Hyphomycetes, whilst acervuli and pycnidia are formed in Coelomycetes.

CLASSIFICATION OF DEUTEROMYCOTINA

Ainsworth (1973) has provided the Key below to the classes of Deuteromycotina. The Blastomycetes (imperfect yeasts) probably include forms with affinities with Ascomycotina and with Basidiomycotina (see p. 495).

Key to classes of Deuteromycotina

1   Budding (yeast or yeast-like) cells with or without pseudomycelium characteristic; true mycelium lacking or not well developed
                                                                    Blastomycetes
     Mycelium well developed, assimilative budding cells absent   2
2   Mycelium sterile or bearing spores directly or on special branches (sporophores) which may be variously aggregated but not in pycnidia or acervuli   **Hyphomycetes**
     Spores in pycnidia or acervuli   **Coelomycetes**

No attempt will be made to provide a more detailed classification of these groups. Kendrick & Carmichael (1973) have given a list of form-genera of Hyphomycetes, together with a Key and figures. Other useful literature to the group are the books by Ellis (1971b, 1976) on Dematiaceous Hyphomycetes, and Barron (1968) on *The Genera of Hyphomycetes from Soil*. Sutton (1973) has similarly treated the Coelomycetes.

Instead of a systematic treatment of Deuteromycotina, it is proposed to deal with three ecological groups of fungi which include a large proportion of Fungi Imperfecti: Aquatic Fungi Imperfecti, Predacious Fungi and Seed-Borne Fungi.

AQUATIC FUNGI IMPERFECTI

If a sample of foam from a well-aerated, rapid stream flowing through deciduous woodland is examined microscopically, especially in the Autumn after leaf-fall, it will be found to contain a rich variety of conidia of unusual shape (Fig. 297). These conidia belong to aquatic Hyphomycetes which grow on decaying leaves and twigs. If decaying leaves are collected from the stream and incubated in a shallow layer of water at a temperature of 10–20 °C, numerous conidiophores will develop. Ingold (1975a) has written an excellent guide to the group, which has a world-wide distribution. Over 100 species are

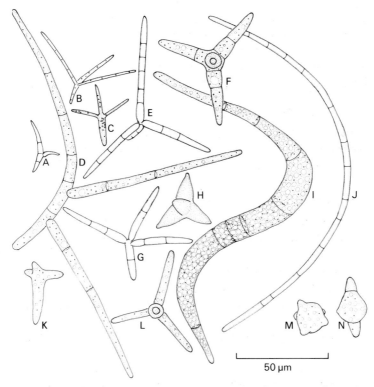

Fig. 297. Conidia of aquatic Hyphomycetes from river foam. A. *Volucrispora* sp. B. *Alatospora acuminata*. C. *Clavatospora longibrachiata*. D. *Tricladium splendens*. E. *Lemonniera aquatica*. F. *Lemonniera terrestris*. G. *Articulospora tetracladia*. H. *Clavatospora stellata*. I. *Anguillospora crassa*. J. *Anguillospora* sp. K. *Heliscus lugdunensis*. L. Unidentified. M. *Margaritispora aquatica*. N. *Pyricularia aquatica*.

known. Two common spore shapes can be recognised – the tetraradiate or branched conidium, and the sigmoid or steep helix type of conidium. Studies of the development of tetraradiate and sigmoid conidia show that they may develop in a variety of different ways, and Ingold (1966b, 1975b) has presented the thesis that both represent the results of a number of lines of convergent evolution. Evidence for the view that tetraradiate propagules have evolved independently several times is provided by:

1. Developmental studies.
2. Knowledge of the perfect states of some tetraradiate and sigmoid conidial forms.
3. The occurrence of tetraradiate propagules in unrelated aquatic organisms.

Development

A few examples of development of branched conidia will illustrate great variation.

PHIALIDIC TETRARADIATE CONIDIA

Lemonniera (Fig. 298A): *Lemonniera* shows phialidic development. Six species have so far been distinguished (Descals *et al.*, 1977), and *L. aquatica* is one of the most common. The conidiophores, which may develop from mycelium embedded in leaf tissues or from chlamydospores or sclerotia, terminate in one to three phialides. From the tip of the phialide, a tetrahedral conidium primordium develops, and the four corners of the tetrahedron expand *simultaneously* to form cylindrical arms which may become septate. The mature conidium is thus attached centrally to the phialide at the point of divergence of the arms. When the first phialoconidium is detached, to be carried away by water currents, a second conidium develops, and others follow.

Alatospora (Fig. 298B): *Alatospora acuminata* is another phialidic aquatic Hyphomycete. The phialides develop singly at the tips of short, inconspicuous conidiophores. From the tip of the phialide, a curved axis grows, from which two lateral arms arise midway along its length, developing simultaneously.

Heliscus (Fig. 298C): *Heliscus lugdunensis* is a common, early coloniser of twigs and leaves in streams. Its conidiophores often develop in pustules, and branch repeatedly to terminate in phialides. When phialoconidia develop underwater, they are somewhat clove-shaped, with short, conical projections at the upper end. Under aerial conditions, such as an agar culture, the conidia are more cylindrical, and somewhat reminiscent of the conidia of a *Cylindrocarpon*. A perfect state of this fungus has been discovered, a member of the genus *Nectria,* which forms its bright red perithecia singly on half-submerged twigs (Webster, 1959a; Willoughby & Archer, 1973).

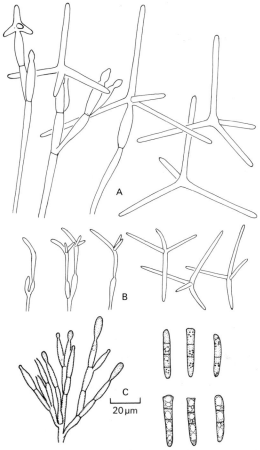

Fig. 298. Three phialidic aquatic Hyphomycetes. A. *Lemonniera aquatica*. B. *Alatospora acuminata*. C. *Heliscus lugdenensis*.

BLASTIC TETRARADIATE CONIDIA

There are numerous examples of blastic conidial development in aquatic Hyphomycetes.

Articulospora tetracladia (Fig. 299A) forms short conidiophores extending from mycelium within a leaf. At the apex of the conidiophore, the first arm develops as a cylindrical bud. At the apex of this first arm, three further cylindrical buds develop in turn. A narrow constriction or joint marks the point of attachment of these later-formed arms to the first (hence the name *Articulospora*, a jointed spore). The mycelium and conidia of *Articulospora* are hyaline.

Clavariopsis aquatica (Fig. 299B) has dark mycelium and conidia. The conidia have a broad, obconical body with a rounded tip bearing three

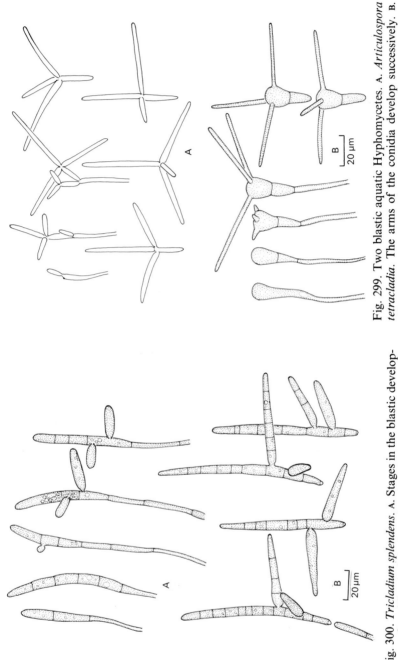

Fig. 299. Two blastic aquatic Hyphomycetes. A. *Articulospora tetracladia*. The arms of the conidia develop successively. B. *Clavariopsis aquatica*. The top-shaped body of the conidium develops first, followed by simultaneous development of the three apical arms.

Fig. 300. *Tricladium splendens*. A. Stages in the blastic development of conidia. A club-shaped main axis develops lateral arms successively from different points along its length. B. Mature detached conidia.

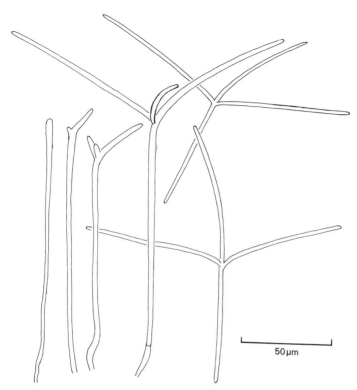

50 µm

Fig. 301. *Tetrachaetum elegans.* The main axis of the conidium bends and, at the point of curvature, two lateral arms arise simultaneously.

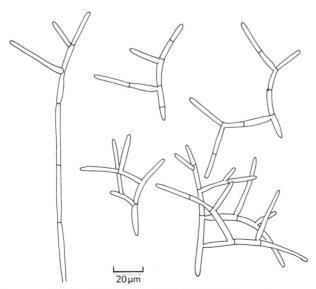

20 µm

Fig. 302. *Varicosporium elodeae*: branched blastoconidia formed by repeated branching of lateral arms which develop mostly from one side of the main axis of the conidium.

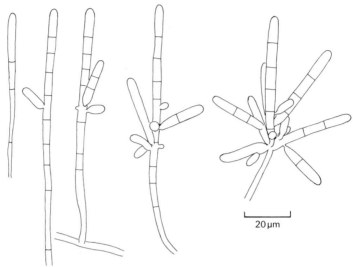

20 μm

Fig. 303. *Dendrospora erecta*: much-branched blastoconidia formed by repeated branching of lateral arms which develop on all sides of the main axis of the conidium.

cylindrical arms which develop simultaneously. The mature conidium usually has a single septum in the central body. In culture, a spermogonial state has been found, a dark-coloured pycnidium containing minute, colourless spermatia. The perfect state is a Loculoascomycete belonging to the genus *Massarina*. The pseudothecia grow on decorticated twigs in streams, and the bitunicate asci contain two-celled, hyaline ascospores (Webster & Descals, 1979).

Tricladium splendens (Fig. 300) has dark-coloured mycelium and conidia. The conidiophore develops as a club-shaped swelling which becomes septate, and forms the main axis of the conidium. A bud develops at one point on the main axis, to be followed by a second, at a different point. The arms taper and are constricted where they join the main axis. Its perfect state is an inoperculate Discomycete.

Tetrachaetum elegans (Fig. 301) has a hyaline mycelium and conidia which are relatively large, spanning up to 200 μm. The conidium develops by the curvature of the first arm which is narrowly cylindrical. Two laterals arise at a common point about half-way along the main axis, and develop simultaneously.

Varicosporium elodeae (Fig. 302) and *Dendrospora erecta* (Fig. 303) bear blastoconidia which are more highly branched.

BRANCHED CONIDIA WITH CLAMP CONNECTIONS

A number of branched conidia found in water or foam have clamp connections at their septa. They include *Leptosporomyces galzinii* (Fig. 304) which has a conidium somewhat resembling a *Tricladium*, but with a single clamp connection at the septum lying between the two arms (Nawawi *et al.*, 1977a).

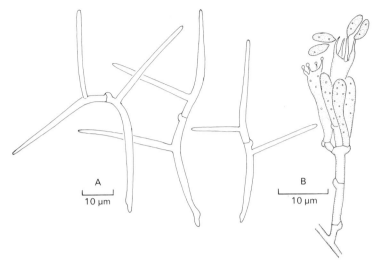

Fig. 304. *Leptosporomyces galzinii.* A. Mature detached conidia. Note the clamp connection between the two lateral arms. B. Basidial state.

The conidium is dikaryotic, and the basidial state has developed in cultures derived from a conidium. Somewhat surprisingly, the basidial state in nature grows in terrestrial rather than aquatic habitats, on rotting wood, ferns, etc. (Nawawi *et al.,* 1977a). *Ingoldiella hamata* (Fig. 305), a tropical aquatic fungus, has large, dikaryotic conidia with numerous clamp connections. The basidial state is a species of *Sistotrema* with eight-spored basidia. Single basidiospores germinate to form monokaryotic mycelia on which monokaryotic conidia develop, closely resembling the dikaryotic conidia but lacking clamp connections (Nawawi, unpublished).

There are probably other aquatic Hyphomycetes with branched conidia with Basidiomycetous affinities. One such is *Dendrosporomyces prolifer* which has conidia with a strong morphological resemblance to *Dendrospora,* but with dolipore septa (Nawawi *et al.*, 1977b).

SIGMOID CONIDIA

A similar range of different types of conidial ontogeny can be demonstrated for sigmoid conidia.

Flagellospora curvula (Fig. 306A) has narrow, sigmoid phialoconidia developing from phialides on a sparsely branched conidiophore. A more richly branched, penicillate arrangement of phialides is found in *F. penicillioides* (Fig. 306B), for which Ranzoni (1956) has described a *Nectria* perfect state.

Anguillospora has blastic, sigmoid conidia, but evidence from the known perfect states indicates that this form genus is very heterogeneous, and includes conidial states from unrelated groups of Ascomycetes. There are also important differences in the mechanism of conidial separation. *A. longissima*

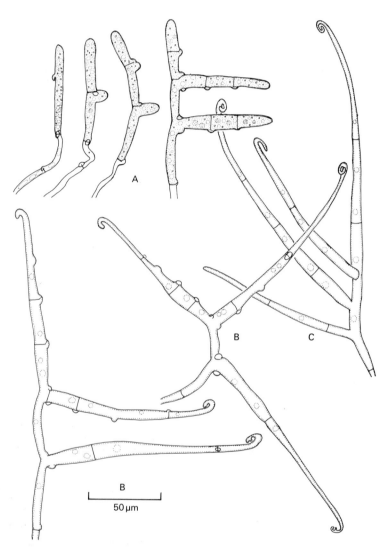

Fig. 305. *Ingoldiella hamata*. A. Developing dikaryotic conidia. Note the clamp connections. B. Mature detached dikaryotic conidium. C. Monokaryotic conidium.

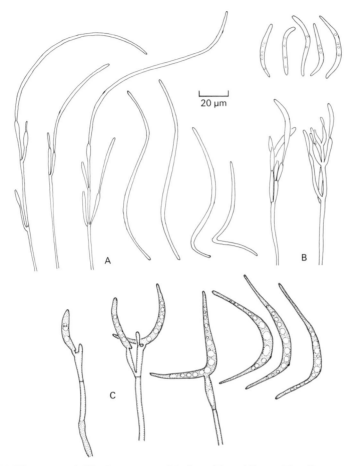

Fig. 306. Three aquatic Hyphomycetes with sigmoid conidia. A. *Flagellospora curvula*: conidiophores with phialides and sigmoid phialoconidia. B. *Flagellospora penicillioides*: conidiophores with phialides and phialoconidia. C. *Lunulospora curvula*, showing blastic development of crescent-shaped conidia.

(Fig. 307) has dark mycelium and conidia. The conidia develop as a club-shaped swelling from the apex of a conidiophore. The conidium becomes septate and helically curved. Conidial separation is brought about by the collapse of a special *separating cell* at the base of the conidium. The contents of the separating cell disintegrate, and the cell wall breaks down at a line of weakness near the middle. When the conidium separates, it carries at its base a little collar which represents half of the empty separating cell. Similarly, the apex of the conidiophore bears a collar after detachment of the first conidium. The conidiophore may develop a second conidium by percurrent extension through the remnants of the first separating cell and, after several conidia have been formed, a succession of collars may be found at the tip of the conidio-

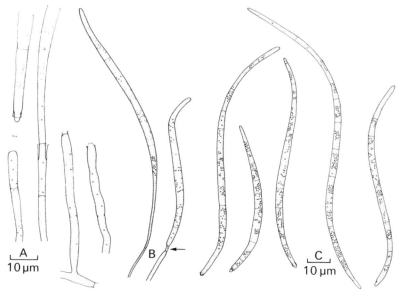

Fig. 307. *Anguillospora longissima*, an aquatic Hyphomycete with blastic sigmoid conidia. A. Detachment of conidium showing the remnants of the separating cell, and percurrent proliferation. B. Developing conidia. The arrow marks a separating cell. C. Mature detached conidia.

phore (Webster & Descals, 1979). The perfect state of *A. longissima* is a species of *Massarina* found on twigs in streams (Willoughby & Archer, 1973). A second species, *A. furtiva*, has conidia which closely resemble those of *A. longissima*, but cultures derived from such conidia have developed apothecia of an inoperculate Discomycete. In *A. furtiva*, there is no separating cell: conidium separation is by dissolution of a septum at its base (Webster & Descals, 1979). In *A. crassa*, which has fatter conidia, separation is also brought about by septum dissolution, and this species also has an inoperculate Discomycete as its perfect state, a species of *Mollisia* (Webster, 1961).

Lunulospora curvula (Fig. 306C) has crescent-shaped blastoconidia which develop from specialised conidiogenous cells at the apex of dark conidiophores. This fungus is more common in warmer countries than in temperate regions, and in Britain its season of maximum abundance is in late summer and autumn (Iqbal & Webster, 1973b).

OTHER TYPES OF SPORE

Not all aquatic Hyphomycetes have branched or sigmoid conidia. *Margaritispora aquatica* forms hyaline, globose phialoconidia bearing a few conical protrusions (Fig. 308A), whilst *Pyricularia aquatica* (Fig. 308B) has pear-shaped or broadly fusiform blastoconidia which separate by septal dissolution. The perfect state of this fungus is *Massarina aquatica* (Webster, 1965) which forms pseudothecia on submerged wood in streams.

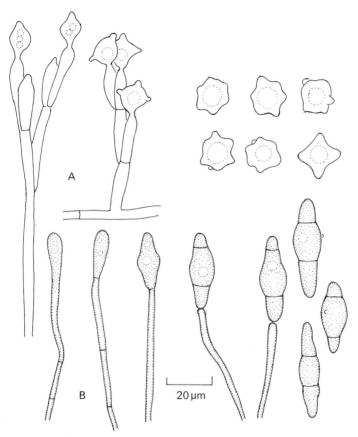

Fig. 308. A. *Margaritispora aquatica*: conidiophores, phialides and phialoconidia. B. *Pyricularia aquatica*: conidiophores and blastoconidia.

THE SIGNIFICANCE OF THE TETRARADIATE PROPAGULE

We have seen that tetraradiate conidia may develop in a variety of ways. There is also evidence that some are conidial Ascomycetes (e.g. *Clavariopsis aquatica*), whilst some are conidial Basidiomycetes (e.g. *Ingoldiella hamata*). Tetraradiate propagules have also been found as secondary conidia of *Entomophthora* spp. attacking aquatic insects (Webster *et al.*, 1978), as basidiospores in the marine fungi *Digitatospora marina* (Doguet, 1962) and *Nia vibrissa* (Gasteromycetes) (Doguet, 1967). The brown alga *Sphacelaria* also forms tetraradiate propagules. There is thus ample evidence for the view that this type of structure has evolved repeatedly in aquatic environments. What is its significance? Ingold (1953) speculated that there might be three reasons:

... in aquatic Hyphomycetes the tetraradiate spore is of such frequent occurrence that it is natural to suppose it has some biological significance in the aquatic environment. What this significance may be is pure speculation. Perhaps a spore of this kind settles relatively slowly in water and so stands a good chance of adequate dispersal; perhaps on the other hand it acts as an anchor and readily becomes entangled in a suitable substratum, for the arrest of a spore in a stream may be a real problem; or perhaps it is not so easily devoured by small aquatic animals as a spherical or oval spore would be.

Experimental studies by Webster (1959b) suggest that the most important advantage of the tetraradiate propagule is that it is more effectively trapped by impaction on to underwater objects than spores of more conventional shape. Possibly this is because when a tetraradiate propagule makes contact with a surface, it makes a three-point landing, a very stable form of attachment. At the point of contact, the tips of the conidial arms quickly develop 'appressoria' and germ tubes, but the fourth arm not in contact with a surface fails to develop in this way. There is no support for the view that slow sedimentation rates are of any great advantage to fungi with tetraradiate propagules. The rate of sedimentation is very low (of the order of 0·1 mm/second) in comparison with flow rates in streams which are commonly 1,000 mm/second and may reach 5,000 mm/second or more. Whether spore shape is a deterrent to animal feeding is doubtful because it is possible to rear animals such as *Asellus* on a diet of high spore concentrations of aquatic fungi such as *Tricladium splendens.*

Bandoni (1974) has suggested that trapping efficiency may not be the only advantage of a tetraradiate propagule. He and others have shown (Bandoni, 1972; Park, 1974) that tetraradiate spores may be found among leaves in terrestrial habitats, and has advanced the idea that tetraradiate spores may be adapted to movement in surface films of water.

Foam is an effective trap for both tetraradiate and sigmoid spores. In experiments in which air bubbles were passed through concentrated suspensions of conidia, it has been found that the concentration fell very rapidly. Tetraradiate conidia were removed more readily than conidia of sigmoid or other shape (Iqbal & Webster, 1973a), and comparison of spores collected in foam or by Millipore filtration from the same stream indicates that the spore content of foam tends to over-represent the tetraradiate type of conidium in relation to other spore-types known to be present.

The concentrations of aquatic Hyphomycete spores in streams can be readily estimated by filtration of water samples through a Millipore filter (preferably with 8 μm pore size, to facilitate rapid filtration). Not surprisingly, the concentrations reach a peak which coincides with or follows soon after the peak period of deciduous tree leaf fall in temperate countries. Concentrations as high as 10^4/litre have been found in October and November in Britain (Iqbal & Webster, 1973b).

It is now becoming clear from the work of Kaushik & Hynes (1968, 1971), Bärlocher & Kendrick (1973a,b) and Suberkropp & Klug (1977, 1980) that the aquatic Hyphomycetes play an important rôle in the cycling of nutrients. Many streams receive the bulk of their fixed carbon input not from attached

Table 9. *Growth of* Gammarus pseudolimnaeus *on fungal mycelium and on leaf litter*

Source of food	Average weight of animals at start (mg)	Average weight of surviving animals after 62 days (mg)	Average wet weight increase (mg)	Dry matter consumed per animal per day (mg)
Anguillospora	7·22	12·18	4·96	0·084
Clavariopsis	7·34	11·82	4·48	0·092
Tricladium	7·26	11·74	4·48	0·086
Elm leaves	7·32	9·50	2·18	1·07
Sugar maple leaves	7·34	8·42	1·08	0·98

From Bärlocher & Kendrick (1973a).

macrophytes or algae, but from the leaf, twig and other debris such as bud-scales, catkins, fruits and fallen logs shed by trees and other plants which border the streams. However, such material may be relatively poor in nitrogen or be otherwise unpalatable or un-nutritious to the invertebrate animal population. The colonisation of leaf litter by aquatic Hyphomycetes and bacteria is an important part of the 'processing' which is required to make the leaf litter available to the animals (Berrie, 1975). There are two main reasons for this. Firstly, the protein content of leaf material is enhanced by microbial colonisation. The fungi can concentrate inorganic nitrogen present in solution in the water, in low concentration but in relatively large total amounts and, making use of the organic matter in the leaves, manufacture microbial protein. Secondly, the aquatic animals may not have the enzyme equipment to soften and break up leaf tissue into digestible morsels. Spectacular differences in the growth rate of *Gammarus* fed on uncolonised leaf litter or on fungal mycelium, as shown in table 9, demonstrate these effects. Column three shows that animals fed on a fungus diet make a much greater wet weight increase than those fed on a leaf diet. In order to sustain this restricted growth rate on leaf diets, the animals consume about ten times the dry weight of leaf material than animals fed on fungus diets.

AERO-AQUATIC FUNGI IMPERFECTI

If leaves and twigs from the mud-surface of stagnant pools or slow-running ditches are incubated at room temperature in a humid environment (e.g. a Petri-dish or plastic lunch-box lined with wet blotting paper), a group of fungi with very characteristic conidia usually develops within a few days. The common feature of the conidia of these fungi is that, as they develop, they trap

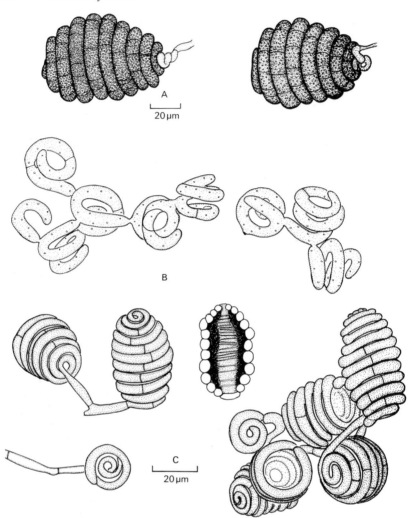

Fig. 309. Some helicosporous fungi. A. *Helicoon richonis*. B. *Helicodendron triglitziense*. C. *Helicodendron conglomeratum*. The central spore is drawn in optical section to show the trapped air bubble.

air, which assists in floating the conidia off if the substratum is submerged in water. It seems likely that such fungi can grow vegetatively on leaves and twigs, often in water with quite low amounts of dissolved oxygen. Under submerged conditions, these fungi do not sporulate, but only do so if a moist interface between air and water is provided, as might happen at a pond margin as the water dried up. There are various ways in which the conidia may develop. In *Helicoon* (Fig. 309), they develop as cylindrical or beehive-shaped spirals. The conidia vary in colour from hyaline to dark black. The direction

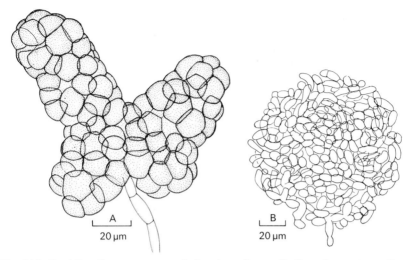

Fig. 310. Conidia of two aero-aquatic fungi. A. *Beverwijkella pulmonaria.* B. *Spirosphaera* sp.

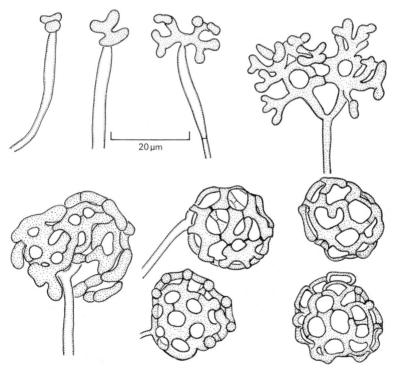

Fig. 311. *Clathrosphaerina zalewskii*: conidial development.

of coiling of the spirals (looking upwards from the apex of the conidiophore) is clockwise in *H. richonis*. In some other helicosporous fungi, the direction of coiling is counter-clockwise, and the direction appears to be constant for a given species. In *Helicoon*, the conidia themselves do not branch, but in *Helicodendron* the conidia may bear further conidia as lateral branches (Fig. 309B, C). *Beverwijkella pulmonaria* (Fig. 310A) forms bi-lobed aggregates of dark, thick-walled, subglobose cells with air trapped in intracellular spaces, whilst *Spirosphaera* achieves the same end by the formation of globose propagules made up of richly branched, incurved hyphae (Fig. 310B). Yet another way of entrapping air within the propagule is shown by *Clathrosphaerina zalewskii* which forms hollow, spherical propagules with a lattice wall, resembling plastic practice golf balls. These clathrate structures are formed by the repeated dichotomy of the arms of the developing conidium, which then curve inwards and join firmly where the tips of the arms touch (Fig. 311). This fungus has a minute inoperculate Discomycete perfect state, a species of *Hyaloscypha* (Descals & Webster, 1976). These methods of conidial development do not exhaust the range of possibilities for enclosing air, because several other propagules of this kind are known with different structure.

Although the taxonomy of the group is fairly well known (Linder, 1929; Glen-Bott, 1951, 1955; van Beverwijk, 1953, 1954; Moore, 1955; Hennebert, 1968; Tubaki, 1975a,b), much still remains to be discovered about their ecology, but simple techniques exist for their study (Fisher, 1977; Fisher & Webster, 1980).

PREDACIOUS FUNGI IMPERFECTI

The most exciting group of Fungi Imperfecti are the predacious Hyphomycetes which ensnare or attack living nematodes and consume them. This habit is not confined to the Fungi Imperfecti, because examples are known amongst the Zoopagales and Entomophthorales (Zygomycotina), Chytridiomycetes and Oomycetes. Within the Basidiomycotina, the genus *Nematoctonus* is made up of forms with conidia borne on a clamped mycelium, and it has recently been shown that the perfect state of one species is the agaric genus *Hohenbuehelia* (Barron, 1977; Barron & Dierkes, 1977). The predacious Hyphomycetes are easy to study. A small pinch of soil is added to a Petri-dish containing weak corn-meal agar and incubated for a few weeks at room temperature. Nematodes present in the soil crawl out over the agar surface, feeding on bacteria. If predacious fungi are present in the soil, they develop structures for trapping nematodes, and the trapped dead and dying animals are easy to see under a dissecting microscope. Over 150 species of nematode-destroying fungi have been described. An excellent account of the group has been written by Barron (1977). The taxonomy is based largely on the pioneering studies of Drechsler, and a valuable Key to species has been provided by Cooke & Godfrey (1964).

Barron (1977) has distinguished between **predatory** and **endoparasitic** nematode-destroying fungi. The predatory forms produce an extensive hyphal system in the soil and, at intervals along the length of the hyphae, trapping devices are formed which capture living nematodes. In the endoparasitic forms, there is no extensive mycelial development outside the host, and the fungus exists in the soil as a spore which may either become attached to the body of its host or be ingested. The spore then germinates and penetrates the animal, either from outside, or through the gut wall, and develops a mycelium within the body of the nematode. Only reproductive hyphae (conidiophores) penetrate to the outside of the dead nematode, although chlamydospores formed on the mycelium of the fungus within the body of its host may survive in the soil as the body decays.

The distinction between these two habits is not clear-cut. This is well shown by the fact that, within the single genus *Nematoctonus,* some species are predatory and others endoparasitic.

PREDATORY NEMATODE-DESTROYING HYPHOMYCETES

Various types of trapping organs are developed by predatory Hyphomycetes. In many cases, they do not develop in pure culture, but the addition of nematodes or nematode extracts will induce their development within 24–48 hours. Other substances of biological origin, such as horse serum or yeast extract, are effective in inducing trap formation. Pramer & Stoll (1959) coined the name nemin for the substance or substances active in trap induction. Later work suggested that the effective substance might be one or more amino acids or a peptide of relatively low molecular weight. Nordbring-Herz (1973) showed that peptides were more effective than amino acids in inducing trap formation in *Arthrobotrys oligospora*, and that valyl peptides were especially active. The range of trapping structures in predatory Hyphomycetes includes:

1. *Adhesive knobs:* single-celled, sessile or stalked, globose knobs, covered by a sticky secretion, and spaced at intervals along a hypha, form the trapping organs in several predatory Hyphomycetes – e.g. *Dactylaria candida.* Ultrastructural studies of the sticky knobs of *Monacrosporium drechsleri* show dense inclusions not found in unmodified cells of the mycelium, and well-developed, rough and smooth endoplasmic reticulum. Such structures are possibly involved in the secretion of the adhesive, but there are no channels in the hyphal wall through which adhesive material might be conducted (Heintz & Pramer, 1972). Sometimes, after a nematode has become attached to an adhesive knob, its violent movements may cause the knob to be detached from the mycelium. In such cases, it is possible that penetration of the host by the fungus may occur later (see below). Barron (1977) has suggested that detachable knobs may provide an effective means of dispersal.

2. *Adhesive lateral branches:* short lateral branches, sometimes composed of

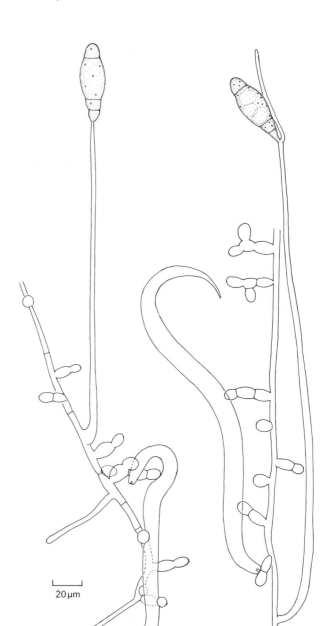

Fig. 312. *Monacrosporium* sp. Hyphae showing short lateral branches modified as sticky knob traps, with nematodes attached at several points. There is a single conidium at the end of the conidiophore.

only a single swollen cell, but often of a few cells, become covered by a film of a very tenacious adhesive and are held in an upright position above the general level of the mycelium. These traps take the form of sticky knobs spaced apart in such a way that a nematode may become attached to several of them. Such traps are illustrated in *Monacrosporium* sp. (Fig. 312). In some species, e.g. *M. cionopagum* and *M. gephyropagum,* the lateral branches may develop sufficiently close together for anastomosis to take place, resulting in the formation of a primitive network.

3. *Adhesive nets:* one of the most common types of trap is an adhesive network formed by anastomosis of the recurved branch tips of a lateral branch

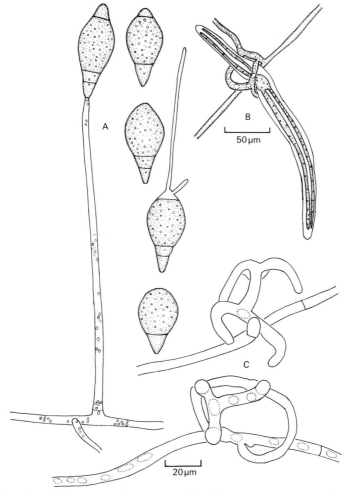

Fig. 313. *Monacrosporium eudermatum.* A. Conidiophore with its single attached conidium, and several detached conidia. B. Trapped nematode showing infection bulb and assimilative hyphae. C. Adhesive trapping networks.

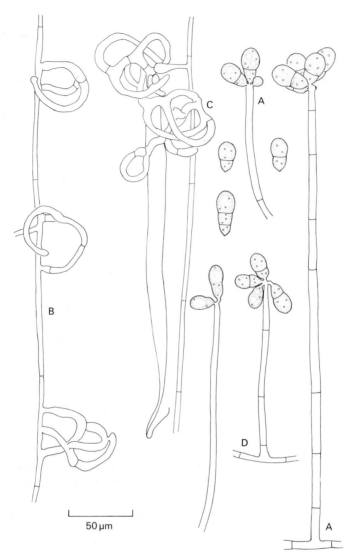

50 μm

Fig. 314. *Arthrobotrys* spp. A. *A. robusta*, conidiophores and conidia. B. *A. robusta*, mycelium with anastomosing traps. C. *A. robusta*, nematode caught in trap. D. *A. oligospora*, two conidiophores showing the sequence of conidial development.

system. The network is lifted above the general level of the mycelium. Traps of this kind are illustrated in *M. eudermatum* (Fig. 313) and *Arthrobotrys robusta* (Fig. 314). The entire surface of the network is lined by an adhesive film, and a nematode which thrusts its body into the network is quickly immobilised. Nordbring-Herz (1972) has prepared scanning electron micrographs showing the distribution of adhesive over the network, and, in an excellent film by

Comandon & de Fonbrune (*Champignons prédateurs de nématodes du sol:* Institut Pasteur, Paris), the adhesive has been demonstrated on the outside of the net by means of a micromanipulator.

4. *Non-constricting rings:* a number of predacious Hyphomycetes ensnare their prey by three-celled rings formed by recurvature and anastomosis of the tip of a lateral branch with itself. A nematode thrusting its body into such a loop may become tightly wedged inside it, and may find it impossible to retract. The point of junction of the loop to the hypha which bears it is often weak, and the struggles of the nematode may detach the loop. Occasionally, a single nematode bearing several loops may be found. The detached loops are still capable of penetrating and killing the nematode. The action of this type of trap is, however, essentially passive, and there is no inflation of the cells of the trap to constrict the nematodes. Non-constricting rings occur in *Dactylaria candida*, which also forms sticky knobs (Fig. 316) (Dowsett & Reid, 1977a,b).

5. *Constricting rings:* the most dramatic type of trap is the constricting ring trap, which develops in the same way as the non-constricting ring trap, but differs from it in that, when the inner surface of the ring is stimulated by contact, the individual cells rapidly inflate, to occlude the lumen of the trap, and thereby to constrict severely a nematode which it surrounds. Such traps have been found in the three main genera of predacious Hyphomycetes, *Arthrobotrys* (e.g. *A. anchonia, A. dactyloides*), *Monacrosporium* (e.g. *M. bembicodes*) and *Dactylaria* (e.g. *D. brochopaga*). This type of trap in a species of *Monacrosporium* is illustrated in Fig. 315.

The mechanism of ring closure has aroused much interest. The insertion of a fine glass rod into a ring by means of a micromanipulator, followed by gentle friction of one of the cells, can trigger off the closure of the trap. Other stimuli, such as heat or a stream of dry air, are also effective. The enlargement of the cells is accompanied by vacuolation of their contents, and by elastic stretching of the *inside* wall of the ring, whilst the outer part does not change shape (Estey & Tzean, 1976). Ring closure is complete within 0·1 second. Enlargement of the three cells making up the ring is not simultaneous, but one cell inflates a fraction of a second before the others. By immersing rings in 0·3–0·5 M sucrose solutions, and inducing trap closure by heat, Muller (1958) succeeded in slowing down the rate of ring closure by a factor of 100, so that closure took about 10 seconds. Estimations of the volume change of the cells during closure showed a three-fold increase. It is likely that several physiological changes take place during closure:

(i) Change in membrane permeability permitting rapid uptake of water.

(ii) Water uptake. Muller has estimated that, in the constricting ring of *Monacrosporium doedycoides*, there must be an uptake of 18,000 μm^3 of water in 0·1 second. It seems unlikely that this water could pass by conduction through the stalk of the constricting ring and the three septa dividing the cells, so that water uptake over the whole surface of the ring may take place.

(iii) Wall changes. Since the inner part of the ring wall rapidly changes

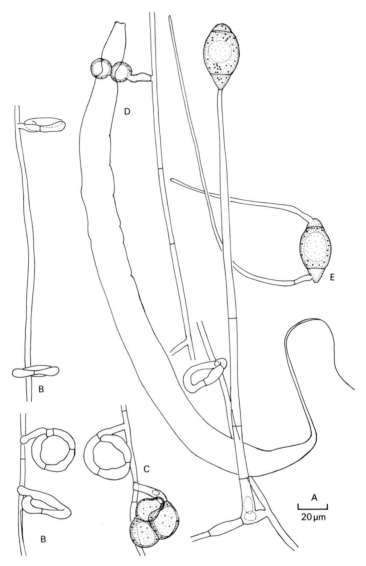

Fig. 315. *Monacrosporium* sp. A. Conidiophore with a single terminal conidium. B. Three-celled constricting ring traps. C. Two traps, one inflated. D. Nematode caught in constricting ring trap. E. Germinating conidium.

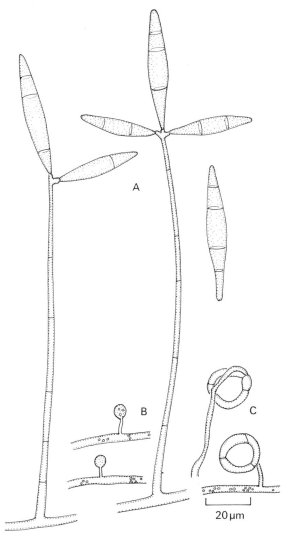

Fig. 316. *Dactylaria candida*. A. Conidiophores and a detached conidium. B. Stalked unicellular knob traps. C. Non-constricting ring traps.

shape, it is likely that there is a slippage of the microfibrils making up the wall, and this part of the wall also becomes thinner. Working with *Dactylaria brochopaga,* Dowsett *et al.* (1977) showed that the inner or luminal wall of the ring cells is thicker than the outer wall. The luminal wall is four-layered before expansion. During expansion, the two external wall layers of the luminal wall rupture.

(iv) Changes in osmotic concentration. The rapid uptake of water is possibly due to hydrolysis of polymers within the cell leading to an increase in osmotic concentration of the cell sap. However, the water taken up would presumably dilute the osmotically active substances. This may explain why Muller could not detect changes in the osmotic concentration of cells before and after closure.

(v) Rearrangement of membranes. The three-fold increase in cell volume and the commensurate increase in surface area, together with the process of vacuolation, clearly necessitates a rapid rearrangement of membrane material within the cell. It is significant that Heintz & Pramer (1972) discovered, in their fine-structural investigation of constricting rings of *Arthrobotrys dactyloides*, a labyrinth of membrane-bound material close to the plasma membrane on the inside of the ring cells, and a wide zone of electron-lucent material. In expanded rings, a more usual type of plasma membrane organisation was found, suggesting that the membrane-bound inclusions had contributed to the formation of the enlarged plasma membrane. (See also Dowsett *et al.*, 1977.)

The development of all these different types of trap can be elicited by the presence of nematodes, and presumably by their bodily secretions. Contact with traps is not entirely due to chance, but may be the result of chemotactic movement of nematodes towards them. If nematodes are placed equidistantly between disks from an agar culture of a predacious Hyphomycete, one disk having been induced to form traps, and the other without traps, a significantly larger number of nematodes moves towards the stimulated disk (Field & Webster, 1977).

The capture of a nematode is quickly followed by its death, although it may struggle vigorously for some time. In the case of the constricting ring trap, the stricture of the body of the nematode may well be a contributory cause of death. There is, however, evidence that toxins may also be produced by certain fungi. Olthof & Estey (1963) showed that sterile extracts of the *mycelium* of *Arthrobotrys oligospora* (which forms adhesive nets) had no apparent effect on the vitality of the nematode *Rhabditis* sp., nor had sterile extracts from crushed nematodes. When, however, extracts were prepared from crushed nematodes which had been parasitised by the fungus, living nematodes placed in the filtrate became inactive and were killed. It was suggested that the extracts contained a nematotoxic principle. Balan & Gerber (1972) have shown that culture filtrates from the unstimulated mycelium of *Arthrobotrys dactyloides* are nematocidal, and that the active principle is ammonia. Stable nematotoxins have also been shown to be produced by germinating conidia of the endoparasitic *Nematoctonus haptocladus* and *N. concurrens*. Here the conidia adhere externally to the host cuticle, and the secretion of toxins (probably, in this case, a polysaccharide) may help to immobilise the host before penetration (Giuma & Cooke, 1971; Giuma *et al.*, 1973).

The penetration of the host by predatory nematophagous fungi is by a haustorium which pierces the cuticle and body wall. Within the body cavity of the host, the fungus swells up to produce a globose vesicle which has been given various names, e.g. the mortiferous excrescence, infection bulb or post-infection bulb. It was once thought that death of the host was brought about by 'internal strangulation' of the body contents as the bulb expanded, but Shepherd (1955) showed that the host becomes inactive before the bulb has fully expanded. From the bulb, assimilative hyphae develop in both directions, consuming the body contents. Finally, the fungus may re-emerge to form conidiophores and conidia.

REPRODUCTION

The predatory nematophagous fungi mostly belong to four genera. In *Arthrobotrys* (Fig. 314), the conidia are two-celled and produced either in a single whorl at the apex or in a series of whorls along the length of the conidiophore. The conidia develop holoblastically. Although a young conidium may first appear somewhat below the point of insertion of the next oldest conidium (Fig. 314D), further growth of the conidiophore apex thrusts the older conidia downwards so that the youngest conidium is towards the apex of the conidiophore. The conidia of *Trichothecium,* although superficially similar, develop in a different sequence, in basipetal succession, so that the youngest conidium is clearly at the base of the group of terminal conidia. This is illustrated in Fig. 317 for *T. roseum* which is *not* a predatory fungus but a common saprophyte. During the development of successive conidia, the conidiophore apex actually becomes shorter, and it has been inferred that the wall material of the

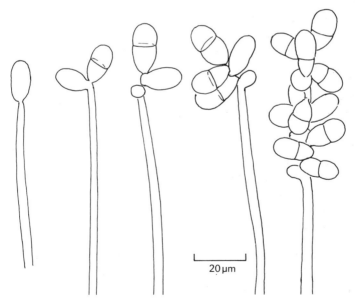

Fig. 317. *Trichothecium roseum*: conidiophores showing successive development of conidia. The second conidium is formed beneath the first, and the conidiophore becomes shorter as conidial development proceeds.

conidiophore is *converted* into material from which the spores develop (Kendrick & Cole, 1969). Although *T. roseum* is not nematophagous, some other species are, e.g. *T. flagrans* and *T. cystosporium*. The two other most common genera are *Monacrosporium* in which a single hyaline phragmospore is borne at the tip of a conidiophore (Fig. 315), and *Dactylaria* which forms a succession of similar conidia so that an 'ear' of several conidia is found at the apex of the conidiophore (Fig. 316). Many species now classified in *Monacrosporium* were once placed in the genus *Dactylella* (Cooke & Dickinson, 1965). Schenck *et al.* (1977) have argued that conidial septation is not a good criterion for separating species of nematophagous fungi, and have transferred the nematophagous species of *Dactylaria* (i.e. those forms with a plurality of phragmoconidia) to the genus *Arthrobotrys*.

ENDOPARASITIC NEMATODE-DESTROYING HYPHOMYCETES

Several genera of Hyphomycetes include forms which are endoparasitic. In some species, the conidia attach themselves to the cuticle of the host and, on germination, penetrate into the body cavity and eventually fill it with hyphae. *Meria coniospora* (Fig. 318) has conical conidia which, at maturity, bear a globose, adhesive knob at the narrow end. Such conidia readily adhere to the cuticle of a nematode which brushes past them, and they are also commonly found attached to the buccal end of nematodes which have attempted to ingest them. In *Nematoctonus* (Fig. 318) which has a clamped mycelium, erect conidia with tapering ends, which are sometimes bent at an angle, are attached by a sticky secretion to the cuticle of a nematode which brushes them. The conidia secrete nematotoxins which immobilise the nematode (Giuma & Cooke, 1971; Giuma *et al.*, 1973). Penetration and digestion then follow. In some other forms, infection follows ingestion of a conidium. *Harposporium anguillulae* has crescent-shaped conidia and, when these are ingested, the pointed end of the spore becomes lodged in the wall of the oesophagus and, on germination, penetrates the body cavity from within (Aschner & Kohn, 1958). The conidiophores which emerge from the dead host bear sessile, subglobose phialides (Fig. 319). The mycelium within the host may form chlamydospores which presumably survive in the soil when the body of the nematode decays.

BIOLOGICAL CONTROL OF NEMATODE PATHOGENS

Some nematodes are serious pathogens of plants and animals. Attempts have been made to control the populations of parasitic nematodes in the soil by the addition of cultures of nematophagous fungi. Although, in some experiments, some success in reducing nematodes and improving crop yields has been reported, so far the technique is not reproducible and cannot be regarded as commercially practicable (Duddington, 1957).

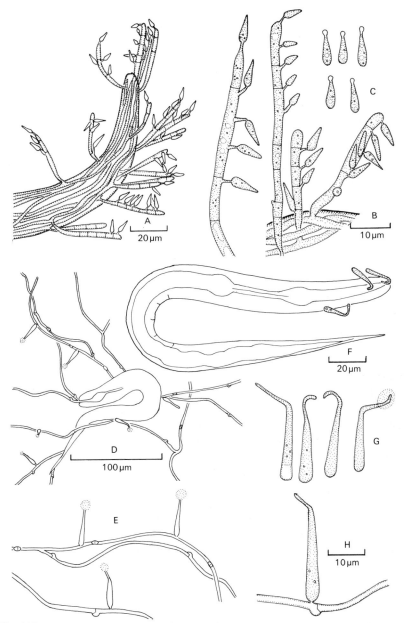

Fig. 318. *Meria coniospora*: A. Dead nematode containing hyphae, and bearing emergent conidiophores. B. Conidiophores. C. Mature conidia showing a terminal adhesive bulb.

Nematoctonus leptosporus: D. Dead nematode with radiating clamped hyphae and conidia. E. Clamped hyphae bearing erect conidia ending in adhesive globules. F. Nematode with three conidia attached at the buccal end, one inside the mouth. G. Detached conidia showing the tapering extensions which are sometimes recurved. H. Attached conidium.

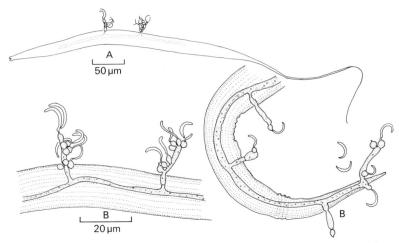

Fig. 319. *Harposporium anguillulae.* A. Dead nematode showing extent of internal mycelium, and external conidiophores. B. Enlargement to show details of conidiophores, bearing subglobose phialides and sickle-shaped phialoconidia.

SEED-BORNE FUNGI IMPERFECTI

Seeds of many plants bear a distinctive flora of fungi. In some cases, the fungi are pathogenic. The presence of pathogenic fungi in a seed sample of an economically important crop may be a means by which the disease may survive from one season to the next, or indeed for several seasons, and may also be a means by which the disease is spread into previously trouble-free areas. Pathogenic fungi may impair the viability of the seed or lower its quality. Seed-borne fungi may also be a cause of animal disease, e.g. toxicosis or hormonal imbalance leading to loss of fertility or milk production. Many, but not all, seed-borne fungi belong to the Deuteromycotina, and include not only Hyphomycetes, but also Coelomycetes, forms with pycnidia and acervuli as their conidial fruit-body. Because of their potential significance to plant pathology, seed-borne fungi have been the subject of much study and, in most countries with well-developed agricultural and horticultural industries, seed-testing stations carry out routine tests for the incidence of pathogens in seed-crops. There is thus an extensive literature on seed-borne fungi (Neergaard, 1977).

Seed-borne fungi are relatively easy to study. During development, the floral parts and ovaries of seed-plants become exposed to colonisation by common, air-borne, saprophytic fungi such as *Aureobasidium, Cladosporium, Alternaria, Epicoccum, Aspergillus* and *Penicillium,* so that, if seed samples are incubated in moist chambers (e.g. a Petri-dish lined by moist blotting paper) or are placed directly on a nutrient agar surface, extensive development of these common moulds may occur, and may obscure the presence of the potentially more serious pathogens which may be present in the deeper-lying layers of the seed. This problem can be avoided by rinsing the seed in antiseptic (e.g. a dilute solution of sodium hypochlorite for 10 minutes) to surface-sterilise it. If the seed is then dried and placed on a suitable agar medium (e.g. corn-meal agar, V8-juice agar) or incubated on moist filter paper, fruiting of the pathogens may take place on the surface of the seed or on the medium. Fruiting is often stimulated by using plastic dishes and irradiating them with near-ultra-violet light (Leach, 1962, 1963). Some of the genera commonly encountered, following such treatment, are:

Dematiaceous Hyphomycetes: *Alternaria, Stemphylium, Epicoccum, Nigrospora, Drechslera, Cercospora, Curvularia, Pyricularia.*

Coelomycetes: *Colletotrichum, Phoma, Ascochyta, Septoria.*

SEED-BORNE DEMATIACEOUS HYPHOMYCETES

Alternaria (Figs. 320, 321)

The conidia of *Alternaria* are highly characteristic. They are yellowish-brown, beaked, and possess transverse and longitudinal septa. Spores with this type of septation are described as **muriform:** the term **dictyospore** is also used. They develop by apical budding of a conidiogenous cell, or by budding at the tip of a spore. If a spore buds to produce more than a single spore, branching of the spore chain takes place. Where perfect states of *Alternaria* are known, they belong mostly to the Loculoascomycete genus *Pleospora* (p. 390). The form-genus is a large one (Joly, 1964), and contains species which are host-specific, and also others which are found on a wide range of host plants. *Alternaria alternata* (often called *A. tenuis*) is abundant on moribund and senescent plant material. Its spores form a very significant element of the air-borne spora, especially in the late summer and autumn in temperate countries (Hyde & Williams, 1946). It is known that *Alternaria* conidia are allergenic to man, and may cause respiratory disorders. In wheat seeds, *Alternaria* may form mycelium in the pericarp, and systemic infection of the shoots has been reported. *A. alternata* is associated with 'black-point', a deterioration of wheat grains, and *A. triticina* with a leaf blight (Bhowmik, 1969). *Brassica* seeds may be infected by two species of *Alternaria, A. brassicicola* (Fig. 320) and *A. brassicae* (Fig. 321). The two species are readily distinguishable by the shape of the conidia, those of *A. brassicae* being extended into a long, tapering beak. These species are associated with various disorders of brassicas. *A. brassicicola* causes a disease known as dark leaf spot, which causes a substantial lowering of yield, whilst *A. brassicae* causes grey leaf spot, and is a major cause of loss of seed production. It is probable that seed infection takes place through the ovary wall, from infections which originated on the decaying petals.

The structure of developing, mature and germinating conidia of *A. brassicicola* has been studied by Campbell (1969, 1970). Transverse septa are first laid down in a developing spore by annular ingrowths. A pore in the centre of the septum allows conduction of cytoplasm to later-developing spores, but even-

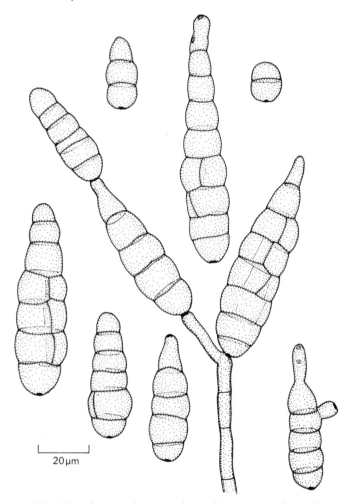

20 µm

Fig. 320. *Alternaria brassicicola*: conidiophores and conidia.

tually the septa of the basal spore become plugged, thus preventing further extension of the spore chain. The wall surrounding the spore is distinguishable into two layers, an outer, melanised layer, and an inner, hyaline layer. When a spore buds at its apex, a pore develops in the outer layer, through which the inner layer extends. A non-pigmented (i.e. non-melanised) mutant was isolated. It was found that its spores were much less resistant to chemical degradation than those of the pigmented wild-type (Campbell *et al.*, 1968).

Stemphylium (Fig. 216)

The conidia of *Stemphylium* are dark- and usually rough-walled, and have

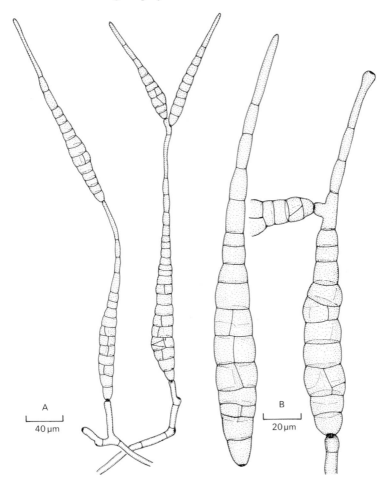

Fig. 321. *Alternaria brassicae.* A. Conidiophores and conidia. B. Conidia.

transverse and longitudinal septa. In contrast with *Alternaria,* they develop singly, not in chains. There are several species, distinguished from each other on conidial size and shape. In addition to their occurrence on seeds, some are weak parasites of shoots causing leaf-spots. *S. radicinum* causes a root-rot of carrots. The perfect states of *Stemphylium* are species of *Pleospora*, a Loculoascomycete with brownish-yellow, muriform ascospores (p. 393) (Simmons, 1969).

Epicoccum (Fig. 322)

Epicoccum is a ubiquitous saprophyte on senescent plant parts, and is also parasitic on some hosts – e.g. millet, apple and cotton bolls (Mulder & Pugh,

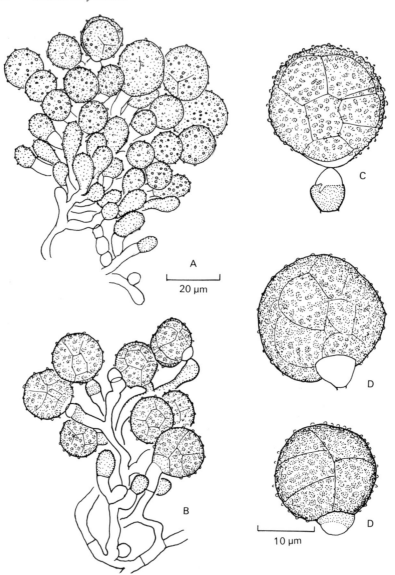

Fig. 322. *Epicoccum nigrum.* A. Young sporodochium. B. Conidiophores and conidia. C. Conidium almost separated from conidiophore. Note the bulging septum of the conidiophore. D. Two detached conidia.

1971). The most common species is *E. nigrum* (syn. *E. purpurascens*) (Schol-Schwartz, 1959). Although often found on seeds, especially of cereals and grasses, there is little evidence, in most cases, that it is parasitic. The conidial fructifications take the form of cushion-shaped sporodochia, which are black or purplish-red in colour, and covered with rough, warted, segmented, brownish-red conidia (Fig. 322A). The conidia may be violently projected from the sporodochium, probably by the rounding off of the two turgid cells, on either side of the septum which separates the conidium from its conidiophore (Fig. 322C) (Webster, 1966). It is possible that conidial separation is stimulated by drying, because the peak concentration of spores in air occurs shortly before noon (Meredith, 1966). A pycnidial state, *Phoma epicoccina,* has been reported (Punithalingam *et al.*, 1972). The fine structure of conidia has been studied by Brushaber & Haskins (1973) and Duncan & Herald (1974).

Nigrospora (Fig. 323)

Species of *Nigrospora* are common on plants such as rice, sugar cane, maize and other Monocotyledonous hosts. *N. oryzae* is a saprophyte or weak parasite found on grains of rice, sorghum and maize. Its perfect state is *Khuskia oryzae* (Hudson, 1963), a unitunicate Pyrenomycete. The black, shiny, depressed-globose conidia are distinguished by size from those of *N. sphaerica*. The conidia of both species are common components of the air spora in Jamaican banana plantations (Meredith, 1961, 1962), reaching a maximum in the early daylight hours. In *N. sphaerica*, the conidia are discharged explosively by the discharge of a jet of cytoplasm from a swollen, ampulliform cell. The conidium is held in place over the orifice of the jet by a thin-walled, supporting collar (Webster, 1952).

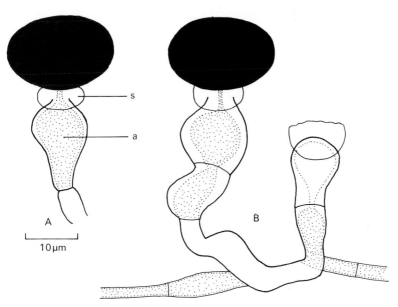

Fig. 323. *Nigrospora sphaerica.* A. Conidiophores consisting of a swollen ampulliform cell, a, full of cytoplasm, and a supporting collar, s, through which a narrow neck of cytoplasm passes. B. One undischarged and one discharged conidiophore. Note that the discharged conidiophore is void of cytoplasm.

Drechslera (Fig. 324)

Species of this form-genus are to be found on seeds of cereals, grasses and other hosts, and are pathogenic. Many of them were previously classified in the form-genus *Helminthosporium*. However, the type species of this form-genus is *H. velutinum*, in which conidial development is tretic (Fig. 292, p. 520). The conidia develop *terminally* or *laterally* through very narrow channels in the thick wall of the conidiophore. The formation of a terminal conidium in *H. velutinum* prevents further elongation of the conidiophore. The sequence of conidium development in *Drechslera* is not of this type. Here the conidia develop *apically* from a pore at the tip of the conidiophore. After the first conidium has developed, the conidiophore may proliferate, either by growing through the scar of the first conidium (percurrent proliferation) or by forming a new apex laterally to the first conidium, which grows past the point of attachment to form a second conidium (lateral proliferation). The process may be repeated. Lateral proliferation in *D. sorokiniana* is illustrated in Fig. 324B. The conidia of *Drechslera* are cylindrical or club-shaped and transversely septate. Luttrell (1963) has distinguished two types of septation.

Euseptate conidia are surrounded by a single wall and have true septa formed as inward extensions of the lateral walls. The septum probably develops as a closing diaphragm, and usually a pore remains in the center connecting the two cells separated by the septum.

Distoseptate conidia . . . have a common outer wall enclosing more or less spherical cells each of which is surrounded by an individual wall. Apparently the wall of the young conidium is double. Invaginations of the inner wall divide the protoplast into a series of cells resembling peas in a pod . . . Pores in the inner walls connect the protoplasts of adjoining cells.

Septation of this type has been reported in *D. sorokiniana* (Fig. 324c), *D. avenacea* and *D. oryzae*. The term pseudoseptate is sometimes used for this condition.

There are differing accounts of the fine structural details of conidial development in *Drechslera* spp. For *D. sorokiniana*, Cole (1973) has shown that conidial development is tretic. The newly formed conidium develops enteroblastically through a channel in the wall of the conidiogenous cell, probably as a result of autolysis of the outer wall layers. In *D. maydis*, Brotzman *et al.* (1975) have shown that development is holoblastic, and all the layers of the wall of the conidium are involved in development.

Perfect states of several *Drechslera* species are known. They are all Loculoascomycetes, and include the following genera: *Cochliobolus, Pyrenophora, Pleospora* and *Trichometasphaeria* (Ellis, 1971b).

Important pathogenic, seed-borne *Drechslera* species include: *D. sorokiniana* (conidial state of *Cochliobolus sativus*), the cause of seedling blight, foot rot and ear blight of barley and other hosts; *D. teres* (conidial state of *Pyrenophora teres*), the cause of net blotch of barley; and *D. graminea* (conidial state of *Pyrenophora graminea*), the cause of leaf-stripe of barley.

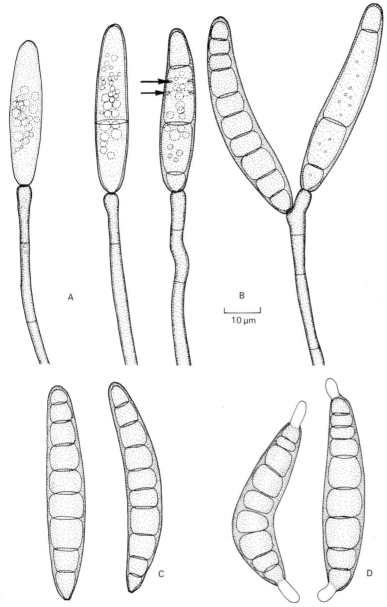

Fig. 324. *Drechslera sorokiniana*, the conidial state of *Cochliobolus sativus*. A. Developing conidia. The arrows point to developing septa. B. Conidiophore showing the development of a second conidium lateral to the first. C. Mature conidia. D. Two germinating conidia showing emergence of germ tubes from each end of the conidium.

Violent discharge of conidia has been reported in *Drechslera turcica* (conidial state of *Trichometasphaeria turcica*), the cause of northern leaf blight of maize. It should be noted that this fungus has not yet been found to be seed-borne on maize (Neergaard, 1977), but it does occur on seeds of *Sorghum*. Meredith (1965b) showed that discharge was associated with drying, and believed that this caused the cells of the conidiophore to shrink.

The walls, being elastic, tend to resist this inward pull and so are brought into a state of tension. A point is reached when the forces of adhesion or cohesion, or both, are overcome; a rupture occurs in the cell solution and a gas phase suddenly develops. Tension is suddenly released, causing a jolt which, in turn, if it is of sufficient magnitude, causes the conidium to be projected.

Leach (1976), working with the same fungus, confirmed that spore discharge was associated with drying, but showed that it was also associated with increasing humidity. Irradiation of lesions on maize leaves with red-infra-red light also resulted in spore discharge. Irradiation was accompanied by an increase in voltage measurements across the lesions. Leach has suggested that the conidia are discharged by electrostatic repulsion.

Violent spore discharge probably does not occur in all species of *Drechslera*. In fact, considerable force may be required to separate the conidia of *D. maydis* from their conidiophores. Studies of the centrifugal force, or the wind speed, required to bring about separation have been made by Aylor (1975) and Aist *et al.* (1976). Spores became detached at wind speeds above 5 m/second, and at speeds of 6 to 7 m/second liberation was almost complete 3–5 mm from the edge of a maize leaf, whilst 10 mm from the edge removal was nil. The mature conidium is attached to its conidiophore only by a narrow isthmus of cytoplasm, surrounding which is a circular, raised hilum. Laterally applied forces may lever the spore away from its conidiophore, with the hilum as a pivot.

Cercospora (Fig. 325)

About 2,000 species names have been included in this form-genus, but many are probably synonyms. The multitude of names reflects the tendency of plant pathologists to describe, as new species, forms of *Cercospora* on previously unrecorded host plants. However, Johnson & Valleau (1949) isolated *Cercospora* from twenty-eight host plants in sixteen families, and all seemed to belong to the same species. There is great variation in the dimensions of conidiophores and conidia, produced in response to changes in humidity. Most species of *Cercospora* are weak parasites. *C. beticola* (Fig. 325) (regarded by Ellis (1971b) as probably a synonym of *C. apii*) is the cause of leaf spot of sugar beet, and is seed-borne. The conidia are narrow and tapering, and contain numerous transverse septa. They are effectively dispersed by rain-splash.

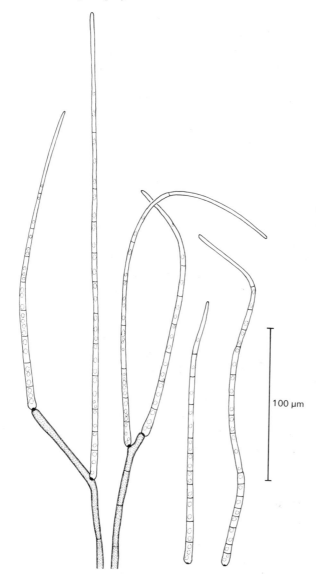

Fig. 325. *Cercospora beticola*. Conidiophores and conidia from sugar-beet seed.

Curvularia (Fig. 326)

The form-genus *Curvularia* includes over thirty species, not all of which are seed-borne (Ellis, 1966, 1971b). An account of seed-borne species of *Curvularia* on agricultural seeds in Canada has been given by Groves & Skolko (1944–1945), whilst those on rice grain have been monographed by Benoit & Mathur (1970). The characteristic features of the form-genus are the forma-

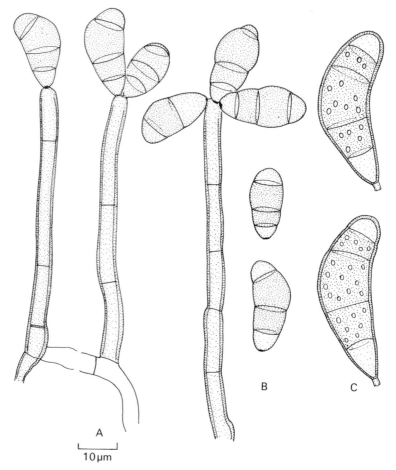

Fig. 326. *Curvularia* spp. A. *C. lunata*, conidiophores showing sequence of conidial development. B. *C. lunata*, mature detached conidia. Note the paler end cells. C. *C. cymbopogonis*, mature detached conidia showing the protruberant hilum.

tion of macronematous, mononematous, erect conidiophores (and occasionally stromata), bearing spores spirally or in whorls. The spores are usually curved; the third cell from the base of the spore is larger than the rest, and the end cells are paler (see Fig. 326A,B, of *C. lunata*). In some species (e.g. *C. cymbopogonis,* Fig. 326C), the base of the conidium bears a protruberant hilum. Kendrick & Cole (1968) have described, and photographed by time-lapse cinematography, the events associated with conidium development in *C. inaequalis.* The first conidium develops tretically, i.e. as a poroconidium at the apex of the elongating conidiophore. A tiny apical pore forms at the tip of the conidiophore by dissolution of the outer wall, and a spherical, cytoplasmic bubble is blown out through the pore. The first conidium assumes an obovoid shape and, after it has matured, the conidiophore develops a new subterminal growing point, from which a second conidium initial arises. The process is repeated so that a succession of new apices, each terminated by a conidium, is formed. The term sympodula has been applied to this type of conidiophore apex.

A range of different disease symptoms is associated with *Curvularia* spp. On rice and other crops, they have been reported as causal agents of leaf spots, leaf blights, kernel rot, root rot, seedling blights, grain discolorations, grain lesions and grain deformations (Benoit & Mathur, 1970).

The perfect states of *Curvularia*, where known, are species of *Cochliobolus*.

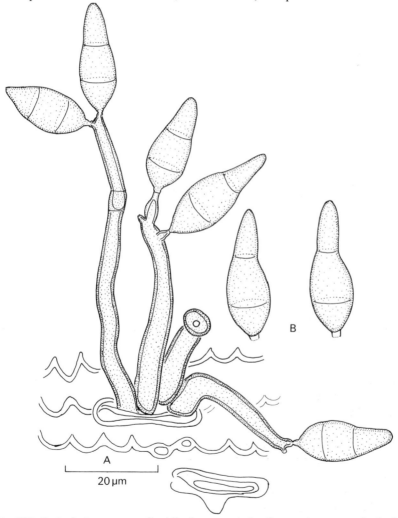

Fig. 327. *Pyricularia oryzae*. A. Conidiophores emerging from a stoma on a rice leaf. B. Detached conidia.

Pyricularia (Fig. 327)

There are two common species of *Pyricularia* associated with diseases of cereals and grasses in the tropics and occasionally elsewhere. *P. grisea* is associated with leaf spots of many grasses. *P. oryzae* is the cause of the serious disease, rice blast, and is frequently seed-borne in all rice-producing countries. Whether *P. grisea* and *P. oryzae* should be regarded as distinct species is a matter of opinion: some would regard them as synonyms. *P. oryzae* is the principal, most destructive disease of rice. Rice seedlings or plants at the tillering stage are often completely killed. Heavy infections of the panicles are often detrimental to yield (Ou, 1972; I.R.R.I., 1963). The conidiophores are slender, and extend from the mycelium within the diseased leaf through stromata. A succession of obpyriform, two-septate, pale-brown conidia is produced, each attached to the conidiophore by a protruberant hilum. The conidia develop blastically. According to Ingold (1964), they are discharged violently, possibly by explosion of the hilum (which he termed the stalk cell). Discharge is at a maximum during the night. coinciding with maximum humidity. On germination, the conidia develop short germ tubes bearing appressoria.

The perfect state of *P. grisea* is a heterothallic, unitunicate Pyrenomycete, *Magnaporthe grisea* (Hebert, 1971; Yaegashi & Udagawa, 1978). It is obvious that *P. oryzae* and *P. grisea* are not related to *P. aquatica* (see p. 536) which has as its perfect state the Loculoascomycete *Massarina*.

SEED-BORNE COELOMYCETES

Several form-genera belonging to the Coelomycetes are seed-borne. They may not always produce their conidial fructifications on the seed, but may have mycelium within the seed from which the seedling can become infected. The conidial fructifications then develop later on the diseased host. There are two main types of conidial fructification in Coelomycetes, acervuli and pycnidia, and these terms are defined on p. 526.

Colletotrichum (Fig. 328)

In *Colletotrichum*, the conidial fructification is an acervulus. Von Arx (1957) has distinguished eleven species, and has treated as synonyms a large number of named forms found on different host plants. Some serious plant diseases are caused by *Colletotrichum* spp., including:

C. dematium, on legumes, spinach.

C. gloeosporoides, on a wide range of hosts; the perfect state is *Glomerella cingulata* (Sphaeriales).

C. gossypii, causing pink boll rot and anthracnose (i.e. a disease with limited lesions) of cotton; its perfect state is *G. gossypii*.

C. graminicola, common on seeds of many grasses and cereals; its perfect state is *G. graminicola.*

C. lindemuthianum, causing anthracnose of *Phaseolus* beans; its perfect state is *G. cingulata* f. sp. *phaseoli.*

C. lini, causing stem anthracnose and canker of flax.

For a more detailed account of these diseases, see Neergaard (1977).

Colletotrichum graminicola is illustrated in Fig. 328, growing on grains of *Sorghum.* The acervulus is saucer-shaped and surrounded by stiff, black, unbranched hairs. The conidiogenous cells form a closely packed palisade of phialides. The phialoconidia accumulate under moist conditions as glistening pustules held in place by the setae.

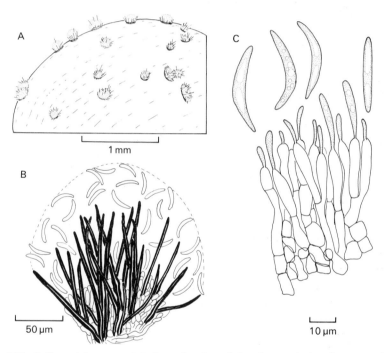

Fig. 328. *Colletotrichum graminicola.* A. Portion of *Sorghum* grain bearing acervuli. B. An acervulus. C. Phialides and phialoconidia.

Phoma (Fig. 329)

The form-genus *Phoma* includes a large number of species names, but it is artificial because perfect states are found in different genera of the Pseudo-sphaeriales. The conidial fructification is a pycnidium, a dark, flask-shaped structure opening usually by a single, circular ostiole, and lined by a hymenium of conidiogenous cells from which numerous, one-celled, hyaline pycnospores (pycnidiospores) develop. The pycnospores often ooze out from

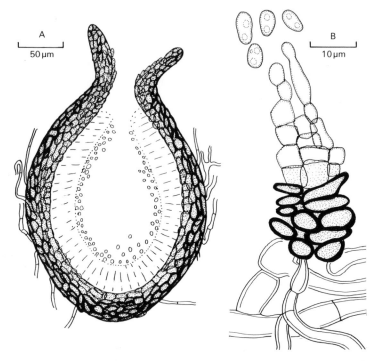

Fig. 329. *Phoma betae*. A. L.S. pycnidium. B. Portion of pycnidium wall, showing conidiogenous cells and conidia (phialides and phialospores).

the ostiole as a tendril or cirrhus. The conidiogenous cells are very small, and the details of conidium development are difficult to discern with the light microscope. Brewer & Boerema (1965) and Boerema (1965) have studied spore development using the electron microscope. They describe the process of spore formation as a monopolar, repetitive budding of the small, undifferentiated cells of the pycnidial wall. As repeated spore formation occurs, the apex of the conidiogenous cell develops a thickened rim which resembles a phialide or an annellophore (annellide). Sutton (1964) interprets the conidia as phialospores.

Important seed-borne pathogens belonging to the form-genus *Phoma* include:

Phoma betae (Fig. 329), the cause of black-leg and damping-off of beet; its perfect state is *Pleospora bjorlingii*.

Phoma lingam (Plenodomus lingam), the cause of black-leg of crucifers; the perfect state is *Leptosphaeria maculans*.

Phoma medicaginis var *pinodella*, the cause of foot rot and collar rot of peas.

For a fuller account of these pathogens, see Neergaard (1977).

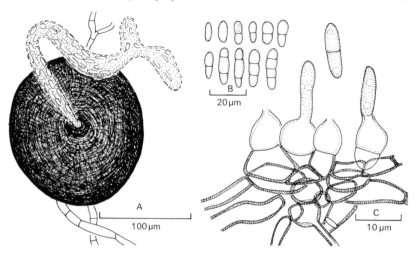

Fig. 330. *Ascochyta pisi*. A. Pycnidium seen from above, showing a cirrhus of spores oozing from the ostiole. B. Pycnospores. C. Portion of pycnidium wall in section, showing origin of pycnospores.

Ascochyta (Fig. 330)

This form-genus contains about 500 species names, but many of these are synonyms. It is probably heterogeneous, and critical studies are needed to determine details of conidium ontogeny. The pycnidia contain hyaline, two-celled conidia which, according to Brewer & Boerema (1965), arise by septation from the conidiogenous cell. A few conidia with no, two or three septa

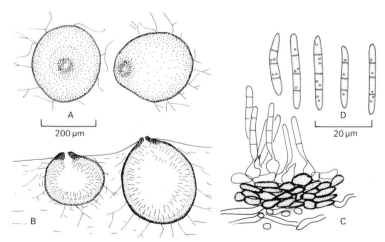

Fig. 331. *Septoria nodorum*. A. Pycnidia seen from above. B. Pycnidia in section in agar culture. C. Portion of wall of pycnidium showing origin of conidia. D. Conidia. A, B to same scale; C, D to same scale.

may occasionally be produced. Seed-borne species of *Ascochyta* of pathogenic significance include:

A. fabae, the cause of leaf and pod spot of broad beans, *Vicia faba.*

A. gossypii, the cause of *Ascochyta* blight or wet-weather canker of cotton.

A. pinodes, the cause of foot rot or blight of pea; its perfect state is *Mycosphaerella pinodes* (Pseudosphaeriales).

A. pisi (Fig. 330), the cause of leaf and pod spot of pea. For a fuller account of these pathogens, see Neergaard (1977).

Septoria (Fig. 331)

This is another large form-genus containing over 1,000 species names, many representing collections found on different host plants, but not necessarily different taxa. The pycnidia are immersed in the host tissues and contain filiform conidia with several transverse septa. Some important seed-borne plant pathogens include:

S. apiicola, the cause of late blight of celery, which is a destructive leaf disease.

S. avenae, which causes leaf blotch, stem break and black stem of oats; its perfect state is *Leptosphaeria avenaria.*

S. nodorum (Fig. 331), glume blotch of wheat, which is common in regions of high rainfall; the perfect state is *Leptosphaeria nodorum* (Loculoascomycetes). The pycnidia develop beneath the epidermis of the leaves and glumes of wheat, and give rise to cylindrical, three-septate pycnospores. These accumulate in a tendril or cirrhus and are dispersed by rain-splash. A second pycnidial state containing minute, unicellular conidia has also been discovered (Shaw, 1953; Harrower, 1976). The microspores are capable of infecting host leaves via germ tubes which penetrate the stomata. There is no evidence that they play a spermatial role. Low temperatures (5–10 °C) and irradiation with near-ultra-violet light favour their production.

References

Abbayes, H. des (1951). Traité de lichénologie. *Encyclopédie Biologique*, **41**, 217 pp. Paris: Lechevalier.

Abrams, E. (1950). Microbiological deterioration of cellulose during the first 72 hours of attack. *Textile Research Journal*, **20**, 71.

Agarwal, P. N., Verma, G. M., Verma, R. K. & Sahgal, D. D. (1963a). Decomposition of cellulose by the fungus *Chaetomium globosum*: Part I. Studies on enzyme activity. *Indian Journal of Experimental Biology*, **1**, 46–50.

Agarwal, P. N., Verma, G. M., Verma, R. K. & Rastogi, V. K. (1963b). Decomposition of cellulose by the fungus *Chaetomium globosum*. Part III. Factors affecting elaboration of cellulolytic enzymes. *Indian Journal of Experimental Biology*, **1**, 229–230.

Ahlquist, C. N. & Gamow, R. I. (1973). *Phycomyces*. Mechanical behaviour of stage II and stage IV. *Plant Physiology*, **51**, 586–587.

Ahmad, M. (1965). Incompatibility in yeasts. In *Incompatibility in Fungi*, 13–23. Editors: K. Esser & J. R. Raper. Berlin: Springer.

Ahmadjian, V. (1965). Lichens. *Annual Review of Microbiology*, **19**, 1–20.

Ahmadjian, V. (1967a). *The Lichen Symbiosis*. 152 pp. Waltham, Mass.: Blaisdell.

Ahmadjian, V. (1967b). A guide to the algae occurring as lichen symbionts: Isolation, culture, cultural physiology and identification. *Phycologia*, **6**, 127–160.

Ahmadjian, V. (1973). Resynthesis of lichens. In *The Lichens*, 565–579. Editors: V. Ahmadjian & M. E. Hale. New York and London: Academic Press.

Ahmadjian, V. & Hale, M. E. (eds.) (1973). *The Lichens*. 697 pp. New York and London: Academic Press.

Ainsworth, G. C. (1952). *Medical Mycology. An Introduction to its Problems*. 105 pp. London: Pitman.

Ainsworth, G. C. (1965). Longevity of *Schizophyllum commune*. II. *Nature, London*, **195**, 1120–1121.

Ainsworth, G. C. (1973). Introduction and Keys to higher taxa. In *The Fungi: An Advanced Treatise*, **IVB**, 1–7. Editors: G. C. Ainsworth, F. K. Sparrow & A. S. Sussman. New York and London: Academic Press.

Ainsworth, G. C. & Austwick, P. K. C. (1973). *Fungal Diseases of Animals*. 216 pp. Farnham Royal, England: Commonwealth Agricultural Bureaux.

Ainsworth, G. C., James, P. W. & Hawksworth, D. L. (1971). *Ainsworth & Bisby's Dictionary of the Fungi*, 6th edition. 663 pp. Kew, Surrey: Commonwealth Mycological Institute.

Ainsworth, G. C. & Sampson, K. (1950). *The British Smut Fungi (Ustilaginales)*. 137 pp. Kew, Surrey: Commonwealth Mycological Institute.

Ainsworth, G. C. & Sussman, A. S. (eds.) (1965). *The Fungi: An Advanced Treatise*. **I.** *The Fungal Cell*. 748 pp. New York and London: Academic Press.

Aist, J. R., Aylor, D. E. & Parlange, J.-Y. (1976). Ultrastructure and mechanics of the conidium–conidiophore attachment of *Helminthosporium maydis*. *Phytopathology*, **66**, 1050–1055.

Aist, J. R. & Williams, P. H. (1971). The cytology and kinetics of cabbage root hair penetration by *Plasmodiophora brassicae. Canadian Journal of Botany*, **49**, 2023–2034.

Aist, J. R. & Williams, P. H. (1972). Ultrastructure and time course of mitosis in the fungus *Fusarium oxysporum. Journal of Cell Biology*, **55**, 368–389.

Ajello, L. (1977). Taxonomy of dermatophytes: A review of their imperfect and perfect states. In *Recent Advances in Medical and Veterinary Mycology*, 289–297. Editor: K. Iwata. Baltimore: University Park Press.

Alasoadura, S. O. (1963). Fruiting in *Sphaerobolus* with special reference to light. *Annals of Botany, London*, N.S., **27**, 123–145.

Aldrich, H. C. (1967). The ultrastructure of meiosis in three species of *Physarum. Mycologia*, **59**, 127–148.

Aldrich, H. C. (1968). The development of flagella in swarm cells of the myxomycete *Physarum flavicomum. Journal of General Microbiology*, **50**, 217–222.

Aldrich, H. C. & Carroll, G. (1972). Synaptonemal complexes and meiosis in *Didymium iridis*: a re-investigation. *Mycologia*, **63**, 308–316.

Alexopoulos, C. J. (1960a). Morphology and laboratory cultivation of *Echinostelium minutum* de Bary. *American Journal of Botany*, **47**, 37–43.

Alexopoulos, C. J. (1960b). Gross morphology of the plasmodium and its possible significance in the relationships among the Myxomycetes. *Mycologia*, **52**, 1–20.

Alexopoulos, C. J. (1969). The experimental approach to the taxonomy of the Myxomycetes. *Mycologia*, **61**, 219–239.

Alexopoulos, C. J. (1973). Myxomycetes. In *The Fungi: An Advanced Treatise*, **IVB**, 39–60. Editors: G. C. Ainsworth, F. K. Sparrow & A. S. Sussman. New York and London: Academic Press.

Alléra, A. & Wohlfarth-Bottermann, K.-E. (1972). Weitreichende fibrilläre Protoplasmadifferenzierungen und ihre Bedeutung für die Protoplasmaströmung. IX. Aggregationszustände des Myosins und Bedingungen zur Enstehung von Myosinfilamenten in den Plasmodien von *Physarum polycephalum. Cytobiologie*, **6**, 261–286.

Alvin, K. L. (1977). *The Observer's Book of Lichens*. 188 pp. London: Warne.

Ames, L. M. (1934). Hermaphroditism involving self-sterility and cross-fertility in the Ascomycete *Pleurage anserina. Mycologia*, **26**, 392–414.

Ames, L. M. (1963). A monograph of the Chaetomiaceae. *U.S. Army Research and Development Series, No. 2*. 125 pp.

Amon, J. P. & Perkins, F. O. (1968). Structure of *Labyrinthula* sp. zoospores. *Journal of Protozoology*, **15**, 543–546.

Andrus, C. F. & Harter, L. L. (1937). Organization of the unwalled ascus in two species of *Ceratostomella. Journal of Agricultural Research*, **54**, 19–46.

Anon (1967a). *Fomes annosus*. A fungus causing butt rot, root rot and death of conifers. *Forestry Commission Leaflet No. 5*. 11 pp. London: H.M.S.O.

Anon (1967b). Silver Leaf Disease of Fruit Trees. *Advisory Leaflet No. 246, Ministry of Agriculture, Fisheries and Food*. 7 pp. London: H.M.S.O.

Anon (1973). Wart Disease of Potatoes. *Advisory Leaflet No. 274, Ministry of Agriculture, Fisheries and Food*. 2nd edition. 5 pp. London: H.M.S.O.

Aoshima, K. (1953). Sexuality of *Elfingia applanata (Fomes applanatus). Mycological Journal of the Nagao Institute*, **3**, 5–11.

Apinis, A. E. (1964). Revision of British Gymnoascaceae. *Mycological Paper No. 96, Commonwealth Mycological Institute*. 56 pp.

Aragno, M. (1968). Formation et évolution de l'asque chez *Rhytisma acerinum* (Pers.) Fr. *Bericht der Schweizerischen Botanischen Gesellschaft*, **77**, 173–186.

Arnold, C. A. (1928). The development of the perithecium and spermagonium of *Sporormia leporina. American Journal of Botany*, **15**, 241–245.

Aronson, J. M. (1965). The cell wall. In *The Fungi: An Advanced Treatise*, **I**, 49–76.

Editors: G. C. Ainsworth & A. S. Sussman. New York and London: Academic Press.

Aronson, J. M., Cooper, B. A. & Fuller, M. S. (1967). Glucans of oomycete cell walls. *Science, N.Y.*, **155**, 332–335.

Aronson, J. M. & Preston, R. D. (1960). An electron microscopic and X-ray analysis of the walls of selected lower Phycomycetes. *Proceedings of the Royal Society, Series B*, **152**, 346–352.

Arx, J. A. von (1957). Die Arten der Gattung *Colletotrichum* Corda. *Phytopathologische Zeitschrift*, **29**, 413–468.

Arx, J. A. von & Müller, E. (1954). Die Gattungen der amerosporen Pyrenomyceten. *Beiträge der Kryptogamenflora der Schweiz*, Bd. **11**(1), 1–434.

Arx, J. A. von & Müller, E. (1975). A re-evaluation of the bitunicate Ascomycetes with Keys to families and genera. *Studies in Mycology*, **9**, 159 pp. Baarn, Holland: Centraalbureau voor Schimmelcultures.

Aschner, M. & Kohn, S. (1958). The biology of *Harposporium anguillulae. Journal of General Microbiology*, **19**, 182–189.

Ashworth, D. (1931). *Puccinia malvacearum* in monosporidial culture. *Transactions of the British Mycological Society*, **16**, 177–202.

Ashworth, J. M. & Sackin, M. J. (1969). Role of aneuploid cells in cell differentiation in the cellular slime mould *Dictyostelium discoideum. Nature, London*, **224**, 817–818.

Ashworth, J. M. & Watts, D. J. (1970). Metabolism of the cellular slime mould *Dictyostelium discoideum* grown in axenic culture. *Biochemical Journal*, **119**, 175–182.

Assche, C. van & Vanachter, A. (1970). Systemic fungicides to control fungal diseases in vegetables. *Parasitica*, **26**, 117–125.

Atkins, F. C. (1966). *Mushroom Growing Today*. 188 pp. London: Faber & Faber.

Austin, W. L., Lafayette, F. & Roth, I. L. (1974). Scanning electron microscope studies on ascospores of homothallic species of *Neurospora. Mycologia*, **66**, 130–138.

Ave, R. & Müller, E. (1967). Vergleichende Untersuchungen an einigen Chaetomium-arten. *Berichte der Schweizerischen Botanische Gesellschaft*. **77**, 187–207.

Aylor, D. E. (1975). Force required to detach conidia of *Helminthosporium maydis. Plant Physiology*, **55**, 99–101.

Bachmann, B. J. & Strickland, W. N. (1965). *Neurospora Bibliography and Index*. 225 pp. New Haven and London: Yale University Press.

Backus, M. P. (1939). The mechanics of conidial fertilization in *Neurospora sitophila. Bulletin of the Torrey Botanical Club*, **66**, 63–76.

Bacon, C. W., Porter, J. K., Robbins, J. D. & Luttrell, E. S. (1977). *Epichloe typhina* from toxic tall fescue grasses. *Applied and Environmental Microbiology*, **34**, 576–581.

Bagchee, K. (1954). *Merulius lacrymans* (Wulf.) Fr. in India. *Sydowia*, **8**, 80–85.

Bailiss, K. W. & Wilson, I. M. (1967). Growth hormones and the creeping thistle rust. *Annals of Botany, London*, N.S., **31**, 195–211.

Baker, H. G. (1947). Infection of species of *Melandrium* by *Ustilago violacea* and the transmission of the resultant disease. *Annals of Botany, London*, N.S., **11**, 333–348.

Bakshi, B. K. (1951). Development of perithecia and reproductive structures in two species of *Ceratocystis. Annals of Botany, London*, N.S., **15**, 53–61.

Balan, J. & Gerber, N. N. (1972). Attraction and killing of the nematode *Panagrellus redivivus* by the predaceous fungus *Arthrobotrys dactyloides. Nematologica*, **18**, 163–173.

Banbury, G. H. (1955). Physiological studies in the Mucorales. iI. The zygotropism of zygophores of *Mucor mucedo* Brefeld. *Journal of Experimental Botany*, **6**, 235–244.

Bandoni, R. J. (1963). Conjugation in *Tremella mesenterica. Canadian Journal of Botany*, **41**, 467–474.

Bandoni, R. J. (1965). Secondary control of conjugation in *Tremella mesenterica*. *Canadian Journal of Botany*, **43**, 627–630.

Bandoni, R. J. (1972). Terrestrial occurrence of some aquatic hyphomycetes. *Canadian Journal of Botany*, **50**, 2283–2288.

Bandoni, R. J. (1974). Mycological observations on the aqueous films covering decaying leaves and other litter. *Transactions of the Mycological Society of Japan*, **15**, 309–315.

Bandoni, R. J. & Bisalputra, A. A. (1971). Budding and fine structure of *Tremella mesenterica* haplonts. *Canadian Journal of Botany*, **49**, 27–30.

Bandoni, R. J., Lobo, K. J. & Brezden, S. A. (1971). Conjugation and chlamydospores in *Sporobolomyces odorus*. *Canadian Journal of Botany*, **49**, 683–686.

Banerjee, S. (1956). Heterothallism in *Auricularia auricula-judae* (Linn.) Schroet. *Science and Culture*, **21**, 549–550.

Banno, I. (1967). Studies on the sexuality of *Rhodotorula*. *Journal of General and Applied Microbiology*, **13**, 167–196.

Barger, G. (1931). *Ergot and ergotism*. London and Edinburgh: Gurney & Jackson.

Barksdale, A. W. (1960). Inter-thallic sexual reactions in *Achlya*, a genus of the aquatic fungi. *American Journal of Botany*, **47**, 14–23.

Barksdale, A. W. (1962). Effect of nutritional deficiency on growth and sexual reproduction of *Achlya ambisexualis*. *American Journal of Botany*, **49**, 633–638.

Barksdale, A. W. (1963a). The uptake of exogenous hormone A by certain strains of *Achlya*. *Mycologia*, **55**, 164–171.

Barksdale, A. W. (1963b). The role of hormone A during sexual conjugation in *Achlya ambisexualis*. *Mycologia*, **55**, 627–632.

Barksdale, A. W. (1965). *Achlya ambisexualis* and a new cross-conjugating species of *Achlya*. *Mycologia*, **57**, 493–501.

Barksdale, A. W. (1966). Segregation of sex in the progeny of a selfed heterozygote of *Achlya ambisexualis*. *Mycologia*, **58**, 802–804.

Barksdale, A. W. (1969). Sexual hormones in *Achlya* and other fungi. *Science*, **166**, 831–837.

Barksdale, A. W. (1970). Nutrition and antheridiol-induced branching in *Achlya ambisexualis*. *Mycologia*, **62**, 411–420.

Bärlocher, F. & Kendrick, W. B. (1973a). Fungi in the diet of *Gammarus pseudolimnaeus* (Amphipoda). *Oikos*, **24**, 295–300.

Bärlocher, F. & Kendrick, W. B. (1973b). Fungi and food preference of *Gammarus pseudolimnaeus*. *Archiv für Hydrobiologie*, **72**, 501–516.

Barnett, H. L. (1937). Studies in the sexuality of the Heterobasidiae. *Mycologia*, **29**, 626–649.

Barnett, J. A. & Pankhurst, R. J. (1974). *A new key to the yeasts*. 273 pp. Amsterdam and London: North-Holland Publishing Company.

Barratt, R. W. (1974). *Neurospora crassa*. In *Handbook of Genetics*, **I**, 511–529. Editor: R. C. King. New York and London: Plenum Press.

Barron, G. L. (1968). *The Genera of Hyphomycetes from Soil*. Baltimore: Williams & Wilkins.

Barron, G. L. (1977). *The Nematode-destroying Fungi*. Guelph, Ontario: Canadian Biological Publications Ltd.

Barron, G. L. & Dierkes, Y. (1977). Nematophagous fungi: *Hohenbuehelia*, the perfect state of *Nematoctonus*. *Canadian Journal of Botany*, **55**, 3054–3062.

Barron, J. L. & Hill, E. P. (1974). Ultrastructure of zoosporogenesis in *Allomyces macrogynus*. *Journal of General Microbiology*, **80**, 319–327.

Bartnicki-Garcia, S. (1968). Cell wall chemistry, morphogenesis, and taxonomy of fungi. *Annual Review of Microbiology*, **22**, 87–108.

Bartnicki-Garcia, S. (1970). Cell wall composition and other biochemical markers in fungal taxonomy. In *Phytochemical Phylogeny*, 81–103. Editor: J. B. Harborne. New York and London: Academic Press.

Bartnicki-Garcia, S. (1973). Fundamental aspects of hyphal morphogenesis. *Symposium of the Society for General Microbiology*, **23**, 245–267.

Bartnicki-Garcia, S., Bracker, C. E., Reyes, E. & Ruiz-Herrera, J. (1978). Isolation of chitosomes from taxonomically diverse fungi and synthesis of chitin microfibrils *in vitro*. *Experimental Mycology*, **2**, 173–192.

Bartnicki-Garcia, S. & Hemmes, D. E. (1976). Some aspects of the form and function of oomycete spores. In *The Fungus Spore. Form and Function*, 593–641. Editors: W. M. Hess & D. J. Weber. New York: John Wiley & Sons.

Bartnicki-Garcia, S. & McMurrough, I. (1971). Biochemistry of morphogenesis in yeasts. In *The Yeasts*, **2**, 441–491. Editors: A. H. Rose & J. S. Harrison. London and New York: Academic Press.

Bartnicki-Garcia, S., Nelson, N. & Cota-Robles, E. (1968). Electron microscopy of spore germination and cell wall formation in *Mucor rouxii*. *Archiv für Mikrobiologie*, **63**, 242–255.

Bartnicki-Garcia, S. & Nickerson, W. J. (1962a). Induction of yeast-like development in *Mucor* by carbon-dioxide. *Journal of Bacteriology*, **84**, 829–840.

Bartnicki-Garcia, S. & Nickerson, W. J. (1962b). Nutrition, growth and morphogenesis of *Mucor rouxii*. *Journal of Bacteriology*, **84**, 841–858.

Bartnicki-Garcia, S. & Nickerson, W. J. (1962c). Isolation, composition and structure of cell walls of filamentous and yeast-like forms of *Mucor rouxii*. *Biochimica et Biophysica Acta*, **58**, 102–119.

Bartnicki-Garcia, S. & Reyes, E. (1968). Polyuronides in the cell wall of *Mucor rouxii*. *Biochimica et Biophysica Acta*, **170**, 54–62.

Bashi, E. & Fokkema, N. J. (1976). Scanning electron microscopy of *Sporobolomyces roseus* on wheat leaves. *Transactions of the British Mycological Society*, **67**, 500–505.

Bashi, E. & Fokkema, N. J. (1977). Environmental factors limiting growth of *Sporobolomyces roseus*, an antagonist of *Cochliobolus sativus*, on wheat leaves. *Transactions of the British Mycological Society*, **68**, 17–25.

Basith, M. & Madelin, M. F. (1968). Studies on the production of perithecial stromata by *Cordyceps militaris* in artificial culture. *Canadian Journal of Botany*, **46**, 473–480.

Basu, S. N. (1951). Significance of calcium in the fruiting of *Chaetomium* species, particularly *Chaetomium globosum*. *Journal of General Microbiology*, **5**, 231–238.

Batra, L. R. (1963). Contributions to our knowledge of Ambrosia fungi. II. *Endomycopsis fasciculata* nom. nov. (Ascomycetes). *American Journal of Botany*, **50**, 481–487.

Batts, C. C. V. (1951). Physiologic specialization of *Puccinia graminis* Pers. in South-East Scotland. *Transactions of the British Mycological Society*, **34**, 533–538.

Batts, C. C. V. (1955). Observations on the infection of wheat by loose smut (*Ustilago tritici* (Pers.) Rostr.). *Transactions of the British Mycological Society*, **38**, 465–475.

Bayliss-Elliott, J. S. (1926). Concerning fairy rings in pastures. *Annals of Applied Biology*, **13**, 277–288.

Beadle, G. W. (1959). Genes and chemical reactions in *Neurospora*. *Science, N.Y.*, **129**, 1715–1719.

Beadle, G. W. & Coonradt, V. L. (1944). Heterokaryosis in *Neurospora crassa*. *Genetics*, **29**, 291–308.

Beakes, G. W. & Gay, J. L. (1978a). Light and electron microscopy of oospore maturation in *Saprolegnia furcata*. 1. Cytoplasmic changes. *Transactions of the British Mycological Society*, **71**, 11–24.

Beakes, G. W. & Gay, J. L. (1978b). Light and electron microscopy of oospore maturation in *Saprolegnia furcata*. 2. Wall development. *Transactions of the British Mycological Society*, **71**, 25–35.

Beaumont, A. (1947). Dependence on the weather of the date of potato blight epidemics. *Transactions of the British Mycological Society*, **31**, 45–53.

Becker, G. (1975). Initiation aux Bolets. *Bulletin Trimestriel de la Société Mycologique de France*, **91**, 191–196.

Beckett, A., Barton, R. & Wilson, I. M. (1968). Fine structure of the wall and appendage formation in ascospores of *Podospora anserina*. *Journal of General Microbiology*, **53**, 89–94.

Beckett, A. & Crawford, R. M. (1973). The development and fine structure of the ascus apex and its role during spore discharge in *Xylaria longipes*. *New Phytologist*, **72**, 357–369.

Beckett, A., Heath, I. B. & McLauglin, D. J. (1974). *An Atlas of Fungal Ultrastructure*. 221 pp. London: Longman.

Beckett, A. R. & Wilson, I. M. (1968). Ascus cytology of *Podospora anserina*. *Journal of General Microbiology*, **53**, 81–87.

Bega, R. V. & Scott, H. A. (1966). Ultrastructure of the sterigma and sporidium of *Cronartium ribicola*. *Canadian Journal of Botany*, **44**, 1726–1727.

Bellemère, A. & Melendez-Howell, L.-M. (1976). Étude ultrastructurale comparée de l'ornamentation externe de la paroi des ascospores de deux Pezizales: *Peziza fortini* n.sp., récoltée au Mexique, et *Aleuria aurantia* (Oed. ex Fr.) Fuck. *Revue de Mycologie*, **40**, 3–19.

Beneke, E. S. (1963). Calvatia, calvacin and cancer. *Mycologia*, **55**, 257–270.

Benjamin, C. R. (1955). Ascocarps of *Aspergillus* and *Penicillium*. *Mycologia*, **47**, 669–687.

Benjamin, C. R. & Hesseltine, C. W. (1959). Studies on the genus *Phycomyces*. *Mycologia*, **51**, 751–771.

Benjamin, R. K. (1956). A new genus of the Gymnoascaceae with a review of the other genera. *Aliso*, **3**, 301–328.

Benjamin, R. K. (1959). The merosporangiferous Mucorales. *Aliso*, **4**, 321–433.

Benjamin, R. K. (1962). A new *Basidiobolus* that forms microspores. *Aliso*, **5**, 223–233.

Benjamin, R. K. (1966). The merosporangium. *Mycologia*, **58**, 1–42.

Benjamin, R. K. (1973). Laboulbeniomycetes. In *The Fungi: An Advanced Treatise*, **IVA**, 223–246. Editors: G. C. Ainsworth, F. K. Sparrow & A. S. Sussman. New York and London: Academic Press.

Benjamin, R. K. & Mehrotra, B. S. (1963). Obligate azygospore formation in two species of *Mucor* (Mucorales). *Aliso*, **5**, 235–245.

Benoit, M. A. & Mathur, S. B. (1970). Identification of species of *Curvularia* on rice seed. *Proceedings of the International Seed Testing Association*, **35**, 99–119.

Bent, K. J. (1970). Fungitoxic action of dimethirimol and ethirimol. *Annals of Applied Biology*, **66**, 103–113.

Bent, K. J. (1978). Chemical control of powdery mildews. In *The Powdery Mildews*, 259–282. Editor: D. M. Spencer. London, New York, San Francisco: Academic Press.

Ben-Yephet, Y., Dinoor, A. & Henis, Y. (1975). The physiological basis of carboxin sensitivity and tolerance in *Ustilago hordei*. *Phytopathology*, **65**, 936–942.

Bergman, K., Burke, P. V., Cerdá-Olmedo, E., David, C. N., Delbrück, M., Foster, K. W., Goodell, E. W., Heisenberg, M., Meissner, G., Zalokar, M., Dennison, D. S. & Shropshire, W. (1969). Phycomyces. *Bacteriological reviews*, **33**, 99–157.

Berkson, B. M. (1966). Cytomorphological studies of the ascogenous hyphae in four species of *Chaetomium*. *Mycologia*, **58**, 125–130.

Berlin, J. D. & Bowen, C. C. (1964). The host–parasite interface of *Albugo candida* on *Raphanus sativus*. *American Journal of Botany*, **51**, 445–452.

Berliner, M. D. (1971). Induction of protoplasts of *Schizosaccharomyces octosporus* by magnesium sulfate and 2 deoxy-d-glucose. *Mycologia*, **63**, 819–825.

Berrie, A. D. (1975). Detritus, micro-organisms and animals in fresh water. In *The Role of Terrestrial and Aquatic Organisms in Decomposition Processes. British Ecological Society Symposium*, **17**, 323–338. Editors: J. M. Anderson & A. MacFadyen. Oxford: Blackwell Scientific Publications Ltd.

Berry, C. R. & Barnett, H. L. (1957). Mode of parasitism and host range of *Piptocephalis virginiana. Mycologia*, **49**, 374–386.

Bessey, E. A. (1950). *Morphology and taxonomy of fungi.* 791 pp. Philadelphia and Toronto: Blakiston.

Beuchat, L. R. & Toledo, R. T. (1977). Behaviour of *Byssochlamys nivea* ascospores in fruit syrups. *Transactions of the British Mycological Society*, **68**, 65–71.

Bevan, R. J. & Greenhalgh, G. N. (1976). *Rhytisma acerinum* as a biological indicator of pollution. *Environmental Pollution*, **10**, 271–285.

Beverwijk, A. L. van (1953). Helicosporous Hyphomycetes. I. *Transactions of the British Mycological Society*, **36**, 111–124.

Beverwijk, A. L. van (1954). Three new fungi: *Helicoon pluriseptatum* n.sp., *Papulospora pulmonaria* n.sp. and *Tricellula inaequalis* n.gen., n.sp. *Antonie van Leeuwenhoek*, **20**, 1–16.

Bhowmik, T. P. (1969). *Alternaria* seed infection of wheat. *Plant Disease Reporter*, **53**, 77–80.

Bistis, G. N. (1956). Sexuality in *Ascobolus stercorarius*. I. Morphology of the ascogonium; plasmogamy; evidence for a sexual hormonal mechanism. *American Journal of Botany*, **43**, 389–394.

Bistis, G. N. (1957). Sexuality in *Ascobolus stercorarius*. II. Preliminary experiments on various aspects of the sexual process. *American Journal of Botany*, **44**, 436–443.

Bistis, G. N. (1970). Dikaryotization in *Clitocybe truncicola. Mycologia*, **62**, 911–923.

Bistis, G. N. & Raper, J. R. (1963). Heterothallism and sexuality in *Ascobolus stercorarius. American Journal of Botany*, **50**, 880–891.

Black, W. (1952). A genetical basis for the classification of strains of *Phytophthora infestans. Proceedings of the Royal Society of Edinburgh, Series B*, **65**, 36–51.

Blackwell, E. M. (1943a). The life history of *Phytophthora cactorum* (Leb. & Cohn) Schroet. *Transactions of the British Mycological Society*, **26**, 71–89.

Blackwell, E. M. (1943b). Presidential Address. On germinating the oospores of *Phytophthora cactorum. Transactions of the British Mycological Society*, **26**, 93–103.

Blakeslee, A. F. (1906). Zygospore germinations in the Mucorineae. *Annales Mycologici*, **4**, 1–28.

Bland, C. E. & Charles, T. M. (1972). Fine structure of *Pilobolus*: surface and wall structure. *Mycologia*, **64**, 774–785.

Blaskovics, J. C. & Raper, K. B. (1957). Encystment stages of *Dictyostelium. Biological Bulletin*, **113**, 58–88.

Blumer, S. (1967). *Echte Mehltaupilze (Erysiphaceae). Ein Bestimmungsbuch für die in Europa vorkommenden Arten.* 436 pp. Jena: Fischer.

Boerema, G. H. (1965). Spore development in the form-genus *Phoma. Persoonia*, **3**, 413–417.

Bohus, M. G. (1962). Der Formenkreis des *Agaricus (Psalliota) bisporus* (Lange) Treschow und die Benützung der wildwachsenden Formen (Sorten) beim Veredelungsverfahren. *Schweizerische Zeitschrift für Pilzkunde*, **40**, 1–7.

Bond, T. E. T. (1952). A further note on size and form in agarics. *Transactions of the British Mycological Society*, **35**, 190–194.

Bonner, J. T. (1944). A descriptive study of the development of the slime mold *Dictyostelium discoideum. American Journal of Botany*, **31**, 175–182.

Bonner, J. T. (1967). *The cellular slime molds.* 2nd edition. 205 pp. Princeton University Press.

Bonner, J. T. (1969). Hormones in social amoebae and mammals. *Scientific American*, **220**, June, 78–91.

Bonner, J. T. (1971). Aggregation and differentiation in the cellular slime molds. *Annual Review of Microbiology*, **25**, 75–92.

Bonner, J. T., Kane, K. K. & Levey, R. H. (1956). Studies on the mechanics of growth in the common mushroom, *Agaricus campestris. Mycologia*, **48**, 13–19.

Booth, C. (1959). Studies of Pyrenomycetes. IV. *Nectria* (Part I). *Mycological Paper No. 73, Commonwealth Mycological Institute*. 115 pp.

Booth, C. (1960). Studies of Pyrenomycetes. V. Nomenclature of some *Fusaria* in relation to their Nectrioid perithecial states. *Mycological Paper No. 74, Commonwealth Mycological Institute*. 16 pp.

Booth, C. (1966a). The genus *Cylindrocarpon. Mycological Paper No. 104, Commonwealth Mycological Institute*. 56 pp.

Booth, C. (1966b). Fruit bodies in Ascomycetes. In *The Fungi: An Advanced Treatise*, **II**, 133–150. Editors: G. C. Ainsworth & A. S. Sussman. New York and London: Academic Press.

Booth, C. (1978). Presidential Address. Do you believe in genera? *Transactions of the British Mycological Society*, **71**, 1–9.

Borkowski, M. (1969). Über die axiale Anordnung der Sporen in Zoosporangien von *Saprolegnia. Flora*, **A, 160**, 158–163.

Borriss, H. (1934). Beiträge zur Wachstums- und Entwicklungsphysiologie der Fruchtkörper von *Coprinus lagopus. Planta*, **22**, 28–69.

Bourke, P. M. A. (1970). Use of weather in the prediction of plant disease epiphytotics. *Annual Review of Plant Pathology*, **8**, 345–370.

Bové, F. J. (1970). *The Story of Ergot*. 297 pp. Switzerland: Karger.

Bowden, J., Gregory, P. H. & Johnson, C. G. (1971). Possible wind transport of coffee leaf rust across the Atlantic Ocean. *Nature, London*, **229**, 500–501.

Bracker, C. E. (1966). Ultrastructural aspects of sporangiophore formation in *Gilbertella persicaria*. In *The Fungus Spore*, 39–58. Editor: M. F. Madelin. Colston Papers No. 18. London: Butterworths.

Bracker, C. E. (1967). Ultrastructure of fungi. *Annual Review of Phytopathology*, **5**, 343–374.

Bracker, C. E. (1968). Ultrastructure of the haustorial apparatus of *Erysiphe graminis* and its relationship to the epidermal cell of Barley. *Phytopathology*, **58**, 12–30.

Bracker, C. E. & Butler, E. E. (1963). The ultrastructure and development of septa in hyphae of *Rhizoctonia solani. Mycologia*, **55**, 35–58.

Bracker, C. E. & Butler, E. E. (1964). Function of the septal pore apparatus in *Rhizoctonia solani* during streaming. *Journal of Cell Biology*, **21**, 152–157.

Bracker, C. E. & Littlefield, L. J. (1973). Structural concepts of host–pathogen interfaces. In *Fungal Pathogenicity and the Plant's Response*, 159–318. Editors: R. J. W. Byrde & C. V. Cutting. New York and London: Academic Press.

Bracker, C. E., Ruiz-Herrera, J. & Bartnicki-Garcia, S. (1976). Structure and transformation of chitin synthetase particles (chitosomes) during microfibril synthesis *in vitro. Proceedings of the National Academy of Sciences of the United States of America*, **73**, 4570–4574.

Bradshaw, A. D. (1959). Population differentiation in *Agrostis tenuis* Sibth. II. The incidence and significance of infection by *Epichloe typhina. New Phytologist*, **58**, 310–315.

Brandenburger, W. & Schwinn, F. J. (1974). Oberflächenstrukturen der Sporen des Antherenbrandes der Caryophyllaceen in Raster-Elektronmikroskop. *Nova Hedwigia*, **22**, 879–897.

Brasier, C. M. (1972). Observations on the sexual mechanism in *Phytophthora palmi-*

vora and related species. *Transactions of the British Mycological Society*, **58**, 237–251.

Brasier, C. M. (1975a). Stimulation of sex organ formation in *Phytophthora* by antagonistic species of *Trichoderma*. I. The effect *in vitro*. *New Phytologist*, **74**, 183–194.

Brasier, C. M. (1975b). Stimulation of sex organ formation in *Phytophthora* by antagonistic species of *Trichoderma*. II. Ecological implications. *New Phytologist*, **74**, 195–198.

Brasier, C. M. & Gibbs, J. N. (1978). Origin and development of the current Dutch elm disease epidemic. In *Plant Disease Epidemiology*, 31–39. Editors: P. R. Scott & A. Bainbridge. Oxford, London, Edinburgh, Melbourne: Blackwell Scientific Publications.

Brefeld, O. (1876). Ueben die Zygosporenbildung bei *Mortierella rostafinskii* nebst Bermerkungen über die Systematik der Zygomycoten. *Sitzungsberichte der Gesellschaft Naturforschender Freunde zu Berlin*, 91.

Brefeld, O. (1877). *Botanische Untersuchungen über Schimmelpilze. Untersuchungen aus dem Gesammtgebiet der Mykologie*. III. Basidiomyceten. I. Leipzig.

Brefeld, O. (1881). *Botanische Untersuchungen über Schimmelpilze. Untersuchungen aus dem Gesammtgebiet der Mykologie*. IV. 4. *Pilobolus*. Leipzig.

Brewer, J. G. & Boerema, G. H. (1965). Electron microscope observations on the development of pycnidiospores in *Phoma* and *Ascochyta* spp. *Proceedings of the Academy of Sciences, Amsterdam*, **C, 68**, 86–97.

Brian, P. W. (1960). Presidential address. Griseofulvin. *Transactions of the British Mycological Society*, **43**, 1–13.

Brian, P. W. (1967). Obligate parasitism in fungi. *Proceedings of the Royal Society, Series B*, **168**, 101–118.

Bridgmon, G. H. (1959). Production of new races of *Puccinia graminis* var. *tritici* by vegetative fusion. *Phytopathology*, **49**, 386–388.

Brierley, W. B. (1918). The microconidia of *Botrytis cinerea*. *Kew Bulletin*, 129–146.

Brodie, H. J. (1931). The oidia of *Coprinus lagopus* and their relation with insects. *Annals of Botany*, **45**, 315–344.

Brodie, H. J. (1932). Oidial mycelia and the diploidization process in *Coprinus lagopus*. *Annals of Botany*, **46**, 727–732.

Brodie, H. J. (1936). The occurrence and function of oidia in the Hymenomycetes. *American Journal of Botany*, **23**, 309–327.

Brodie, H. J. (1951). The splash-cup dispersal mechanism in plants. *Canadian Journal of Botany*, **29**, 224–234.

Brodie, H. J. (1956). The structure and function of the funiculus of the Nidulariaceae. *Svensk Botanisk Tidskrift*, **50**, 142–162.

Brodie, H. J. (1963). Twenty years of Nidulariology. *Mycologia*, **54**, 713–726.

Brodie, H. J. (1975). *The Bird's Nest Fungi*. Toronto and Buffalo: University of Toronto Press.

Brodie, H. J. & Gregory, P. H. (1953). The action of wind in the dispersal of spores from cup-shaped plant structures. *Canadian Journal of Botany*, **31**, 402–410.

Brooks, D. H. (1972a). Observations on the effects of mildew, *Erysiphe graminis*, on growth of spring and winter Barley. *Annals of Applied Biology*, **70**, 149–156.

Brooks, D. H. (1972b). Results in practice. I. Cereals. In *Systemic Fungicides*, 186–205. Editor: R. W. Marsh. London: Longman.

Brotzman, H. G., Calvert, O. H., Brown, M. F. & White, J. A. (1975). Holoblastic conidiogenesis in *Helminthosporium maydis*. *Canadian Journal of Botany*, **53**, 813–817.

Brown, A. H. S. & Smith, G. (1957). The genus *Paecilomyces* Bainier and its perfect stage *Byssochlamys* Westling. *Transactions of the British Mycological Society*, **40**, 17–89.

Brummelen, J. van (1967). A world monograph of the genera *Ascobolus* and *Saccobolus* (Ascomycetes, Pezizales). *Persoonia*, Supplement Vol. **1**. 260 pp.

Brushaber, J. A. & Haskins, R. H. (1973). Cell wall structures of *Epicoccum nigrum* (Hyphomycetes). *Canadian Journal of Botany*, **51**, 1071–1073.

Bryant, T. R. & Howard, F. L. (1969). Meiosis in the Oomycetes. I. A microspectrophotometric analysis of nuclear deoxyribonucleic acid in *Saprolegnia terrestris*. *American Journal of Botany*, **56**, 1075–1083.

Buchenau, G. W. (1975). Relationship between yield loss and area under the wheat stem rust and leaf rust progress curves. *Phytopathology*, **65**, 1317–1318.

Bucholtz, F. (1897). Zur Entwicklungsgeschichte der Tuberaceen. *Bericht der Deutschen Botanischen Gesellschaft*, **15**, 211–226.

Buczacki, S. T., Toxopeus, H., Mattusch, P., Johnston, T. D., Dixon, G. R. & Holbolth, L. A. (1975). Study of physiologic specialization in *Plasmodiophora brassicae*: proposals for attempted rationalization through an international approach. *Transactions of the British Mycological Society*, **65**, 295–303.

Bulat, T. J. (1953). Cultural studies of *Dacrymyces ellisii*. *Mycologia*, **45**, 40–45.

Buller, A. H. R. (1909). *Researches on Fungi*, **1**. 274 pp. London: Longmans, Green & Co.

Buller, A. H. R. (1922). *Researches on Fungi*, **2**. 492 pp. London: Longmans, Green & Co.

Buller, A. H. R. (1924). *Researches on Fungi*, **3**. 611 pp. London: Longmans, Green & Co.

Buller, A. H. R. (1931). *Researches on Fungi*, **4**. 329 pp. London: Longmans, Green & Co.

Buller, A. H. R. (1933). *Researches on Fungi*. **5**. 416 pp. London: Longmans, Green & Co.

Buller, A. H. R. (1934). *Researches on Fungi*, **6**. 513 pp. London: Longmans, Green & Co.

Buller, A. H. R. (1950). *Researches on Fungi*, **7**. 458 pp. Toronto University Press.

Buller, A. H. R. & Vanterpool, T. C. (1933). The violent discharge of the basidiospores (secondary conidia) of *Tilletia tritici*. In *Researches on Fungi*, **5**. A. H. R. Buller. London: Longmans, Green & Co.

Bulmer, G. S. (1964). Spore germination of forty-two species of puffballs. *Mycologia*, **56**, 630–632.

Bulmer, G. S. & Li, Y.-T. (1966). Enzymic activities in *Calvatia cyathiformis* during and after meiosis. *Mycologia*, **58**, 555–561.

Bu'Lock, J. D., Jones, B. E., Taylor, D., Winskill, N. & Quarrie, S. A. (1974a). Sex hormones in the Mucorales. The incorporation of C_{20} and C_{18} precursors into trisporic acids. *Journal of General Microbiology*, **80**, 301–306.

Bu'Lock, J. D., Jones, B. E. & Winskill, N. (1974b). Structures of the mating-type-specific prohormones of Mucorales. *Chemical Communications*, 708–709.

Burchill, R. T., Fricke, E. L. & Swait, A. A. J. (1976). The control of peach leaf curl with off-shoot T. *Annals of Applied Biology*, **82**, 379–380.

Burdekin, D. A. & Gibbs, J. N. (1974). The control of Dutch Elm disease. *Forestry Commission Leaflet No. 54*. 7 pp. London: H.M.S.O.

Burgeff, H. (1914). Untersuchungen über Variabilität, Sexualität und Erblichkeit bei *Phycomyces nitens*. I. *Flora*, **107**, 259–316.

Burgeff, H. (1920). Über den Parasitismus des *Chaetocladium* und die heterokaryotische Natur der von ihm auf Mucorineen erzeugten Gallen. *Zeitschrift für Botanik*, **12**, 1–35.

Burgeff, H. (1924). Untersuchungen über Sexualität und Parasitismus bei Mucorineen. I. *Botanische Abhandlungen*, **4**, 1–135.

Burgeff, H. (1925). Über die Arten und Artkreuzung in der Gattung *Phycomyces*. *Flora*, **18**, 40–46.

Burnett, J. H. (1965). The natural history of recombination systems. In *Incompatibility in Fungi*, 98–113. Editors: K. Esser & J. R. Raper. Berlin: Springer.

Burnett, J. H. (1975). *Mycogenetics. An Introduction to the General Genetics of Fungi.* 375 pp. London, New York, Sydney, Toronto: John Wiley & Sons.

Burnett, J. H. (1976). *Fundamentals of Mycology.* 673 pp. London: Edward Arnold.

Burnett, J. H. & Boulter, M. E. (1963). The mating systems of fungi. II. Mating systems of the Gasteromycetes *Mycocalia denudata* and *M. duriaeana. New Phytologist*, **62**, 217–236.

Burnett, J. H. & Evans, E. J. (1966). Genetical homogeneity and the stability of mating-type factors of fairy rings of *Marasmius oreades. Nature, London*, **210**, 1368–1369.

Burnett, J. H. & Partington, D. (1957). Spatial distribution of fungal mating-type factors. *Proceedings of the Royal Physical Society of Edinburgh*, **26**, 61–68.

Bushnell, W. R. (1972). Physiology of fungal haustoria. *Annual Review of Phytopathology*, **10**, 151–176.

Bushnell, W. R. & Gay, J. (1978). Accumulation of solutes in relation to the structure and function of haustoria in powdery mildews. In *The Powdery Mildews*, 183–235. Editor: D. M. Spencer. London, New York, San Francisco: Academic Press.

Buston, H. W. & Khan, A. H. (1956). The influence of certain micro-organisms on the formation of perithecia by *Chaetomium globosum. Journal of General Microbiology*, **14**, 655–660.

Buston, H. W., Jabbar, A. & Etheridge, D. E. (1953). The influence of hexose phosphates, calcium and jute extract on the formation of perithecia by *Chaetomium globosum. Journal of General Microbiology*, **8**, 302–306.

Butcher, D. N., Sayadat El-Tigani & Ingram, D. S. (1974). The rôle of indole glucosinolates in the club root disease of Cruciferae. *Physiological Plant Pathology*, **4**, 127–141.

Butler, E. E. (1960). Pathogenicity and taxonomy of *Geotrichum candidum. Phytopathology*, **50**, 665–672.

Butler, E. E. & Petersen, L. J. (1972). *Endomyces geotrichum*, a perfect state of *Geotrichum candidum. Mycologia*, **64**, 365–374.

Butler, E. E., Webster, R. K. & Eckert, J. W. (1965). Taxonomy, pathogenicity and physiological properties of the fungus causing sour rot of citrus. *Phytopathology*, **55**, 1262–1268.

Butler, G. M. (1957). The development and behaviour of mycelial strands in *Merulius lacrymans* (Wulf.) Fr. I. Strand development during growth from a food-base through a non-nutrient medium. *Annals of Botany, London*, N.S., **21**, 523–537.

Butler, G. M. (1958). The development and behaviour of mycelial strands in *Merulius lacrymans* (Wulf.) Fr. II. Hyphal behaviour during strand formation. *Annals of Botany, London*, N.S., **22**, 219–236.

Butler, G. M. (1961). Growth of hyphal branching systems in *Coprinus disseminatus. Annals of Botany, London*, N.S., **25**, 341–352.

Butler, G. M. (1966). Vegetative structure. In *The Fungi: An Advanced Treatise*, **II**, 83–112. Editors: G. C. Ainsworth & A. S. Sussman. New York and London: Academic Press.

Butt, D. J. (1978). Epidemiology of powdery mildews. In *The Powdery Mildews*, 51–81. Editor: D. M. Spencer. London, New York, San Francisco: Academic Press.

Buxton, E. W. (1960). Heterokaryosis, saltation and adaptation. In *Plant Pathology; An Advanced Treatise*, **2**, 359–405. Editors: J. G. Horsfall & A. E. Dimond. New York and London: Academic Press.

Cabib, E. (1975). Molecular aspects of yeast morphogenesis. *Annual Review of Microbiology*, **29**, 191–214.

Cain, R. F. (1972). Evolution of the fungi. *Mycologia*, **64**, 1–14.

Callaghan, A. A. (1969a). Light and spore discharge in Entomophthorales. *Transactions of the British Mycological Society*, **53**, 87–97.

Callaghan, A. A. (1969b). Morphogenesis in *Basidiobolus ranarum*. *Transactions of the British Mycological Society*, **53**, 99–108.

Callaghan, A. A. (1969c). Secondary conidium formation in *Basidiobolus ranarum*. *Transactions of the British Mycological Society*, **53**, 132–137.

Callaghan, A. A. (1974). Effect of pH and light on conidium germination in *Basidiobolus ranarum*. *Transactions of the British Mycological Society*, **63**, 13–18.

Calleja, G. B. & Johnson, B. F. (1971). Flocculation in a fission yeast: an initial step in the conjugation process. *Canadian Journal of Microbiology*, **17**, 1175–1177.

Callow, J. A. (1975). Endopolyploidy in maize smut neoplasms induced by the maize smut fungus, *Ustilago maydis*. *New Phytologist*, **75**, 253–257.

Callow, J. A. & Ling, I. T. (1973). Histology of neoplasms and chlorotic lesions in maize seedlings following the injection of sporidia of *Ustilago maydis* (DC) Corda. *Physiological Plant Pathology*, **3**, 489–494.

Calonge, F. D. (1970). Notes on the ultrastructure of the microconidium and stroma in *Sclerotinia sclerotiorum*. *Archiv für Mikrobiologie*, **71**, 191–195.

Campbell, R. (1969). An electron microscope study of spore structure and development in *Alternaria brassicicola*. *Journal of General Microbiology*, **54**, 381–392.

Campbell, R. (1970). An electron microscope study of exogenously dormant spores, spore germination, hyphae and conidiophores of *Alternaria brassicicola*. *New Phytologist*, **69**, 287–293.

Campbell, R., Larner, R. W. & Madelin, M. F. (1968). Notes on an albino mutant of *Alternaria brassicicola*. *Mycologia*, **60**, 1122–1125.

Campbell, R. N. & Fry, P. R. (1966). The nature of the association between *Olpidium brassicae* and lettuce big vein and tobacco necrosis viruses. *Virology*, **29**, 222–233.

Campbell, W. A. & Hendrix, F. F. (1967). A new heterothallic *Pythium* from Southern United States. *Mycologia*, **59**, 274–278.

Campbell, W. P. (1957). Studies on ergot infection in gramineous hosts. *Canadian Journal of Botany*, **35**, 315–320.

Campbell, W. P. (1958). Infection of barley by *Claviceps purpurea*. *Canadian Journal of Botany*, **36**, 615–619.

Campbell, W. P. (1959). Inhibition of starch formation by *Claviceps purpurea*. *Phytopathology*, **49**, 451–452.

Campbell, W. P. & Tyner, L. E. (1959). Comparison of degree and duration of susceptibility of barley to ergot and true loose smut. *Phytopathology*, **49**, 348–349.

Canham, S. C. (1969). Taxonomy and morphology of *Hypocrea citrina*. *Mycologia*, **61**, 315–331.

Canter, H. M. & Willoughby, L. G. (1964). A parasitic *Blastocladiella* from Windermere plankton. *Journal of the Royal Microscopical Society*, **83**, 365–372.

Cantino, E. C. (1950). Nutrition and phylogeny in the water molds. *Quarterly Review of Biology*, **25**, 269–277.

Cantino, E. C. (1955). Physiology and phylogeny in the water-molds – a re-evaluation. *Quarterly Review of Biology*, **30**, 138–149.

Cantino, E. C. (1970). Germination of a resistant sporangium of *Blastocladiella britannica*: bearing on its taxonomic status. *Transactions of the British Mycological Society*, **54**, 303–307.

Cantino, E. C. & Horenstein, E. A. (1954). Cytoplasmic exchange without gametic copulation in the water mold *Blastocladiella emersonii*. *American Naturalist*, **88**, 143–154.

Cantino, E. & Hyatt, M. T. (1953). Carotenoids and oxidative enzymes in the aquatic Phycomycetes *Blastocladiella* and *Rhizophlyctis*. *American Journal of Botany*, **40**, 688–694.

Cantino, E. C. & Lovett, J. S. (1964). Non-filamentous aquatic fungi model systems for biochemical studies of morphological differentiation. *Advances in Morphogenesis*, **3**, 33–93.

Cantino, E. C., Lovett, J. S., Leak, L. V. & Lythgoe, J. (1963). The single mitochondrion, fine structure, and germination of the spore of *Blastocladiella emersonii*. *Journal of General Microbiology*, **31**, 393–404.

Cantino, E. C. & Mack, J. P. (1969). Form and function in the zoospore of *Blastocladiella emersonii*. I. The γ particle and satellite ribosome package. *Nova Hedwigia*, **18**, 115–158.

Cantino, E. C., Truesdell, L. C. & Shaw, D. S. (1968). Life history of the motile spore of *Blastocladiella emersonii*: a study in cell differentiation. *Journal of the Elisha Mitchell Scientific Society*, **84**, 125–146.

Cantino, E. C. & Turian, G. F. (1959). Physiology and development of lower fungi (Phycomycetes). *Annual Review of Microbiology*, **13**, 97–124.

Caporali, L. (1964). La biologie du *Taphrina deformans* (Berk.) Tul: relations entre l'hôte et le parasite. *Revue Générale de Botanique*, **71**, 241–282.

Carlile, M. J. (1965). The photobiology of fungi. *Annual Review of Plant Physiology*, **16**, 175–202.

Carlile, M. J. (1971). Myxomycetes and other slime moulds. In *Methods in Microbiology*, 237–265. Editor: C. Booth. New York and London: Academic Press.

Carlile, M. J. (1972). The lethal interaction following plasmodial fusion between two strains of the myxomycete *Physarum polycephalum*. *Journal of General Microbiology*, **71**, 581–590.

Carlile, M. J. (1973). Cell fusion and somatic incompatibility in Myxomycetes. *Bericht der Deutschen Botanischen Gesellschaft*, **86**, 123–139.

Carlile, M. J. & Dee, J. (1967). Plasmodial and lethal interaction between strains in a myxomycete. *Nature, London*, **215**, 832–834.

Carlile, M. J. & Machlis, L. (1965a). The response of male gametes of *Allomyces* to the sexual hormone sirenin. *American Journal of Botany*, **52**, 478–483.

Carlile, M. J. & Machlis, L. (1965b). A comparative study of the chemotaxis of the motile phases of *Allomyces*. *American Journal of Botany*, **52**, 484–486.

Carmichael, J. W. (1957). *Geotrichum candidum*. *Mycologia*, **49**, 820–830.

Carr, A. J. H. & Olive, L. S. (1958). Genetics of *Sordaria fimicola*. II. Cytology. *American Journal of Botany*, **45**, 142–150.

Carroll, F. E. & Carroll, G. C. (1971). Fine structural studies on "poroconidium" formation in *Stemphylium botryosum*. In *Taxonomy of Fungi Imperfecti*, 75–91. Editor: W. B. Kendrick. University of Toronto Press.

Carroll, G. C. (1967). The ultrastructure of ascospore delimitation in *Saccobolus kerverni*. *Journal of Cell Biology*, **33**, 218–224.

Carroll, G. C. & Dykstra, R. (1966). Synaptinemal complexes in *Didymium iridis*. *Mycologia*, **58**, 166–169.

Carroll, G. C. & Munk, A. (1964). Studies on lignicolous Sordariaceae. *Mycologia*, **56**, 77–98.

Cartwright, K. St.G. & Findlay, W. P. K. (1958). *Decay of Timber and its Prevention*. 2nd edition. 332 pp. London: H.M.S.O.

Castle, E. S. (1942). Spiral growth and reversal of spiraling in *Phycomyces*, and their bearing on primary wall structure. *American Journal of Botany*, **29**, 664–672.

Castle, E. S. (1953). Problems of oriented growth and structure in *Phycomyces*. *Quarterly Review of Biology*, **28**, 364–372.

Castle, E. S. (1966). Light responses of *Phycomyces*. *Science, N.Y.*, **154**, 1416–1420.

Catcheside, D. G. (1951). *Genetics of micro-organisms*. 223 pp. London: Pitman.

Caten, C. E. & Jinks, J. L. (1966). Heterokaryosis: its significance in wild homothallic ascomycetes and fungi imperfecti. *Transactions of the British Mycological Society*, **49**, 81–93.

Cerdá-Olmedo, E. (1974). *Phycomyces*. In *Handbook of Genetics*, **I**, 343–357. Editor: R. C. King. New York and London: Plenum Press.

Chadefaud, M. (1960). Les végétaux non vasculaires (Cryptogamie), Tome I. In *Traité de Botanique Systématique*. M. Chadefaud & L. Emberger. Paris: Masson.

Chadefaud, M. (1973). Les asques et la systématique des Ascomycètes. *Bulletin Trimestriel de la Société Mycologique de France*, **89**, 127–170.

Chambers, T. C., Markus, K. & Willoughby, L. G. (1967). The fine structure of the mature zoosporangium of *Nowakowskiella profusa*. *Journal of General Microbiology*, **46**, 135–141.

Chambers, T. C. & Willoughby, L. G. (1964). The fine structure of *Rhizophlyctis rosea*, a soil Phycomycete. *Journal of the Royal Microscopical Society*, **83**, 355–364.

Chang, S. T. & Hayes, W. A. (1978). *The Biology and Cultivation of Edible Mushrooms*. 819 pp. London, New York, San Francisco: Academic Press.

Chapman, J. A. & Vujičić, R. (1965). The fine structure of sporangia of *Phytophthora erythroseptica* Pethybr. *Journal of General Microbiology*, **41**, 275–282.

Chen, K.-C. (1965). The genetics of *Sordaria brevicollis*. I. Determination of seven linkage groups. *Genetics*, **51**, 509–517.

Chen, K.-C. & Olive, L. S. (1965). The genetics of *Sordaria brevicollis*. II. Biased segregation due to spindle overlap. *Genetics*, **51**, 761–766.

Chester, K. S. (1946). *The Nature and Prevention of the Cereal Rusts as Exemplified in the Leaf Rust of Wheat*. 269 pp. Waltham, Mass.: Chronica Botanica Co.

Chesters, C. G. C. & Greenhalgh, G. N. (1964). *Geniculosporium serpens* gen. et sp. nov., the imperfect state of *Hypoxylon serpens*. *Transactions of the British Mycological Society*, **47**, 393–401.

Chet, I., Henis, Y. & Kislev, N. (1969). Ultrastructure of sclerotia and hyphae of *Sclerotium rolfsii* Sacc. *Journal of General Microbiology*, **57**, 143–147.

Chiang, M. S. & Crête, R. (1970). Inheritance of clubroot resistance in cabbage (*Brassica oleracea* L. var. *capitata* L.). *Journal of Genetics and Cytology*, **12**, 253–256.

Chien, C-Y., Kuhlman, E. G. & Gams, W. (1974). Zygospores of two *Mortierella* species with stylospores. *Mycologia*, **66**, 114–121.

Christen, J. & Hohl, H. R. (1972). Growth and ultrastructural differentiation of sporangia in *Phytophthora palmivora*. *Canadian Journal of Microbiology*, **18**, 1959–1964.

Christensen, C. M. (1965). *The Molds and Man*. 3rd edition. 284 pp. University of Minnesota Press.

Christensen, J. J. (1932). Studies on the genetics of *Ustilago zeae*. *Phytopathologische Zeitschrift*, **4**, 129–188.

Christensen, J. J. (1963). Corn smut caused by *Ustilago maydis*. *American Phytopathological Society Monograph* No. 2. 41 pp.

Chu, F. S. & Chang, C. C. (1973). Pecteolytic enzymes of eight *Byssochlamys fulva* isolates. *Mycologia*, **65**, 920–924.

Ciferri, R. & Tomaselli, R. (1955). The symbiotic fungi of lichens and their nomenclature. *Taxon*, **4**, 190–192.

Clark, J. & Collins, O. R. (1973). Directional cytotoxic reactions between incompatible plasmodia of *Didymium iridis*. *Genetics*, **73**, 247–257.

Clemençon, H. (1969). Reifung und endoplasmatisches Retikulum der Agaricales-Basidie. *Zeitschrift für Pilzkunde*, **35**, 295–304.

Clemençon, H. (1970). Bau der Wände der Basidiosporen und ein Vorschlag zur Benennung ihren Schichten. *Zeitschrift für Pilzkunde*, **36**, 113–133.

Clowes, F. A. L. (1951). The structure of mycorrhizal roots of *Fagus sylvatica*. *New Phytologist*, **50**, 1–16.

Clutterbuck, A. J. (1974). *Aspergillus nidulans*. In *Handbook of Genetics*, **I**, 447–510. Editor: R. C. King. New York and London: Plenum Press.

Cochrane, V. W. (1958). *Physiology of Fungi*. 524 pp. New York and London: Wiley. Chapman & Hall.

Coffey, M. D. (1975). Obligate parasites of higher plants, particularly rust fungi. *Symposium of the Society for Experimental Biology*, **29**, 297–323.

Cole, G. T. (1973). Ultrastructure of conidiogenesis in *Drechslera sorokiniana*. *Canadian Journal of Botany*, **51**, 629–638.

Cole, G. T. (1975). The thallic mode of conidiogenesis in the Fungi Imperfecti. *Canadian Journal of Botany*, **53**, 2983–3001.

Cole, G. T. & Kendrick, W. B. (1969a). Conidium ontogeny in hyphomycetes. The phialides of *Phialophora*, *Penicillium* and *Ceratocystis*. *Canadian Journal of Botany*, **47**, 779–789.

Cole, G. T. & Kendrick, W. B. (1969b). Conidium ontogeny in hyphomycetes. The annellophores of *Scopulariopsis brevicaulis*. *Canadian Journal of Botany*, **47**, 925–929.

Cole, G. T. & Kendrick, W. B. (1969c). Conidium ontogeny in hyphomycetes. The arthrospores of *Oidiodendron* and *Geotrichum*, and the endoarthrospores of *Sporendonema*. *Canadian Journal of Botany*, **47**, 1773, 1780.

Cole, G. T. & Samson, R. A. (1979). *Patterns of Development in Conidial Fungi*. 336 pp. London: Pitman Publishing Ltd.

Coley-Smith, J. R. & Cooke, R. C. (1971). Survival and germination of fungal sclerotia. *Annual Review of Phytopathology*, **9**, 65–92.

Colhoun, J. (1958). Club root disease of crucifers, caused by *Plasmodiophora brassicae* Woron. *Phytopathological Paper No. 3, Commonwealth Mycological Institute*. 108 pp.

Colhoun, J. (1966). The biflagellate zoospore of aquatic Phycomycetes with particular reference to *Phytophthora* spp. In *The Fungus Spore*, 85–92. Editor: M. F. Madelin. Colston Papers No. 18. London: Butterworths.

Collins, O. R. (1963). Multiple alleles at the incompatibility locus in the myxomycete *Didymium iridis*. *American Journal of Botany*, **50**, 477–480.

Collins, O. R. (1965). Evidence for a mutation at the incompatibility locus in the slime mold *Didymium iridis*. *Mycologia*, **57**, 314–315.

Collins, O. R. (1966). Plasmodial compatibility in heterothallic and homothallic isolates of *Didymium iridis*. *Mycologia*, **58**, 362–372.

Collins, O. R. & Clark, J. (1968). Genetics of plasmodial compatibility and heterokaryosis in *Didymium iridis*. *Mycologia*, **60**, 90–103.

Collins, O. R. & Haskins, E. F. (1970). Evidence for polygenic control of plasmodial fusion in *Physarum polycephalum*. *Nature, London*, **226**, 279–280.

Collins, O. R. & Ling, H. (1964). Further studies in multiple allelomorph heterothallism in the myxomycete *Didymium iridis*. *American Journal of Botany*, **51**, 315–317.

Collins, O. R. & Ling, H. (1972). Genetics of somatic cell fusion in two isolates of *Didymium iridis*. *American Journal of Botany*, **59**, 337–340.

Colotelo, N. (1974). A scanning electron microscope study of developing sclerotia of *Sclerotinia sclerotiorum*. *Canadian Journal of Botany*, **52**, 1127–1130.

Colson, B. (1934). The cytology and morphology of *Neurospora tetrasperma*. *Annals of Botany, London*, **48**, 211–225.

Conti, S. F. & Naylor, H. B. (1959). Electron microscopy of ultrathin sections of *Schizosaccharomyces octosporus*. I. Cell division. *Journal of Bacteriology*, **78**, 868–877.

Conti, S. F. & Naylor, H. B. (1960). Electron microscopy of ultrathin sections of *Schizosaccharomyces octosporus*. II. Morphological and cytological changes preceding ascospore formation. *Journal of Bacteriology*, **79**, 331–340.

Cook, A. H. (1958). *The Chemistry and Biology of yeasts.* 763 pp. New York and London: Academic Press.

Cooke, J. C. (1969a). Morphology of *Chaetomium funicolum. Mycologia,* **61,** 1060–1065.

Cooke, J. C. (1969b). Morphology of *Chaetomium erraticum. American Journal of Botany,* **56,** 335–340.

Cooke, J. C. (1970). Morphology of *Chaetomium trilaterale. Mycologia,* **62,** 282–288.

Cooke, R. C. (1969). Changes in soluble carbohydrates during sclerotium formation by *Sclerotinia sclerotiorum* and *S. trifoliorum. Transactions of the British Mycological Society,* **53,** 77–86.

Cooke, R. C. (1970). Physiological aspects of sclerotium growth in *Sclerotinia sclerotiorum. Transactions of the British Mycological Society,* **54,** 361–365.

Cooke, R. C. (1971a). Physiology of sclerotia of *Sclerotinia sclerotiorum* during growth and maturation. *Transactions of the British Mycological Society,* **56,** 51–59.

Cooke, R. C. (1971b). Uptake of [¹⁴C] glucose and loss of water by sclerotia of *Sclerotinia sclerotiorum* during development. *Transactions of the British Mycological Society,* **57,** 379–384.

Cooke, R. C. & Dickinson, C. (1965). Nematode-trapping species of *Dactylella* and *Monacrosporium. Transactions of the British Mycological Society,* **48,** 621–629.

Cooke, R. C. & Godfrey, B. E. S. (1964). A key to the nematode-destroying fungi. *Transactions of the British Mycological Society,* **47,** 61–74.

Cooke, R. C. & Mitchell, D. T. (1966). Sclerotium size and germination in *Claviceps purpurea. Transactions of the British Mycological Society,* **49,** 95–100.

Cooke, R. C. & Mitchell, D. T. (1967). Germination pattern and capacity for repeated stroma formation in *Claviceps purpurea. Transactions of the British Mycological Society,* **50,** 275–283.

Cooke, R. C. & Mitchell, D. T. (1969). Sugars and polyols in sclerotia of *Claviceps purpurea, C. nigricans* and *Sclerotinia curreyana* during germination. *Transactions of the British Mycological Society,* **52,** 365–372.

Cooke, R. C. & Mitchell, D. T. (1970). Carbohydrate physiology of sclerotia of *Claviceps purpurea* during dormancy and germination. *Transactions of the British Mycological Society,* **54,** 93–99.

Cooke, W. B. (1961). The genus *Schizophyllum. Mycologia,* **53,** 575–599.

Cooke, W. B. & Hawksworth, D. L. (1970). A preliminary list of the families proposed for fungi (including the lichens). *Mycological Paper No. 121, Commonwealth Mycological Institute.* 86 pp.

Corlett, M. (1966). Perithecium development in *Chaetomium trigonosporum. Canadian Journal of Botany,* **44,** 155–162.

Corlett, M. (1973). Observations and comments on the *Pleospora* centrum type. *Nova Hedwigia,* **24,** 347–360.

Corner, E. J. H. (1929). Studies in the morphology of Discomycetes. I–II. *Transactions of the British Mycological Society,* **14,** 263–275; 275–291.

Corner, E. J. H. (1932). A *Fomes* with two systems of hyphae. *Transactions of the British Mycological Society,* **17,** 51–81.

Corner, E. J. H. (1948). Studies in the basidium. I. The ampoule effect, with a note on nomenclature. *New Phytologist,* **47,** 22–51.

Corner, E. J. H. (1950). *A Monograph of Clavaria and Allied Genera.* 740 pp. Oxford University Press.

Corner, E. J. H. (1953). The construction of polypores. I. Introduction: *Polyporus sulphureus, P. squamosus, P. betulinus* and *Polystictus microcyclus. Phytomorphology,* **3,** 152–167.

Corner, E. J. H. (1954). The classification of the higher fungi. *Proceedings of the Linnean Society of London,* **165,** 4–6.

Corner, E. J. H. (1966). *A Monograph of Cantharelloid Fungi*. 255 pp. Oxford University Press.

Couch, J. N. (1926). Heterothallism in *Dictyuchus*, a genus of the water moulds. *Annals of Botany, London*, **40**, 849–881.

Couch, J. N. (1939). Heterothallism in the Chytridiales. *Journal of the Elisha Mitchell Scientific Society*, **55**, 409–414.

Cox, A. E. & Large, E. C. (1960). Potato blight epidemics throughout the world. *U.S. Department of Agriculture Handbook No. 174*. 230 pp. Washington.

Cox, G. & Sanders, F. (1974). Ultrastructure of the host interface in vesicular–arbuscular mycorrhiza. *New Phytologist*, **73**, 901–912.

Cox, R. J. & Niederpruem, D. J. (1975). Differentiation in *Coprinus lagopus*. III. Expansion of excised fruit bodies. *Archives of Microbiology*, **105**, 257–260.

Crady, E. E. & Wolf, F. T. (1959). The production of indole acetic acid by *Taphrina deformans* and *Dibotryon morbosum*. *Physiologia Plantarum*, **12**, 526–533.

Craigie, J. H. (1927a). Experiments on sex in rust fungi. *Nature, London*, **120**, 116–117.

Craigie, J. H. (1927b). Discovery of the function of the pycnia of the rust fungi. *Nature, London*, **120**, 765–767.

Craigie, J. H. (1945). Epidemiology of stem rust in Western Canada. *Scientific Agriculture*, **25**, 285–401.

Craigie, J. H. & Green, G. J. (1962). Nuclear behaviour leading to conjugate association in haploid infections of *Puccinia graminis*. *Canadian Journal of Botany*, **40**, 163–178.

Crane, J. L. & Schoknecht, J. D. (1973). Conidiogenesis in *Ceratocystis ulmi*, *Ceratocystis piceae* and *Graphium penicillioides*. *American Journal of Botany*, **60**, 346–354.

Croft, J. H. & Jinks, J. L. (1977). Aspects of population genetics of *Aspergillus nidulans*. In *Genetics and Physiology of Aspergillus*, 339–360. Editors: J. E. Smith & J. A. Pateman. London, New York, San Francisco: Academic Press.

Croxall, H. E., Collingwood, C. A. & Jenkins, J. E. E. (1951). Observations on brown rot (*Sclerotinia fructigena*) of apples in relation to injury by earwigs (*Forficula auricularia*). *Annals of Applied Biology*, **38**, 833–843.

Crump, E. & Branton, D. (1966). Behavior of primary and secondary zoospores of *Saprolegnia* sp. *Canadian Journal of Botany*, **44**, 1393–1400.

Cullum, F. J. & Webster, J. (1977). Cleistocarp dehiscence in *Phyllactinia*. *Transactions of the British Mycological Society*, **68**, 316–320.

Cummins, G. B. (1959). *Illustrated Genera of Rust Fungi*. 129 pp. Minneapolis: Burgess Publishing Company.

Curtis, K. M. (1921). The life history and cytology of *Synchytrium endobioticum* (Schilb.) Perc., the cause of wart disease in potato. *Philosophical Transactions of the Royal Society, Series B*, **210**, 409–478.

Curtis, F. C., Evans, G. H., Lillis, V., Lewis, D. H. & Cooke, R. C. (1978). Studies on Mucoralean mycoparasites. I. Some effects of *Piptocephalis* species on host growth. *New Phytologist*, **80**, 157–165.

Cutter, V. M. (1942a). Nuclear behavior in the Mucorales. I. The *Mucor* pattern. *Bulletin of the Torrey Botanical Club*, **69**, 480–508.

Cutter, V. M. (1942b). Nuclear behavior in the Mucorales. II. The *Rhizopus*, *Phycomyces* and *Sporodinia* patterns. *Bulletin of the Torrey Botanical Club*, **69**, 592–616.

Daniels, J. (1961). *Chaetomium piluliferum* sp. nov., the perfect state of *Botryotrichum piluliferum*. *Transactions of the British Mycological Society*, **44**, 79–86.

D'Arcy Thompson, W. (1961). *On Growth and Form*. Abridged edition. Editor: J. T. Bonner. 346 pp. Cambridge University Press.

Dargent, R, R., Darnaud, M. & Montant, C. (1973). Sur l'ultrastructure des hyphes en croissance d'*Achlya bisexualis* Coker. Localisation des organites cytoplasmiques et étude de la morphologie des mitochondries. *Compte Rendu Hebdomadaire des Séances de l'Académie des Sciences, D*, **277**, 1141–1144.

Davey, C. B. & Papavizas, G. C. (1962). Growth and sexual reproduction of *Aphanomyces euteiches* as affected by the oxidation state of sulfur. *American Journal of Botany*, **49**, 400–404.

Davies, B. H. (1961). The carotenoids of *Rhizophlyctis rosea*. *Phytochemistry*, **1**, 25–29.

Davis, E. E. (1967). Zygospore formation in *Syzygites megalocarpus*. *Canadian Journal of Botany*, **45**, 531–532.

Davis, R. H. (1966). Mechanisms of inheritance. 2. Heterokaryosis. In *The Fungi: An Advanced Treatise*, **II**, 567–588. Editors: G. C. Ainsworth & A. S. Sussman. New York and London: Academic Press.

Day, A. W. (1972). Genetic implications of current models of somatic nuclear division in fungi. *Canadian Journal of Botany*, **50**, 1337–1347.

Day, A. W. & Poon, N. H. (1975). Fungal fimbriae. II. Their role in conjugation in *Ustilago violacea*. *Canadian Journal of Microbiology*, **21**, 547–557.

Day, P. R. (1963). The structure of the A mating-type factor in *Coprinus lagopus*: wild alleles. *Genetical Research, Cambridge*, **4**, 323–325.

Day, P. R. (1974). *Genetics of Host–Parasite Interaction*. 238 pp. San Francisco: W. H. Freeman & Co.

De Bary, A. (1887). *Comparative Morphology and Biology of the Fungi, Mycetozoa and Bacteria*. English translation. 525 pp. Oxford: Clarendon Press.

Decaris, B., Girard, J. & Leblon, G. (1974). *Ascobolus. Handbook of Genetics*, **I**. Editor: R. C. King. New York and London: Plenum Press.

Dee, J. (1960). A mating type system in an acellular slime mould. *Nature, London*, **185**, 780–781.

Dee, J. (1966). Multiple alleles and other factors affecting plasmodium formation in the true slime mould *Physarum polycephalum* Schw. *Journal of Protozoology*, **13**, 610–616.

Dekhuizen, H. M. & Overeem, J. C. (1971). The role of cytokinins in club root formation. *Physiological Plant Pathology*, **1**, 151–161.

Delamater, E. D., Yaverbaum, S. & Schwartz, L. (1953). The nuclear cytology of *Eremascus albus*. *American Journal of Botany*, **40**, 475–492.

Delay, C. (1966). Étude de l'infrastructure de l'asque d'*Ascobolus immersus* Pers., pendant la maturation des spores. *Annales des Sciences Naturelles (Botanique)*, Sér. 12, **7**, 361–420.

Delmas, J. (1976). La truffe et sa culture. 54 pp. Étude No. 60. Institut National de la Recherche Agronomique. Editors: S.E.I. C.N.R.A., Versailles.

Dengler, I. (1937). Entwicklungsgeschichtliche Untersuchungen an *Sordaria macrospora* Auersw., *S. uvicola* Viala et Mars., und *S. brefeldii* Zopf. *Jahrbuch für Wissenschaftliche Botanik* **84**, 427–448.

Dengler, R. E., Filosa, M. F. & Shao, Y. Y. (1970). Ultrastructural aspects of macrocyst development in *Dictyostelium mucoroides*. (Abstract). *American Journal of Botany*, **57**, 737.

Dennis, R. L. (1970). A middle Pennsylvanian Basidiomycete mycelium with clamp connections. *Mycologia*, **62**, 578–584.

Dennis, R. W. G. (1956). A revision of the British Helotiaceae in the herbarium of the Royal Botanic Gardens, Kew, with notes on related European species. *Mycological Paper No. 62, Commonwealth Mycological Institute*. 216 pp.

Dennis, R. W. G. (1858a). *Xylaria* versus *Hypoxylon* and *Xylosphaera*. *Kew Bulletin*, **13**, 101–106.

Dennis, R. W. G. (1958b). Some Xylosphaeras from tropical Africa. *Revista di Biologia, Lisboa*, **1**, 175–208.

Dennis, R. W. G. (1968). *British Ascomycetes*. 455 pp. Stuttgart: Cramer.

Dennis, R. W. G., Orton, P. D. & Hora, F. B. (1960). New check list of British Agarics and Boleti. *Supplement to Transactions of the British Mycological Society*, **43**. 225 pp.

Denward, T. (1970). Differentiation in *Phytophthora infestans*. II. Somatic recombination in vegetative mycelium. *Hereditas*, **66**, 35–48.

Derx, H. G. (1930). Étude sur les Sporobolomycètes. *Annales Mycologici*, **28**, 1–23.

Derx, H. G. (1948). *Itersonilia*, nouveau genre de sporobolomycètes à mycelium bouclé. *Bulletin of the Buitenzorg Botanical Gardens*, III, **18**, 465–472.

Descals, E. & Webster, J. (1976). *Hyaloscypha*: perfect state of *Clathrosphaerina zalewskii*. *Transactions of the British Mycological Society*, **67**, 525–528.

Descals, E., Webster, J. & Dyko, B. S. (1977). Taxonomic studies on aquatic Hyphomycetes. I. *Lemonniera* de Wildeman. *Transactions of the British Mycological Society*, **69**, 89–109.

Dick, M. W. (1969). Morphology and taxonomy of the Oomycetes, with special reference to the Saprolegniaceae, Leptomitaceae and Pythiaceae. *New Phytologist*, **68**, 751–775.

Dick, M. W. (1970). Saprolegniaceae on insect exuviae. *Transactions of the British Mycological Society*, **55**, 449–459.

Dick, M. W. (1971). Oospore structure in *Aphanomyces*. *Mycologia*, **63**, 686–688.

Dick, M. W. (1972). Morphology and taxonomy of the Oomycetes, with special reference to Saprolegniaceae, Leptomitaceae and Pythiaceae. II. Cytogenetic systems. *New Phytologist*, **71**, 1151–1159.

Dick, M. W. (1973a). Leptomitales. In *The Fungi: An Advanced Treatise*, **IVB**, 145–158. Editors: G. C. Ainsworth, F. K. Sparrow & A. S. Sussman. New York and London: Academic Press.

Dick, M. W. (1973b). Saprolegniales. In *The Fungi: An Advanced Treatise*, **IVB**, 113–144. Editors: G. C. Ainsworth, F. K. Sparrow & A. S. Sussman. New York and London: Academic Press.

Dick, M. W. (1976). The ecology of aquatic Phycomycetes. In *Recent Advances in Aquatic Mycology*, 513–542. Editor: E. B. Gareth Jones. London: Elek Science.

Dick, M. W. & Win-Tin (1973). The development of cytological theory in the Oomycetes. *Biological Reviews, Cambridge*, **48**, 133–158.

Dick, S. (1965). Physiological aspects of tetrapolar incompatibility. In *Incompatibility in Fungi*, 72–80. Editors: K. Esser & J. R. Raper. Berlin: Springer.

Dodge, B. O. (1924). Aecidiospore discharge as related to the character of the spore wall. *Journal of Agricultural Research*, **27**, 749–756.

Dodge, B. O. (1927). Nuclear phenomena associated with heterothallism and homothallism in the ascomycete *Neurospora*. *Journal of Agricultural Research*, **35**, 289–305.

Dodge, B. O. (1937). The perithecial cavity formation in a *Leptosphaeria* on *Opuntia*. *Mycologia*, **29**, 707–716.

Dodge, B. O. (1942). Heterocaryotic vigor in *Neurospora*. *Bulletin of the Torrey Botanical Club*, **69**, 75–91.

Douget, G. (1957). Organogénie de *Creopus spinulosus* (Fuck.) Moravec. Organogénie comparée de quelques Hypocréales du même type. *Bulletin de la Société Mycologique de France*, **73**, 144–164.

Doguet, G. (1960). Morphologie, organogénie et evolution nucléaire de l'*Epichloe typhina*. La place des Clavicipitaceae dans la classification. *Bulletin de la Société Mycologique de France*, **76**, 171–203.

Doguet, G. (1962). *Digitatospora marina* n.g., n.sp., basidiomycète marin. *Compte Rendu Hebdomadaire des Séances de l'Académie des Sciences, Paris*, Sér. D, **254**, 4336–4338.

Doguet, G. (1967). *Nia vibrissa* Moore et Meyers, remarquable basidiomycète marin.

Compte Rendu Hebdomadaire des Séances de l'Académie des Sciences, Paris, Sér. D, **265,** 1780–1783.

Doguet, G. (1968). *Nia vibrissa* Moore et Meyers, Gasteromycète marin. I. Conditions générale de formation des carpophores en culture. *Bulletin Trimestriel de la Société Mycologique de France,* **84,** 343–351.

Doguet, G. (1969). *Nia vibrissa* Moore et Meyers, Gasteromycète marin. II. Développement des carpophores et des basides. *Bulletin Trimestriel de la Société Mycologique de France,* **85,** 93–104.

Doling, D. A. (1966). Loose smut in wheat and barley. *Agriculture,* **73,** 523–527.

Donk, M. A. (1964). A conspectus of the families of Aphyllophorales. *Persoonia,* **3,** 199–324.

Donk, M. A. (1972a). The Heterobasidiomycetes: a reconnaissance. I. A restricted emendation. *Proceedings of the Konlike Nederlandse Akademie van Wetenschappie, Amsterdam, C,* **75,** 365–375.

Donk, M. A. (1972b). The Heterobasidiomycetes: a reconnaissance. II. Some problems connected with the restricted emendation. *Proceedings of the Konlike Nederlandse Akademie van Wetenschappie, Amsterdam, C,* **75,** 376–390.

Douwes, G. A. C. & Arx, J. A. von (1965). Das hymenophorale Trama bei den Agaricales. *Acta Botanica Neerlandica,* **14,** 197–217.

Dowding, E. S. (1931). The sexuality of *Ascobolus stercorarius* and the transportation of oidia by mites and flies. *Annals of Botany, London,* **45,** 621–637.

Dowding, E. S. (1958). Nuclear streaming in *Gelasinospora. Canadian Journal of Microbiology,* **4,** 295–301.

Dowding, E. S. & Bakerspigel, A. (1954). The migrating nucleus. *Canadian Journal of Microbiology,* **1,** 68–78.

Dowding, E. S. & Bulmer, G. S. (1964). Notes on the cytology and sexuality of puffballs. *Canadian Journal of Microbiology,* **10,** 783–789.

Dowsett, J. A. & Reid, J. (1977a). Light microscope observations on the trapping of nematodes by *Dactylaria candida. Canadian Journal of Botany,* **55,** 2956–2962.

Dowsett, J. A. & Reid, J. (1977b). Transmission and scanning electron microscope observations on the trapping of nematodes by *Dactylaria candida. Canadian Journal of Botany,* **55,** 2963–2970.

Dowsett, J. A., Reid, J. & van Caeseele, L. (1977). Transmission and scanning electron microscope observations on the trapping of nematodes by *Dactylaria brochopaga. Canadian Journal of Botany,* **55,** 2945–2955.

Drayton, F. L. (1934). The sexual mechanism of *Sclerotinia gladioli. Mycologia,* **26,** 46–72.

Drayton, F. L. & Groves, J. W. (1952). *Stromatinia narcissi,* a new, sexually dimorphic discomycete. *Mycologia,* **44,** 119–140.

Drechsler, C. (1956). Supplementary developmental stages of *Basidiobolus ranarum* and *Basidiobolus haptosporus. Mycologia,* **48,** 655–676.

Drechsler, C. (1960). Two root rot fungi closely related to *Pythium ultimum. Sydowia,* **14,** 107–115.

Dring, D. M. (1973). Gasteromycetes. In *The Fungi: An Advanced Treatise,* **IVB,** 451–478. Editors: G. C. Ainsworth, F. K. Sparrow & A. S. Sussman. New York and London: Academic Press.

Duddington, C. L. (1957). *The Friendly Fungi. A New Approach to the Eelworm Problem.* 188 pp. London: Faber & Faber.

Duddington, C. L. (1973). Zoopagales. In *The Fungi: An Advanced Treatise,* **IVB,** 231–234. Editors: G. C. Ainsworth, F. K. Sparrow & A. S. Sussman. New York and London: Academic Press.

Duncan, B. & Herald, A. C. (1974). Some observations on the ultrastructure of *Epicoccum nigrum. Mycologia,* **66,** 1022–1029.

Duncan, E. G. & Galbraith, M. H. (1972). Post-meiotic events in the Homobasidio-mycetidae. *Transactions of the British Mycological Society*, **58**, 387–392.

Duncan, E. G. & MacDonald, J. A. (1967). Micro-evolution in *Auricularia auricula*. *Mycologia*, **59**, 803–818.

Duncan, U. K. (1959). *A Guide to the Study of Lichens*. 164 pp. Arbroath: Buncle.

Duncan, U. K. (1963). *Lichen Illustrations. Supplement to A Guide to the Study of Lichens*. 144 pp. Arbroath: Buncle.

Duntze, W., Mackay, V. & Manney, T. R. (1970). *Saccharomyces cerevisiae*: A diffusible sex factor. *Science*, **168**, 1472–1473.

Durán, R. (1973). Ustilaginales. In *The Fungi: An Advanced Treatise*, **IVB**, 281–300. Editors: G. C. Ainsworth, F. K. Sparrow & A. S. Sussman. New York and London: Academic Press.

Durrieu, G. & Zambettakis, C. (1973). Les *Ustilago* parasites des Caryophyllacées. Apports de la microscopie électronique. *Bulletin Trimestriel de la Société Mycologique de France*, **89**, 283–290.

Dykstra, M. J. (1974). An ultrastructural examination of the structure and germination of asexual propagules of four mucoralean fungi. *Mycologia*, **66**, 477–489.

Edwards, H. H. (1970). A basic staining material associated with the penetration process in resistant and susceptible powdery mildewed barley. *New Phytologist*, **69**, 299–301.

Edwards, H. H. & Allen, P. J. (1970). A fine-structure study of the primary infection process during infection of barley by *Erysiphe graminis* f. sp. *hordei*. *Phytopathology*, **60**, 1504–1509.

Egel, R. (1971). Physiological aspects of conjugation in fission yeast. *Planta*, **98**, 89–96.

Eger, G. (1963). Untersuchungen zur Fruchtkörperbildung des Kulturchampignons. *Mushroom Science*, **5**, 314.

Ehrlich, H. G. & Ehrlich, M. A. (1963). Electron microscopy of the host–parasite relationships in stem rust of wheat. *American Journal of Botany*, **50**, 123–130.

Ehrlich, H. G. & McDonough, E. S. (1949). The nuclear history in the basidia and basidiospores of *Schizophyllum commune* Fries. *American Journal of Botany*, **36**, 360–363.

Ehrlich, M. A. & Ehrlich, H. G. (1966). Ultrastructure of the hyphae and haustoria of *Phytophthora infestans* and hyphae of *P. parasitica*. *Canadian Journal of Botany*, **44**, 1495–1503.

Ehrlich, M. A. & Ehrlich, H. G. (1970). Electron microscope radioautography of ^{14}C transfer from rust uredospores to wheat host cells. *Phytopathology*, **60**, 1850–1851.

Ehrlich, M. A. & Ehrlich, H. G. (1971). Fine structure of the host–parasite interfaces in mycoparasitism. *Annual Review of Phytopathology*, **9**, 155–184.

Ehrlich, M. A., Ehrlich, H. G. & Schafer, J. F. (1968). Septal pores in the Heterobasidiomycetidae, *Puccinia graminis* and *P. recondita*. *American Journal of Botany*, **55**, 1020–1027.

Ekundayo, J. A. (1966). Further studies on germination of spores of *Rhizopus arrhizus*. *Journal of General Microbiology*, **42**, 283–291.

El-Abyad, M. S. H. & Webster, J. (1968a). Studies on pyrophilous Discomycetes. I. Comparative physiological studies. *Transactions of the British Mycological Society*, **51**, 353–367.

El-Abyad, M. S. H. & Webster, J. (1968b). Studies on pyrophilous Discomycetes. II. Competition. *Transactions of the British Mycological Society*, **51**, 369–375.

El-Ani, A. S. (1971). Chromosome numbers in the Hypocreales. II. Ascus development in *Nectria cinnabarina*. *American Journal of Botany*, **58**, 56–60.

El Shafie, A. K. & Webster, J. (1980). Ascospore liberation in *Cochliobolus cymbopogonis*. *Transactions of the British Mycological Society* (in the press).

Ellingboe, A. H. (1961). Somatic recombination in *Puccinia graminis* var. *tritici*. *Phytopathology*, **51**, 13–15.

Ellingboe, A. H. (1972). Genetics and physiology of primary infection by *Erysiphe graminis*. *Phytopathology*, **62**, 401–406.

Ellingboe, A. H. & Raper, J. R. (1962). The Buller Phenomenon in *Schizophyllum commune*: nuclear selection in fully compatible dikaryotic-homokaryotic matings. *American Journal of Botany*, **49**, 454–459.

Elliott, C. G. (1972). Sterols and the production of oospores by *Phytophthora cactorum*. *Journal of General Microbiology*, **72**, 321–327.

Elliott, C. G. & MacIntyre, D. (1973). Genetical evidence on the life history of *Phytophthora*. *Transactions of the British Mycological Society*, **60**, 311–316.

Elliott, E. W. (1949). The swarm cells of Myxomycetes. *Mycologia*, **41**, 141–170.

Elliott, M. E. (1964). Self-fertility in *Botryotrinia porri*. *Canadian Journal of Botany*, **42**, 1393–1395.

Ellis, M. B. (1966). Dematiaceous Hyphomycetes. VII. *Curvularia, Brachysporium*, etc. *Mycological Paper No. 106, Commonwealth Mycological Institute*. 43 pp.

Ellis, M. B. (1971a). Porospores. In *Taxonomy of Fungi Imperfecti*, 71–74. Editor: W. B. Kendrick. University of Toronto Press.

Ellis, M. B. (1971b). *Dematiaceous Hyphomycetes*. 608 pp. Kew, Surrey, England: Commonwealth Mycological Institute.

Ellis, M. B. (1976). *More Dematiaceous Hyphomycetes*. 505 pp. Kew, Surrey, England: Commonwealth Mycological Institute.

Ellis, T. T., Rogers, M. A. & Mims, C. W. (1972). The fine structure of the septal pore cap in *Coprinus stercorarius*. *Mycologia*, **64**, 681–688.

Ellis, T. T., Scheetz, R. W. & Alexopoulos, C. J. (1973). Ultrastructural observations on capillitial types in the Trichiales (Myxomycetes). *Transactions of the American Microscopical Society*, **92**, 65–79.

Ellzey, J. T. (1974). Ultrastructural observations of meiosis within antheridia of *Achlya ambisexualis*. *Mycologia*, **66**, 32–47.

Elsner, P. R., Vandermolen, G. E., Horton, J. C. & Bowen, C. C. (1970). Fine structure of *Phytophthora infestans* during sporangial differentiation and germination. *Phytopathology*, **60**, 1765–1772.

Emecz, T. I. & Jones, D. G. (1970). Effect of gibberellic acid on inflorescence production in cocksfoot plants infected with choke (*Epichloe typhina*). *Transactions of the British Mycological Society*, **55**, 77–82.

Emerson, R. (1941). An experimental study of the life cycles and taxonomy of *Allomyces*. *Lloydia*, **4**, 77–144.

Emerson, R. (1954). The biology of water molds. In *Aspects of Synthesis and Order in Growth*, 171–208. Editor: D. Rudnick. Princeton University Press.

Emerson, R. (1958). Mycological organization. *Mycologia*, **50**, 589–621.

Emerson, R. (1973). Mycological relevance in the nineteen seventies. *Transactions of the British Mycological Society*, **60**, 363–387.

Emerson, R. & Whisler, H. (1968). Cultural studies of *Oedogoniomyces* and *Harpochytrium*, and a proposal to place them in a new order of aquatic Phycomycetes. *Archiv für Mikrobiologie*, **61**, 195–211.

Emerson, R. & Wilson, C. M. (1954). Interspecific hybrids and the cytogenetics and cytotaxonomy of *Eu-Allomyces*. *Mycologia*, **46**, 393–434.

Emmons, C. W. & Bridges, C. H. (1961). *Entomophthora coronata*, the etiologic agent of a Phycomycosis of horses. *Mycologia*, **53**, 307–312.

Ende, H. van den (1976). *Sexual Interactions in Plants. The Role of Specific Substances in Sexual Reproduction*. 186 pp. London, New York, San Francisco: Academic Press.

Ende, H. van den & Stegwee, D. (1971). Physiology of sex in Mucorales. *Botanical Review*, **37**, 22–36.

Engel, H. & Friederichsen, I. (1964). Der Abschluss der Sporangiolen von *Sphaerobolus stellatus* (Thode) Pers., in kontinuerlicher Dunkelheit. *Planta*, **61**, 361–370.

Engel, H. & Schneider, J. C. (1963). Die Umwandlung von Glykogen in Zucker in den Fruchtkörpern von *Sphaerobolus stellatus* (Thode) Pers., vor ihrem Abschluss. *Bericht der Deutschen Botanischen Gesellschaft*, **75**, 397–400.

Ennis, H. L. & Sussman, M. (1958). The initiator cell for slime mold aggregation. *Proceedings of the National Academy of Sciences of the United States of America*, **44**, 407–411.

Erdos, G. W., Nickerson, A. W. & Raper, K. B. (1972). Fine structure of macrocysts in *Polysphondylium violaceum*. *Cytobiologie*, **6**, 351–366.

Erdos, G. W., Nickerson, A. W. & Raper, K. B. (1973a). The fine structure of macrocyst germination in *Dictyostelium mucoroides*. *Developmental Biology*, **32**, 321–330.

Erdos, G. W., Raper, K. B. & Vogen, L. K. (1973b). Mating types and macrocyst formation in *Dictyostelium discoideum*. *Proceedings of the National Academy of Sciences of the United States of America*, **70**, 1828–1830.

Erdos, G. W., Raper, K. B. & Vogen, L. K. (1975). Sexuality in the cellular slime mold *Dictyostelium giganteum*. *Proceedings of the National Academy of Sciences of the United States of America*, **72**, 970–973.

Eriksson, J. & Henning, E. (1894). Die Hauptresultate einer neuen Untersuchung über die Getreideroste. *Zeitschrift für Pflanzenkrankheiten*, **4**, 66–73.

Erlenmeyer, H. & Geiger-Huber, M. (1935). Notiz über die durch einen Brandpilz verursachte Geschlechtsummstimmung bei *Melandrium album*. *Helvetica Chimica Acta*, **18**, 921–923.

Escobar, G., McCabe, D. E. & Harpel, C. W. (1976). *Limnoperdon*, a floating Gasteromycete isolated from marshes. *Mycologia*, **68**, 874–880.

Esser, K. (1965). Heterogenic incompatibility. In *Incompatibility in Fungi*, 6–13. Editors: K. Esser & J. R. Raper. Berlin: Springer.

Esser, K. (1971). Breeding systems in fungi and their significance for genetic recombination. *Molecular and General Genetics*, **110**, 86–100.

Esser, K. (1974). *Podospora anserina*. In *Handbook of Genetics*, **I**, 531–551. Editor: R. C. King. New York and London: Plenum Press.

Esser, K. (1978). Genetics of the ergot fungus *Claviceps purpurea*. I. Proof of a monoecious life cycle and segregation patterns for mycelial morphology and alkaloid production. *Theoretical and Applied Genetics*, **53**, 145–149.

Esser, K. & Kuehnen, R. (1967). *Genetics of Fungi*. Translated by E. Steiner. 500 pp. Berlin: Springer.

Essig, F. M. (1922). The morphology, development and economic aspects of *Schizophyllum commune* Fries. *University of California Publications in Botany*, **7**, No. 14, 447–498.

Estey, R. H. & Tzean, S. S. (1976). Scanning electron microscopy of fungal nematode-trapping devices. *Transactions of the British Mycological Society*, **66**, 520–522.

Evans, R. G. (1965). *Sporobolomyces* as a cause of respiratory allergy. *Acta Allergologica*, **20**, 197–205.

Evans, S. M. & Wilson, I. M. (1971). The anther smut of sea campion. A study of the role of growth regulators in the dwarfing symptom. *Annals of Botany*, **35**, 543–553.

Eversmeyer, M. G. & Browder, L. E. (1974). Effect of leaf and stem rust on 1973 Kansas wheat yields. *Plant Disease Reporter*, **58**, 469–471.

Eversmeyer, M. G., Burleigh, J. R. & Roelfs, A. P. (1973). Equations for predicting wheat stem rust development. *Phytopathology*, **63**, 348–351.

Faro, S. (1971). Utilization of certain amino acids and carbohydrates as carbon sources of *Achlya heterosexualis*. *Mycologia*, **63**, 1234–1237.

Fassi, B. & Fontana, A. (1969). Sintesi micorrizici tra "*Pinus strobus*" e "*Tuber maculatum"* II. Sviluppo dei semenzali trapiantati e produzione di ascocarpi. *Allionia*, **15**, 115–120.

Fassi, B., Fontana, A. & Trappe, J. M. (1969). Ectomycorrhizae formed by *Endogone lactiflua* with species of *Pinus* and *Pseudotsuga*. *Mycologia*, **61**, 412–414.

Federici, B. A. (1977). Differential pigmentation in the sexual phase of *Coelomomyces*. *Nature, London*, **267**, 514–515.

Fell, J. W. (1976). Yeasts in oceanic regions. In *Recent Advances in Aquatic Mycology*, 93–124. Editor: E. B. Gareth Jones. London: Elek Science.

Fell, J. W., Hunter, I. L. & Tallman, A. S. (1973). Marine basidiomycetous yeasts (*Rhodosporidium* spp.n.) with tetrapolar and multiple allelic bipolar mating systems. *Canadian Journal of Microbiology*, **19**, 643–657.

Fell, J. W., Statzell, A. C., Hunter, I. L. & Phaff, H. J. (1969). *Leucosporidium* gen.n., the heterobasidiomycetous stage of several yeasts of the genus *Candida*. *Antonie van Leeuwenhoek*, **35**, 433–462.

Fennell, D. I. (1973). Plectomycetes; Eurotiales. In *The Fungi: An Advanced Treatise*, **IVA**, 45–68. Editors: G. C. Ainsworth, F. K. Sparrow & A. S. Sussman. New York and London: Academic Press.

Fergus, C. L. (1960). *Illustrated Genera of Wood Decay Fungi*. 132 pp. Minneapolis: Burgess Publishing Company.

Field, J. I. & Webster, J. (1977). Traps of predacious fungi attract nematodes. *Transactions of the British Mycological Society*, **68**, 467–470.

Fields, W. G. & Maniotis, J. (1963). Some cultural and genetic aspects of a new heterothallic *Sordaria*. *American Journal of Botany*, **50**, 80–85.

Filer, T. H. (1966). Effect on grass and cereal seedlings of hydrogen cyanide produced by mycelium and sporophores of *Marasmius oreades*. *Plant Disease Reporter*, **50**, 264–266.

Filosa, M. F. & Chan, M. (1972). Isolations from soil of macrocyst-forming strains of the cellular slime mould *Dictyostelium mucoroides*. *Journal of General Microbiology*, **71**, 413–414.

Fincham, J. R. S. (1971). Using fungi to study genetic recombination. *Oxford Biology Readers*, **2**. 16 pp. Oxford University Press.

Fincham, J. R. S. & Day, P. R. (1971). *Fungal Genetics*. 3rd edition. 402 pp. Oxford and Edinburgh: Blackwell Scientific Publications.

Findlay, W. P. K. (1940). Studies in the physiology of wood-destroying fungi. III. Progress of decay under natural and controlled conditions. *Annals of Botany, London*, N.S., **4**, 701–712.

Findlay, W. P. K. & Savory, J. G. (1960). *Dry Rot in Wood*. 6th edition. 36 pp. London: H.M.S.O.

Fischer, A. (1892). Die Pilze. IV. Abt. Phycomycetes. In *Kryptogamenflora von Deutschland, Oesterreich und der Schweiz*. Editor: L. Rabenhorst. Leipzig.

Fischer, E. (1938). Tuberineae. In *Die Natürlichen Pflanzenfamilien*. 2nd edition, 5b, VIII, 1–42. Editors: A. Engler & K. Prantl. Leipzig: Engelmann.

Fischer, F. G. & Werner, G. (1955). Eine Analyse des Chemotropismus einiger Pilze, insbesondere der Saprolegniaceen. *Hoppe-Seyler's Zeitschrift für Physiologische Chemie*, **300**, 211–236.

Fischer, F. G. & Werner, G. (1958a). Die Chemotaxis der Schwärmsporen von Wasserpilzen (Saprolegniaceen). *Hoppe-Seyler's Zeitschrift für Physiologische Chemie*, **310**, 65–91.

Fischer, F. G. & Werner, G. (1958b). Über die Wirkungen von Nicotinsäureamid auf die Schwärmsporen wasserbewohnende Pilze. *Hoppe-Seyler's Zeitschrift für Physiologische Chemie*, **310**, 92–96.

Fischer, G. W. & Holton, C. S. (1957). *Biology and Control of the Smut Fungi*. 622 pp. New York: Ronald Press Company.

Fisher, P. J. (1977). New methods of detecting and studying saprophytic behaviour of aero-aquatic hyphomycetes from stagnant water. *Transactions of the British Mycological Society*, **68**, 407–411.

Fisher, P. J. & Webster, J. (1980). Ecological studies on aero-aquatic Hyphomycetes. In *Fungal Ecology*. Editors: D. T. Wicklow & G. C. Carroll. New York: Marcel Dekker, Inc.

Flanagan, P. W. (1970). Meiosis and mitosis in Saprolegniaceae. *Canadian Journal of Botany*, **48**, 2069–2076.

Fleming, A. (1944). The discovery of penicillin. *British Medical Bulletin*, **2**, 4–5.

Fletcher, H. J. (1969). The development and tropisms of the sporangiophores of *Pilaira anomala*. *Transactions of the British Mycological Society*, **53**, 130–132.

Fletcher, H. J. (1973). The sporangiophore of *Pilaira* species. *Transactions of the British Mycological Society*, **61**, 553–568.

Fletcher, J. (1971). Conidium ontogeny in *Penicillium*. *Journal of General Microbiology*, **67**, 207–214.

Fletcher, J. (1972). Fine structure of developing merosporangia and sporangiospores of *Syncephalastrum racemosum*. *Archiv für Mikrobiologie*, **87**, 269–284.

Fletcher, J. (1973a). The distribution of cytoplasmic vesicles, multivesicular bodies and paramural bodies in elongating sporangiophores and swelling sporangia of *Thamnidium elegans* Link. *Annals of Botany*, **37**, 955–961.

Fletcher, J. (1973b). Ultrastructural changes associated with spore formation in sporangia and sporangiola of *Thamnidium elegans* Link. *Annals of Botany*, **37**, 963–972.

Fletcher, J. (1976). Electron microscopy of genesis, maturation and wall structure of conidia of *Aspergillus terreus*. *Transactions of the British Mycological Society*, **66**, 27–34.

Flor, H. H. (1971). Current status of the gene-for-gene concept. *Annual Review of Phytopathology*, **9**, 275–296.

Flores de Cunha, M. (1970). Mitotic mapping of *Schizosaccharomyces pombe*. *Genetical Research*, **16**, 127–144.

Fontana, A. & Palenzona, M. (1969). Sintesi micorrizica di "*Tuber albidum*" in coltura pure, con *Pinus strobus* è Pioppo euroamericano. *Allionia*, **15**, 99–104.

Foster, J. W. (1949). *Chemical Activities of Fungi*. 648 pp. New York and London: Academic Press.

Fothergill, P. G. & Child, J. H. (1964). Comparative studies of the mineral nutrition of three species of *Phytophthora*. *Journal of General Microbiology*, **36**, 67–78.

Fothergill, P. G. & Hide, D. (1962). Comparative nutritional studies of *Pythium* spp. *Journal of General Microbiology*, **29**, 325–334.

Fowell, R. R. (1969). Sporulation and hybridization of yeasts. In *The Yeasts*, **I**, 303–383. Editors: A. H. Rose & J. S. Harrison. New York and London: Academic Press.

Fox, R. A. (1953). Heterothallism in *Chaetomium*. *Nature, London*, **172**, 165–166.

Franke, G. (1957). Die Zytologie der Ascusentwicklung von *Podospora anserina*. *Zeitschrift für Induktive Abstammungs-u. Vererbungslehre*, **88**, 159–160.

Fraymouth, J. (1956). Haustoria of the Peronosporales. *Transactions of the British Mycological Society*, **39**, 79–107.

Fricke, S. & Handke, H. H. (1962). Untersuchungen zur Offnungswerke der Geastracee-Fruchtkorper. *Zeitschrift für Pilzkunde*, **27**, 113–122.

Fries, N. (1942). Einspormyzelien einiger Basidiomyceten als Mykorrhizabildner von Kiefer und Fichte. *Svensk Botanisk Tidskrift*, **36**, 151–156.

Fries, N. (1948). Heterothallism in some Gasteromycetes and Hymenomycetes. *Svensk Botanisk Tidskrift*, **42**, 158–168.

Fukui, Y. & Takeuchi, I. (1971). Drug resistant mutants and appearance of heterozygotes in the cellular slime mould *Dictyostelium discoideum. Journal of General Microbiology*, **67**, 307–317.

Fuller, J. G. (1969). *The Day of Antony's Fire*. 310 pp. New York: Macmillan.

Fuller, M. S. (1966). Structure of the uniflagellate zoospores of aquatic Phycomycetes. In *The Fungus Spore*, 67–84. Editor: M. F. Madelin. Colston Papers No. 18. London: Butterworths.

Fuller, M. S. (1976). The zoospore, hallmark of aquatic fungi. *Mycologia*, **69**, 1–20.

Fuller, M. S. & Olson, L. W. (1971). The zoospore of *Allomyces. Journal of General Microbiology*, **66**, 171–183.

Fuller, M. S. & Reichle, R. (1965). The zoospore and early development of *Rhizidiomyces apophysatus. Mycologia*, **57**, 946–961.

Fuller, M. S. & Reichle, R. E. (1968). The fine structure of *Monoblepharella* sp. zoospores. *Canadian Journal of Botany*, **46**, 279–283.

Fullerton, R. A. (1970). An electron microscope study of the intracellular hyphae of some smut fungi (Ustilaginales). *Australian Journal of Botany*, **18**, 285–292.

Fulton, I. W. (1950). Unilateral nuclear migration and the interactions of haploid mycelia in the fungus *Cyathus stercoreus. Proceedings of the National Academy of Sciences of the United States of America*, **36**, 306–312.

Furtado, J. S. (1962). Structure of the spore of the Ganodermoideae Donk. *Rickia, Archivos de Botanica do Estado de São Paulo*, **1**, 227–241.

Furtado, J. S. (1965). Relation of microstructures to the taxonomy of the Ganodermoideae (Polyporaceae) with special reference to the cover of the pilear surface. *Mycologia*, **57**, 588–611.

Furtado, J. S. (1966). Significance of the clamp-connection in the Basidiomycetes. *Persoonia*, **4**, 125–144.

Furtado, J. S. (1971). The septal pore and other ultrastructural features of the Pyrenomycete *Sordaria fimicola. Mycologia*, **63**, 104–113.

Furtado, J. S. & Olive, L. S. (1970). Ultrastructure of ascospore development in *Sordaria fimicola. Journal of the Elisha Mitchell Scientific Society*, **86**, 131–138.

Furtado, J. S. & Olive, L. S. (1971). Ultrastructural evidence of meiosis in *Ceratiomyxa fruticulosa. Mycologia*, **63**, 413–416.

Furtado, J. S., Olive, L. S. & Jones, S. B. (1971). Ultrastructural studies of protostelids: the fruiting stage of *Cavostelium bisporum. Mycologia*, **63**, 132–143.

Galindo, J. & Gallegly, M. E. (1960). The nature of sexuality in *Phytophthora infestans. Phytopathology*, **50**, 123–128.

Galindo, J. A. & Zentmeyer, G. A. (1967). Genetical and cytological studies of *Phytophthora* strains pathogenic to pepper plants. *Phytopathology*, **57**, 1300–1304.

Gallegly, M. E. & Galindo, J. (1958). Mating types and oospores of *Phytophthora infestans* in nature in Mexico. *Phytopathology*, **48**, 274–277.

Gams, W., Chien, C.-Y. & Domsch, K. H. (1972). Zygospore formation by the heterothallic *Mortierella elongata* and a related homothallic species *Mortierella epigama* n.spec. *Transactions of the British Mycological Society*, **58**, 5–13.

Gams, W. & Williams, S. T. (1963). Heterothallism in *Mortierella parvispora* Linnemann. I. Morphology and development of zygospores and some factors influencing their formation. *Nova Hedwigia*, **5**, 347–357.

Gamundi, I. J. & Ranalli, M. E. (1963). Apothecial development in *Ascobolus stercorarius. Transactions of the British Mycological Society*, **46**, 393–400.

Ganesan, A. T. (1959). The cytology of *Saccharomyces. Comptes Rendus des Travaux du Laboratoire Carlsberg*, **31**, 149–174.

Garay, A. & Sagi, F. (1960). Untersuchungen über die Geschlechtsumwandlung bei *Melandrium album* Mill., nach Infektion mit *Ustilago violacea* (Pers.) Fckl., unter besonder Berücksichtung der Auxinoxydase und Flavonoide. *Phytopathologische Zeitschrift*, **38**, 201–208.

Garnjobst, L. (1955). Further analysis of genetic control of heterokaryosis in *Neurospora crassa*. *American Journal of Botany*, **42**, 444–448.

Garnjobst, L. & Wilson, J. F. (1956). Heterokaryosis and protoplasmic incompatibility in *Neurospora crassa*. *Proceedings of the National Academy of Sciences of the United States of America*, **42**, 613–618.

Garraway, M. O. (1970). Rhizomorph initiation and growth in *Armillaria mellea* promoted by *o*-aminobenzoic and *p*-aminobenzoic acids. *Phytopathology*, **60**, 861–865.

Garraway M. O. & Weinhold, A. A. (1968). Period of access to ethanol in relation to carbon utilization and rhizomorph initiation and growth in *Armillaria mellea*. *Phytopathology*, **58**, 1190–1191.

Garrett, R. G. & Tomlinson, J. A. (1967). Isolate differences in *Olpidium brassicae*. *Transactions of the British Mycological Society*, **50**, 429–435.

Garrett, S. D. (1953). Rhizomorph behaviour in *Armillaria mellea* (Vahl) Quél. I. Factors controlling rhizomorph initiation by *Armillaria mellea* in pure culture. *Annals of Botany, London*, N.S., **17**, 63–79.

Garrett, S. D. (1954). Function of the mycelial strands in substrate colonization by the cultivated mushroom *Psalliota hortensis*. *Transactions of the British Mycological Society*, **37**, 51–57.

Garrett, S. D. (1956). *Biology of Root-Infecting Fungi*. 293 pp. Cambridge University Press.

Garrett, S. D. (1960). Inoculum potential. In *Plant Pathology: An Advanced Treatise*, **3**, 23–56. Editors: J. G. Horsfall & A. E. Dimond. New York and London: Academic Press.

Garrett, S. D. (1970). *Pathogenic Root-Infecting Fungi*. 294 pp. Cambridge University Press.

Garrod, D. & Ashworth, J. M. (1973). Development of the cellular slime mould *Dictyostelium discoideum*. *23rd Symposium of the Society for General Microbiology*, 407–435.

Gauger, W. L. (1961). The germination of zygospores of *Rhizopus stolonifer*. *American Journal of Botany*, **48**, 427–429.

Gauger, W. L. (1965). The germination of zygospores of *Mucor hiemalis*. *Mycologia*, **57**, 634–641.

Gauger, W. L. (1966). Sexuality in an azygosporic strain of *Mucor hiemalis*. I. Breakdown of the azygosporic component. *American Journal of Botany*, **53**, 751–755.

Gauger, W. L. (1975). Further studies on sexuality in azygosporic strains of *Mucor hiemalis*. *Transactions of the British Mycological Society*, **64**, 113–118.

Gäumann, E. (1959). Die Rostpilze Mitteleuropas. *Beiträge der Kryptogamenflora der Schweiz*, Bd. **XII**. 1407 pp. Bern: Büchler.

Gäumann, E. (1964). *Die Pilze. Grundzuge ihrer Entwicklungsgeschichte und Morphologie*. 541 pp. Basel and Stuttgart: Birkhauser.

Gaunt, R. E. & Manners, J. G. (1971a). Host–parasite relations in loose smut of wheat. I. The effect of infection on host growth. *Annals of Botany*, **35**, 1131–1140.

Gaunt, R. E. & Manners, J. G. (1971b). Host–parasite relations in loose smut of wheat. II. Distribution of ^{14}C-labelled assimilates. *Annals of Botany*, **35**, 1141–1150.

Gaunt, R. E. & Manners, J. G. (1971c). Host–parasite relations in loose smut of wheat. III. Utilization of ^{14}C-labelled assimilate. *Annals of Botany*, **35**, 1151–1161.

Gay, J. L. & Greenwood, A. D. (1966). Structural aspects of zoospore production in *Saprolegnia ferax* with particular reference to the cell and vacuolar membranes. In *The Fungus Spore*, 95–108. Editor: M. F. Madelin. Colston Papers No. 18. London: Butterworths.

Gay, J. L., Greenwood, A. D. & Heath, I. B. (1971). The formation and behaviour of vacuoles (vesicles) during oosphere development and zoospore germination in *Saprolegnia. Journal of General Microbiology*, **65**, 233–241.

Gentles, J. C. & La Touche, C. J. (1969). Yeasts as human and animal pathogens. In *The Yeasts*, **1**, 107–182. Editors: A. H. Rose & J. S. Harrison. New York and London: Academic Press.

George, R. P., Hohl, H. R. & Raper, K. B. (1972). Ultrastructural development of stalk-producing cells in *Dictyostelium discoideum*, a cellular slime mould. *Journal of General Microbiology*, **70**, 477–489.

Gerdemann, J. W. & Trappe, J. M. (1974). The Endogonaceae of the Pacific Northwest. *Mycologia Memoir No. 5*. 76 pp.

Gibbons, J. R. & Grimstone, A. V. (1960). On flagellar structure in certain flagellates. *Journal of Biophysical and Biochemical Cytology*, **7**, 697–716.

Gibbs, J. N. (1974). Biology of Dutch Elm Disease. *Forestry Commission Forest Record* **94**. 9 pp. London: H.M.S.O.

Gibbs, J. N. (1978). Intercontinental epidemiology of Dutch Elm Disease. *Annual Review of Phytopathology*, **16**, 287–307.

Gibbs, J. N. & Brasier, C. M. (1973a). Correlation between cultural characters and pathogenicity in *Ceratocystis ulmi* from Britain, Europe and America. *Nature, London*, **241**, 381–383.

Gibbs, J. N. & Brasier, C. M. (1973b). Origin of the Dutch Elm disease epidemic in Britain. *Nature, London*, **242**, 607–609.

Gibbs, J. N., Brasier, C. M., McNabb, H. S. & Heybroek, H. M. (1975). Further studies on pathogenicity in *Ceratocystis ulmi. European Journal of Forest Pathology*, **5**, 161–174.

Giesy, R. M. & Day, P. R. (1965). The septal pores of *Coprinus lagopus* (Fr.) sensu Buller in relation to nuclear migration. *American Journal of Botany*, **52**, 287–293.

Gilbert, H. C. (1935). Critical events in the life history of *Ceratiomyxa. American Journal of Botany*, **22**, 52–74.

Gilbert, O. L. (1973). Lichens and air pollution. In *The Lichens*, 443–472. Editors: V. Ahmadjian & M. E. Hale, New York and London: Academic Press.

Gingold, E. & Ashworth, J. M. (1974). Evidence for mitotic crossing-over during the parasexual cycle of the cellular slime mold *Dictyostelium discoideum. Journal of General Microbiology*, **84**, 70–78.

Girbardt, M. (1958). Über die Substruktur von *Polyporus versicolor* L. *Archiv für Mikrobiologie*, **28**, 255–269.

Girbardt, M. (1961). Licht- und elektronenoptische Untersuchungen an *Polystictus versicolor* (L). VII. Lebendbeobachtung und Zeitdauer der Teilung des vegetativen Kernes. *Experimental Cell Research*, **23**, 181–194.

Girbardt, M. (1968). The ultrastructure and dynamics of the moving nucleus. In *Aspects of Cell Motility*, 249–259. Editor: P. L. Miller. *Symposia of the Society for Experimental Biology*, **22**.

Giuma, A. Y. & Cooke, R. C. (1971). Nematotoxin production by *Nematoctonus haptocladus* and *N. concurrens. Transactions of the British Mycological Society*, **56**, 89–94.

Giuma, A. Y., Hackett, A. M. & Cooke, R. C. (1973). Thermostable nematotoxins produced by germinating conidia of some endozoic fungi. *Transactions of the British Mycological Society*, **60**, 49–56.

Gleason, F. H. (1972). Lactate dehydrogenases in Oomycetes. *Mycologia*, **64**, 663–666.

Gleason, F. H. (1973). Uptake of amino acids by *Saprolegnia. Mycologia*, **65**, 465–468.

Gleason, F. H. (1976). The physiology of the lower freshwater fungi. In *Recent Advances in Aquatic Mycology*, 543–572. Editor: E. B. Gareth Jones. London: Elek Science.

Gleason F. H., Rudolph, C. R. & Price, J. R. (1970a). Growth of certain aquatic

Oomycetes on amino acids. I. *Saprolegnia, Achlya, Leptolegnia* and *Dictyuchus. Physiologia Plantarum*, **23**, 513–516.

Gleason, F. H., Stuart, T. D., Price, J. S. & Nelbach, E. T. (1970b). Growth of certain aquatic Oomycetes on amino acids. II. *Apodachlya, Aphanomyces* and *Pythium. Physiologia Plantarum*, **23**, 769–774.

Glen-Bott, J. I. (1951). *Helicodendron giganteum* n.sp. and other aerial-sporing Hyphomycetes of submerged dead leaves. *Transactions of the British Mycological Society*, **34**, 275–279.

Glen-Bott, J. I. (1955). On *Helicodendron tubulosum* and some similar species. *Transactions of the British Mycological Society*, **38**, 17–30.

Goldie-Smith, E. K. (1954). The position of *Woronina polycystis* in the Plasmodiophoraceae. *American Journal of Botany*, **41**, 441–448.

Goldstein, B. (1923). Resting spores of *Empusa muscae. Bulletin of the Torrey Botanical Club*, **50**, 317–327.

Goldstein, S. (1960a). Physiology of aquatic fungi. I. Nutrition of two monocentric chytrids. *Journal of Bacteriology*, **80**, 701–707.

Goldstein, S. (1960b). Factors affecting the growth and pigmentation of *Cladochytrium replicatum. Mycologia*, **52**, 490–498.

Goldstein, S. (1961). Studies of two polycentric chytrids in pure culture. *American Journal of Botany*, **48**, 294–298.

Gooday, G. W. (1971). An autoradiographic study of hyphal growth of some fungi. *Journal of General Microbiology*, **67**, 125–133.

Gooday, G. W. (1973). Differentiation in the Mucorales. *Symposia of the Society for General Microbiology*, **23**, 269–294.

Gooday, G. W. (1974a). Fungal sex hormones. *Annual Review of Biochemistry*, **43**, 35–49.

Gooday, G. W. (1974b). Control of development of excised fruit bodies and stipes of *Coprinus cinereus. Transactions of the British Mycological Society*, **62**, 391–399.

Gooday, G. W., De Rousset-Hall, A. & Hunsley, D. (1976). Effect of polyoxin D on chitin synthesis in *Coprinus cinereus. Transactions of the British Mycological Society*, **67**, 193–200.

Gooday, G. W., Fawcett, P., Green, D. & Shaw, G. (1973). The formation of fungal sporopollenin in the zygospore wall of *Mucor mucedo:* a role for the sexual carotenogenesis in the Mucorales. *Journal of General Microbiology*, **74**, 233–239.

Goodell, E. W. (1971). "Apical dominance" in the sporangiophore of the fungus *Phycomyces. Planta*, **98**, 63–75.

Goodman, E. M. (1972). Axenic culture of myxamoebae of the Myxomycete *Physarum polycephalum. Journal of Bacteriology*, **111**, 242–247.

Gordon, C. C. (1966). A re-interpretation of the ontogeny of the ascocarp of species of the Erysiphaceae. *American Journal of Botany*, **53**, 652–662.

Gottsberger, C. (1967). Geisseln bei Myxomyceten (Elektronenoptische Studie). *Nova Hedwigia*, **13**, 235–243.

Graham, K. M. (1955). Distribution of physiological races of *Phytophthora infestans* (Mont.) de Bary in Canada. *American Potato Journal*, **32**, 277–282.

Grainger, J. (1962). Vegetative and fructifying growth in *Phallus impudicus. Transactions of the British Mycological Society*, **45**, 147–155.

Gray, W. D. (1941). Studies on alcohol tolerance of yeasts. *Journal of Bacteriology*, **42**, 561–574.

Gray, W. D. (1959). *The Relation of Fungi to Human Affairs.* 510 pp. New York: Henry Holt & Company.

Gray, W. D. & Alexopoulos, C. J. (1968). *Biology of the Myxomycetes.* 288 pp. New York: The Ronald Press Company.

Green, G. J. (1971a). Physiologic races of wheat stem rust in Canada from 1919 to 1969. *Canadian Journal of Botany*, **49**, 1575–1588.

Green, G. J. (1971b). Hybridization between *Puccinia graminis tritici* and *Puccinia graminis secalis* and its evolutionary implications. *Canadian Journal of Botany*, **49**, 2089–2095.

Greenhalgh, G. N. & Bevan, R. J. (1978). Response of *Rhytisma acerinum* to air pollution. *Transactions of the British Mycological Society*, **71**, 491–523.

Greenhalgh, G. N. & Evans, L. V. (1967). The structure of the ascus apex in *Hypoxylon fragiforme* with reference to ascospore release in this and related species. *Transactions of the British Mycological Society*, **50**, 183–188.

Greer, D. L. & Friedman, L. (1966). Studies on the genus *Basidiobolus* with reclassification of the species pathogenic for man. *Sabouraudia*, **4**, 231–241.

Gregg, J. H. (1966). Organization and synthesis in the cellular slime molds. In *The Fungi: An Advanced Treatise*, **II**, 235–281. Editors: G. C. Ainsworth & A. S. Sussman. New York and London: Academic Press.

Gregory, P. H. (1949). The operation of the puff-ball mechanism of *Lycoperdon perlatum* by raindrops shown by ultra-high-speed Schlieren cinematography. *Transactions of the British Mycological Society*, **32**, 11–15.

Gregory, P. H. (1957). Electrostatic charges on spores of fungi in air. *Nature, London*, **180**, 330.

Gregory, P. H. (1973). *The Microbiology of the Atmosphere*. 2nd edition. Aylesbury: Leonard Hill.

Gregory, P. H., Guthrie, E. J. & Bunce, M. E. (1952). Experiments on splash dispersal of fungus spores. *Journal of General Microbiology*, **20**, 328–354.

Gregory, P. H. & Henden, D. R. (1976). Terminal velocity of basidiospores of the giant puffball (*Lycoperdon giganteum*). *Transactions of the British Mycological Society*, **67**, 399–407.

Greis, H. (1938). Die sexualvorgange bei *Tuber aestivum* und *T. brumale*. *Biologisches Zentralblatt*, **58**, 617–631.

Greis, H. (1939). Ascusentwicklung von *Tuber aestivum* und *T. brumale*. *Zeitschrift für Botanik*, **34**, 129–178.

Gremmen, J. (1955). Über Apothezienbildung bein *Ascobolus stercorarius* (Bull.) Schroet. *Schweizerische Zeitschrift für Pilzkunde*, **33**, 42–45.

Griffiths, E. (1959). The cytology of *Gloeotinia temulenta* (blind seed disease of rye-grass). *Transactions of the British Mycological Society*, **42**, 132–148.

Griffiths, H. B. (1973). Fine structure of seven unitunicate pyrenomycete asci. *Transactions of the British Mycological Society*, **60**, 261–271.

Grove, S. N. & Bracker, C. E. (1970). Protoplasmic organization of hyphal tips among fungi: vesicles and Spitzenkörper. *Journal of Bacteriology*, **104**, 989–1009.

Grove, S. N. & Bracker, C. E. (1978). Protoplasmic changes during zoospore encystment and cyst germination in *Pythium aphanidermatum*. *Experimental Mycology*, **2**, 51–98.

Grove, S. N., Bracker, C. E. & Morré, D. J. (1970). An ultrastructural basis for hyphal tip growth in *Pythium ultimum*. *American Journal of Botany*, **57**, 245–266.

Groves, J. W. & Drayton, F. L. (1939). The perfect stage of *Botrytis cinerea*. *Mycologia*, **31**, 485–489.

Groves, J. W. & Loveland, C. A. (1953). The connection between *Botryotinia fuckeliana* and *Botrytis cinerea*. *Mycologia*, **45**, 415–425.

Groves, J. W. & Skolko, A. J. (1944–1945). Notes on seed-borne fungi. III. *Curvularia*. *Canadian Journal of Research*, **C**, 22–33.

Gruen, H. E. (1959). Auxins and fungi. *Annual Review of Plant Physiology*, **10**, 405–440.

Gruen, H. E. (1963). Endogenous growth regulations in carpophores of *Agaricus bisporus*. *Plant Physiology*, **38**, 652–666.

Gruen, H. E. (1969). Growth and rotation of *Flammulina velutipes* and the dependence of stipe elongation on the cap. *Mycologia*, **61**, 149–166.

Gruen, H. E. & Wu, S. (1972a). Promotion of stipe elongation in isolated *Flammulina velutipes* fruit bodies by carbohydrates, natural extracts, and amino acids. *Canadian Journal of Botany*, **50**, 803–818.

Gruen, H. E. & Wu, S. (1972b). Dependence of fruit body elongation on the mycelium in *Flammulina velutipes*. *Mycologia*, **64**, 995–1007.

Guerdoux, J. L. (1974). *Coprinus*. In *Handbook of Genetics*, **I**, 627–636. Editor: R. C. King. New York and London: Plenum Press.

Gull, K. (1976). Differentiation of septal ultrastructure according to cell type in the Basidiomycete, *Agrocybe praecox*. *Journal of Ultrastructure Research*, **54**, 89–94.

Gull, K. & Trinci, A. P. J. (1974). Detection of areas of wall differentiation in fungi using fluorescent staining. *Archiv für Mikrobiologie*, **96**, 53–57.

Gustafsson, M. (1965). On species of the genus *Entomophthora* Fres. in Sweden. II. Cultivation and physiology. *Lantbrukshögskolans Annaler*, **31**, 405–457.

Gustafsson, M. (1969). On the species of the genus *Entomophthora* Fres. in Sweden. III. Possibility of usage in biological control. *Lantbrukshögskolans Annaler*, **35**, 235–274.

Guttenberg, H. von & Schmoller, H. (1958). Kulturversuche mit *Peronospora brassicae* Gäum. *Archiv für Mikrobiologie*, **30**, 268–279.

Guttes, E., Guttes, S. & Rusch, H. P. (1961). Morphological observations on growth and differentiation of *Physarum polycephalum* grown in pure culture. *Developmental Biology*, **3**, 588–614.

Gutz, H. & Doe, F. J. (1975). On homo- and heterothallism in *Schizosaccharomyces pombe*. *Mycologia*, **67**, 748–759.

Gutz, H., Heslot, H., Leupold, U. & Loprieno, N. (1974). *Schizosaccharomyces pombe*. In *Handbook of Genetics*, **I**, 395–446. Editor: R. D. King. New York and London: Plenum Press.

Gwynne-Vaughan, H. C. I. & Broadhead, Q. E. (1936). Contributions to the study of *Ceratostomella fimbriata*. *Annals of Botany, London*, **50**, 747–758.

Gwynne-Vaughan, H. C. I. & Williamson, H. S. (1930). Contributions to the study of *Humaria granulata* Quél. *Annals of Botany, London*, **44**, 127–145.

Gwynne-Vaughan, H. C. I. & Williamson, H. S. (1932). The cytology and development of *Ascobolus magnificus*. *Annals of Botany, London*, **46**, 653–670.

Haber, J. E. & Halvorson, H. O. (1975). Methods in sporulation and germination of yeasts. In *Methods in Cell Biology*, **9**, 45–69. Editor: D. M. Prescott. New York, San Francisco, London: Academic Press.

Hadley, G. (1968). Development of stromata in *Claviceps purpurea*. *Transactions of the British Mycological Society*, **51**, 763–769.

Hagimoto, H. (1963). Studies on the growth of fruit body of fungi. IV. The growth of the fruit body of *Agaricus bisporus* and the economy of the mushroom growth hormone. *Botanical Magazine, Tokyo*, **76**, 256–263.

Hagimoto, H. & Konishi, M. (1959). Studies on the growth of fruit body of fungi. I. Existence of a hormone active to the growth of fruit body in *Agaricus bisporus* (Lange) Sing. *Botanical Magazine, Tokyo*, **72**, 359–366.

Hagimoto, H. & Konishi, M. (1960). Studies on the growth of fruit body of fungi. II. Activity and stability of growth hormone in the fruit body of *Agaricus bisporus* (Lange) Sing. *Botanical Magazine, Tokyo*, **73**, 283–287.

Hale, M. E. (1967). *The Biology of Lichens*. 176 pp. London: Edward Arnold.

Halisky, P. M. (1965). Physiologic specialization and genetics of the smut fungi. III. *Botanical Review*, **31**, 114–150.

Hall, I. M. & Halfhill, J. C. (1959). The germination of resting spores of *Entomophthora virulenta* Hall and Dunn. *Journal of Economic Entomology*, **52**, 30–35.

Hammett, K. R. W. & Manners, J. G. (1971). Conidium liberation in *Erysiphe graminis*. I. Visual and statistical analysis of spore trap records. *Transactions of the British Mycological Society*, **56**, 387–401.

Hammett, K. R. W. & Manners, J. G. (1973). Conidium liberation in *Erysiphe graminis*. II. Conidial chain and pustule structure. *Transactions of the British Mycological Society*, **61**, 121–133.

Hammett, K. R. W. & Manners, J. G. (1974). Conidium liberation in *Erysiphe graminis*. III. Wind tunnel studies. *Transactions of the British Mycological Society*, **62**, 267–282.

Hammill, T. M. (1971). Fine structure of annellophores. I. *Scopulariopsis brevicaulis* and *S. koningii*. *American Journal of Botany*, **58**, 88–97.

Hammill, T. M. (1972). Fine structure of annellophores. II. *Doratomyces nanus*. *Transactions of the British Mycological Society*, **59**, 249–253.

Hanlin, R. T. (1961). Studies in the genus *Nectria*. II. Morphology of *N. gliocladioides*. *American Journal of Botany*, **48**, 900–908.

Hanlin, R. T. (1965). Morphology of *Hypocrea schweinitzii*. *American Journal of Botany*, **52**, 570–579.

Hanlin, R. T. (1971). Morphology of *Nectria haematococca*. *American Journal of Botany*, **58**, 105–116.

Hanlin, R. T. (1976). Phialide and conidium development in *Aspergillus clavatus*. *American Journal of Botany*, **63**, 144–155.

Hansen, L. (1958). On the anatomy of the Danish species of *Ganoderma*. *Botanisk Tidsskrift*, **54**, 333–352.

Harley, J. L. (1969). *The Biology of Mycorrhiza*. 2nd edition. 334 pp. London: Leonard Hill.

Harris, J. S. & Taber, W. A. (1973). Ultrastructure and morphogenesis of the synnema of *Ceratocystis ulmi*. *Canadian Journal of Botany*, **51**, 1565–1571.

Harrison, K. A. (1971). The evolutionary lines in the fungi with spines supporting the hymenium. In *Evolution in the Higher Basidiomycetes*, 375–392. Editor: R. H. Petersen. Knoxville: University of Tennessee Press.

Harrison, K. A. (1973). Aphyllophorales. III. Hydnaceae and Echinodontiaceae. In *The Fungi: An Advanced Treatise*, **IVB**, 369–395. Editors: G. C. Ainsworth, F. K. Sparrow & A. S. Sussman. New York and London: Academic Press.

Harrold, C. E. (1950). Studies in the genus *Eremascus*. I. The re-discovery of *Eremascus albus* Eidam and some new observations concerning its life history and cytology. *Annals of Botany, London*, N.S., **14**, 127–148.

Harrower, K. M. (1976). The micropycnidiospores of *Leptosphaeria nodorum*. *Transactions of the British Mycological Society*, **76**, 335–336.

Hart, J. H. & Landis, W. R. (1971). Rate and extent of colonization of naturally and artificially inoculated American Elms by *Ceratocystis ulmi*. *Phytopathology*, **61**, 1456–1458.

Hartley, M. J. & Williams, P. G. (1971). Genotypic variation within a phenotype as a possible basis for somatic hybridization in rust fungi. *Canadian Journal of Botany*, **49**, 1085–1087.

Hartwell, L. H. (1974). *Saccharomyces cerevisiae* cell cycle. *Bacteriological Reviews*, **38**, 164–198.

Harvey, R. (1958). Sporophore development and proliferation in *Hydnum auriscalpium* Fr. *Transactions of the British Mycological Society*, **41**, 325–334.

Haskins, R. H. (1939). Cellulose as a substratum for saprophytic Chytrids. *American Journal of Botany*, **26**, 635–639.

Haskins, R. H. & Weston, W. H. (1950). Studies in the lower Chytridiales. I. Factors affecting pigmentation, growth and metabolism in a strain of *Karlingia (Rhizophlyctis) rosea*. *American Journal of Botany*, **37**, 739–750.

Hassan, A. & Macdonald, J. A. (1971). *Ustilago violacea* on *Silene dioica*. *Transactions of the British Mycological Society*, **56**, 451–461.

Hawker, L. E. (1954). British hypogeous fungi. *Philosophical Transactions of the Royal Society, Series B*, **237**, 429–546.

Hawker, L. E. (1955). Hypogeous fungi. *Biological Reviews*, **30**, 127–158.

Hawker, L. E. (1957). *The Physiology of Reproduction in Fungi*. 128 pp. Cambridge University Press.

Hawker, L. E. (1974). Revised annotated list of British hypogeous fungi. *Transactions of the British Mycological Society*, **63**, 67–76.

Hawker, L. E. & Abbott, P. McV. (1963a). The fine structure of vegetative hyphae of *Rhizopus*. *Journal of General Microbiology*, **30**, 401–408.

Hawker, L. E. & Abbott, P. McV. (1963b). An electron microscope study of maturation and germination of sporangiospores of two species of *Rhizopus*. *Journal of General Microbiology*, **32**, 295–298.

Hawker, L. E. & Beckett, A. (1971). Fine structure and development of the zygospore of *Rhizopus sexualis* (Smith) Callen. *Philosophical Transactions of the Royal Society, Series B*, **263**, 71–100.

Hawker, L. E. & Gooday, M. (1967). Delimitation of the gametangia of *Rhizopus sexualis* (Smith) Callen: an electron microscope study of septum formation. *Journal of General Microbiology*, **49**, 371–376.

Hawker, L. E. & Gooday, M. A. (1968). Development of the zygospore wall in *Rhizopus sexualis* (Smith) Callen. *Journal of General Microbiology*, **54**, 13–120.

Hawker, L. E. & Linton, A. H. (1971). *Micro-organisms. Function, Form and Environment*. 727 pp. London: Edward Arnold.

Hawker, L. E., Thomas, B. & Beckett, A. (1970). An electron microscope study of structure and germination of *Cunninghamella elegans* Lendner. *Journal of General Microbiology*, **60**, 181–189.

Hawksworth, D. L. (1971). A revision of the genus *Ascotricha* Berk. *Mycological Paper No. 126, Commonwealth Mycological Institute*. 28 pp.

Hawksworth, D. L. & Rose, F. (1976). Lichens as pollution monitors. *Studies in Biology No. 66, Institute of Biology*. 59 pp. London: Edward Arnold.

Hawksworth, D. L. & Wells, H. (1973). Ornamentation on the terminal hairs in *Chaetomium* Kunze ex Fr. and some allied genera. *Mycological Paper No. 134, Commonwealth Mycological Institute*. 24 pp.

Hayes, W. A. (1972). Nutritional factors in relation to mushroom production. *Mushroom Science*, **8**, 663–674.

Hayes, W. A., Randle, P. E. & Last, F. T. (1969). The nature of the microbial stimulus affecting sporophore formation in *Agaricus bisporus* (Lange) Sing. *Annals of Applied Biology*, **64**, 177–187.

Heath, I. B. (1974). Mitosis in the fungus *Thraustotheca clavata*. *Journal of Cell Biology*, **60**, 204–220.

Heath, I. B. (1976). Ultrastructure of freshwater phycomycetes. In *Recent Advances in Aquatic Mycology*, 603–650. Editor: E. B. Gareth Jones. London: Elek Science.

Heath, I. B., Gay, J. L. & Greenwood, A. D. (1971). Cell wall formation in the Saprolegniales: cytoplasmic vesicles underlying developing walls. *Journal of General Microbiology*, **65**, 225–232.

Heath, I. B. & Greenwood, A. D. (1970a). The structure and formation of lomasomes. *Journal of General Microbiology*, **62**, 129–137.

Heath, I. B. & Greenwood, A. D. (1970b). Centriole replication and nuclear division in *Saprolegnia*. *Journal of General Microbiology*, **62**, 139–148.

Heath, I. B. & Greenwood, A. D. (1970c). Wall formation in the Saprolegniales. II.

Formation of cysts by the zoospores of *Saprolegnia* and *Dictyuchus*. *Achiv für Mikrobiologie*, **75**, 67–79.

Heath, I. B. & Greenwood, A. D. (1971). Ultrastructural observations on the kinetosomes and Golgi bodies during the asexual life cycle of *Saprolegnia*. *Zeitschrift für Zellforschung und Mikroskopische Anatomie*, **112**, 371–389.

Heath, I. B., Greenwood, A. D. & Griffiths, B. (1970). The origin of Flimmer in *Saprolegnia, Dictyuchus, Synura* and *Cryptomonas*. *Journal of Cell Science*, **7**, 445–451.

Heath, M. C. (1976). Hypersensitivity, the cause or the consequence of rust resistance? *Phytopathology*, **66**, 935–936.

Hebert, T. T. (1971). The perfect state of *Pyricularia grisea (Ceratosphaeria grisea)*. *Phytopathology*, **61**, 339–417.

Heim, P. (1955). Le noyau dans le cycle évolutif de *Plasmodiophora brassicae* Woron. *Revue de Mycologie, Paris*, **20**, 131–157.

Heim, P. (1956a). Remarques sur le cycle évolutif du *Synchytrium endobioticum*. *Compte Rendu Hebdomadaire des Séances de l'Académie des Sciences, Paris*, **42**, 2759–2761.

Heim, P. (1956b). Remarques sur le développement, les divisions nucléaires et sur le cycle évolutif du *Synchytrium endobioticum* (Schilb.) Perc. *Revue de Mycologie, Paris*, **21**, 93–100.

Heim, P. (1960). Évolution du *Spongospora* parasite des racines du Cresson. *Revue de Mycologie, Paris*, **25**, 3–12.

Heim, R. (1948). Phylogeny and natural classification of macro-fungi. *Transactions of the British Mycological Society*, **30**, 161–178.

Heim, R. (1962). L'organisation architecturale des spores de Ganodermes. *Revue de Mycologie, Paris*, **27**, 199–212.

Heim, R. (1963). *Les Champignons Toxiques et Hallucinogènes*. Paris: Boubée.

Heim, R. (1971). The interrelationships between the Agaricales and the Gasteromycetes. In *Evolution in the Higher Basidiomycetes*, 505–534. Editor: R. H. Petersen. Knoxville, U.S.A.: The University of Tennessee Press.

Heim, R. & Wasson, R. G. (1958). Les champignons hallucinogènes du Mexique. *Archives du Muséum National d'Histoire Naturelle, Paris*, 7 sér **6**, IV–VIII, 322 pp.

Heintz, C. E. & Niederpruem, D. J. (1970). Ultrastructure and respiration of oidia and basidiospores of *Coprinus lagopus* (sensu Buller). *Canadian Journal of Microbiology*, **16**, 481–484.

Heintz, C. E. & Pramer, D. (1972). Ultrastructure of nematode-trapping fungi. *Journal of Bacteriology*, **110**, 1163–1170.

Hemmes, D. E. & Bartnicki-Garcia, S. (1975). Electron microscopy of gametangial interaction and oospore development in *Phytophthora capsici*. *Archives of Microbiology*, **103**, 91–112.

Hemmes, D. E. & Hohl, H. R. (1969). Ultrastructural changes in directly germinating sporangia of *Phytophthora parasitica*. *American Journal of Botany*, **56**, 300–313.

Henderson, D. M., Orton, P. D. & Watling, R. (1969). *British Fungus Flora. Agarics and Boleti: Introduction*. 58 pp. Edinburgh: H.M.S.O.

Hendrix, J. W. (1970). Sterols in growth and reproduction of fungi. *Annual Review of Phytopathology*, **8**, 111–130.

Hennebert, G. L. (1968). New species of *Spirosphaera*. *Transactions of the British Mycological Society*, **51**, 13–24.

Herman, A. I. (1971). Sex-specific growth responses in yeasts. *Antonie van Leeuwenhoek*, **37**. 379–384.

Herrick, J. A. (1953). An unusually large *Fomes*. *Mycologia*, **45**, 622–624.

Heslop-Harrison, J. (1957). The experimental modification of sex expression in flowering plants. *Biological Reviews*, **32**, 38–90.

Heslot, H. (1958). Contribution a l'étude cytogenetique et genetique des Sordariacées. *Revue de Cytologie et Biologie Végétale*, **19**, Suppl. 2, 1–235.

Hess, W. M. & Weber, D. J. (1976). Form and function in Basidiomycete spores. In *The Fungal Spore: Form and Function*, 643–713. Editors: D. J. Weber & W. M. Hess. New York, London, Sydney, Toronto: John Wiley & Sons.

Hesseltine, C. W. (1953). A revision of the Choanephoraceae. *American Midland Naturalist*, **50**, 248–256.

Hesseltine, C. W. (1957). The genus *Syzygites* (Mucoraceae). *Lloydia*, **20**, 228–237.

Hesseltine, C. W. (1960). *Gilbertella* gen. nov. (Mucorales). *Bulletin of the Torrey Botanical Club*, **87**, 21–30.

Hesseltine, C. W. (1961). Carotenoids in the fungi Mucorales: special reference to Choanephoraceae. *United States Department of Agriculture Technical Bulletin No. 1245*, 1–33.

Hesseltine, C. W. (1965). A millenium of fungi, food and fermentation. *Mycologia*, **57**, 149–197.

Hesseltine, C. W. & Anderson, P. (1956). The genus *Thamnidium* and a study of the formation of its zygospores. *American Journal of Botany*, **43**, 696–703.

Hesseltine, C. W. & Anderson, P. (1957). Two genera of molds with low temperature requirements. *Bulletin of the Torrey Botanical Club*, **84**, 31–45.

Hesseltine, C. W. & Benjamin, C. R. (1957). Notes on the Choanephoraceae. *Mycologia*, **49**, 723–733.

Hesseltine, C. W., Benjamin, C. R. & Mehrotra, B. S. (1959). The genus *Zygorhynchus*. *Mycologia*, **51**, 173–194.

Hesseltine, C. W. & Ellis, J. J. (1973). Mucorales. In *The Fungi: An Advanced Treatise*, **IVB**, 187–217. Editors: G. C. Ainsworth, F. K. Sparrow & A. S. Sussman. New York and London: Academic Press.

Hesseltine, C. W., Whitehill, A. R., Pidacks, C., Tenhagen, M., Bohonos, M., Hutchings, B. L. & Williams, J. H. (1953). Coprogen, a new growth factor present in dung required by *Pilobolus*. *Mycologia*, **45**, 7–19.

Hewitt, W. B. & Grogan, R. G. (1967). Unusual vectors of plant viruses. *Annual Review of Microbiology*, **21**, 205–224.

Hickman, C. J. (1958). *Phytophthora* – plant destroyer. *Transactions of the British Mycological Society*, **41**, 1–13.

Hill, E. P. (1969). The fine structure of the zoospores and cysts of *Allomyces macrogynus*. *Journal of General Microbiology*, **56**, 125–130.

Hintikka, V. (1973). A note on the polarity of *Armillariella mellea*. *Karstenia*, **13**, 32–39.

Hiratsuka, Y. (1973). The nuclear cycle and terminology of spore states in Uredinales. *Mycologia*, **65**, 432–443.

Hiratsuka, Y. (1975). Recent controversies on the terminology of rust fungi. *Reports of the Tottori Mycological Institute (Japan)*, **12**, 99–104.

Hirsch, H. E. (1950). No brachymeiosis in *Pyronema confluens*. *Mycologia*. **42**, 301–305.

Hirst, J. M. (1955). The early history of a potato blight epidemic. *Plant Pathology*, **4**, 44–50.

Hirst, J. M. & Stedman, O. J. (1960a). The epidemiology of *Phytophthora infestans*. I. Climate, ecoclimate and the phenology of disease outbreak. *Annals of Applied Biology*, **48**, 471–488.

Hirst, J. M. & Stedman, O. J. (1960b). The epidemiology of *Phytophthora infestans*. II. The source of inoculum. *Annals of Applied Biology*, **48**, 489–517.

Hiura, U. (1978). Genetic basis of *formae speciales* in *Erysiphe graminis* DC. In *The Powdery Mildews*, 101–128. Editor: D. M. Spencer. London, New York, San Francisco: Academic Press.

Ho, H. H., Hickman, C. J. & Telford, R. W. (1968a). The morphology of zoospores of *Phytophthora megasperma* var. *sojae* and other Phycomycetes. *Canadian Journal of Botany*, **46**, 88–89.

Ho, H. H., Zachariah, K. & Hickman, C. J. (1968b). The ultrastructure of zoospores of *Phytophthora megasperma* var. *sojae*. *Canadian Journal of Botany*, **46**, 37–41.

Hoch, H. C. & Mitchell, J. E. (1972a). The ultrastructure of *Aphanomyces euteiches* during asexual spore formation. *Phytopathology*, **62**, 149–160.

Hoch, H. C. & Mitchell, J. E. (1972b). The ultrastructure of zoospores of *Aphanomyces euteiches* and of their encystment and subsequent germination. *Protoplasma*, **75**, 113–138.

Hocking, D. (1963). β-carotene and sexuality in the Mucoraceae. *Nature, London*, **197**, 404.

Hocking D. (1967). Zygospore initiation, development and germination in *Phycomyces blakesleeanus*. *Transactions of the British Mycological Society*, **50**, 207–220.

Hodgetts, W. J. (1917). On the forcible discharge of spores of *Leptosphaeria acuta*. *New Phytologist*, **16**, 139–146.

Hohl, H. R. (1966). The fine structure of the slimeways in *Labyrinthula*. *Journal of Protozoology*, **13**, 41–43.

Hohl, H. R. & Streit, W. (1975). Ultrastructure of ascus, ascospore and ascocarp in *Neurospora lineolata*. *Mycologia*, **67**, 367–381.

Holliday, R. (1961a). The genetics of *Ustilago maydis*. *Genetical Research, Cambridge*, **2**, 204–230.

Holliday, R. (1961b). Induced mitotic crossing-over in *Ustilago maydis*. *Genetical Research, Cambridge*, **2**, 231–248.

Holliday, R. (1974). *Ustilago maydis*. In Handbook of Genetics, **I**, 575–595. Editor: R. C. King. New York and London: Plenum Press.

Holloway, S. A. & Heath, I. B. (1974). Observations on the mechanism of flagellar retraction in *Saprolegnia terrestris*. *Canadian Journal of Botany*, **52**, 939–942.

Holloway, S. A. & Heath, I. B. (1977a). Morphogenesis and the role of microtubules in synchronous populations of *Saprolegnia* zoospores. *Experimental Mycology*, **1**, 9–29.

Holloway, S. A. & Heath, I. B. (1977b). An ultrastructural analysis of changes in organelle arrangement and structure between the various spore types of *Saprolegnia*. *Canadian Journal of Botany*, **55**, 1328–1339.

Holm, L. (1959). Some comments on the ascocarps of the Pyrenomycetes. *Mycologia*, **50**, 777–788.

Holm, L. & Müller, E. (1965). Nomina conservanda proposita. II. Proposals in Fungi. *Xylaria* Hill & Greville. *Regnum Vegetabile*, **40**, 13.

Holton, C. S. (1936). Origin and production of morphologic and pathogenic strains of the oat fungi by mutation and hybridization. *Journal of Agricultural Research*, **52**, 311–317.

Holton, C. S., Hoffmann, J. A. & Duran, R. (1968). Variation in the smut fungi. *Annual Review of Phytopathology*, **6**, 213–242.

Hooker, A. L. (1967). The genetics and expression of resistance in plants to rusts of the genus *Puccinia*. *Annual Review of Phytopathology*, **5**, 163–182.

Horgen, P. A. & Griffin, D. H. (1969). Structure and germination of *Blastocladiella emersonii* resistant sporangia. *American Journal of Botany*, **56**, 22–25.

Howard, F. L. (1971). Oospore types in the Saprolegniaceae. *Mycologia*, **63**, 679–686.

Howard, K. L. & Bigelow, H. E. (1969). Nutritional studies on two Gasteromycetes: *Phallus ravenelii* and *Crucibulum levis*. *Mycologia*, **61**, 606–613.

Howard, K. L. & Bryant, T. R. (1971). Meiosis in the Oomycetes. II. A microspectrophotometric analysis of DNA in *Apodachlya brachynema*. *Mycologia*, **63**, 58–68.

Howard, K. L. & Moore, R. T. (1970). Ultrastructure of oogenesis in *Saprolegnia terrestris*. *Botanical Gazette*, **131**, 311–336.

Huber, J. (1958). Untersuchungen zur Physiologie insektentötender Pilze. Archiv für Mikrobiologie, **29**, 257–276.

Hudson, H. J. (1963). The perfect state of *Nigrospora oryzae*. *Transactions of the British Mycological Society*, **46**, 355–360.

Huffman, D. M. & Olive, L. S. (1964). Engulfment and anastomosis in the cellular slime molds (Acrasiales). *American Journal of Botany*, **51**, 465–471.

Hughes, S. J. (1953). Conidiophores, conidia, and classification. *Canadian Journal of Botany*, **31**, 577–659.

Hughes, S. J. (1970). Ontogeny of spore forms in the Uredinales. *Canadian Journal of Botany*, **48**, 2147–2157.

Hughes, S. J. (1971). Annellophores. In *Taxonomy of Fungi Imperfecti*, 132–140. Editor: W. B. Kendrick. Toronto and Buffalo: University of Toronto Press.

Hunsley, D. (1973). Apical wall structure in hyphae of *Phytophthora parasitica*. *New Phytologist*, **72**, 980–990.

Hunsley, D. & Burnett, J. H. (1970). The ultrastructural architecture of the walls of some hyphal fungi. *Journal of General Microbiology*, **62**, 203–218.

Hunt, J. (1956). Taxonomy of the genus *Ceratocystis*. *Lloydia*, **19**, 1–58.

Hüttig, W. (1931). Über den Einfluss der Temperatur auf die Keimung und Geschlechtsverteilung bei Brandpilzen. *Zeitschrift für Botanik*, **24**, 529–577.

Hyde, H. A. & Williams, D. A. (1946). A daily census of *Alternaria* spores caught from the atmosphere at Cardiff in 1942 and 1943. *Transactions of the British Mycological Society*, **29**, 78–85.

Illingworth, R. F., Rose, A. H. & Beckett, A. (1973). Changes in the lipid composition and fine structure of *Saccharomyces cerevisiae* during ascus formation. *Journal of Bacteriology*, **113**, 373–386.

Indira, P. U. (1964). Swarmer formation from plasmodia of Myxomycetes. *Transactions of the British Mycological Society*, **47**, 531–533.

Ing, B. (1968). *A Census Catalogue of British Myxomycetes*. 24 pp. British Mycological Society Foray Committee.

Ingold, C. T. (1939). *Spore Discharge in Land Plants*. 178 pp. Oxford University Press.

Ingold, C. T. (1946a). Spore discharge in *Daldinia concentrica*. *Transactions of the British Mycological Society*, **29**, 43–51.

Ingold, C. T. (1946b). Size and form in agarics. *Transactions of the British Mycological Society*, **29**, 108–113.

Ingold, C. T. (1948). The water relations of spore discharge in *Epichloe*. *Transactions of the British Mycological Society*, **31**, 277–280.

Ingold, C. T. (1953). *Dispersal in Fungi*. 197 pp. Oxford: Clarendon Press.

Ingold, C.T. (1954a). The ascogenous hyphae in *Daldinia*. *Transactions of the British Mycological Society*, **37**, 108–110.

Ingold, C. T. (1954b). Fungi and water. *Transactions of the British Mycological Society*, **37**, 97–107.

Ingold, C. T. (1956). A gas phase in viable fungal spores. *Nature, London*, **177**, 1242–1243.

Ingold, C. T. (1957). Spore liberation in higher fungi. *Endeavour*, **16**, 78–83.

Ingold, C. T. (1959). Jelly as a water reserve in fungi. *Transactions of the British Mycological Society*, **42**, 475–478.

Ingold, C. T. (1964). Possible spore discharge mechanism in *Pyricularia*. *Transactions of the British Mycological Society*, **47**, 573–575.

Ingold, C. T. (1965). *Spore Liberation*. 210 pp. Oxford University Press.

Ingold, C. T. (1966a) Aspects of spore liberation: violent discharge. In *The Fungus Spore*, 113–132. Editor: M. F. Madelin. Colston Papers No. 18. London: Butterworths.

Ingold, C. T. (1966b). The tetraradiate aquatic fungal spore. *Mycologia*, **58**, 43–56.

Ingold, C. T. (1971). *Fungal Spores, Their Liberation and Dispersal*. 302 pp. Oxford: Clarendon Press.

Ingold, C. T. (1972). Presidential address. *Sphaerobolus:* the story of a fungus. *Transactions of the British Mycological Society*, **58**, 179–195.

Ingold, C. T. (1975a). *An Illustrated Guide to Aquatic and Water-Borne Hyphomycetes (Fungi Imperfecti) with Notes on their Biology*. 96 pp. Scientific Publication No. 30. Ambleside: Freshwater Biological Association.

Ingold, C. T. (1975b). Hooker Lecture 1974. Convergent evolution in aquatic fungi: the tetraradiate spore. *Biological Journal of the Linnean Society*, **7**, 1–25.

Ingold, C. T. & Cox, V. J. (1955). Periodicity of spore discharge in *Daldinia*. *Annals of Botany, London*, N.S., **19**, 201–209.

Ingold, C. T. & Dann, V. (1968). Spore discharge in fungi under very high surrounding air-pressure, and the bubble-theory of ballistospore release. *Mycologia*, **60**, 285–289.

Ingold, C. T. & Dring, V. J. (1957). An analysis of spore discharge in *Sordaria*. *Annals of Botany, London*, N.S., **21**, 465–477.

Ingold, C. T. & Hadland, S. A. (1959a). The ballistics of *Sordaria*. *New Phytologist*, **58**, 46–57.

Ingold, C. T. & Hadland, S. A. (1959b). Phototropism and pigment production in *Sordaria* in relation to quality of light. *Annals of Botany, London*, N.S., **23**, 425–429.

Ingold, C. T. & Zoberi, M. H. (1963). The asexual apparatus of Mucorales in relation to spore liberation. *Transactions of the British Mycological Society*, **46**, 115–134.

Ingram, D. S. (1971). An attempt to establish dual cultures of *Synchytrium endobioticum* and *Solanum tuberosum* callus. *Phytopathologische Zeitschrift*, **71**, 21–24.

Ingram, D. S. & Joachim, I. (1971). The growth of *Peronospora farinosa* f.sp. *betae* and sugar beet callus tissues in dual culture. *Journal of General Microbiology*, **69**, 211–220.

Ingram, D. S., Tommerup, I. C. & Dixon, G. R. (1975). The occurrence of oospores in lettuce cultivars infected with *Bremia lactucae* Regel. *Transactions of the British Mycological Society*, **64**, 149–153.

Ingram, M. (1955). *An Introduction to the Biology of Yeasts*. 273 pp. London: Pitman.

Iqbal, S. H. & Webster, J. (1973a). The trapping of aquatic hyphomycete spores by air bubbles. *Transactions of the British Mycological Society*, **60**, 37–48.

Iqbal, S. H. & Webster, J. (1973b). Aquatic Hyphomycete spora of the River Exe and its tributaries. *Transactions of the British Mycological Society*, **61**, 331–346.

I.R.R.I. (1963). *The Rice Blast disease*. 507 pp. Proceedings of a Symposium at the International Rice Research Institute. Baltimore, Maryland, U.S.A.: The Johns Hopkins Press.

Iten, W. (1969–1970). Zur Funktion hydrolytischer Enzyme bei der Autolyse von *Coprinus*. *Bericht der Schweizerischen Botanischen Gesellschaft*, **79**, 175–197.

Iten, W. & Matile, P. (1970). Role of chitinase and other lysosomal enzymes of *Coprinus lagopus* in the autolysis of fruiting bodies. *Journal of General Microbiology*, **61**, 301–309.

Jackson, G. V. H. & Gay, J. L. (1976). Perennation of *Sphaerotheca mors-uvae* as cleistothecia with particular reference to microbial activity. *Transactions of the British Mycological Society*, **66**, 463–471.

Jackson, G. V. H. & Wheeler, B. E. J. (1974). Perennation of *Sphaerotheca mors-uvae* as cleistocarps. *Transactions of the British Mycological Society*, **62**, 73–87.

Jackson, H. S. (1931). Present evolutionary tendencies and the origin of life cycles in the Uredinales. *Memoirs of the Torrey Botanical Club*, **18**, 1–108.

Jacobsen, B. J. & Williams, P. H. (1970). Control of cabbage clubroot using benomyl fungicide. *Plant Disease Reporter*, **54**, 456–460.

Jaffe, L. F. (1968). Localization in the developing *Fucus* egg and the general role of localizing currents. *Advances in Morphogenesis*, **7**, 295–328.

James, P. W. (1965). A new check-list of British lichens. *Lichenologist*, **3**, 95–153.

Janszen, F. H. A. & Wessels, J. G. H. (1970). Enzymic dissolution of hyphal septa in a Basidiomycete. *Antonie van Leeuwenhoek*, **36**, 255–257.

Jenkins, W. A. (1934). The development of *Cordyceps agariciformia*. *Mycologia*, **26**, 220–243.

Jenkyn, J. F. & Bainbridge, A. (1977). Biology and pathology of cereal powdery mildews. In *The Powdery Mildews*, 283–321. Editor: D. M. Spencer. London, New York, San Francisco: Academic Press.

Jersild, R., Mishkin, S. & Niederpruem, D. J. (1967). Origin and ultrastructure of complex septa in *Schizophyllum commune* development. *Archiv für Mikrobiologie*, **57**, 22–32.

Ji, Thakur & Dayal, R. (1971). Studies in the life cycle of *Allomyces javanicus* Kniep. *Hydrobiologia*, **37**, 245–251.

Jinks, J. L. (1952). Heterokaryosis: a system of adaptation in wild fungi. *Proceedings of the Royal Society, London, Series B*, **140**, 83–99.

Jinks, J. L. & Simchen, G. (1966). A consistent nomenclature for the nuclear status of fungal cells. *Nature, London*, **210**, 778–780.

John, B. & Lewis, K. R. (1973). The meiotic mechanism. *Oxford Biology Readers*, **65**. 32 pp. Oxford University Press.

Johns, R. M. & Benjamin, R. K. (1954). Sexual reproduction in *Gonapodya*. *Mycologia*, **46**, 202–208.

Johnson, B. F., Yoo, B. Y. & Calleja, G. B. (1973). Cell division in yeasts: movement of organelles associated with cell plate growth of *Schizosaccharomyces pombe*. *Journal of Bacteriology*, **115**, 358–366.

Johnson, E. M. & Valleau, W. D. (1949). Synonymy in some common species of *Cercospora*. *Phytopathology*, **39**, 763–770.

Johnson, T. (1953). Variation in the rusts of cereals. *Biological Reviews*, **28**, 105–157.

Johnson, T. (1960). Genetics of pathogenicity. In *Plant Pathology: An Advanced Treatise*, **2**, 407–459. Editors: J. G. Horsfall & A. E. Dimond. New York and London: Academic Press.

Johnson, T. (1961). Rust research in Canada and related plant-disease investigations. *Publication 1098, Research Branch, Canada Department of Agriculture*, 1–69.

Johnson, T., Green, G. J. & Samborski, D. J. (1967). The world situation of the cereal rusts. *Annual Review of Phytopathology*, **5**, 183–200.

Johnson, T. W. (1956). *The Genus Achlya: Morphology and Taxonomy*. 180 pp. Ann Arbor: The University of Michigan Press.

Johnson, T. W. (1969). Aquatic fungi of Iceland: *Olpidium* (Braun) Rabenhorst. *Archiv. für Mikrobiologie*, **69**, 1–11.

Johnson, T. W. (1973). Aquatic fungi from Iceland: some polycentric species. *Mycologia*, **65**, 1337–1355.

Johnston, T. D. (1970). A new factor for resistance to club root in *Brassica napus* L. *Plant Pathology*, **19**, 156–159.

Joly, P. (1964). *Le Genre Alternaria*. Paris: Editions Paul Chevalier.

Jones, D., Bacon, J. S. D., Farmer, V. C. & Webley, D. M. (1968). Lysis of cell walls of *Mucor ramannianus* Möller by a *Streptomyces* sp. *Antonie van Leeuwenhoek*, **34**, 173–182.

Jones, R. A. C. & Harrison, B. D. (1969). The behaviour of potato mop-top virus in soil and the evidence for its transmission by *Spongospora subterranea* (Wallr.) Lagenh. *Annals of Applied Biology*, **63**, 1–17.

Jones, R. A. C. & Harrison, B. D. (1972). Ecological studies on potato mop-top virus in Scotland. *Annals of Applied Biology*, **71**, 47–57.

Jones, S. G. (1925). Life-history and cytology of *Rhytisma acerinum* (Pers.) Fries. *Annals of Botany, London*, **39**, 41–73.

Jump, J. A. (1954). Studies on sclerotization in *Physarum polycephalum*. *American Journal of Botany*, **41**, 561–567.

Jurand, M. K. & Kemp, R. F. O. (1972). Surface ultrastructure of oidia in the Basidiomycete *Psathyrella coprophila*. *Journal of General Microbiology*, **72**, 575–579.

Jurand, M. K. & Kemp, R. F. O. (1973). An incompatibility system determined by three factors in a species of *Psathyrella* (Basidiomycetes). *Genetical Research, Cambridge*, **22**, 125–134.

Kane, B. E., Reiskind, J. B. & Mullins, J. T. (1973). Hormonal control of sexual morphogenesis in *Achlya:* dependence on protein and ribonucleic acid syntheses. *Science, N.Y.*, **180**, 1192–1193.

Karling, J. S. (1958). *Synchytrium fulgens* Schroeter. *Mycologia*, **50**, 373–375.

Karling, J. S. (1964). *Synchytrium*. 470 pp. New York and London: Academic Press.

Karling, J. S. (1968). *The Plasmodiophorales*. 256 pp. New York and London: Hafner Publishing Company.

Karling, J. S. (1969). Zoosporic fungi of Oceania. VII. Fusions in *Rhizophlyctis*. *American Journal of Botany*, **56**, 211–221.

Karling, J. S. (1973). A note on *Blastocladiella* (Blastocladiaceae). *Mycopathologia et Mycologia Applicata*, **49**, 169–172.

Kaushik, N. K. & Hynes, H. B. N. (1968). Experimental study on the role of autumn-shed leaves in aquatic environments. *Journal of Ecology*, **56**, 229–243.

Kaushik, N. K. & Hynes, H. B. N. (1971). The fate of dead leaves that fall into streams. *Archiv für Hydrobiologie*, **68**, 465–515.

Kemp, R. F. O. (1975a). Breeding biology of *Coprinus* species in the section *Lanatuli*. *Transactions of the British Mycological Society*, **65**, 375–388.

Kemp, R. F. O. (1975b). Oidia, plasmogamy and speciation in Basidiomycetes. In *The Biology of the Male Gamete*, 57–69. Editors: J. G. Duckett & P. A. Racey. Supplement No. 1, Biological Journal of the Linnean Society, **7**, Academic Press.

Kemp, R. F. O. (1976). A new interpretation of unilateral nuclear migration in fungi with special reference to *Podospora anserina*. *Transactions of the British Mycological Society*, **66**, 1–5.

Kendrick, W. B. (ed.) (1971a). *Taxonomy of Fungi Imperfecti*. 309 pp. University of Toronto Press.

Kendrick, W. B. (1971b). Arthroconidia and meristem arthroconidia. In *Taxonomy of Fungi Imperfecti*, 160–175. Editor: W. B. Kendrick. University of Toronto Press.

Kendrick, W. B. (ed.) (1979). *The Whole Fungus: The Sexual–Asexual Synthesis*. 793 pp. 2 vols. Ottawa: National Museums of Canada.

Kendrick, W. B. & Carmichael, J. W. (1973). Hyphomycetes. In *The Fungi: An Advanced Treatise*, **IVA**, 323–509. Editors: G. C. Ainsworth, F. K. Sparrow & A. S. Sussman. New York and London: Academic Press.

Kendrick, W. B. & Cole, G. T. (1968). Conidium ontogeny in Hyphomycetes. The sympodulae of *Beauveria* and *Curvularia*. *Canadian Journal of Botany*, **46**, 1297–1301.

Kendrick, W. B. & Cole, G. T. (1969). Conidium ontogeny in hyphomycetes. *Trichothecium roseum* and its meristem arthrospores. *Canadian Journal of Botany*, **47**, 345–350.

Kern, H. & Naef-Roth, S. (1975). Zur Bildung von Auxinen und Cytokininen durch *Taphrina*-Arten. *Phytopathologische Zeitschrift*, **83**, 193–222.

Kerr, N. S. (1960). Flagella formation by myxamoebae of the true slime mold, *Didymium nigripes*. *Journal of Protozoology*, **7**, 103–108.

Kerr, N. S. (1963). The growth of myxamoebae of the true slime mold, *Didymium nigripes*, in axenic culture. *Journal of General Microbiology*, **32**, 409–416.

Kerr, N. S. (1967). Plasmodium formation by a minute mutant of the true slime mold, *Didymium nigripes*. *Experimental Cell Research*, **45**, 646–655.

Kerr, S. (1968). Ploidy level in the true slime mould, *Didymium nigripes*. *Journal of General Microbiology*, **53**, 9–15.

Keskin, B. (1964). *Polymyxa betae* n.sp. ein Parasite in den Wurzeln von *Beta vulgaris* Tournefort, besonders während der Jugendentwickelung der Zuckerrübe. *Archiv für Mikrobiologie*, **49**, 348–374.

Keskin, B. & Fuchs, W. H. (1969). Der Infektions vorgang bei *Polymyxa betae*. *Archiv für Mikrobiologie*, **68**, 218–226.

Kevorkian, A. G. (1937). Studies in the Entomophthoraceae. I. Observations on the genus *Conidiobolus*. *Journal of the Agricultural University of Puerto Rico*, **21**, 191–200.

Khairi, S. M. & Preece, T. F. (1978a). Hawthorn powdery mildew: diurnal and seasonal distribution of conidia in air near infected plants. *Transactions of the British Mycological Society*, **71**, 395–397.

Khairi, S. M. & Preece, T. F. (1978b). Hawthorn powdery mildew: overwintering mycelium in buds and the effect of clipping hedges on disease epidemiology. *Transactions of the British Mycological Society*, **71**, 399–404.

Killian, C. (1919). Sur la sexualité de l'ergot de Seigle, le *Claviceps purpurea* (Tulasne). *Bulletin de la Société Mycologique de France*, **35**, 182–197.

Kimbrough, J. W. (1970). Current trends in the classification of Discomycetes. *Botanical Review*, **36**, 91–161.

Kinugawa, K. (1965). On the growth of *Dictyophora indusiata*. II. Relations between the change in osmotic value of expressed sap and the conversion of glycogen to reducing sugar in tissues during receptaculum elongation. *Botanical Magazine, Tokyo*, **78**, 171–176.

Kinugawa, K. & Furukawa, H. (1965). The fruit-body formation in *Collybia velutipes* induced by the lower temperature treatment of one short duration. *Botanical Magazine, Tokyo*, **78**, 240–244.

Kirby, E. J. M. (1961). Host–parasite relations in the choke disease of grasses. *Transactions of the British Mycological Society*, **44**, 493–503.

Kirk, D., McKeen, W. E. & Smith, R. (1971). Cytoplasmic connections between *Dictyostelium discoideum* cells. *Canadian Journal of Botany*, **49**, 19–20.

Klein, D. T. (1960). Interrelations between growth rate and nuclear ratios in heterokaryons of *Neurospora crassa*. *Mycologia*, **52**, 137–147.

Kniep, H. (1919). Untersuchungen über den Antherenbrand (*Ustilago violacea* Pers.). Ein Beitrag zum Sexualitätsproblem. *Zeitschrift für Botanik*, **11**, 275–284.

Koch, W. J. (1951). Studies in the genus *Chytridium*, with observations on a sexually reproducing species. *Journal of the Elisha Mitchell Scientific Society*, **67**, 267–278.

Koch, W. J. (1956). Studies of the motile cells of chytrids. I. Electron microscope observations of the flagellum, blepharoplast and rhizoplast. *American Journal of Botany*, **43**, 811–819.

Koch, W. J. (1958). Studies of the motile cells of chytrids. II. Internal structure of the body observed with light microscopy. *American Journal of Botany*, **45**, 59–72.

Koch, W. J. (1961). Studies of the motile cells of chytrids. III. Major types. *American Journal of Botany*, **48**, 786–788.

Koch, W. J. (1968). Studies of the motile cells of chytrids. V. Flagellar retraction in posteriorly uniflagellate fungi. *American Journal of Botany*, **55**, 841–859.

Koevenig, J. L. (1964). Life cycle of *Physarum gyrosum* and other Myxomycetes. *Mycologia*, **56**, 170–184.

Köhler, E. (1923). Über den derseitigen Stand der Erforschung des Kartoffelkrebses.

Arbeiten aus der Biologischen Bundesanstalt für Land- und Forstwirtschaft, **11**, 289–315.

Köhler, E. (1931a). Der Kartoffelkrebs und sein Erreger (*Synchytrium endobioticum* (Schilb.) Perc.). *Landwirtschaftliche Jahrbücher*, **74**, 729–806.

Köhler, E. (1931b). Zur Biologie und Cytologie von *Synchytrium endobioticum* (Schilb.) Perc. *Phytopathologische Zeitschrift*, **4**, 43–55.

Köhler, E. (1956). Zur Kenntniss der Sexualität bei *Synchytrium. Bericht der Deutschen Botanischen Gesellschaft*, **69**, 121–127.

Köhler, F. (1935). Genetische Studien an *Mucor mucedo* Brefeld I–III. *Zeitschrift für Induktive Abstammungs- und Vererbungslehre*, **70**, 1–54.

Kohlmeyer, J. & Kohlmeyer, E. (1974). Distribution of *Epichloe typhina* (Ascomycetes) and its parasitic fly. *Mycologia*, **66**, 77–86.

Kole, A. P. (1954). *A contribution to the knowledge of Spongospora subterranea (Wallr.) Lagerh., the cause of powdery scab of potatoes*. 65 pp. Thesis. University of Wageningen.

Kole, A. P. (1965). Resting-spore germination in *Synchytrium endobioticum. Netherlands Journal of Plant Pathology*, **71**, 72–78.

Kole, A. P. & Gielink, A. J. (1961). Electron microscope observations on the flagella of the zoosporangial zoospores of *Plasmodiophora brassicae* and *Spongospora subterranea. Proceedings. Koniklijke Nederlandse Akademie van Wetenschappen*, C, **64**, 157–161.

Kole, A. P. & Gielink, A. J. (1962). Electron microscope observations on the resting spore germination of *Plasmodiophora brassicae. Proceedings. Koniklijke Nederlandse Akademie van Wetenschappen*, C, **65**, 117–121.

Kole, A. P. & Gielink, A. J. (1963). The significance of the zoosporangial stage in the life cycle of the Plasmodiophorales. *Netherlands Journal of Plant Pathology*, **69**, 258–262.

Koltin, Y. & Flexer, A. S. (1969). Alteration of nuclear distribution in B mutants of *Schizophyllum commune. Journal of Cell Science*, **4**, 739–749.

Koltin, Y., Raper, J. R. & Simchen, G. (1967). Genetics of the incompatibility factors of *Schizophyllum commune:* the B factor. *Proceedings of the National Academy of Sciences of the United States of America*, **57**, 55–62.

Konijn, T. M., Van de Meene, J. G. C., Bonner, J. T. & Barkley, D. S. (1967). The acrasin activity of adenosine-3′-5′ cyclic phosphate. *Proceedings of the National Academy of Sciences of the United States of America*, **58**, 1152–1154.

Korf, R. P. (1973). Discomycetes and Tuberales. In *The Fungi: An Advanced Treatise*, **IVA**, 249–319. Editors: G. C. Ainsworth, F. K. Sparrow & A. S. Sussman. New York and London: Academic Press.

Korhonen, K. & Hintikka, V. (1974). Cytological evidence for somatic diploidization in dikaryotic cells of *Armillariella mellea. Archiv für Mikrobiologie*, **95**, 187–192.

Korohoda, W., Rakoczy, L. & Walczak, T. (1970). On the control mechanism of protoplasmic streaming in the plasmodia of Myxomycetes. *Acta Protozoologica*, **7**, 363–373.

Kosasih, B. D. & Willetts, H. J. (1975). Ontogenetic and histochemical studies of the apothecium of *Sclerotinia sclerotiorum. Annals of Botany*, **39**, 185–191.

Kramer, C. L. (1961). Morphological development and nuclear behaviour in the genus *Taphrina. Mycologia*, **52**, 295–320.

Kramer, C. L. (1973). Protomycetales and Taphrinales. In *The Fungi: An Advanced Treatise*, **IVA**, 33–41. Editors: G. C. Ainsworth, F. K. Sparrow & A. S. Sussman. New York and London: Academic Press.

Kreger, D. R. (1954). Observations on cell walls of yeasts and some other fungi by X-ray diffraction and solubility tests. *Biochimica et Biophysica Acta*, **13**, 1–9.

Kreger-van Rij, N. J. W. (1969). Taxonomy and systematics of yeasts. In *The Yeasts*, **I**,

3–78. Editors: A. H. Rose & J. S. Harrison. New York and London: Academic Press.

Kreger-van Rij, N. J. W. (1970). *Endomycopsis* Dekker. In *The Yeasts, a Taxonomic Study*, 166–208. Editor: J. Lodder. Amsterdam and London: North-Holland Publishing Company.

Kreger-van Rij, N. J. W. (1973). Endomycetales, Basidiomycetous yeasts, and related fungi. In *The Fungi: An Advanced Treatise*, **IVA**, 11–32. Editors: G. C. Ainsworth, F. K. Sparrow & A. S. Sussman. New York and London: Academic Press.

Kreger-van Rij, N. J. W. & Veenhuis, M. (1971). A comparative study of the cell wall structure of basidiomycetous and related yeasts. *Journal of General Microbiology,* **68,** 87–95.

Kreisel, H. (1961). Die Lycoperdaceae der Deutschen Demokratischen Republik. *Repertorium Novarum Specierum Regni Vegetabilis,* **64,** 89–201.

Krenner, J. A. (1961). Studies in the field of microscopic fungi. III. On *Entomophthora aphidis* H. Hoffm. with special regard to the family of the Entomophthoraceae in general. *Acta Botanica Hungarica,* **7,** 345–376.

Kuehn, H. H. (1956). Observations on the Gymnoascaceae. III. Developmental morphology of *Gymnoascus reessii,* a new species of *Gymnoascus* and *Eidamella deflexa. Mycologia,* **48,** 805–820.

Kuehn, H. H. (1958). A preliminary survey of the Gymnoascaceae. I. *Mycologia,* **50,** 417–439.

Kuehn, H. H. (1959). A preliminary survey of the Gymnoascaceae. II. *Mycologia,* **51,** 665–692.

Kuehn, H. H., Orr, G. F. & Ghosh, G. R. (1964). Pathological implications of Gymnoascaceae. *Mycopathologia et Mycologia Applicata,* **24,** 35–46.

Kuhlman, E. G. (1972). Variation in zygospore formation among species of *Mortierella. Mycologia,* **64,** 325–341.

Kuhlman, E. G. (1975). Zygospore formation in *Mortierella alpina* and *M. spinosa. Mycologia,* **67,** 678–681.

Kühner, R. (1973). Architecture de la paroi sporique des Hyménomycètes et de ses différenciations. *Persoonia,* **7,** 217–248.

Kühner, R. (1977). Variation of nuclear behaviour in the Homobasidiomycetes. *Transactions of the British Mycological Society,* **68,** 1–16.

Kühner, R. & Romagnesi, H. (1953). *Flore Analytique des Champignons Supérieurs (Agarics, Bolets, Chanterelles).* 556 pp. Paris: Masson.

Kulkarni, U. K. (1963). Initiation of the dikaryon in *Claviceps microcephala* (Wallr.) Tul. *Mycopathologia et Mycologia Applicata,* **21,** 19–22.

Kusano, S. (1930a). The life-history and physiology of *Synchytrium fulgens* Schroet., with special reference to its sexuality. *Japanese Journal of Botany,* **5,** 35–132.

Kusano, S. (1930b). Cytology of *Synchytrium fulgens* Schroet. *Journal of the College of Agriculture, Imperial University of Tokyo,* **10,** 347–388.

Kwon, K. J. & Raper, K. B. (1967). Sexuality and cultural characteristics of *Aspergillus heterothallicus. American Journal of Botany,* **54,** 36–48.

Kwon-Chung, K. J. (1975). A new genus *Filobasidiella,* the perfect state of *Cryptococcus neoformans. Mycologia,* **67,** 1197–1200.

Kwon-Chung, K. J. (1976). Morphogenesis of *Filobasidiella neoformans,* the sexual state of *Cryptococcus neoformans. Mycologia,* **68,** 821–833.

Laane, M. M. (1974). Nuclear behaviour during vegetative stage and zygospore formation in *Absidia glauca. Norwegian Journal of Botany,* **21,** 125–135.

Lacoste, L. (1965). *Biologie naturelle et culturale du genre Leptosphaeria Cesati & de Notaris. Determinisme de la reproduction sexuelle.* Thèse de Doctorat ès Sciences. Toulouse, 1965. 230 pp.

Laffin, R. J. & Cutter, V. M. (1959). Investigations on the life cycle of *Sporidiobolus johnsonii*. I. Irradiation and cytological studies. *Journal of the Elisha Mitchell Scientific Society*, **75**, 89–96.

Laibach, F. (1927). Zytologische Untersuchungen über die Monoblepharideen. *Jahrbuch für Wissenschaftliche Botanik*, **66**, 596–630.

Laidlaw, W. M. R. (1961). Extracting barley embryos for loose smut examination. *Plant Pathology*, **10**, 63–65.

Lakon, G. (1963). Entomophthoraceae. *Nova Hedwigia*, **5**, 7–26.

Lammerink, J. (1970). Interspecific transfer of clubroot resistance from *Brassica campestris* L. to *Brassica napus* L. *New Zealand Journal of Agricultural Research*, **13**, 105–110.

Langdon, R. F. N. & Fullerton, R. A. (1975). Sorus ontogeny and sporogenesis in some smut fungi. *Australian Journal of Botany*, **23**, 915–930.

Lange, J. E. (1935–1940). *Flora Agaricina Danica*. I–V. Copenhagen.

Lange, M. & Hora, F. B. (1963). *Collins Guide to Mushrooms and Toadstools*. 257 pp. London: Collins.

Lara, S. L. & Bartnicki-Garcia, S. (1974). Cytology of budding in *Mucor rouxii:* Wall ontogeny. *Archiv für Mikrobiologie*, **97**, 1–16.

Large, E. C. (1952). Surveys for choke *(Epichloe typhina)* in cocksfoot seed crops, 1951. *Plant Pathology*, **1**, 23–28.

Large, E. C. (1958). *The Advance of the Fungi*. 488 pp. London: Jonathan Cape.

Last, F. T. (1955). Spore content of air within and above mildew-infected cereal crops. *Transactions of the British Mycological Society*, **38**, 453–464.

Last, F. T. & Deighton, F. C. (1965). The non-parasitic microflora on the surfaces of living leaves. *Transactions of the British Mycological Society*, **48**, 83–99.

Laundon, G. F. (1973). Uredinales. In *The Fungi: An Advanced Treatise*, **IVB**, 247–279. Editors: G. C. Ainsworth, F. K. Sparrow & A. S. Sussman. New York and London: Academic Press.

Leach, C. M. (1962). Sporulation of diverse species of fungi under near ultraviolet radiation. *Canadian Journal of Botany*, **40**, 151–161.

Leach, C. M. (1963). The qualitative and quantitative relationship of monochromatic radiation to sexual and asexual reproduction of *Pleospora herbarum*. *Mycologia*, **55**, 151–163.

Leach, C. M. (1968). An action spectrum for light inhibition of the "terminal phase" of photosporogenesis in the fungus *Stemphylium botryosum*. *Mycologia*, **60**, 532–546.

Leach, C. M. (1971). Regulation of perithecium development and maturation in *Pleospora herbarum* by light and temperature. *Transactions of the British Mycological Society*, **57**, 295–315.

Leach, C. M. (1976). An electrostatic theory to explain violent spore liberation by *Drechslera turcica* and other fungi. *Mycologia*, **68**, 63–86.

Leadbeater, G. & Mercer, C. (1956). Zygospores in *Piptocephalis cylindrospora* Bain. *Transactions of the British Mycological Society*, **39**, 17–20.

Leadbeater, G. & Mercer, C. (1957a). Zygospores in *Piptocephalis*. *Transactions of the British Mycological Society*, **40**, 109–116.

Leadbeater, G. & Mercer, C. (1957b). *Piptocephalis virginiana* sp.nov. *Transactions of the British Mycological Society*, **40**, 461–471.

Lebeau, J. B. & Hawn, E. J. (1963). Formation of hydrogen cyanide by the mycelial stage of a fairy ring fungus. *Phytopathology*, **53**, 1395–1396.

Lentz, P. L. (1954). Modified hyphae of Hymenomycetes. *Botanical Review*, **20**, 135–199.

Lentz, P. L. (1973). Analysis of modified hyphae as a tool in taxonomic research in the Higher Basidiomycetes. In *Evolution in the Higher Basidiomycetes*, 99–127. Editor: R. H. Petersen. Knoxville, U.S.A.: University of Tennessee Press.

Lerbs, V. (1971). Licht- und elektronen-mikroskopische Untersuchungen an meiotischen Basidien von *Coprinus radiatus* (Bolt.) Fr. *Archiv für Mikrobiologie*, **77**, 308–330.

Lesemann, D. E. & Fuchs, W. H. (1970a). Elektronenmikroskopische Untersuchung über die Vorbereitung der Infektion in encystierten Zoosporen von *Olpidium brassicae. Archiv für Mikrobiologie*, **71**, 9–19.

Lesemann, D. E. & Fuchs, W. H. (1970b). Die Ultrastruktur des Penetrationsvorganges von *Olpidium brassicae* an Kohlrabi-Wurzeln. *Archiv für Mikrobiologie*, **71**, 20–30.

Lessie, P. E. & Lovett, J. S. (1968). Ultrastructural changes during sporangium formation and zoospore differentiation in *Blastocladiella emersonii. American Journal of Botany*, **55**, 220–236.

Leupold, U. (1970). Genetical methods for *Schizosaccharomyces pombe*. In *Methods in Cell Physiology*, **4**, 169–177. Editor: D. M. Prescott. New York and London: Academic Press.

Levi, M. P. & Preston, R. D. (1965). A chemical and microscopic examination of the action of the soft-rot fungus *Chaetomium globosum* on Beechwood *(Fagus sylv.). Holzforschung*, **19**, 183–190.

Levisohn, I. (1927). Beitrag zur Entwicklungsgesichte und Biologie von *Basidiobolus ranarum* Eidam. *Jahrbuch für Wissenschaftliche Botanik*, **66**, 513–555.

Lewis, D. H. (1973). Concepts in fungal nutrition and the origin of biotrophy. *Biological Reviews, Cambridge*, **48**, 261–278.

Lewis, D. H. (1974). Micro-organisms and plants: the evolution of parasitism and mutualism. *Symposium of the Society for General Microbiology*, **24**, 367–392.

Lichtwardt, R. W. (1973a). The Trichomycetes: what are their relationships? *Mycologia*, **65**, 1–20.

Lichtwardt, R. W. (1973b). Trichomycetes. In *The Fungi: An Advanced Treatise*, **IVB**, 237–243. Editors: G. C. Ainsworth, F. K. Sparrow & A. S. Sussman. New York and London: Academic Press.

Lichtwardt, R. W. (1976). Trichomycetes. In *Recent Advances in Aquatic Mycology*, 651–671. Editor: E. B. Gareth Jones. London: Elek Science.

Lilly, V. G. & Barnett, H. L. (1951). *Physiology of the Fungi*. 464 pp. New York: McGraw Hill.

Lin, C. C. & Aronson, J. M. (1970). Chitin and cellulose in the cell walls of the oomycete *Apodachlya* sp. *Archiv für Mikrobiologie*, **72**, 111–114.

Lin, M.-R. & Edwards, H. H. (1974). Primary penetration process in powdery mildewed barley related to host cell age, cell type, and occurrence of basic staining material. *New Phytologist*, **73**, 131–137.

Lindegren, C. C. (1949). *The Yeast Cell, its Genetics and Cytology*. St. Louis, U.S.A.: Educational Publishers.

Lindegren, C. C. & Lindegren, G. (1943). Segregation, mutation and copulation in *Saccharomyces cerevisiae. Annals of Missouri Botanical Garden*, **30**, 453–468.

Linder, D. H. (1929). A monograph of the helicosporous Fungi Imperfecti. *Annals of Missouri Botanical Garden*, **16**, 227–388.

Linder, D. H. (1933). The genus *Schizophyllum*. I. Species of the Western Hemisphere. *American Journal of Botany*, **20**, 552–564.

Lingappa, B. T. (1958a). Development and cytology of the evanescent prosori of *Synchytrium brownii* Karling. *American Journal of Botany*, **45**, 116–123.

Lingappa, B. T. (1958b). The cytology of development and germination of resting spores of *Synchytrium brownii. American Journal of Botany*, **45**, 613–620.

Lingappa, B. T. & Sussman, A. S. (1959). Endogenous substrates of dormant activated and germinating ascospores of *Neurospora tetrasperma. Plant Physiology*, **34**, 466–472.

Lissouba, P., Mousseau, J., Rizet, G. & Rossignol, J. L. (1962). Fine structure of genes in the Ascomycete *Ascobolus immersus. Advances in Genetics*, **11**, 343–380.

List, P. H. & Freund, B. (1968). Geruchstoffe der Stinkmorchel, *Phallus impudicus* L. 18. Mitteilung über Pilzinhaltstoffe. *Planta Medical Supplement*, **1968**, 123–132.

Littlefield, L. J. (1973). Histological evidence for diverse mechanisms of resistance to flax rust, *Melampsora lini* (Ehrenb.) Lév. *Physiological Plant Pathology*, **3**, 241–247.

Lodder, J. (1970). *The Yeasts*. 2nd edition. 1385 pp. Amsterdam and London: North-Holland Publishing Company.

Lodha, B. C. (1964a). Studies on coprophilous fungi. I. *Chaetomium*. *Journal of the Indian Botanical Society*, **43**, 121–140.

Lodha, B. C. (1964b). Studies on coprophilous fungi. II. *Chaetomium*. *Antonie van Leeuwenhoek*, **20**, 163–167.

Long, P. E. & Jacobs, L. (1974). Aseptic fruiting of the cultivated mushroom, *Agaricus bisporus*. *Transactions of the British Mycological Society*, **63**, 99–107.

Loomis, W. F. (1975). *Dictyostelium discoideum. A Developmental System*. 214 pp. New York, San Francisco, London: Academic Press.

Lopez-Real, J. M. (1975). Formation of pseudosclerotia ('zone lines') in wood decayed by *Armillaria mellea* and *Stereum hirsutum*. I. Morphological aspects. *Transactions of the British Mycological Society*, **64**, 465–471.

Lopez-Real, J. M. & Swift, M. J. (1975). The formation of pseudosclerotia ('zone lines') in wood decayed by *Armillaria mellea* and *Stereum hirsutum*. II. Formation in relation to the moisture content of the wood. *Transactions of the British Mycological Society*, **64**, 473–481.

Lorenz, F. (1933). Beiträge zur Entwicklungsgeschichte von *Sphaerobolus*. *Archiv für Protistenkunde*, **81**, 361–398.

Loveless, A. R. (1971). Conidial evidence for host restriction in *Claviceps purpurea*. *Transactions of the British Mycological Society*, **56**, 419–434.

Loveless, A. R. & Peach, J. M. (1974). Evidence for the genotypic control of spore size in *Claviceps purpurea*. *Transactions of the British Mycological Society*, **63**, 612–616.

Lowry, R. J., Durkee, T. L. & Sussman, A. S. (1967). Ultrastructural studies of microconidium formation in *Neurospora crassa*. *Journal of Bacteriology*, **94**, 1757–1763.

Lowry, R. J. & Sussman, A. S. (1958). Wall structure of ascospores of *Neurospora tetrasperma*. *American Journal of Botany*, **45**, 397–403.

Lowry, R. J. & Sussman, A. S. (1968). Ultrastructural changes during germination of ascospores of *Neurospora tetrasperma*. *Journal of General Microbiology*, **51**, 403–409.

Lowy, B. (1951). A morphological basis for classifying the species of *Auricularia*. *Mycologia*, **43**, 351–358.

Lowy, B. (1952). The genus *Auricularia*. *Mycologia*, **44**, 656–692.

Lu, B. C. (1964). Polyploidy in the Basidiomycete *Cyathus stercoreus*. *American Journal of Botany*, **51**, 343–347.

Lu, B. C. & Brodie, H. J. (1964). Preliminary observation of meiosis in the fungus *Cyathus*. *Canadian Journal of Botany*, **42**, 307–310.

Lu, S.-H. (1973). Effect of calcium on fruiting of *Cyathus stercoreus*. *Mycologia*, **65**, 329–334.

Lucas, M. T. & Webster, J. (1967). Conidial states of British species of *Leptosphaeria*. *Transactions of the British Mycological Society*, **50**, 85–121.

Lundqvist, N. (1972). Nordic Sordariaceae. *Symbolae Botanicae Upsaliensis*, **20**, 1–374.

Lunney, C. Z. & Bland, C. E. (1976). An ultrastructural study of zoosporogenesis in *Pythium proliferum* de Bary. *Protoplasma*, **88**, 85–100.

Luttrell, E. S. (1951). Taxonomy of the Pyrenomycetes. *University of Missouri Studies*, **24**(3), 1–20.

Luttrell, E. S. (1955). The ascostromatic ascomycetes. *Mycologia*, **47**, 511–532.

Luttrell, E. S. (1963). Taxonomic criteria in *Helminthosporium*. *Mycologia*, **55**, 643–674.

Luttrell, E. S. (1965b). Paraphysoids, pseudoparaphyses, and apical paraphyses. *Transactions of the British Mycological Society*, **48**, 135–144.

Luttrell, E. S. (1973). Loculoascomycetes. In *The Fungi: An Advanced Treatise*, **IVA**, 135–219. Editors: G. C. Ainsworth, F. K. Sparrow & A. S. Sussman. New York and London: Academic Press.

Lyr, H. (1953). Zur Kenntniss der Ernahrungsphysiologie der Gattung *Pilobolus*. *Archiv. für Mikrobiologie*, **19**, 402–434.

Lythgoe, J. N. (1958). Taxonomic notes on the genera *Helicostylum* and *Chaetostylum* (Mucoraceae). *Transactions of the British Mycological Society*, **41**, 135–141.

Lythgoe, J. N. (1961). Effect of light and temperature on growth and development in *Thamnidium elegans* Link. *Transactions of the British Mycological Society*, **44**, 199–213.

Lythgoe, J. N. (1962). Effect of light and temperature on sporangium development in *Thamnidium elegans* Link. *Transactions of the British Mycological Society*, **45**, 161–168.

Maas-Geesteranus, R. A. (1963a). Hyphal structures in Hydnums. II. *Proceedings. Koniklijke Nederlandse Akademie van Wetenschappen, C*, **66**, 426–436.

Maas-Geesteranus, R. A. (1963b). Hyphal structures in Hydnums. IV. Proceedings. *Koniklijke Nederlandse Akademie van Wetenschappen, C*, **66**, 447–457.

Maas-Geesteranus, R. A. (1975). *Die Terrestrischen Stachelpilze Europas*. Amsterdam and London: North-Holland Publishing Company.

McClary, D. O., Williams, M. A., Lindegren, C. C. & Ogur, M. (1957). Chromosome counts in a polyploid series of *Saccharomyces*. *Journal of Bacteriology*, **73**, 360–364.

McClintock, B. (1945). *Neurospora*. I. Preliminary observations on the chromosomes of *Neurospora crassa*. *American Journal of Botany*, **32**, 671–678.

McCranie, J. (1942). Sexuality in *Allomyces cystogenus*. *Mycologia*, **34**, 209–213.

McCully, E. K. & Robinow, C. F. (1971). Mitosis in the fission yeast *Schizosaccharomyces pombe:* a comparative study with light and electron microscopy. *Journal of Cell Science*, **9**, 475–507.

Macdonald, J. A. (1937). A study of *Polyporus betulinus* (Bull.) Fr. *Annals of Applied Biology*, **24**, 289–310.

Macer, R. C. F. (1959). Pathology. *Annual Report. Plant Breeding Institute, Cambridge*, 1958–1959, 60–63.

McEwen, F. L. (1963). *Cordyceps* infections. In *Insect Pathology: An Advanced Treatise*, **2**, 273–290. Editor: E. A. Steinhaus. New York and London: Academic Press.

Macfarlane, I. (1952). Factors affecting the survival of *Plasmodiophora brassicae* Wor. in the soil and its assessment by a host test. *Annals of Applied Biology*, **39**, 239–256.

Macfarlane, I. (1959). A solution-culture technique for obtaining root-hair, or primary, infection by *Plasmodiophora brassicae*. *Journal of General Microbiology*, **18**, 720–732.

Macfarlane, I. (1968). Problems in the systematics of the Olpidiaceae. In *Marine Mykologie (Symposium über Niedere Pilze im Küstenbereich)*, Editor: A. Gaertner. *Veröffentlichungen des Instituts für Meeresforschung in Bremerhaven*, Sonderband 3, 39–58.

Macfarlane, I. (1970). Germination of resting spores of *Plasmodiophora brassicae*. *Transactions of the British Mycological Society*, **55**, 97–112.

Macfarlane, I. & Last, F. T. (1959). Some effects of *Plasmodiophora brassicae* Woron. on the growth of the young cabbage plant. *Annals of Botany, London*, N.S., **23**, 547–570.

Macfarlane, T. D., Kuo, J. & Hilton, R. N. (1978). Structure of the giant sclerotium of

Polyporus mylittae. Transactions of the British Mycological Society, **71**, 359–365.

Machlis, L. (1958a). Evidence for a sexual hormone in *Allomyces. Physiologia Plantarum,* **11**, 181–192.

Machlis, L. (1958b). A study of sirenin, the chemotactic sexual hormone from the watermold *Allomyces. Physiologia Plantarum,* **11**, 845–854.

Machlis, L. (1972). The coming of age of sex hormones in plants. *Mycologia,* **64**, 235–247.

MacInnes, M. & Francis, D. (1974). Meiosis in *Dictyostelium mucoroides. Nature,* **251**, 321–324.

McIntosh, D. L. (1954). A cytological study of ascus development in *Pyronema confluens* Tul. *Canadian Journal of Botany,* **32**, 440–446.

McIntosh, R. A. (1978). Breeding for resistance to powdery mildew in the temperate cereals. In *The Powdery Mildews,* 237–257. Editor: D. M. Spencer. London, New York, San Francisco: Academic Press.

McKay, R. (1957). The longevity of the oospores of onion downy mildew, *Peronospora destructor* (Berk.) Casp. *Scientific Proceedings of the Royal Dublin Society,* N.S., **27**, 295–307.

McKeen, W. E. (1962). The flagellation, movement and encystment of some Phycomycetous zoospores. *Canadian Journal of Microbiology,* **8**, 897–904.

McKeen, W. E. (1970). Lipid in *Erysiphe graminis hordei* and its possible role during germination. *Canadian Journal of Microbiology,* **16**, 1041–1044.

McKeen, W. E., Mitchell, N. & Smith, R. (1967). The *Erysiphe cichoracearum* conidium. *Canadian Journal of Botany,* **45**, 1489–1496.

McKeen, W. E., Smith, R. & Mitchell, N. (1966). The haustorium of *Erysiphe cichoracearum* and the host–parasite interface on *Helianthus annuus. Canadian Journal of Botany,* **44**, 1299–1306.

McLaughlin, D. J. (1973). Ultrastructure of sterigma growth and basidiospore formation in *Coprinus* and *Boletus. Canadian Journal of Botany,* **51**, 145–150.

McLaughlin, J. L., Goyan, J. E. & Paul, A. G. (1964). Thin layer chromatography of ergot alkaloids. *Journal of Pharmaceutical Sciences,* **53**, 306–310.

MacLean, N. (1964). Electron microscopy of a fission yeast, *Schizosaccharomyces pombe. Journal of Bacteriology,* **88**, 1459–1466.

MacLeod, D. M. (1963). Entomophthorales infections. In *Insect Pathology: An Advanced Treatise,* **2**, 189–231. Editor: E. A. Steinhaus. New York and London: Academic Press.

MacLeod, D. M. & Müller-Kögler, E. (1973). Entomogenous fungi: *Entomophthora* species with pear-shaped to almost spherical conidia (Entomophthorales: Entomophthoraceae). *Mycologia,* **65**, 823–893.

McManus, M. A. (1958). *In vivo* studies of plasmogamy in *Ceratiomyxa. Bulletin of the Torrey Botanical Club,* **85**, 28–37.

McMeekin, D. (1960). The role of the oospores of *Peronospora parasitica* in downy mildew of crucifers. *Phytopathology,* **50**, 93–97.

McMorris, T. C. & Barksdale, A. W. (1967). Isolation of a sex-hormone from the water-mould *Achlya bisexualis. Nature, London,* **215**, 302–321.

McMorris, T. C., Seshadri, R., Weihe, G. R., Arsenault, G. P. & Barksdale, A. W. (1975). Structure of oogoniol-1, -2, and -3, steroidal sex hormones of the water mould *Achlya. Journal of the American Chemical Society,* **97**, 2544–2545.

McNabb, R. F. R. (1965). Taxonomic studies in the Dacrymycetaceae. II. *Calocera* (Fries) Fries. *New Zealand Journal of Botany,* **3**, 31–58.

McNabb, R. F. R. (1973). Phragmobasidiomycetidae: Tremellales, Auriculariales, Septobasidiales. In *The Fungi: An Advanced Treatise,* **IVB**, 303–316. Editors: G. C. Ainsworth, F. K. Sparrow & A. S. Sussman. New York and London, Academic Press.

McNabb, R. F. R. & Talbot, P. H. B. (1973). Holobasidiomycetidae. In *The Fungi: An Advanced Treatise*, **IVB**, 317–325. Editors: G. C. Ainsworth, F. K. Sparrow & A. S. Sussman. New York and London: Academic Press.

Madelin, M. F. (1956). The influence of light and temperature on fruiting of *Coprinus lagopus* Fr. in pure culture. *Annals of Botany, London*, N.S., **20**, 467–480.

Mai, S. H. (1976). Morphological studies in *Podospora anserina*. *American Journal of Botany*, **63**, 821–825.

Mainwaring, H. R. (1972). The fine structure of ascospore wall formation in *Sordaria fimicola*. *Archiv für Mikrobiologie*, **81**, 126–135.

Malcolmson, J. F. (1969). Races of *Phytophthora infestans* occurring in Great Britain. *Transactions of the British Mycological Society*, **53**, 417–423.

Malcolmson, J. F. (1970). Vegetative hybridity in *Phytophthora infestans*. *Nature*, **225**, 971–972.

Malençon, G. (1938). Les truffes européenes: historique, morphogenie, organographie, classification, culture. *Revue de Mycologie, Paris* (Mémoire hor série), 1–92.

Malik, M. M. S. (1974). Nuclear behaviour during teliospore germination in *Ustilago tritici* and *U. nuda*. *Pakistan Journal of Botany*, **6**, 59–63.

Manners, J. G. & Hossain, S. M. M. (1963). Effect of temperature and humidity on conidial germination in *Erysiphe graminis*. *Transactions of the British Mycological Society*, **46**, 225–234.

Manocha, M. S. (1965). Fine structure of the *Agaricus* carpophore. *Canadian Journal of Botany*, **43**, 1329–1334.

Manocha, M. S. (1975). Host–parasite relations in a mycoparasite. III. Morphological and biochemical differences in the parasitic- and axenic-culture spores of *Piptocephalis virginiana*. *Mycologia*, **76**, 382–391.

Manocha, M. S. & Shaw, M. (1966). The physiology of host–parasite relations. XVI. Fine structure of the nucleus in rust-infected mesophyll cells of wheat. *Canadian Journal of Botany*, **44**, 669–673.

Mantle, P. G. (1974). Industrial exploitation of ergot fungi. In *The Filamentous Fungi*, **3**, 426–450. Editors: J. E. Smith, & D. R. Berry. London: Edward Arnold.

Mantle, P. G. & Shaw, S. (1976). Role of ascospore production by *Claviceps purpurea* in aetiology of ergot disease in male sterile wheat. *Transactions of the British Mycological Society*, **67**, 17–22.

Manton, I., Clarke, B. & Greenwood, A. D. (1951). Observations with the electron microscope on a species of *Saprolegnia*. *Journal of Experimental Botany*, **2**, 321–331.

Marchant, R. & Robards, A. W. (1968). Membrane systems associated with the plasmalemma of plant cells. *Annals of Botany*, **32**, 457–471.

Marchant, R., Peat, A. & Banbury, G. H. (1967). The ultrastructural basis of hyphal growth. *New Phytologist*, **66**, 623–629.

Marshall, G. M. (1960a). The incidence of certain seed-borne diseases in commercial seed-samples. II. Ergot, *Claviceps purpurea* (Fr.) Tul. in cereals. *Annals of Applied Biology*, **48**, 19–26.

Marshall, G. M. (1960b). The incidence of certain seed-borne diseases in commercial seed-samples. IV. Bunt of wheat, *Tilletia caries* (DC.) Tul. V. Earcockles of wheat, *Anguina tritici* (Stein.) Filipjev. *Annals of Applied Biology*, **48**, 34–38.

Marte, M. & Gargiulo, A. M. (1972). Electron microscopy of peach leaves infected by *Taphrina deformans* (Berk.) Tul. *Phytopathologia Mediterranea*, **11**, 166–179.

Martin, E. (1940). The morphology and cytology of *Taphrina deformans*. *American Journal of Botany*, **27**, 743–751.

Martin, G. W. (1925). Morphology of *Conidiobolus villosus*. *Botanical Gazette*, **80**, 311–318.

Martin, G. W. (1927). Basidia and spores of the Nidulariaceae. *Mycologia*, **19**, 239–247.

Martin, G. W. (1952) Revision of the North Central Tremellales. *University of Iowa Studies in Natural History*, **19** (3), 1–122.

Martin, G. W. (1960). The systematic position of the Myxomycetes. *Mycologia*, **52**, 119–129.

Martin, G. W. (1961). Key to the families of fungi. In *Ainsworth & Bisby's Dictionary of the Fungi*, 5th edition, 497–517. Editor: G. C. Ainsworth. Kew, Surrey: Commonwealth Mycological Institute.

Martin, G. W. & Alexopoulos, C. J. (1969). *Monograph of the Myxomycetes*. 560 pp. Iowa City: University of Iowa Press.

Martin, M., Gay, J. L. & Jackson, G. V. H. (1976). Electron microscopic study of developing and mature cleistothecia of *Sphaerotheca mors-uvae*. *Transactions of the British Mycological Society*, **66**, 473–487.

Martin, P. (1967). Studies in the Xylariaceae. I. New and old concepts. *Journal of South African Botany*, **33**, 205–240.

Masri, S. S. & Ellingboe, A. H. (1966). Primary infection of wheat and barley by *Erysiphe graminis*. *Phytopathology*, **56**, 389–395.

Mather, K. & Jinks, J. L. (1958). Cytoplasm in sexual reproduction. *Nature, London*, **182**, 1188–1190.

Mathew, K. T. (1961). Morphogenesis of mycelial strands in the cultivated mushroom, *Agaricus bisporus*. *Transactions of the British Mycological Society*, **44**, 285–290.

Matile, P., Moor, H. & Robinow, C. F. (1969). Yeast cytology. In *The Yeasts*, **1**, 219–302. Editors: A. H. Rose & J. S. Harrison. New York and London: Academic Press.

Matthews, T. R. & Niederpruem, D. J. (1972). Differentiation in *Coprinus lagopus*. I. Control of fruiting and cytology of initial events. *Archiv für Mikrobiologie*, **87**, 257–268.

Matthews, T. R. & Niederpruem, D. J. (1973). Differentiation in *Coprinus lagopus*. II. Histology and ultrastructural aspects of developing primordia. *Archiv für Mikrobiologie*, **88**, 169–180.

Meier, H. & Webster, J. (1954). An electron microscope study of zoospore cysts in the Saprolegniaceae. *Journal of Experimental Botany*, **5**, 401–409.

Menzies, B. P. (1953). Studies on the systemic fungus *Puccinia suaveolens*. *Annals of Botany, London*, N.S., **17**, 551–568.

Meredith, D. S. (1961). Atmospheric content of *Nigrospora* spores in Jamaican banana plantations. *Journal of General Microbiology*, **26**, 343–349.

Meredith, D. S. (1962). Some components of the air spora in Jamaican banana plantations. *Annals of Applied Biology*, **50**, 577–594.

Meredith, D. S. (1965a). Violent spore release in *Stemphylium botryosum* Wallr. *Plant Disease Reporter*, **49**, 1006.

Meredith, D. S. (1965b). Violent spore release in *Helminthosporium turcicum*. *Phytopathology*, **55**, 1099–1102.

Meredith, D. S. (1966). Diurnal periodicity and violent liberation of conidia in *Epicoccum*. *Phytopathology*, **56**, 988.

Mesland, D. A. M., Huisman, J. G. & van den Ende, H. (1974). Volatile sexual hormones in *Mucor mucedo*. *Journal of General Microbiology*, **80**, 111–117.

Middleton, J. T. (1943). The taxonomy, host range and geographic distribution of the genus *Pythium*. *Memoirs of the Torrey Botanical Club*, **20**, 1–171.

Middleton, J. T. (1952). Generic concepts in the Pythiaceae. *Tijdschrift over Plantenziekten*, **58**, 226–235.

Milburn, J. A. (1970). Cavitation and osmotic potentials of *Sordaria* ascospores. *New Phytologist*, **69**, 133–141.

Millbank, J. W. & Kershaw, K. A. (1973). Nitrogen metabolism. In *The Lichens*,

289—307. Editors: V. Ahmadjian & M. E. Hale. New York and London: Academic Press.

Miller, J. H. (1930). British Xylariaceae. *Transactions of the British Mycological Society*, **15**, 134–154.

Miller, J. H. (1932a). British Xylariaceae. II. *Transactions of the British Mycological Society*, **17**, 125–135.

Miller, J. H. (1932b). British Xylariaceae. III. A revision of specimens in the Herbarium of the Royal Botanic Gardens, Kew. *Transactions of the British Mycological Society*, **17**, 136–146.

Miller, J. H. (1949). A revision of the classification of the Ascomycetes with special emphasis on the Pyrenomycetes. *Mycologia*, **41**, 99–127.

Miller, J. H. (1961). *A Monograph of the World Species of Hypoxylon*. 158 pp. University of Georgia Press.

Miller, R. E. (1971). Evidence of sexuality in the cultivated mushroom, *Agaricus bisporus*. *Mycologia*, **63**, 630–634.

Mills, G. L. & Cantino, E. C. (1978). The lipid composition of the *Blastocladiella emersonii* γ-particle and the function of the γ-particle lipid in chitin formation. *Experimental Mycology*, **2**, 99–109.

Mims, C. W. (1969). Capillitial formation in *Arcyria cinerea*. *Mycologia*, **61**, 784–798.

Mirza, J. H. & Cain, R. F. (1969). Revision of the genus *Podospora*. *Canadian Journal of Botany*, **47**, 1999–2048.

Mitchell, D. T. & Cooke, R. C. (1968). Some effects of temperature on germination and longevity of sclerotia in *Claviceps purpurea*. *Transactions of the British Mycological Society*, **51**, 721–729.

Mitchell, M. B. (1965). Characteristics of developing asci of *Neurospora crassa*. *Canadian Journal of Botany*, **43**, 933–938.

Mitchison, J. D. (1970). Physiological and cytological methods for *Schizosaccharomyces pombe*. In *Methods in Cell Physiology*, **4**, 131–168. Editor: D. M. Prescott. New York and London: Academic Press.

Mix, A. J. (1949). A monograph of the genus *Taphrina*. *Kansas University Scientific Bulletin*, **33**, 3–167.

Moens, P. B. & Rapport, E. (1971). Spindles, spindle plaques and meiosis in the yeast *Saccharomyces cerevisiae* (Hansen). *Journal of Cell Biology*, **50**, 344–361.

Moore, D. & Jirjis, R. I. (1976). Regulation of sclerotium production by primary metabolites in *Coprinus cinereus* (=*C. lagopus* sensu Lewis). *Transactions of the British Mycological Society*, **66**, 377–382.

Moore, E. J. (1963). The ontogeny of the apothecia in *Pyronema domesticum*. *American Journal of Botany*, **50**, 37–44.

Moore, E. J. & Korf, R. P. (1963). The genus *Pyronema*. *Bulletin of the Torrey Botanical Club*, **90**, 33–42.

Moore, R. T. (1955). Index to the Helicosporae. *Mycologia*, **47**, 90–103.

Moore, R. T. (1965). The ultrastructure of fungal cells. In *The Fungi: An Advanced Treatise*, **1**, 95–118. Editors: G. C. Ainsworth & A. S. Sussman. New York and London: Academic Press.

Moore, R. T. (1972). Ustomycota, a new division of higher fungi. *Antonie van Leeuwenhoek Journal of Microbiology and Serology*, **38**, 567–584.

Moore, R. T. (1975). Early ontogenetic stages in dolipore/parenthesome formation in *Polyporus biennis*. *Journal of General Microbiology*, **87**, 251–259.

Moore, R. T. & McAlear, J. H. (1962). Fine structure of mycota. 7. Observations on septa of Ascomycetes and Basidiomycetes. *American Journal of Botany*, **49**, 86–94.

Moore, R. T. & Marchant, R. (1972). Ultrastructural characterization of the Basidiomycete septum of *Polyporus biennis*. *Canadian Journal of Botany*, **50**, 2463–2469.

Moore, W. C. (1959). *British Parasitic Fungi*. 430 pp. Cambridge University Press.

Moore-Landecker, E. (1975). A new pattern of reproduction in *Pyronema domesticum*. *Mycologia*, **67**, 1119–1127.

Moreau, C. (1953). Les genres *Sordaria* et *Pleurage*, leurs affinités systématiques. *Encyclopédie Mycologique*, **15**, 1–330. Paris: Lechevalier.

Morisset, E. (1963). Recherches sur le Pyrénomycète *Sporormia leporina* Niessl (Pléosporale Sordarioide). *Revue Générale de Botanique*, **70**, 69–106.

Morrison, D. J. (1976). Vertical distribution of *Armillaria mellea* rhizomorphs in soil. *Transactions of the British Mycological Society*, **66**, 393–399.

Mortimer, A. M. & Shaw, D. S. (1975). Cytofluorimetric evidence for meiosis in gametangial nuclei of *Phytophthora drechsleri*. *Genetical Research, Cambridge*, **25**, 201–205.

Mortimer, R. K. & Hawthorne, D. C. (1969). Yeast genetics. In *The Yeasts*, **1**, 385–460. Editors: A. H. Rose & J. S. Harrison. New York and London: Academic Press.

Morton, D. J. (1961). Trypan blue and boiling lactophenol for staining and clearing barley tissues infected with *Ustilago nuda*. *Phytopathology*, **51**, 27–29.

Morton, F. J. & Smith, G. (1963). The genera *Scopulariopsis* Bainer, *Microascus* Zukal and *Doratomyces* Corda. *Mycological Paper No. 86, Commonwealth Mycological Institute*. 96 pp.

Moseman, J. G. (1966). Genetics of powdery mildews. *Annual Review of Plant Pathology*, **4**, 269–290.

Moseman, J. G. & Powers, H. R. (1957). Function and longevity of cleistothecia of *Erysiphe graminis* f. sp. *hordei*. *Phytopathology*, **47**, 53–56.

Moser, M. (1967). *Die Röhrlinge und Blatterpilze (Agaricales)*. Kleine Kryptogamenflora. Bd. IIb2. Basidiomyceten II Teil. Editor: H. Gams. Stuttgart: Fischer.

Moses, M. J. (1968). Synaptinemal complex. *Annual Review of Genetics*. **2**, 363–412.

Mosse, B. (1973). Advances in the study of vesicular–arbuscular mycorrhiza. *Annual Review of Phytopathology*, **11**, 171–196.

Mosse, B. & Bowen, G. D. (1968). A key to the recognition of some *Endogone* spore types. *Transactions of the British Mycological Society*, **51**, 469–483.

Motta, J. (1967). A note on the mitotic apparatus in the rhizomorph meristem of *Armillaria mellea*. *Mycologia*, **59**, 370–375.

Mower, R. L. & Hancock J. G. (1975a). Sugar composition of ergot honeydews. *Canadian Journal of Botany*, **53**, 2813–2825.

Mower, R. L. & Hancock J. G. (1975b). Mechanism of honeydew formation by *Claviceps* species. *Canadian Journal of Botany*, **53**, 2826–2834.

Mulder, J. L. & Pugh, G. J. F. (1971). Fungal biological flora. II. *Epicoccum nigrum* Link. *International Biodeterioration Bulletin*, **7**, 67–71.

Müller, F. (1954). Die Abschleuderung der Sporen von *Sporobolomyces* – Spiegelhefe – gefilmt. *Friesia*, **6**, 65–74.

Müller, E. & Arx, J. A. von (1962). Die Gattungen der didymosporen Pyrenomyceten. *Beiträge der Kryptogamenflora der Schweiz*, Bd. **11**(2), 1–922.

Müller, E. & Arx, J. A. von (1973). Pyrenomycetes: Meliolales, Coronophorales, Sphaeriales. In *The Fungi: An Advanced Treatise*, IVA, 87–132. Editors: G. C. Ainsworth, F. K. Sparrow & A. S. Sussman. New York and London: Academic Press.

Müller, E. & Tomasevic, M. (1957). Kulturversuche mit einigen Arten der Gattung *Leptosphaeria* Ces. & de Not. *Phytopathologische Zeitschrift*, **29**, 287–294.

Müller, G. (1964). Zur Kenntniss der Gattung *Endomycopsis* Stelling-Dekker. *Zentralblatt für Bakteriologie, Parasitenkunde, Infektionskrankheiten und Hygiene*, Abteilung II, **118**, 40–43.

Muller, H. G. (1958) The constricting ring mechanism of two predacious Hyphomycetes. *Transactions of the British Mycological Society*, **41**, 341–364.

Müller, K. O. (1959). Hypersensitivity. In *Plant Pathology: An Advanced Treatise*, **1**,

469–519. Editors: J. G. Horsfall & A. E. Dimond. New York and London: Academic Press.

Müller-Kögler, E. (1959). Zur Isolierung und Kultur insektpathogener Entomophthoraceen. *Entomophaga*, **4**, 261–274.

Müller- Kögler, E. (1965). *Pilzkrankheiten bei Insekten*. 444 pp. Berlin and Hamburg: Paul Parey.

Mullins, J. T. (1973). Lateral branch formation and cellulase production in the water molds. *Mycologia*, **65**, 1007–1014.

Mullins, J. T. & Raper, J. R. (1965). Heterothallism in biflagellate aquatic fungi: preliminary genetic analysis. *Science. N.Y.*, **150**, 1174–1175.

Munnecke, D. E., Wilbur, W. D. & Kolbezen, M. J. (1970). Dosage response of *Armillaria mellea* to methyl bromide. *Phytopathology*, **60**, 992–993.

Myers, R. B. & Cantino, E. C. (1974). The gamma particle. A study of cell-organelle interactions in the development of the water mold *Blastocladiella emersonii*. *Monographs in Developmental Biology*, **8**. 117 pp. S. Karger.

Nakai, Y. (1975). Fine structure of shiitake, *Lentinus edodes* (Berk.) Sing. IV. External and internal features of the hilum in relation to basidiospore discharge. *Reports of the Tottori Mycological Institute, Japan*, **12**, 41–45.

Nakai, Y. & Ushiyama, R. (1974a). Fine structure of the shiitake, *Lentinus edodes* (Berk.) Sing. *Report of the Tottori Mycological Institute, Japan*, **11**, 1–6.

Nakai, Y. & Ushiyama, R. (1974b). Fine structure of shittake, *Lentinus edodes* (Berk.) Sing. II. Development of basidia and basidiospores. *Report of the Tottori Mycological Institute, Japan*, **11**, 7–15.

Nakai, Y. & Ushiyama, R. (1978). Fine structure of shiitake, *Lentinus edodes*. VI. Cytoplasmic microtubules in relation to nuclear movement. *Canadian Journal of Botany*, **56**, 1206–1211.

Nannfeldt, J. A. (1932). Studien über die Morphologie und Systematik der nichtlichenisierten inoperculaten Discomyceten. *Nova Acta Regiae Societatis Scientiarum Upsaliensis*, IV, **8**(2), 1–368.

Nannfeldt, J. A. (1942). The Geoglossaceae of Sweden. *Arkiv för Botanik*, Bd. 30A, No. 4, 67 pp.

Nawawi, A., Descals, E. & Webster, J. (1977a). *Leptosporomyces galzinii*, the basidial state of a clamped branched conidium from fresh water. *Transactions of the British Mycological Society*, **68**, 31–36.

Nawawi, A., Webster, J. & Davey, R. A. (1977b). *Dendrosporomyces prolifer* gen. et sp.nov., a Basidiomycete with branched spores. *Transactions of the British Mycological Society*, **68**, 59–63.

Neergaard, P. (1977). *Seed Pathology*, 2 vols. London and Basingstoke: Macmillan.

Nelson, A. C. & Backus, M. P. (1968). Ascocarp development in two homothallic Neurosporas. *Mycologia*, **60**, 16–28.

Nelson, R. K. & Scheetz, R. W. (1975). Swarm cell ultrastructure in *Ceratiomyxa fruticulosa*. *Mycologia*, **67**, 733–740.

Nelson, R. K. & Scheetz, R. W. (1976). Thread phase ultrastructure in *Ceratiomyxa fruticulosa*. *Mycologia*, **68**, 144–150.

Newell, P. C. (1971). The development of the cellular slime mould *Dictyostelium discoideum*: A model system for the study of cellular differentiation. *Essays in Biochemistry*, **7**, 87–126.

Newell, P. C. (1975). Cellular communication during aggregation of the slime mold *Dictyostelium*. In *Microbiology – 1975*, 426–433. Editor: D. Schlessinger. Washington, D.C., U.S.A.: American Society for Microbiology.

Nickerson, A. W. & Raper, K. B. (1973a). Macrocysts in the life cycle of the Dictyosteliaceae. I. Formation of the macrocysts. *American Journal of Botany*, **60**, 190–197.

Nickerson, A. W. & Raper, K. B. (1973b). Macrocysts in the life cycle of the Dictyosteliaceae. II. Germination of the macrocysts. *American Journal of Botany*, **60**, 247–254.

Niederpruem, D. J. & Wessels, J. G. H. (1969). Cytodifferentiation and morphogenesis in *Schizophyllum commune*. *Bacteriological Reviews*, **33**, 505–535.

Niel, C. B. van, Garner, G. E. & Cohen, A. L. (1972). On the mechanism of ballistospore discharge. *Archiv für Mikrobiologie*, **84**, 129–140.

Nielsen, R. I. (1978). Sexual mutants of a heterothallic *Mucor* species, *Mucor pusillus*. *Experimental Mycology*, **2**, 193–197.

Noble, M. & Glynne, M. D. (1970). Wart disease of potatoes. *F.A.O. Plant Protection Bulletin*, **18**, 125–135.

Nobles, M. K. (1958a). A rapid test for extracellular oxidase in cultures of wood-inhabiting Hymenomycetes. *Canadian Journal of Botany*, **36**, 91–99.

Nobles, M. K. (1958b). Cultural characters as a guide to the taxonomy and phylogeny of the Polyporaceae. *Canadian Journal of Botany*, **36**, 883–926.

Nobles, M. K. (1965). Identification of cultures of wood-inhabiting Hymenomycetes. *Canadian Journal of Botany*, **43**, 1097–1139.

Nolan, R. A. & Lewis, J. D. (1974). Studies on *Pythiopsis cymosa* from Newfoundland. *Transactions of the British Mycological Society*, **62**, 163–179.

Nordbring-Herz, B. (1972). Scanning electron microscopy of the nematode-trapping organs in *Arthrobotrys oligospora*. *Physiologia Plantarum*, **26**, 279–284.

Nordbring-Herz, B. (1973). Peptide-induced morphogenesis in the nematode-trapping fungus *Arthrobotrys oligospora*. *Physiologia Plantarum*, **29**, 223–233.

Novaes-Ledieu, M., Jiménez-Martínez, A. & Villanueva, J. R. (1967). Chemical composition of hyphal wall of Phycomycetes. *Journal of General Microbiology*, **47**, 237–245.

O'Donnell, K. L., Fields, W. G. & Hooper, G. R. (1974). Scanning ultrastructural ontogeny of cleistohymenial apothecia in the operculate Discomycete *Ascobolus furfuraceus*. *Canadian Journal of Botany*, **52**, 1653–1656.

Oduro, K. A., Munnecke, D. E., Sims, J. J. & Keen, N. T. (1976). Isolation of antibiotics produced in culture by *Armillaria mellea*. *Transactions of the British Mycological Society*, **66**, 195–199.

Ogilvie, L. & Thorpe, I. G. (1966). Black stem rust of wheat in Great Britain. In *Cereal Rust Conferences, Cambridge, 1964*, 172–176. Cambridge: Plant Breeding Institute.

Ohr, H. D. & Munnecke, D. E. (1974). Effects of methyl bromide on antibiotic production by *Armillaria mellea*. *Transactions of the British Mycological Society*, **62**, 65–72.

Ojha, M. & Turian, G. (1971). Interspecific transformation and DNA characteristics in *Allomyces*. *Molecular and General Genetics*, **112**, 49–59.

Olchowecki, A. & Reid, J. (1974). Taxonomy of the genus *Ceratocystis* in Manitoba. *Canadian Journal of Botany*, **52**, 1675–1711.

Olive, L. S. (1952). Studies on the morphology and cytology of *Itersonilia perplexans* Derx. *Bulletin of the Torrey Botanical Club*, **79**, 126–138.

Olive, L. S. (1953). The structure and behavior of fungus nuclei. *Botanical Review*, **19**, 439–586.

Olive, L. S. (1956). Genetics of *Sordaria fimicola*. I. Spore color mutants. *American Journal of Botany*, **43**, 97–107.

Olive, L. S. (1963a). Genetics of homothallic fungi. *Mycologia*, **55**, 93–103.

Olive, L. S. (1963b). The question of sexuality in cellular slime molds. *Bulletin of the Torrey Botanical Club*, **90**, 144–153.

Olive, L. S. (1964). Spore discharge mechanism in Basidiomycetes. *Science, N.Y.*, **146**, 542–543.

Olive, L. S. (1965). Nuclear behavior during meiosis. In *The Fungi: An Advanced*

Treatise, **I**, 143–161. Editors: G. C. Ainsworth & A. S. Sussman. New York and London: Academic Press.

Olive, L. S. (1967). The Protostelida – a new order of the Mycetozoa. *Mycologia*, **59**, 1–29.

Olive, L. S. (1970). The Mycetozoa: a revised classification. *Botanical Review*, **36**, 59–89.

Olive, L. S. & Fantini, A. A. (1961). A new heterothallic species of *Sordaria*. *American Journal of Botany*, **48**, 124–128.

Olive, L. S. & Stoianovitch, C. (1960). Two new members of the Acrasiales. *Bulletin of the Torrey Botanical Club*, **87**, 1–20.

Olive, L. S. & Stoianovitch, C. (1966). A simple new mycetozoan with ballistospores. *American Journal of Botany*, **53**, 344–349.

Olive, L. S. & Stoianovitch, C. (1971). A new genus of protostelids showing affinities with *Ceratiomyxa*. *American Journal of Botany*, **58**, 32–40.

Oliver, P. T. P. (1972). Conidiophore and spore development in *Aspergillus nidulans*. *Journal of General Microbiology*, **73**, 45–54.

Olson, L. W. & Fuller, M. S. (1968). Ultrastructural evidence for the biflagellate origin of the uniflagellate fungal zoospore. *Archiv für Mikrobiologie*, **62**, 237–250.

Olthof, T. H. A. & Estey, R. H. (1963). A nematotoxin produced by the nematophagous fungus *Arthrobotrys oligospora* Fresenius. *Nature, London*, **197**, 514–515.

Orr, G. F., Kuehn, H. H. & Plunkett, O. A. (1963). The genus *Gymnoascus* Baranetzky. *Mycopathologia et Mycologia Applicata*, **21**, 1–18.

Oso, B. (1969). Electron microscopy of ascus development in *Ascobolus*. *Annals of Botany*, N.S., **33**, 205–209.

Ou, S. H. (1972). *Rice Diseases*. 368 pp. Kew, Surrey, England: Commonwealth Mycological Institute.

Padhye, A. A. & Carmichael, J. W. (1971). The genus *Arthroderma* Berkeley. *Canadian Journal of Botany*, **49**, 1525–1540.

Pady, S. M. (1974). Sporobolomycetaceae in Kansas. *Mycologia*, **66**, 333–338.

Pady, S. M. & Johnston, C. O. (1955). The concentration of airborne rust spores in relation to epidemiology of wheat rusts in Kansas in 1954. *Plant Disease Reporter*, **39**, 463–466.

Page, R. M. (1952). Studies on the development of asexual reproductive structures in *Pilobolus*. *Mycologia*, **48**, 206–224.

Page, R. M. (1959). Stimulation of asexual reproduction of *Pilobolus* by *Mucor plumbeus*. *American Journal of Botany*, **46**, 579–585.

Page, R. M. (1960). The effect of ammonia on growth and reproduction of *Pilobolus kleinii*. *Mycologia*, **52**, 480–489.

Page, R. M. (1964). Sporangium discharge in *Pilobolus*: a photographic study. *Science, N.Y.*, **146**, 925–927.

Page, R. M. & Curry, G. M. (1966). Studies on phototropism of young sporangiophores of *Pilobolus kleinii*. *Photochemistry and Photobiology*, **5**, 31–40.

Page, R. M. & Humber, R. A. (1973). Phototropism in *Conidiobolus coronatus*. *Mycologia*, **65**, 335–354.

Page, R. M. & Kennedy, D. (1964). Studies on the velocity of discharged sporangia of *Pilobolus kleinii*. *Mycologia*, **56**, 363–368.

Palenzona, M. (1969). Sintesi micorrizica tra *"Tuber aestivum"* Vitt., *"Tuber brumale"* Vitt., *"Tuber melanosporum"* Vitt. e gemenzali di *"Corylus avellana"* L. *Allionia*, **15**, 121–131.

Palmer, J. T. (1955). Observations on Gasteromycetes. 1–3. *Transactions of the British Mycological Society*, **38**, 317–334.

Papa, K. E., Campbell, W. A. & Hendrix, F. F. (1967). Sexuality in *Pythium sylvaticum*: heterothallism. *Mycologia*, **59**, 589–595.

Papavizas, G. C. & Davey, C. B. (1960). Some factors affecting growth of *Aphanomyces euteiches* in synthetic media. *American Journal of Botany*, **47**, 758–765.

Papazian, H. P. (1950). Physiology of the incompatibility factors in *Schizophyllum commune*. *Botanical Gazette*, **112**, 143–163.

Parag, Y. (1965). Genetic investigation into the mode of action of the genes controlling self-incompatibility and heterothallism in Basidiomycetes. In *Incompatibility of Fungi*, 80–98. Editors: K. Esser & J. R. Raper. Berlin: Springer.

Parag, Y. & Koltin, Y. (1971). The structure of the incompatibility factors of *Schizophyllum commune*: Constitution of the three classes of B factors. *Molecular and General Genetics*, **112**, 43–48.

Paravicini, E. (1917). Untersuchungen über das Verhalten der Zellkerne bei der Fortpflanzung der Brandpilze. *Annales Mycologici*, **15**, 57–96.

Parguey-Leduc, A. (1967a). Recherches préliminaires sur l'ontogénie et l'anatomie comparée des ascocarpes des Pyrénomycètes ascohyméniaux. *Revue de Mycologie (Paris)*, **32**, 57–68.

Parguey-Leduc, A. (1967b). Recherches préliminaires sur l'ontogénie et l'anatomie comparée des ascocarpes des Pyrénomycètes ascohyméniaux. III. Les asques des Sordariales et leurs ascothécies, du type "Diaporthe". *Revue de Mycologie (Paris)*, **32**, 369–407.

Park, D. (1974). Aquatic hyphomycetes in non-aquatic habitats. *Transactions of the British Mycological Society*, **63**, 179–183.

Park, J. Y. & Agnihotri, V. P. (1969). Bacterial metabolites trigger sporophore formation in *Agaricus bisporus*. *Nature, London*, **222**, 984.

Parker-Rhodes, A. F. (1955). Fairy ring kinetics. *Transactions of the British Mycological Society*, **38**, 59–72.

Parmeter, J. R., Snyder, W. C. & Reichle, R. E. (1963). Heterokaryosis and variability in plant-pathogenic fungi. *Annual Review of Phytopathology*, **1**, 51–76.

Peace, W. R. (1962). *Pathology of Trees and Shrubs with Special Reference to Britain*. 722 pp. Oxford University Press.

Peat, A. & Banbury, G. H. (1967). Ultrastructure, protoplasmic streaming, growth and tropisms of *Phycomyces* sporangiophores. *New Phytologist*, **66**, 475–484.

Pegler, D. N. (1966). "Polyporaceae" – Part I, with a key to British genera. *News Bulletin of the British Mycological Society*, **26**, 14–28.

Pegler, D. N. (1967). "Polyporaceae" – Part II, with a key to world genera. *Bulletin of the British Mycological Society*, **1**, 17–38.

Pegler, D. N. (1973). Aphyllophorales. IV. Poroid families. In *The Fungi: An Advanced treatise*, **IVB**, 397–420. Editors: G. C. Ainsworth, F. K. Sparrow & A. S. Sussman. New York and London: Academic Press.

Pegler, D. N. & Young, T. W. K. (1971). Basidiospore morphology in the Agaricales. Beihefte, *Nova Hedwigia*, **35**, 210 pp.

Pendergrass, W. R. (1950). Studies on the Plasmodiophoraceous parasite, *Octomyxa brevilegniae*. *Mycologia*, **42**, 279–298.

Pentland, G. D. (1968). The stimulatory effect of *Aureobasidium pullulans* on rhizomorph development of *Armillaria mellea* in autoclaved soil. *Canadian Journal of Microbiology*, **14**, 87–88.

Perera, R. G. & Gay, J. L. (1976). The ultrastructure of haustoria of *Sphaerotheca pannosa* (Wallroth ex Fries) Léveillé and changes in infected and associated cells of rose. *Physiological Plant Pathology*, **9**, 57–65.

Perera, R. G. & Wheeler, B. E. J. (1975). Effect of water droplets on the development of *Sphaerotheca pannosa* on rose leaves. *Transactions of the British Mycological Society*, **64**, 313–319.

Perkins, D. D. & Barry, E. G. (1977). The cytogenetics of *Neurospora*. In *Advances in*

Genetics, 133–285. Editor: E. W. Caspari. New York, London, San Francisco: Academic Press.

Perkins, F. O. (1970). Formation of centriole and centriole-like structures during meiosis and mitosis in *Labyrinthula* sp. (Rhizopodea, Labyrinthulida). An electron microscope study. *Journal of Cell Science*, **6**, 629–653.

Perkins, F. O. (1972). The ultrastructure of holdfasts, "rhizoids" and "slime tracks" in thraustochytriaceous fungi and *Labyrinthula* spp. *Archiv für Mikrobiologie*, **84**, 97–118.

Perkins, F. O. (1973a). Observations of thraustochytriaceous (Phycomycetes) and labyrinthulid (Rhizopodea) ectoplasmic nets on natural and artificial substrates – an electron microscope study. *Canadian Journal of Botany*, **51**, 485–491.

Perkins, F. O. (1973b). A new species of marine Labyrinthulid *Labyrinthuloides yorkensis* gen.nov. sp.nov. Cytology and fine structure. *Archiv für Mikrobiologie*, **90**, 1–17.

Perkins, F. O. (1974a). Phylogenetic considerations of the problematic thraustochytriaceous–labyrinthulid–*Dermocystidium* complex based on observations of fine structure. *Veröffentlichungen des Instituts für Meeresforschung in Bremerhaven*, Supplement 5.

Perkins, F. O. (1974b). Fine structure of lower marine and estuarine fungi. In *Recent Advances in Aquatic Mycology*, 279–312. Editor: E. B. Gareth Jones. London: Elek Science.

Perkins, F. O. & Amon, J. P. (1969). Zoosporulation in *Labyrinthula* sp., an electron microscope study. *Journal of Protozoology*, **16**, 235–257.

Perreau-Bertrand, J. (1967). Recherches sur la différentiation et la structure de la paroi sporale chez les homobasidiomycètes à spores ornées. *Annales des Sciences Naturelles Botanique*, Sér. 12, **8**, 639–746.

Perrott, P. E. (1955). The genus *Monoblepharis*. *Transactions of the British Mycological Society*, **38**, 247–282.

Perrott, P. E. (1958). Isolation and pure culture of *Monoblepharis*. *Nature, London*, **182**, 1322–1324.

Petch, T. (1936). *Cordyceps militaris* and *Isaria farinosa*. *Transactions of the British Mycological Society*, **20**, 216–224.

Petersen, H. E. (1936). The wasting disease of *Zostera marina*. I. A phytological investigation of the diseased plant. *Biological Bulletin*, **70**, 148–158.

Petersen, R. H. (1973). Aphyllophorales. II. The Clavarioid and Cantharelloid Basidiomycetes. In *The fungi: An Advanced Treatise*, **IVB**, 351–368. Editors: G. C. Ainsworth, F. K. Sparrow & A. S. Sussman. New York and London: Academic Press.

Petersen, R. H. (1974). The rust fungus life cycle. *Botanical Review*, **40**, 453–513.

Petrak, F. (1962). Über die Gattungen *Xylosphaera* Dum. und *Xylosphaeria* Otth. *Sydowia*, **15**, 288–290.

Peveling, E. (1973). Fine structure. In *The Lichens*, 147–179. Editors: V. Ahmadjian & M. E. Hale. New York and London: Academic Press.

Peyton, G. A. & Bowen, C. C. (1963). The host–parasite interface of *Peronospora manshurica* on *Glycine max*. *American Journal of Botany*, **50**, 787–797.

Phaff, H. J. (1963). Cell wall of yeasts. *Annual Review of Microbiology*, **17**, 15–30.

Phaff, H. J. (1970). Discussion of the yeast-like genera belonging to the Sporobolomycetaceae. In *The Yeasts. A taxonomic Study*. 2nd edition. Editor: J. Lodder. Amsterdam and London: North-Holland Publishing Company.

Phaff, H. J., Miller, M. W. & Mrak, E. M. (1966). *The Life of Yeasts*. 186 pp. Cambridge, Mass.: Harvard University Press.

Phaff, H. J. & Yoneyama, M. (1961). *Endomycopsis scolyti*, a new heterothallic species of yeast. *Antonie van Leeuwenhoek*, **27**, 196–202.

Piard-Douchez, Y. (1949). Le *Spongospora subterranea* et son action pathogène. *Annales des Sciences Naturelles Botanique*, Sér. 11, 91–122.

Pickett-Heaps, J. D. (1969). The evolution of the mitotic apparatus: an attempt at comparative ultrastructural cytology in dividing plant cells. *Cytobios*, **3**, 257–280.

Pidacks, C., Whitehill, A. R., Pruess, L. M., Hesseltine, C. W., Hutchings, B. C., Bohonos, N. & Williams, J. H. (1953). Coprogen, the isolation of a new growth factor required by *Pilobolus* species. *Journal of the American Chemical Society*, **75**, 6064–6065.

Piehl, A. (1929). The cytology and morphology of *Sordaria fimicola*. *Transactions of the Wisconsin Academy of Science, Arts and Letters*, **24**, 323–341.

Pittenger, R. H. & Atwood, K. C. (1956). Stability of nuclear proportions during growth of *Neurospora* heterokaryons. *Genetics*, **41**, 227–241.

Pittenger, R. H., Kimball, A. W. & Atwood, K. C. (1955). Control of nuclear ratios in *Neurospora* heterokaryons. *American Journal of Botany*, **42**, 954–958.

Plaats-Niterink, A. J. van der (1968). The occurrence of *Pythium* in the Netherlands. I. Heterothallic species. *Acta Botanica Neerlandica*, **17**, 320–329.

Plaats-Niterink, A. J. van der (1969). The occurrence of *Pythium* in the Netherlands. II. Another heterothallic species: *Pythium splendens* Braun. *Acta Botanica Neerlandica*, **18**(4), 489–495.

Plank, J. E. van der (1963). *Plant Diseases: Epidemics and Control.* 349 pp. New York and London: Academic Press.

Plank, J. E. van der (1971). Stability of resistance to *Phytophthora infestans* in cultivars without R genes. *Potato Research*, **14**, 263–270.

Plank, J. E. van der (1975). *Principles of Plant Infection.* 216 pp. New York and London: Academic Press.

Platford, R. G. & Bernier, C. C. (1970). Resistance to *Claviceps purpurea* in spring and durum wheat. *Nature, London*, **226**, 770.

Plattner, J. J. & Rapoport, H. (1971). The synthesis of D- and L-sirenin and their absolute configurations. *Journal of the American Chemical Society*, **93**, 1758–1761.

Plumb, R. & Turner, R. H. (1972). Scanning electron microscopy of *Erysiphe graminis*. *Transactions of the British Mycological Society*, **59**, 149–150.

Plunkett, B. E. (1953). Nutritional and other aspects of fruit body formation in pure cultures of *Collybia velutipes* (Curt.) Fr. *Annals of Botany, London*, N.S., **17**, 193–217.

Plunkett, B. E. (1956). The influence of factors of the aeration complex and light upon fruit-body form in pure culture of an agaric and a polypore. *Annals of Botany, London*, N.S., **20**, 563–586.

Plunkett, B. E. (1958). Translocation and pileus formation in *Polyporus brumalis*. *Annals of Botany, London*, N.S., **22**, 237–249.

Plunkett, B. E. (1961). The change of tropism of *Polyporus brumalis* stipes and the effect of directional stimuli on pileus differentiation. *Annals of Botany, London*, N.S., **25**, 206–223.

Plunkett, B. E. (1966). Morphogenesis in the mycelium: control of lateral hypha frequency in *Mucor hiemalis* by amino-acids. *Annals of Botany, London*, N.S., **30**, 133–151.

Poelt, J. (1973). Classification. In *The Lichens*, 599–632. Editors: V. Ahmadjian & M. E. Hale. New York and London: Academic Press.

Poitras, A. W. (1955). Observations on asexual and sexual reproductive structures of the Choanephoraceae. *Mycologia*, **47**, 702–713.

Pokorny, K. L. (1967). *Labyrinthula. Journal of Protozoology*, **14**, 697–708.

Pomerlau, R. (1970). Pathological anatomy of the Dutch Elm disease. Distribution and development of *Ceratocystis ulmi* in Elm tissues. *Canadian Journal of Botany*, **48**, 2043–2057.

Pontecorvo, G. (1956). The parasexual cycle in fungi. *Annual Review of Microbiology*, **10**, 393–400.

Poon, N. H. & Day, A. W. (1975). Fungal fimbriae. I. Structure, origin and synthesis. *Canadian Journal of Microbiology*, **21**, 537–546.

Poon, N. H., Martin, J. & Day, A. W. (1974). Conjugation in *Ustilago violacea*. I. Morphology. *Canadian Journal of Microbiology*, **20**, 187–191.

Popp, W. (1958). An improved method of detecting loose-smut mycelium in whole embryos of wheat and barley. *Phytopathology*, **48**, 641–643.

Porter, D. (1969). Ultrastructure of *Labyrinthula*. *Protoplasma*, **67**, 1–19.

Poulter, R. T. M. & Dee, J. (1968). Segregation of factors controlling fusion between plasmodia of the true slime mould *Physarum polycephalum*. *Genetical Research, Cambridge*, **12**, 71–79.

Pramer, D. & Stoll, N. R. (1959). Nemin: a morphogenic substance causing trap formation by predaceous fungi. *Science*, **129**, 966–967.

Prasertphon, S. & Tanada, Y. (1969). Mycotoxins of Entomophthoraceous fungi. *Hilgardia*, **39**, 581–600.

Pratt, R. G. & Green, R. J. (1971). The taxonomy and heterothallism of *Pythium sylvaticum*. *Canadian Journal of Botany*, **49**, 273–279.

Pratt, R. G. & Green, R. J. (1973). The sexuality and population structure of *Pythium sylvaticum*. *Canadian Journal of Botany*, **51**, 429–436.

Price, T. V. (1970). Epidemiology and control of powdery mildew (*Sphaerotheca pannosa*) on roses. *Annals of Applied Biology*, **65**, 231–248.

Prince, A. E. (1943). Basidium formation and spore discharge in *Gymnosporangium nidus-avis*. *Farlowia*, **1**, 79–93.

Punithalingam, E., Tulloch, M. & Leach, C. M. (1972). *Phoma epicoccina* sp.nov. on *Dactylis glomerata*. *Transactions of the British Mycological Society*, **59**, 341–345.

Puranik, S. B. & Mathre, D. E. (1971). Biology and control of ergot on male sterile wheat and barley. *Phytopathology*, **61**, 1075–1080.

Quantz, L. (1943). Untersuchungen über die Ernährungsphysiologie einiger niederer Phycomyceten (*Allomyces kniepii, Blastocladiella variabilis* und *Rhizophlyctis rosea*). *Jahrbuch für Wissenschaftliche Botanik*, **91**, 120–168.

Quispel, A. (1959). Lichens. In *Encyclopedia of Plant Physiology*, **11**, 577–604. Editor: W. Ruhland. Berlin: Springer.

Raa, J. (1971). Indole-3-acetic acid levels and the role of indole-3-acetic oxidase in the normal root and club root of cabbage. *Physiologia plantarum*, **25**, 130–134.

Ragab, M. A. (1953). Hyphal systems of *Auriscalpium vulgare*. *Bulletin of the Torrey Botanical Club*, **80**, 21–25.

Ramsbottom, J. (1953). *Mushrooms and Toadstools. A Study of the Activities of Fungi*. 306 pp. London: Collins.

Ranzoni, F. V. (1956). The perfect stage of *Flagellospora penicillioides*. *American Journal of Botany*, **43**, 13–17.

Raper, C. A., Raper, J. R. & Miller, R. E. (1972). Genetical analysis of the life cycle of *Agaricus bisporus*. *Mycologia*, **64**, 1088–1117.

Raper, J. R. (1939). Sexual hormones in *Achlya*. I. Indicative evidence of a hormonal coordinating mechanism. *American Journal of Botany*, **26**, 639–650.

Raper, J. R. (1950). Sexual hormones in *Achlya*. VII. The hormonal mechanism in homothallic species. *Botanical Gazette*, **112**, 1–24.

Raper, J. R. (1952). Chemical regulation of the sexual processes in the Thallophytes. *Botanical Review*, **18**, 447–545.

Raper, J. R. (1954). Life cycles, sexuality, and sexual mechanisms in the fungi. In *Sex in Microorganisms*, 42–81. Editor: D. H. Wenrich. Washington, D.C.: American Association for the Advancement of Science.

Raper, J. R. (1957). Hormones and sexuality in lower plants. *Symposia of the Society for Experimental Biology*, **11**, 143–165.

Raper, J. R. (1959). Sexual versatility and evolutionary processes in fungi. *Mycologia*, **51**, 107–124.

Raper, J. R. (1966a). *Genetics of Sexuality in Higher Fungi*. 283 pp. New York: Ronald Press Co.

Raper, J. R. (1966b). Life cycles, basic patterns of sexuality, and sexual mechanisms. In *The Fungi: An Advanced Treatise*, **II**, 473–511. Editors: G. C. Ainsworth & A. S. Sussman. New York and London: Academic Press.

Raper, J. R. & Flexer, A. S. (1971). Mating systems and evolution of the Basidiomycetes. In *Evolution in the Higher Basidiomycetes*, 149–167. Editor: R. H. Petersen. Knoxville: University of Tennessee Press.

Raper, J. R. & Hoffman, R. M. (1974). *Schizophyllum commune*. In *Handbook of Genetics*, **I**, 597–626. Editor: R. C. King. New York and London: Plenum Press.

Raper, J. R., Krongelb, G. S. & Baxter, M. G. (1958). The number and distribution of incompatibility factors in *Schizophyllum commune*. *American Naturalist*, **92**, 221–232.

Raper, K. B. (1935). *Dictyostelium discoideum*, a new species of slime mould. *Journal of Agricultural Research*, **50**, 135–147.

Raper, K. B. (1951). Isolation, cultivation and conservation of simple slime molds. *Quarterly Review of Biology*, **26**, 169–190.

Raper, K. B. (1952). A decade of antibiotics in America. *Mycologia*, **44**, 1–59.

Raper, K. B. (1957). Nomenclature in *Aspergillus* and *Penicillium*. *Mycologia*, **49**, 644–662.

Raper, K. B. (1973). Acrasiomycetes. In *The Fungi: An Advanced treatise*, **IVB**, 9–36. Editors: G. C. Ainsworth, F. K. Sparrow & A. S. Sussman. New York and London: Academic Press.

Raper, K. B. & Fennell, D. I. (1965). The genus *Aspergillus*. 686 pp. Baltimore: Williams & Wilkins.

Raper, K. B. & Quinlan, M. S. (1958). *Acytostelium leptosomum*. A unique cellular slime mould with an acellular stalk. *Journal of General Microbiology*, **18**, 16–32.

Raper, K. B. & Thom, C. (1949). *A Manual of the Penicillia*. 875 pp. London: Baillière, Tindall & Cox.

Raudaskoski, M. (1972a). Occurrence of microtubules in the hyphae of *Schizophyllum commune* during intercellular nuclear migration. *Archiv für Mikrobiologie*, **86**, 91–100.

Raudaskoski, M. (1972b). Secondary mutations at the B incompatibility locus and nuclear migration in the basidiomycete *Schizophyllum commune*. *Hereditas*, **72**, 175–182.

Rayner, A. D. M. & Todd, N. K. (1977). Intraspecific antagonism in natural populations of wood-decaying Basidiomycetes. *Journal of General Microbiology*, **103**, 85–90.

Rayner, A. D. M. & Todd, N. K. (1978). Polymorphism in *Coriolus versicolor* and its relation to interfertility and intraspecific antagonism. *Transactions of the British Mycological Society*, **71**, 99–106.

Read, D. J. & Armstrong, W. (1972). A relationship between oxygen transport and the formation of the ectotrophic mycorrhizal sheath in conifer seedlings. *New Phytologist*, **71**, 49–53.

Reddy, M. S. & Kramer, C. L. (1975). A taxonomic revision of the Protomycetales. *Mycotaxon*, **3**, 1–50.

Redfern, D. B. (1968). The ecology of *Armillaria mellea* in Britain: Biological control. *Annals of Botany*, N.S., **32**, 293–300.

Redfern, D. B. (1973). Growth and behaviour of *Armillaria mellea* rhizomorphs in soil. *Transactions of the British Mycological Society*, **61**, 569–581.

Redhead, S. A. & Malloch, D. W. (1977). The Endomycetaceae: new concepts, new taxa. *Canadian Journal of Botany*, **55**, 1701–1711.

Reeves, F. (1967). The fine structure of ascospore formation in *Pyronema domesticum*. *Mycologia*, **59**, 1018–1033.

Reeves, F. B. (1971). The structure of the ascus apex in *Sordaria fimicola*. *Mycologia*, **63**, 204–212.

Reichle, R. E. (1969). Fine structure of *Phytophthora parasitica* zoospores. *Mycologia*, **61**, 30–51.

Reichle, R. E. & Fuller, M. S. (1967). The fine structure of *Blastocladiella emersonii* zoospores. *American Journal of Botany*, **54**, 81–92.

Reid, D. A. (1974). A monograph of the British Dacrymycetales. *Transactions of the British Mycological Society*, **62**, 433–494.

Reiff, F., Kautzmann, R., Luers, H. & Lindemann, M. (1960). Editors. *Die Hefen*. Bd. A. *Die Hefen in der Wissenschaft*. 1024 pp. Nürnberg: Hans Carl.

Reijnders, A. F. M. (1963). *Les Problèmes du Développement des Carpophores des Agaricales et de quelques Groupes Voisins*. 412 pp. The Hague: Junk.

Reijnders, A. F. M. (1975). The development of three species of the Agaricaceae and the ontogenetic pattern of this family as a whole. *Persoonia*, **8**, 307–319.

Reischer, H. S. (1951). Growth of Saprolegniaceae in synthetic media. I. Inorganic nutrition. *Mycologia*, **43**, 142–155.

Remacle, J. & Moreau, C. (1962). Le *Pleurage tetraspora* (Wint.) Griffiths est-il une forme à asques tetrasporés du *Pleurage curvula* (de Bary) Kuntze? *Revue de Mycologie, Paris*, **27**, 213–217.

Rennie, W. J. & Seaton, R. D. (1975). Loose smut of barley. The embryo test as a means of assessing loose smut infection in seed stocks. *Seed Science and Technology*, **3**, 697–709.

Reynolds, D. R. (1971). Wall structure of a bitunicate ascus. *Planta*, **98**, 244–257.

Richard, C. (1971). Sur l'activité antibiotique de l'*Armillaria mellea*. *Canadian Journal of Microbiology*, **17**, 1395–1399.

Richardson, D. H. S. (1973). Photosynthesis and carbohydrate movement. In *The Lichens*, 249–288. Editors: V. Ahmadjian & M. E. Hale. New York and London: Academic Press.

Richardson, D. H. S. (1975). *The Vanishing Lichens. Their History, Biology and Importance*. 231 pp. Newton Abbot: David & Charles.

Riddle, L. W. (1906). On the cytology of the Entomophthoraceae. *Proceedings of the American Academy of Arts and Sciences*, **42**, 177–200.

Rifai, M. A. (1969). A revision of the genus *Trichoderma*. *Mycological Paper No. 116, Commonwealth Mycological Institute*. 56 pp.

Rifai, M. & Webster, J. (1966). Culture studies on *Hypocrea* and *Trichoderma*. III. *H. lactea* (= *H. citrina*) and *H. pulvinata*. *Transactions of the British Mycological Society*, **49**, 297–310.

Rijkenberg, F. H. J. & Truter, S. J. (1974). The ultrastructure of sporogenesis in the pycnial stage of *Puccinia sorghi*. *Mycologia*, **66**, 319–326.

Rishbeth, J. (1952). Control of *Fomes annosus* Fr. *Forestry*, **25**, 41–50.

Rishbeth, J. (1961). Inoculation of pine stumps against infection by *Fomes annosus*. *Nature, London*, **191**, 826–827.

Rishbeth, J. (1968). The growth rate of *Armillaria mellea*. *Transactions of the British Mycological Society*, **51**, 575–586.

Rishbeth, J. (1976). Chemical treatment and inoculation of hardwood stumps for control of *Armillaria mellea*. *Annals of Applied Biology*, **82**, 57–70.

Ritchie, D. (1937). The morphology of the perithecium of *Sordaria fimicola*. *Journal of the Elisha Mitchell Scientific Society*, **53**, 334–342.

Ritchie, D. (1948). The development of *Lycoperdon oblongisporum*. *American Journal of Botany*, **35**, 215–219.

Robinow, C. F. (1957a). The structure and behavior of the nuclei in spores and growing hyphae of Mucorales. I. *Mucor hiemalis* and *Mucor fragilis*. *Canadian Journal of Microbiology*, **3**, 771–789.

Robinow, C. F. (1957b). The structure and behavior of the nuclei in spores and growing hyphae of Mucorales. II. *Phycomyces blakesleeanus*. *Canadian Journal of Microbiology*, **3**, 791–798.

Robinow, C. F. (1962). Some observations on the mode of division of somatic nuclei of *Mucor* and *Allomyces*. *Archiv. für Mikrobiologie*, **42**, 369–377.

Robinow, C. F. (1963). Observations on cell growth, mitosis and division in the fungus *Basidiobolus ranarum*. *Journal of Cell Biology*, **17**, 123–152.

Robinson, P. M. & Smith, J. M. (1976). Morphogenesis and growth kinetics of *Geotrichum candidum* in continuous culture. *Transactions of the British Mycological Society*, **66**, 413–420.

Roelofsen, P. A. (1950). The origin of spiral growth in *Phycomyces* sporangiophores. *Receuil des Travaux Botaniques Néerlandaises*, **42**, 73–110.

Roelofsen, P. A. (1959). *The Plant Cell Wall*. Handbuch der Pflanzenanatomie, III(4). Berlin: Gebrüder Borntraeger.

Rogers, J. D. (1965). *Hypoxylon fuscum*. I. Cytology of the ascus. *Mycologia*, **57**, 789–803.

Rogers, J. D. (1968). *Xylaria curta*: cytology of the ascus. *Canadian Journal of Botany*, **46**, 1337–1340.

Rogers, J. D. (1969). *Xylaria polymorpha*. I. Cytology of a form with small stromata from Minnesota. *Canadian Journal of Botany*, **47**, 1315–1317.

Rogers, J. D. (1975a). *Xylaria polymorpha*. II. Cytology of a form with typical robust stromata. *Canadian Journal of Botany*, **53**, 1736–1743.

Rogers, J. D. (1975b). *Hypoxylon serpens*: cytology and taxonomic considerations. *Canadian Journal of Botany*, **53**, 52–55.

Rogerson, C. T. (1970). The Hypocrealean Fungi (Ascomycetes, Hypocreales). *Mycologia*, **62**, 865–910.

Romagnesi. H. (1944). La cystide chez les Agaricacées. *Revue de Mycologie, Paris*, **9**, (Supplement), 4–21.

Romano, A. H. (1966). Dimorphism. In *The fungi: An Advanced Treatise*, **II**, 181–209. Editors: G. C. Ainsworth & A. S. Sussman. New York and London: Academic Press.

Roper, J. A. (1966). Mechanisms of inheritance. 3. The parasexual cycle. In *The Fungi: An Advanced Treatise*, **II**, 589–617. Editors: G. C. Ainsworth & A. S. Sussman. New York and London: Academic Press.

Rose, A. H. & Harrison, J. S. (1969–1970). Editors. *The Yeasts*. 3 vols. New York and London: Academic Press.

Rosinski, M. A. (1961). Development of the ascocarp of *Ceratocystis ulmi*. *American Journal of Botany*, **48**, 285–293.

Ross, I. K. (1957a). Capillitial formation in the Stemonitaceae. *Mycologia*, **49**, 808–819.

Ross, I. K. (1957b). Syngamy and plasmodium formation in the Myxogastres. *American Journal of Botany*, **44**, 843–850.

Ross, I. K. (1960a). Sporangial development in *Lamproderma arcyrionema*. *Mycologia*, **52**, 621–627.

Ross, I. K. (1960b). Studies on diploid strains of *Dictyostelium discoideum*. *American Journal of Botany*, **47**, 54–59.

Ross, I. K. (1961). Further studies on meiosis in the Myxomycetes. *American Journal of Botany*, **48**, 244–248.

Ross, I. K. (1964). Pure cultures of some Myxomycetes. *Bulletin of the Torrey Botanical Club*, **91**, 23–31.

Ross, I. K. (1973). The Stemonitomycetidae, a new subclass of Myxomycetes. *Mycologia*, **65**, 477–485.

Ross, I. K. (1976). Nuclear migration rates in *Coprinus congregatus*: a new record? *Mycologia*, **68**, 418–422.

Ross, I. K. & Cummings, R. J. (1967). Formation of amoeboid cells from the plasmodium of a myxomycete. *Mycologia*, **59**, 725–732.

Rowell, J. B. (1968). Chemical control of the cereal rusts. *Annual Review of Phytopathology*, **6**, 213–242.

Rowell, J. B. (1973). Control of leaf and stem rusts of wheat by seed treatment with oxycarboxin. *Plant Disease Reporter*, **57**, 567–671.

Rowell, J. B. (1976). Control of leaf rust on spring wheat by seed treatment with 4-N butyl-1,2,4-triazole. *Phytopathology*, **66**, 1129–1134.

Sahtiyanci, S. (1962). Studien über einige wurzelparasitäre Olpidiaceen. *Archiv für Mikrobiologie*, **41**, 187–228.

Salaman, R. N. (1949). *The History and Social Influence of the Potato*. 685. pp. Cambridge University Press.

Salvin, S. B. (1941). Comparative studies on the primary and secondary zoospores of the Saprolegniaceae. I. Influence of temperature. *Mycologia*, **53**, 592–600.

Salvin, S. B. (1942). Preliminary report on the intergeneric mating of *Thraustotheca clavata* and *Achlya flagellata*. *American Journal of Botany*, **29**, 674–676.

Sampson, K. (1939). *Olpidium brassicae* (Wor.) Dang. and its connection with *Asterocystis radicis* de Wildeman. *Transactions of the British Mycological Society*, **32**, 199–205.

Samson, R. A. (1974). *Paecilomyces* and some allied Hyphomycetes. *Studies in Mycology*, **6**, 1–119.

Samson, R. A. & Tansey, M. R. (1975). *Byssochlamys verrucosa* sp.nov. *Transactions of the British Mycological Society*, **65**, 512–514.

San Antonio, J. P. (1971). A laboratory method to obtain fruit from cased grain spawn of the cultivated mushroom, *Agaricus bisporus*. *Mycologia*, **63**, 16–21.

Sanders, F. E., Mosse, B. & Tinker, P. B. (1976). *Endomycorrhizas*. 626 pp. New York, San Francisco, London: Academic Press.

Sansome, E. (1961). Meiosis in the oogonium and antheridium of *Pythium debaryanum* Hesse. *Nature, London*, **191**, 827–828.

Sansome, E. (1963). Meiosis in *Pythium debaryanum* Hesse and its significance in the life history of the Biflagellatae. *Transactions of the British Mycological Society*, **46**, 63–72.

Sansome, E. (1976). Gametangial meiosis in *Phytophthora capsici*. *Canadian Journal of Botany*, **54**, 1535–1545.

Sansome, E. & Brasier, C. M. (1973). Intracellular spore formation in *Ceratocystis ulmi*. *Transactions of the British Mycological Society*, **61**, 588–590.

Sansome, E. & Sansome, F. W. (1974). Cytology and life-history of *Peronospora parasitica* on *Capsella bursa pastoris* and of *Albugo candida* on *C. bursa-pastoris* and on *Lunaria annua*. *Transactions of the British Mycological Society*, **62**, 323–332.

Sansome, E. R. (1946). Heterokaryosis, mating-type factors, and sexual reproduction in *Neurospora*. *Bulletin of the Torrey Botanical Club*, **73**, 397–409.

Santesson, R. (1953). The new systematics of lichenized fungi. *Proceedings of the Seventh International Botanical Congress*, 809–810.

Sauer, H. W. (1973). Differentiation in *Physarum polycephalum*. *Symposium of the Society for General Microbiology*, **23**, 375–405.

Savage, E. J., Clayton, C. W., Hunter, J. A., Brenneman, J. A., Laviola, C. & Gallegly, M. E. (1968). Homothallism, heterothallism and interspecific hybridisation in the genus *Phytophthora*. *Phytopathology*, **58**, 1004–1021.

Savile, D. B. O. (1965). Spore discharge in Basidiomycetes: a unified theory. *Science, N.Y.*, **147**, 165–166.

Savile, D. B. O. (1976). Evolution of the rust fungi (Uredinales) as reflected by their ecological problems. *Evolutionary Biology*, **9**, 137–207.

Savile, D. B. O. & Hayhoe, H. N. (1978). The potential effect of drop size on efficiency of splash cup and springboard dispersal devices. *Canadian Journal of Botany*, **56**, 127–128.

Savory, J. G. (1954). Breakdown of timber by Ascomycetes and Fungi Imperfecti. *Annals of Applied Biology*, **41**, 336–347.

Savory, J. G. & Pinion, L. C. (1958). Chemical aspects of decay of beech wood by *Chaetomium globosum*. *Holzforschung*, **12**, 99–103.

Scheetz, R. W. (1972). The ultrastructure of *Ceratiomyxa fruticulosa*. *Mycologia*, **64**, 38–54.

Schenck, S., Kendrick, W. B. & Pramer, D. (1977). A new nematode-trapping hyphomycete and a re-evaluation of *Dactylaria* and *Arthrobotrys*. *Canadian Journal of Botany*, **55**, 977–985.

Schipper, M. A. A. (1978). On certain species of *Mucor*, with a key to all accepted species. *Studies in Mycology*, **17**, 1–52. Baarn: Centraalbureau voor Schimmelcultures.

Schipper, M. A. A., Samson, R. A. & Stalpers, J. A. (1975). Zygospore ornamentation in the genera *Mucor* and *Zygorhynchus*. *Persoonia*, **8**, 321–328.

Schnathorst, W. C. (1965). Environmental relationships in the powdery mildews. *Annual Review of Phytopathology*, **3**, 343–366.

Schneider, A. (1956). Sur les asques de *Taphrina* (Hémiascomycètes). *Compte Rendu Hebdomadaire des Séances de l'Académie des Sciences, Paris*, **243**, 2139–2142.

Schoknecht, J. D. (1975). Structure of the ascus apex and ascospore dispersal mechanisms in *Sclerotinia tuberosa*. *Transactions of the British Mycological Society*, **64**, 358–362.

Schol-Schwartz, M. B. (1959). The genus *Epicoccum* Link. *Transactions of the British Mycological Society*, **42**, 149–173.

Schremmer, F. (1963). Wechselbeziehungen zwischen Pilzen und Insecten. Beobachtungen an der Stinkmorchel, *Phallus impudicus* L. ex Pers. *Osterreichische Botanische Zeitschrift*, **110**, 380–340.

Schuster, F. (1964). Electron microscope observations on spore formation in the true slime mold *Didymium nigripes*. *Journal of Protozoology*, **11**, 207–216.

Schwab-Stey, H. & Schwab, D. (1973). Über die Feinstruktur von *Labyrinthula coenocystis* Schmoller nach Gefrieratzung. *Protoplasma*, **76**, 455–464.

Schwalb, M. & Roth, R. (1970). Axenic growth and development of the cellular slime mould, *Dictyostelium discoideum*. *Journal of General Microbiology*, **60**, 283–286.

Schweizer, G. (1947). Über die Kultur von *Empusa muscae* Cohn und anderen Entomophthoraceen auf kalt sterilisierten nährboden. *Planta*, **35**, 132–176.

Scott, D. B. & Stolk, A. C. (1967). Studies on the genus *Eupenicillium* Ludwig. II. Perfect states of some penicillia. *Antonie van Leeuwenhoek*, **33**, 297–314.

Scott, G. D. (1956). Further investigation of some lichens for fixation of nitrogen. *New Phytologist*, **55**, 111–119.

Scott, K. J. & Maclean, D. J. (1969). Culturing of rust fungi. *Annual Review of Phytopathology*, **7**, 123–146.

Scott, W. W. (1956). A new species of *Aphanomyces*, and its significance in the taxonomy of the watermolds. *Virginia Journal of Science*, N.s., **7**, 170–175.

Scott, W. W. (1961). A monograph of the genus *Aphanomyces*. *Virginia Agricultural Experimental Station Technical Bulletin*, **151**, 95 pp.

Scott, W. W. & O'Bier, A. H. (1962). Aquatic fungi associated with diseased fish and their eggs. *Progressive Fish Culturist*, **24**, 3–15.

Scurfield, G. (1967). Fine structure of the cell walls of *Polyporus myllitae* Cke. et Mass. *Journal of the Linnean Society (Botany)*, **60**, 159–166.

Sedlar, L., Dreyfuss, M. & Müller, E. (1972). Kompatibilitätsverhältnisse in *Chaetomium*. I. Vorkommen von Homo- und Heterothallie in Arten und Stämmen. *Archiv für Mikrobiologie*, **83**, 172–178.

Seth, H. K. (1967). Studies on the genus *Chaetomium*. I. Heterothallism. *Mycologia*, **59**, 580–584.

Seth, H. K. (1972). A monograph of the genus *Chaetomium*. *Beihefte Nova Hedwigia*, **37**, 1–134.

Seymour, R. L. (1970). The genus *Saprolegnia*. *Nova Hedwigia*, 19, 1–124.

Seymour, R. L. & Johnson, T. W. (1973). Saprolegniaceae: a keratinophilic *Aphanomyces* from soil. *Mycologia*, **65**, 1312–1318.

Shaffer, R. L. (1975). The major groups of Basidiomycetes. *Mycologia*, **67**, 1–18.

Shanor, L. (1936). The production of mature perithecia of *Cordyceps militaris* (Linn.) Link. in laboratory culture. *Journal of the Elisha Mitchell Scientific Society*, **52**, 99–103.

Shantz, H. L. & Piemeisel, R. L. (1917). Fungus fairy rings in Eastern Colorado and their effects on vegetation. *Journal of Agricultural Research,* **11**, 191–245.

Sharma, R. & Cammack, R. H. (1976). Spore germination and taxonomy of *Synchytrium endobioticum* and *S. succisae*. *Transactions of the British Mycological Society*, **66**, 137–147.

Shatkin, A. J. & Tatum, E. L. (1959). Electron microscopy of *Neurospora crassa* mycelia. *Journal of Biophysical and Biochemical Cytology*, **6**, 423–426.

Shatla, M. N., Yang, C. Y. & Mitchell, J. E. (1966). Cytological and fine structure studies of *Aphanomyces euteiches*. *Phytopathology*, **56**, 923–928.

Shaw, D. E. (1953). Cytology of *Septoria* and *Selenophoma* spores. *Proceedings of the Linnean Society of New South Wales*, **78**, 122–130.

Shepherd, A. M. (1955). Formation of the infection bulb in *Arthrobotrys oligospora* Fresenius. *Nature, London*, **175**, 475.

Sherman, F. & Lawrence, C. W. (1974). *Saccharomyces*. In *Handbook of Genetics*, **I**, 359–393. Editor: R. C. King. New York and London: Plenum Press.

Sherwood, W. A. (1969). Sexual reactions between clonal subcultures of a strain of *Dictyuchus monosporus*. *Mycologia*, **61**, 251–263.

Sherwood, W. A. (1971). Some observations on the sexual behavior of the progeny of six isolates of *Dictyuchus monosporus*. *Mycologia*, **63**, 22–30.

Shoemaker, R. A. (1955). Biology, cytology and taxonomy of *Cochliobolus sativus*. *Canadian Journal of Botany*, **33**, 562–576.

Shropshire, W. (1963). Photoresponses of the fungus *Phycomyces*. *Physiological Reviews*, **43**, 38–67.

Siegel, M. (1958). Zur ökologie des Anemonbecherlings. *Zeitschrift für Pilzkunde*, **24**, 18–19.

Sietsma, J. H., Child, J. J., Nesbitt, L. R. & Haskins, R. H. (1975). Chemistry and ultrastructure of the hyphal walls of *Pythium acanthicum*. *Journal of General Microbiology*, **86**, 29–38.

Silver, J. C. & Horgen, P. A. (1974). Hormonal regulation of presumptive mRNA in the fungus *Achlya ambisexualis*. *Nature, London*, **249**, 252–254.

Silver-Dowding, E. (1955). *Endogone* in Canadian rodents. *Mycologia*, **47**, 51–57.

Simmons, E. G. (1969). Perfect states of *Stemphylium*. *Mycologia*, **61**, 1–26.

636   References

Singer, R. (1961). *Mushrooms and Truffles. Botany, Cultivation and Utilization*. 272 pp. London: Leonard Hill.

Singer, R. (1970). *Armillariella mellea. Schweizerische Zeitschrift für Pilzkunde*, **48**, 65–69.

Singer, R. (1975). *The Agaricales in Modern Taxonomy*. 3rd edition. Vaduz: J. Cramer.

Singer, R. & Smith, A. H. (1960). Studies on Secotiaceous fungi. XI. The Astrogastraceous series. *Memoirs of the Torrey Botanical Club*, **21**, 1–112.

Singleton, J. R. (1953). Chromosome morphology and the chromosome cycle in the ascus of *Neurospora crassa*. *American Journal of Botany*, **40**, 124–144.

Sinha, U. & Ashworth, J. M. (1969). Evidence for the existence of elements of a para-sexual cycle in the cellular slime mould *Dictyostelium discoideum*. *Proceedings of the Royal Society of London, Series B*, **173**, 531–540.

Siva, B. & Shaw, M. (1969). Nuclei in haustoria of *Phytophthora infestans.Canadian Journal of Botany*, **47**, 1585–1587.

Sjöwall, von M. (1945). Studien über Sexualität, Vererbung an Zytologie bei einiger diozischen Mucoraceen. *Akademisk afhandling*, Lund, 1–97.

Sjöwall, von M. (1946). Über die zytologischen Verhaltnisse in den Keimschlauchen von *Phycomyces blakesleeanus* und *Rhizopus nigricans*. *Botaniska Notiser*, 1946, 331–334.

Skucas, G. P. (1966). Structure and composition of zoosporangial discharge papillae in the fungus *Allomyces*. *American Journal of Botany*, **53**, 1006–1011.

Skucas, G. P. (1967). Structure and composition of the resistant sporangial wall in the fungus *Allomyces*. *American Journal of Botany*, **54**, 1152–1158.

Skucas, G. P. (1968). Changes in wall and internal structure of *Allomyces*-resistant sporangia during germination. *American Journal of Botany*, **55**, 291–295.

Skupienski, F.-X. (1918). Sur la sexualité chez une espèce de Myxomycète Acrasiée, *Dictyostelium mucoroides*. *Compte Rendu Hebdomadaire des Séances de l'Académie des Sciences, Paris*, **167**, 960–962.

Slayman, C. L. & Slayman, C. W. (1962). Measurement of membrane potentials in *Neurospora*. *Science, N.Y.*, **136**, 876–877.

Slesinski, R. S. & Ellingboe, A. H. (1969). The genetic control of primary infection of wheat by *Erysiphe graminis* f.sp. *tritici*. *Phytopathology*, **59**, 1833–1837.

Slooff, W. C. (1970). *Schizosaccharomyces* Lindner. In *The Yeasts: A Taxonomic Study*, 733–755. Editor: J. L. Lodder. Amsterdam and London: North-Holland Publishing Company.

Smith, A. H. (1966). The hyphal structure of the basidiocarp. In *The Fungi: An Advanced Treatise*, **II**, 151–177. Editors: G. C. Ainsworth & A. S. Sussman. New York and London: Academic Press.

Smith, A. H. (1973). Agaricales and related Secotioid Gasteromycetes. In *The Fungi: An Advanced Treatise*, **IVB**, 421–450. Editors: G. C. Ainsworth, F. K. Sparrow & A. S. Sussman. New York and London: Academic Press.

Smith, A. L. (1918). *A Monograph of British Lichens*. Part I, 519 pp. British Museum.

Smith, A. L. (1921). *A Handbook of the British Lichens*. 158 pp. British Museum.

Smith, A. L. (1926). *A Monograph of British Lichens*. Part II, 447 pp. British Museum.

Smith, D. C. (1961). The physiology of *Peltigera polydactyla* (Neck.) Hoffm. *Lichenologist*, **1**, 209–226.

Smith, D. C. (1962). The biology of lichen thalli. *Biological Reviews*, **37**, 537–570.

Smith, D. C. (1963). Experimental studies of lichen physiology. In *Symbiotic Associations, 13th Symposium of the Society for General Microbiology*, 31–50. Editors: P. S. Nutman & B. Mosse. Cambridge University Press.

Smith, D. C. (1973). The lichen symbiosis. *Oxford Biology Reader*, **42**. 16 pp. Oxford University Press.

Smith, D. C. (1978). What can lichens teach us about REAL fungi? *Mycologia*, **70**, 915–934.

Smith, D. C., Muscatine, L. & Lewis, D. (1969). Carbohydrate movement from autotrophs to heterotrophs in parasitic and mutualistic symbiosis. *Biological Reviews*, **44**, 17–90.

Smith, G. (1969). *An Introduction to Industrial Mycology*. 6th edition, 390 pp. London: Edward Arnold.

Snider, P. J. (1959). Stages of development in rhizomorphic thalli of *Armillaria mellea*. *Mycologia*, **51**, 693–707.

Snider, P. J. (1965). Incompatibility and nuclear migration. In *Incompatibility in Fungi*, 52–70. Editors: K. Esser & J. R. Raper. Berlin: Springer.

Snider, P. J. (1968). Nuclear movements in *Schizophyllum*. In *Aspects of Cell Motility*, *Symposium of the Society for Experimental Biology*, **22**, 261–283. Editor: P. L. Miller. Cambridge University Press.

Somers, E. & Horsfall, J. G. (1966). The water content of powdery mildew conidia. *Phytopathology*, **56**, 1031–1035.

Sommer, N. F. (1961). Production by *Taphrina deformans* of substances stimulating cell elongation and division. *Physiologia Plantarum*, **14**, 460–469.

Sortkjaer, O. & Allermann, K. (1972). Rhizomorph formation in fungi. I. Stimulation by ethanol and acetate and inhibition by disulfiram of growth and rhizomorph formation in *Armillaria mellea*. *Physiologia Plantarum*, **26**, 376–380.

Spaeth, H. (1957). Über *Sclerotinia tuberosa*. *Schweizerische Zeitschrift für Pilzkunde*, **23**, 20–21.

Sparrow, F. K. (1939). *Monoblepharis taylori*, a remarkable soil fungus from Trinidad. *Mycologia*, **31**, 737–738.

Sparrow, F. K. (1957). A further contribution to the Phycomycete flora of Great Britain. *Transactions of the British Mycological Society*, **40**, 523–535.

Sparrow, F. K. (1958). Interrelationships and phylogeny of the aquatic Phycomycetes. *Mycologia*, **50**, 797–813.

Sparrow, F. K. (1960). *Aquatic Phycomycetes*, 2nd edition, 1187 pp. Ann Arbor: The University of Michigan Press.

Sparrow, F. K. (1973a). Mastigomycotina (Zoosporic Fungi). In *The Fungi: An Advanced Treatise*, **IVB**, 61–73. Editors: G. C. Ainsworth, F. K. Sparrow & A. S. Sussman. New York and London: Academic Press.

Sparrow, F. K. (1973b). Chytridiomycetes. Hyphochytridiomycetes. In *The Fungi: An Advanced Treatise*, **IVB**, 85–110. Editors: G. C. Ainsworth, F. K. Sparrow & A. S. Sussman. New York and London: Academic Press.

Sparrow, F. K. (1973c). Lagenidiales. In *The Fungi: An Advanced Treatise*, **IVB**, 159–163. Editors: G. C. Ainsworth, F. K. Sparrow & A. S. Sussman. New York and London: Academic Press.

Sparrow, F. K. (1973d). The type of *Chytridium olla* A. Braun. *Taxon*, **22**, 583–586.

Spencer, D. M. (ed.) (1978). *The Powdery Mildews*. 565 pp. London, New York, San Francisco: Academic Press.

Springer, M. E. (1945). A morphologic study of the genus *Monoblepharella*. *American Journal of Botany*, **32**, 259–269.

Srinivasan, M. C., Narasimhan, M. J. & Thirumalachar, M. J. (1964). Artificial culture of *Entomophthora muscae* and morphological aspects for differentiation of the genera *Entomophthora* and *Conidiobolus*. *Mycologia*, **56**, 683–691.

Srinivasan, M. C. & Thirumalachar, M. J. (1964). On the identity of *Entomophthora coronata*. *Mycopathologia et Mycologia Applicata*, **24**, 294–296.

Stakman, E. C. & Christensen, C. M. (1946). Aerobiology in relation to plant disease. *Botanical Review*, **12**, 205–253.

Stakman, E. C. & Harrar, J. G. (1957). *Principles of Plant Pathology.* 581 pp. New York: Ronald Press Company.

Stakman, E. C. & Levine, M. N. (1922). The determination of biologic forms of *Puccinia graminis* on *Triticum* spp. *Minnesota University Agricultural Experiment Station Technical Bulletin*, **8**.

Stakman, E. C., Levine, M. N. & Cotter, R. U. (1930). Origin of physiologic forms of *Puccinia graminis* through hybridization and mutation. *Scientific Agriculture*, **10**, 707–720.

Stamberg, J. & Koltin, Y. (1973). The organization of the incompatibility factors in higher fungi: the effects of structure and symmetry on breeding. *Heredity*, **30**, 15–26.

Stanbridge, B., Gay, J. L. & Wood, R. K. S. (1971). Gross and fine structural changes in *Erysiphe graminis* and barley before and during infection. In *Ecology of Leaf Surface Microorganisms*, 367–379. Editors: T. F. Preece & C. H. Dickinson. New York and London: Academic Press.

Stanghellini, M. E. & Hancock, J. G. (1971a). The sporangium of *Pythium ultimum* as a survival structure in soil. *Phytopathology*, **61**, 157–164.

Stanghellini, M. E. & Hancock, J. G. (1971b). Radial extent of the bean spermosphere and its relation to the behaviour of *Pythium ultimum*. *Phytopathology*, **61**, 165–168.

Stanier, R. Y. (1942). The culture and nutrient requirements of a chytridiaceous fungus. *Journal of Bacteriology*, **43**, 499–520.

Steele, S. D. & Fraser, T. W. (1973a). Ultrastructural changes during germination of *Geotrichum candidum* arthrospores. *Canadian Journal of Microbiology*, **19**, 1031–1034.

Steele, S. D. & Fraser, T. W. (1973b). The ultrastructure of *Geotrichum candidum* hyphae. *Canadian Journal of Microbiology*, **19**, 1507–1512.

Stephenson, L. W. & Erwin, D. C. (1972). Encirclement of the oogonial stalk by the amphigynous antheridium in *Phytophthora capsici*. *Canadian Journal of Botany*, **50**, 2439–2441.

Stey, H. (1968). Nachweis eines bisher unbekannten Organells bei *Labyrinthula*. *Zeitschrift für Naturforschung*, **23**, 567–568.

Stiegel, S. (1952). Über Erregungsvorgänge bei der Einwirkung von photischen und mechanischen Reizen auf *Coprinus* Fruchtkörper. *Planta*, **40**, 301–312.

Stolk, A. C. (1968). Studies on the genus *Eupenicillium* Ludwig. III. Four new species of *Eupenicillium*. *Antonie van Leeuwenhoek*, **34**, 37–53.

Stolk, A. C. & Samson, R. A. (1972). Studies on *Talaromyces* and related genera. I. *Hamigera* gen.nov. and *Byssochlamys* Westling. *Persoonia*, **6**, 341–357.

Storck, R. (1966). Nucleotide composition of nucleic acids of fungi. II. Deoxyribonucleic acids. *Journal of Bacteriology*, **91**, 227–230.

Storck, R., Alexopoulos, C. J. & Phaff, H. J. (1969). Nucleotide composition of deoxyribonucleic acid of some species of *Cryptococcus, Rhodotorula* and *Sporobolomyces*. *Journal of Bacteriology*, **98**, 1069–1072.

Stosch, H. A. von., Vanzul-Pischinger, M. & Dersch, G. (1964). Nuclear phase alternance in the myxomycete *Physarum polycephalum*. *Abstracts, Xth International Botanical Congress, Edinburgh*, 481–482.

Strickmann, E. & Chadefaud, M. (1961). Recherches sur les asques et les périthèces des *Nectria* et réflexions sur l'évolution des Ascomycètes. *Revue Générale de Botanique*, **68**, 725–770.

Stuart, M. R. & Fuller, H. T. (1968). Mycological aspects of diseased Atlantic salmon. *Nature, London*, **217**, 90–92.

Suberkropp, K. & Klug, M. J. (1977). Extracellular hydrolytic capabilities of aquatic Hyphomycetes on leaf litter. *Abstracts, Second International Mycological Congress, Tampa, Florida, U.S.A.*, 639.

Suberkropp, K. & Klug, M. J. (1980). The degradation of leaf litter by aquatic hyphomycetes. In *Fungal Ecology*. Editors: D. T. Wicklow & G. C. Carroll. New York: Marcel Dekker Inc.

Subramanian, C. V. (1971). The phialide. In *Taxonomy of Fungi Imperfecti*, 92–119. Editor: W. B. Kendrick. University of Toronto Press.

Subramanian, C. V. (1972). The perfect states of *Aspergillus*. *Current Science*, **41**, 755–761.

Suminoe, K. & Dukmo, H. (1963). The life cycles of *Schizosaccharomyces* with reference to the mode of conjugation and ascospore formation. *Journal of General and Applied Microbiology, Tokyo*, **9**, 243–247.

Sun, N. C. & Bowen, C. C. (1972). Ultrastructural studies of nuclear division in *Basidiobolus ranarum* Eidam. *Caryologia*, **25**, 471–494.

Sussman, A. S. (1957). Physiological and genetic adaptability in the fungi. *Mycologia*, **49**, 29–43.

Sussman, A. S. (1966). Types of dormancy as represented by conidia and ascospores of *Neurospora*. In *The Fungus Spore*, 235–256. Editor: M. F. Madelin. Colston Papers No. 18. London: Butterworths.

Sussman, A. S. (1968). Longevity and survivability of fungi. In *The Fungi: An Advanced Treatise*, **III**, 447–486. Editors: G. C. Ainsworth & A. S. Sussman. New York and London: Academic Press.

Sussman, A. S. & Halvorson, H. O. (1966). *Spores: Their Dormancy and Germination*. 354 pp. New York and London: Harper and Row.

Sussman, M. & Sussman, R. R. (1962). Ploidal inheritance in *Dictyostelium discoideum*: stable haploid, stable diploid and metastable strains. *Journal of General Microbiology*, **28**, 417–429.

Sussman, R. R. & Sussman, M. (1967). Cultivation of *Dictyostelium discoideum* in axenic medium. *Biochemical and Biophysical Research Communications*, **29**, 53–55.

Sutton, B. C. (1964). *Phoma* and related genera. *Transactions of the British Mycological Society*, **47**, 497–509.

Sutton, B. C. (1973). Coelomycetes. In *The Fungi: An Advanced Treatise*, **IVA**, 513–582. Editors: G. C. Ainsworth, F. K. Sparrow & A. S. Sussman. New York and London: Academic Press.

Swift, M. J. (1968). Inhibition of rhizomorph development by *Armillaria mellea* in Rhodesian soils. *Transactions of the British Mycological Society*, **51**, 241–247.

Swinbank, P., Taggart, J. & Hutchinson, S. A. (1964). The measurement of electrostatic charges on spores of *Merulius lacrymans* (Wulf.) Fr. *Annals of Botany, London*, N.S., **28**, 239–249.

Syers, J. K. & Iskandar, I. K. (1973). Pedogenetic significance of lichens. In *The Lichens*, 225–248. Editors: V. Ahmadjian & M. E. Hale. New York and London: Academic Press.

Syrop, M. (1975a). Leaf curl disease of almond caused by *Taphrina deformans* (Berk.) Tul. I. A light microscope study of the host/parasite relationship. *Protoplasma*, **85**, 39–56.

Syrop, M. (1975b). Leaf curl disease of almond caused by *Taphrina deformans* (Berk.) Tul. II. An electron microscope study of the host/parasite relationship. *Protoplasma*, **85**, 57–69.

Syrop, M. J. & Beckett, A. (1972). The origin of ascospore-delimiting membranes in *Taphrina deformans*. *Archiv für Mikrobiologie*, **86**, 185–191.

Szaniszlo, P. J. (1965). A study of the effect of light and temperature on the formation of oogonia and oospheres in *Saprolegnia diclina*. *Journal of the Elisha Mitchell Scientific Society*, **81**, 10–15.

Sziráki, I., Balázs, E. & Király, Z. (1975). Increased levels of cytokinin and indole-acetic acid in peach leaves infected with *Taphrina deformans*. *Physiological Plant Pathology*, **5**, 45–50.

Talbot, P. H. B. (1973a). Towards uniformity in basidial terminology. *Transactions of the British Mycological Society*, **61**, 497–512.

Talbot, P. H. B. (1973b). Aphyllophorales I: General characteristics; Thelephoroid and cupuloid families. In *The Fungi: An Advanced Treatise*, **IVB**, 327–349. Editors: G. C. Ainsworth, F. K. Sparrow & A. S. Sussman. New York and London: Academic Press.

Tanaka, K. (1970). Mitosis in the fungus *Basidiobolus ranarum* as revealed by electron microscopy. *Protoplasma*, **70**, 423–440.

Tansey, M. R. (1972). Effect of temperature on growth rate and development of the thermophilic fungus *Chaetomium thermophile*. *Mycologia*, **64**, 1290–1299.

Tarr, S. A. J. (1972). *Principles of Plant Pathology*. 632 pp. London: Macmillan.

Tatum, E. L. (1946). *Neurospora* as a biochemical tool. *Federation Proceedings, Federation of American Societies for Experimental Biology*, **5**, 362–365.

Teixeira, A. R. (1962). The taxonomy of the Polyporaceae. *Biological Reviews*, **37**, 51–81.

Teixeira, A. R. & Furtado, J. S. (1963). Anatomical studies on *Amauroderma regulicolor* (Berk. ex Cke). Murrill. *Rickia, Archivos de Botanica do Estado de São Paulo*, **2**, 17–23.

Temmink, J. H. M. (1971). An ultrastructural study of *Olpidium brassicae* and its transmission of Tobacco necrosis virus. *Mededelingen van de Landbouwhoogeschool te Wageningen*, **71**, 135 pp.

Temmink, J. H. M. & Campbell, R. N. (1968). The ultrastructure of *Olpidium brassicae*. I. Formation of sporangia. *Canadian Journal of Botany*, **46**, 951–956.

Temmink, J. H. M. & Campbell, R. N. (1969a). The ultrastructure of *Olpidium brassicae*. II. Zoospores. *Canadian Journal of Botany*, **47**, 227–231.

Temmink, J. H. M. & Campbell, R. N. (1969b). The ultrastructure of *Olpidium brassicae*. III. infection of host roots. *Canadian Journal of Botany*, **47**, 421–424.

Temmink, J. H. M., Campbell, R. N. & Smith, P. R. (1970). Specificity and site of *in vitro* acquisition of Tobacco necrosis virus by zoospores of *Olpidium brassicae*. *Journal of General Virology*, **9**, 201–213.

Teter, H. E. (1944). Isogamous sexuality in a new strain of *Allomyces*. *Mycologia*, **36**, 194–210.

Thaxter, R. (1888). The Entomophthoreae of the United States. *Memoirs of the Boston Society for Natural History*, **4**, 133–201.

Therrien, C. D. (1966). Microspectrophotometric measurement of nuclear deoxyribonucleic acid content in two Myxomycetes. *Canadian Journal of Botany*, **44**, 1667–1675.

Thielke, C. (1972). Die Dolipore der Basidiomyceten. *Archiv für Mikrobiologie*, **82**, 31–37.

Thielke, C. (1976). Intranucleäre Meiose bei *Agaricus bisporus*. *Zeitschrift für Pilzkunde*, **42**, 57–66.

Thirumalachar, M. J. & Dickson, J. G. (1949). Chlamydospore germination, nuclear cycle and artificial culture of *Urocystis agropyri* on red top. *Phytopathology*, **39**, 333–339.

Thomas, D. des S. & Mullins, J. T. (1969). Cellulose induction and wall extension in the water mold *Achlya ambisexualis*. *Physiologia Plantarum*, **22**, 347–353.

Tiffney, W. N. (1939a). The host range of *Saprolegnia parasitica*. *Mycologia*, **31**, 310–321.

Tiffney, W. N. (1939b). The identity of certain species of the Saprolegniaceae parasitic to fish. *Journal of the Elisha Mitchell Scientific Society*, **55**, 134–151.

Tingle, M., Singh Klar, A. J., Henry, S. A. & Halvorson, H. O. (1973). Ascospore formation in yeast. *Symposium of the Society for General Microbiology*, **23**, 209–243.

Tomiyama, S., Sakuma, T., Ishizaka, N., Sato, N., Katsui, N., Takasugi, M. &

Masamune, T. (1968). A new antifungal substance isolated from resistant potato tuber tissue infected by pathogens. *Phytopathology*, **58**, 115–116.

Tomlinson, J. A. (1958a). Crook root of watercress. II. The control of the disease with zinc-fritted glass and the mechanism of its action. *Annals of Applied Biology*, **46**, 608–621.

Tomlinson, J. A. (1958b). Crook root of watercress. III. The causal organism *Spongospora subterranea* (Wallr.) Lagerh. f.sp. *nasturtii* f.sp.nov. *Transactions of the British Mycological Society*, **41**, 491–498.

Tommerup, I. C. & Broadbent, D. (1974). Nuclear fusion, meiosis and the origin of dikaryotic hyphae in *Armillariella mellea*. *Archiv für Mikrobiologie*, **103**, 279–282.

Tommerup, I. C. & Ingram, D. S. (1971). The life cycle of *Plasmodiophora brassicae* Woron. in *Brassica* tissue cultures and in intact roots. *New Phytologist*, **70**, 327–332.

Tommerup, I. C., Ingram, D. S. & Sargent, J. A. (1974). Oospores of *Bremia lactucae*. *Transactions of the British Mycological Society*, **62**, 145–150.

Townsend, B. B. (1954). Morphology and development of fungal rhizomorphs. *Transactions of the British Mycological Society*, **37**, 222–233.

Townsend, B. B. & Willetts, H. J. (1954). The development of sclerotia of certain fungi. *Transactions of the British Mycological Society*, **37**, 213–221.

Trappe, J. M. & Maser, C. (1976). Germination of spores of *Glomus macrocarpus* (Endogonaceae) after passage through a rodent digestive tract. *Mycologia*, **68**, 433–436.

Trevethick, J. & Cooke, R. C. (1973). Water relations of some *Sclerotinia* and *Sclerotium* species. *Transactions of the British Mycological Society*, **60**, 555–558.

Trinci, A. P. J., Peat, A. & Banbury, G. H. (1968). Fine structure of phialide and conidiophore development in *Aspergillus giganteus* "Wehmer." *Annals of Botany, London,* N.S., **32**, 241–249.

Trione, E. J. (1972). Isolation of *Tilletia caries* from infected wheat plants. *Phytopathology*, **62**, 1096–1097.

Tsao, P. H. (1970). Selective media for isolation of pathogenic fungi. *Annual Review of Microbiology*, **8**, 157–186.

Tschierpe, H. J. (1959). Die Bedeutung des Kohlendioxyds für den Kulturchampignon. *Gartenbauwissenschaft*, **24**(1), 18–75.

Tschierpe, H. J. & Sinden, J. W. (1964). Weitere Untersuchungen über die Bedeutung von Kohlendioxyd für die Fruchtifikation des Kulturchampignons, *Agaricus campestris* var. *bisporus* (L) Lge. *Archiv für Mikrobiologie*, **49**, 405–425.

Tubaki, K. (1958). Studies on the Japanese Hyphomycetes. V. Leaf and stem group with a discussion of the classification of Hyphomycetes and their perfect stages. *Journal, Hattori Botanical Laboratory*, **20**, 142–244.

Tubaki, K. (1966). Sporulating structures in Fungi Imperfecti. In *The Fungi: An Advanced Treatise*, **II**, 113–131. Editors: G. C. Ainsworth & A. S. Sussman. New York and London: Academic Press.

Tubaki, K. (1975a). Notes on the Japanese Hyphomycetes. VI. *Candelabrum* and *Beverwykella* gen. nov. *Transactions of the Mycological Society of Japan*, **16**, 132–140.

Tubaki, K. (1975b). Notes on the Japanese Hyphomycetes. VII. *Cancellidium*, a new Hyphomycetes genus. *Transactions of the Mycological Society of Japan*, **16**, 357–360.

Tucker, C. M. (1931). Taxonomy of the genus *Phytophthora* de Bary. *Research Bulletin, Missouri Agricultural Experiment Station*, **153**, 1–208.

Tuominen, Y. & Jaakkola, T. (1973). Absorption and accumulation of mineral elements and radioactive nuclides. In *The Lichens*, 185–223. Editors: V. Ahmadjian & M. E. Hale. New York and London: Academic Press.

Turian, G. (1976). Spores in Ascomycetes, their controlled differentiation. In *The*

Fungal Spore: Form and Function, 715–788. Editors: D. J. Weber & W. M. Hess. New York and London: John Wiley and Sons.

Tveit, M. (1953). Control of a seed-borne disease by *Chaetomium cochliodes* Pall., under natural conditions. *Nature, London*, **172**, 39.

Tveit, M. (1956). Isolation of a chetomin-like substance from oat seedlings infested with *Chaetomium cochliodes*. *Acta Agriculturae Scandinavica*, **6**, 13–16.

Tveit, M. & Moore, M. B. (1954). Isolates of *Chaetomium* that protect oats from *Helminthosporium victoriae*. *Phytopathology*, **44**, 686–689.

Tveit, M. & Wood, R. K. S. (1955). The control of *Fusarium* blight in oat seedlings with antagonistic species of *Chaetomium*. *Annals of Applied Biology*, **43**, 538–552.

Tyler, V. E. (1971). Chemotaxonomy in the Basidiomycetes. In *Evolution in the Higher Basidiomycetes*, 29–62. Editor: R. H. Petersen. Knoxville, U.S.A.: University of Tennessee Press.

Tyrell, D. (1977). Occurrence of protoplasts in the natural life cycle of *Entomophthora egressa*. *Experimental Mycology*, **1**, 259–263.

Tyrell, D. & MacLeod, D. M. (1972). A taxonomic proposal regarding *Delacroixia coronata* (Entomophthoraceae). *Journal of Invertebrate Pathology*, **20**, 11–13.

Uecker, F. A. (1976). Development and cytology of *Sordaria humana*. *Mycologia*, **68**, 30–46.

Ullrich, R. C. & Anderson, J. B. (1978). Sex and diploidy in *Armillaria mellea*. *Experimental Mycology*, **2**, 119–129.

Unestam, T. (1965). Studies on the crayfish plague fungus, *Aphanomyces astaci*. I. Some factors affecting growth *in vitro*. *Physiologia Plantarum*, **18**, 483–505.

Unestam, T. & Gleason, F. H. (1968). Comparative physiology of respiration in aquatic fungi. II. The Saprolegniales, especially *Aphanomyces astaci*. *Physiologia Plantarum*, **21**, 573–588.

Upadhyay, H. P. & Kendrick, W. B. (1975). Prodromus for a revision of *Ceratocystis* (Microascales, Ascomycetes) and its conidial states. *Mycologia*, **67**, 798–808.

Vanterpool, T. C. (1959). Oospore germination in *Albugo candida*. *Canadian Journal of Botany*, **37**, 169–172.

Varitchak, B. (1931). Contribution à l'étude du développement des Ascomycètes. *Botaniste*, Série **23**, 1–182.

Vishniac, H. S. & Nigrelli, R. F. (1957). The ability of the Saprolegniaceae to parasitise platyfish. *Zoologica*, **42**, 131–134.

Vriesinga, J. D. & Honma, S. (1971). Inheritance of seedling resistance to clubroot in *Brassica oleracea* L. *Horticultural Science*, **6**, 395–396.

Vujičić, R., Colhoun, J. & Chapman, J. A. (1968). Some observations on the zoospores of *Phytophthora erythroseptica*. *Transactions of the British Mycological Society*, **51**, 125–127.

Waksman, S. A. & Bugie, E. (1944). Chaetomin, a new antibiotic substance produced by *Chaetomium cochliodes*. I. Formation and properties. *Journal of Bacteriology*, **48**, 527–530.

Walker, L. B. (1920). Development of *Cyathus fascicularis, C. striatus*, and *Crucibulum vulgare*. *Botanical Gazette*, **70**, 1–24.

Walker, L. B. (1927). Development and mechanism of discharge in *Sphaerobolus iowensis* n.sp. and *S. stellatus* Tode. *Journal of the Elisha Mitchell Scientific Society*, **42**, 151–178.

Walkey, D. G. A. & Harvey, R. (1966a). Studies of the ballistics of ascospores. *New Phytologist*, **65**, 59–74.

Walkey, D. G. A. & Harvey, R. (1966b). Spore discharge rhythms in pyrenomycetes. I. A survey of the periodicity of spore discharge in pyrenomycetes. *Transactions of the British Mycological Society*, **49**, 583–592.

Walsh, J. H. & Harley, J. L. (1962). Sugar absorption by *Chaetomium globosum*. *New Phytologist*, **61**, 299–313.

Walt, J. P. van der (1970a). *Saccharomyces* Meyen emend. Reess. In *The Yeasts: A Taxonomic Study*, 2nd revised and enlarged edition, 553–718. Editor: J. Lodder. Amsterdam and London: North-Holland Publishing Company.

Walt, J. P. van der (1970b). The perfect and imperfect states of *Sporobolomyces salmonicolor*. *Antonie van Leeuwenhoek*, **36**, 49–55.

Walt, J. P. van der & Pitout, M. J. (1969). Ploidy differences in *Sporobolomyces salmonicolor* and *Candida albicans*. *Antonie van Leeuwenhoek*, **35**, 227–231.

Wang, D. T. (1932). Observations cytologiques sur *l'Ustilago violacea* (Pers.) Fuckel. *Compte Rendu Hebdomadaire des Séances de l'Académie des Sciences, Paris*, **195**, 1417–1418.

Wang, D. T. (1934). Contribution à l'étude des Ustilaginées (Cytologie du parasite et pathologie de la cellule hôte). *Botaniste*, **26**, 540–672.

Warcup, J. H. (1951). Studies on the growth of Basidiomycetes in soil. *Annals of Botany, London*, N.S., **15**, 305–317.

Warcup, J. H. & Talbot, P. H. B. (1962). Ecology and identity of mycelia isolated from soil. *Transactions of the British Mycological Society*, **45**, 495–518.

Ward, E. W. B. & Thorn, G. D. (1965). The isolation of a cyanogenic fraction from the fairy ring fungus *Marasmius oreades* Fr. *Canadian Journal of Botany*, **43**, 997–998.

Ward, S. V. & Manners, J. G. (1974). Environmental effects on the quantity and viability of conidia produced by *Erysiphe graminis*. *Transactions of the British Mycological Society*, **62**, 119–128.

Warren, R. C. & Colhoun, J. (1975). Viability of sporangia of *Phytophthora infestans* in relation to drying. *Transactions of the British Mycological Society*, **64**, 73–78.

Waterhouse, G. M. (1956). The genus *Phytophthora*. Diagnoses (or descriptions) and figures from original papers. *Miscellaneous Publication No. 12, Commonwealth Mycological Institute*. 120 pp.

Waterhouse, G. M. (1963). Key to the species of *Phytophthora* de Bary. *Mycological Paper No. 92, Commonwealth Mycological Institute*. 22 pp.

Waterhouse, G. M. (1967). Key to *Pythium* Pringsheim. *Mycological Paper No. 109, Commonwealth Mycological Institute*. 15 pp.

Waterhouse, G. M. (1968). The genus *Pythium* Pringsheim. Diagnoses (or descriptions) and figures from the original papers. *Mycological Paper No. 110, Commonwealth Mycological Institute*. 71 pp.

Waterhouse, G. M. (1973a). Plasmodiophoromycetes. In *The Fungi: An Advanced Treatise*, **IVB**, 75–82. Editors: G. C. Ainsworth, F. K. Sparrow & A. S. Sussman. New York and London: Academic Press.

Waterhouse, G. M. (1973b). Peronosporales. In *The Fungi: An Advanced Treatise*, **IVB**, 165–183. Editors: G. C. Ainsworth, F. K. Sparrow & A. S. Sussman. New York and London: Academic Press.

Waterhouse, G. M. (1973c). Entomophthorales. In *The Fungi: An Advanced Treatise*, **IVB**, 219–229. Editors: G. C. Ainsworth, F. K. Sparrow & A. S. Sussman. New York and London: Academic Press.

Waterhouse, G. M. (1975). Key to the species of *Entomophthora* Fries. *Bulletin of the British Mycological Society*, **9**, 15–41.

Waters, H., Butler, R. D. & Moore, D. (1975a). Structure of aerial and submerged sclerotia of *Coprinus lagopus*. *New Phytologist*, **74**, 199–205.

Waters, H., Moore, D. & Butler, R. D. (1975b). Morphogenesis of aerial sclerotia of *Coprinus lagopus*. *New Phytologist*, **74**, 207–213.

Watkinson, S. C. (1971). The mechanism of mycelial strand induction in *Serpula lacrimans*: a possible effect of nutrient distribution. *New Phytologist*, **70**, 1079–1088.

Watling, R. (1970). *British Fungus Flora. Agarics and Boleti.* I. *Boletaceae: Gomphidiaceae: Paxillaceae.* 125 pp. Edinburgh: H.M.S.O.

Watson, A. G. & Baker, K. F. (1969). Possible gene centers for resistance in the genus *Brassica* to *Plasmodiophora brassicae. Economic Botany,* **23,** 245–252.

Watson, I. A. (1957). Further studies on the production of new races from mixtures of races of *Puccinia graminis* var. *tritici* on wheat seedlings. *Phytopathology,* **47,** 510–512.

Watson, I. A. & Luig, N. H. (1958). Somatic hybridization in *Puccinia graminis* var. *tritici. Proceedings of the Linnean Society of New South Wales,* **83,** 190–195.

Watson, I. A. & Luig, N. H. (1959). Somatic hybridization between *Puccinia graminis* var. *tritici* and *Puccinia graminis* var. *secalis.* Proceedings of the Linnean Society of New South Wales, **84,** 207–208.

Watson, S. W. & Raper, K. B. (1957). *Labyrinthula minuta* sp. nov. *Journal of General Microbiology,* **17,** 368–377.

Watts, D. J. & Ashworth, J. M. (1970). Growth of myxamoebae of the cellular slime mould *Dictyostelium discoideum* in axenic culture. *Biochemical Journal,* **119,** 171–174.

Webster, J. (1952). Spore projection in the Hyphomycete *Nigrospora sphaerica. New Phytologist,* **51,** 229–235.

Webster, J. (1959a). *Nectria lugdunensis* sp.nov., the perfect stage of *Heliscus lugdunensis. Transactions of the British Mycological Society,* **42,** 322–327.

Webster, J. (1959b). Experiments with spores of aquatic Hyphomycetes. I. Sedimentation, and impaction on smooth surfaces. *Annals of Botany, London,* N.S., **23,** 595–611.

Webster, J. (1961). The *Mollisia* perfect stage of *Anguillospora crassa. Transactions of the British Mycological Society,* **44,** 559–564.

Webster, J. (1964). Culture studies on *Hypocrea* and *Trichoderma.* I. Comparison of perfect and imperfect states of *Hypocrea gelatinosa, H. rufa* and *Hypocrea* sp.1. *Transactions of the British Mycological Society,* **47,** 75–96.

Webster, J. (1965). The perfect state of *Pyricularia aquatica.Transactions of the British Mycological Society,* **48,** 449–452.

Webster, J. (1966). Spore projection in *Epicoccum* and *Arthrinium. Transactions of the British Mycological Society,* **49,** 339–343.

Webster, J. (1979). Cleistocarps of *Phyllactinia* as shuttlecocks. *Transactions of the British Mycological Society,* **72,** 489–490.

Webster, J. & Dennis, C. (1967). The mechanism of sporangial discharge in *Pythium middletonii. New Phytologist,* **66,** 307–313.

Webster, J. & Descals, E. (1979). Perfect states of aquatic Hyphomycetes. In *The Whole Fungus – The Sexual–Asexual Synthesis.* 793 pp. 2 vols. Editor: W. B. Kendrick, Ottawa: National Museums of Canada.

Webster, J., Sanders, P. F. & Descals, C. (1978). Tetraradiate aquatic propagules in two species of *Entomophthora. Transactions of the British Mycological Society,* **70,** 472–479.

Wehmeyer, L. E. (1926). A biologic and phylogenetic study of stromatic Sphaeriales. *American Journal of Botany,* **13,** 575–645.

Wehmeyer, L. E. (1955). Development of the ascostroma in *Pleospora armeriae* of the *Pleospora herbarum* complex. *Mycologia,* **47,** 821–834.

Wehmeyer, L. E. (1961). *A World Monograph of the Genus* Pleospora *and its Segregates.* 451 pp. Ann Arbor: University of Michigan Press.

Wells, K. (1964a). The basidia of *Exidia nucleata.* I. Ultrastructure. *Mycologia,* **56,** 327–341.

Wells, K. (1964b). The basidia of *Exidia nucleata.* II. Development. *American Journal of Botany,* **51,** 360–370.

Wells, K. (1965). Ultrastructural features of developing and mature basidia and basidiospores of *Schizophyllum commune*. *Mycologia*, **57**, 236–261.

Wells, K. (1970). Light and electron microscopic studies of *Ascobolus stercorarius*. I. Nuclear divisions in the ascus. *Mycologia*, **62**, 761–790.

Wells, K. (1972). Light and electron microscopic studies of *Ascobolus stercorarius*. II. Ascus and ascospore ontogeny. *University of California Publications in Botany*, **62**, 1–93.

Wells, K. (1977). Meiotic and mitotic divisions in the Basidiomycotina. In *Mechanisms and Control of Cell Division*, 337–374. Editors: T. L. Rost & E. M. Gifford. Stroudsberg, Pennsylvania: Dowden, Hutchinson & Ross.

Werkman, B. A. & Ende, H. van den (1973). Trisporic acid synthesis in *Blakeslea trispora*. Interactions between *plus* and *minus* mating types. *Archiv für Mikrobiologie*, **90**, 365–374.

Werkman, B. A. & Ende, H. van den (1974). Trisporic acid synthesis in homothallic and heterothallic Mucorales. *Journal of General Microbiology*, **82**, 273–278.

Wessels, J. G. H. (1965). Morphogenesis and biochemical processes in *Schizophyllum commune* Fr. *Wentia*, **13**, 1–113.

Wessels, J. G. H. & Koltin, Y. (1972). R-glucanase activity and susceptibility of hyphal walls to degradation in mutants of *Schizophyllum* with disrupted nuclear migration. *Journal of General Microbiology*, **71**, 471–475.

Wessels, J. G. H., Kreger, D. R., Marchant, R., Regensburg, B. A. & De Vries, O. M. H. (1972). Chemical and morphological characterization of the hyphal wall surface of the basidiomycete *Schizophyllum commune*. *Biochimica et Biophysica Acta*, **273**, 346–358.

Western, J. H. & Cavett, J. J. (1959). The choke disease of cocksfoot (*Dactylis glomerata*) caused by *Epichloe typhina* (Fr.) Tul. *Transactions of the British Mycological Society*, **42**, 298–307.

Whalley, A. J. S. & Greenhalgh, G. N. (1973a). Numerical taxonomy of *Hypoxylon*. I. A comparison of classifications of the cultural and the perfect states. *Transactions of the British Mycological Society*, **61**, 435–454.

Whalley, A. J. S. & Greenhalgh, G. N. (1973b). Numerical taxonomy of *Hypoxylon*. II. A key to the identification of British species of *Hypoxylon*. *Transactions of the British Mycological Society*, **61**, 455–459.

Wheals, A. E. (1970). A homothallic strain of the Myxomycete *Physarum polycephalum*. *Genetics*, **66**, 623–633.

Whetzel, H. H. (1945). A synopsis of the genera and species of the Sclerotiniaceae, a family of stromatic inoperculate Discomycetes. *Mycologia*, **37**, 648–714.

Whetzel, H. H. (1946). The cypericolous and juncicolous species of *Sclerotinia*. *Farlowia*, **2**, 385–437.

Whiffen, A. J. (1944). A discussion of taxonomic criteria in the Chytridiales. *Farlowia*, **1**, 583–597.

Whisler, H. C., Zebold, S. L. & Shemanchuk, J. A. (1975). Life cycle of *Coelomomyces psophorae*. *Proceedings of the National Academy of Sciences of the United States of America*, **72**, 693–696.

White, G. A. & Thorn, G. D. (1975). Structure–activity relationships of carboxamide fungicides and the succinic dehydrogenase complex of *Cryptococcus laurentii* and *Ustilago maydis*. *Pesticide Biochemistry and Physiology*, **5**, 380–395.

Whitehouse, H. L. K. (1949a). Heterothallism and sex in the fungi. *Biological Reviews*, **24**, 411–447.

Whitehouse, H. L. K. (1949b). Multiple-allelomorph heterothallism in the fungi. *New Phytologist*, **48**, 212–244.

Whitehouse, H. L. K. (1951). A survey of heterothallism in the Ustilaginales. *Transactions of the British Mycological Society*, **34**, 340–355.

Whiteside, W. C. (1957). Perithecial initials of *Chaetomium*. *Mycologia*, **49**, 420–425.
Whiteside, W. C. (1961). Morphological studies in the Chaetomiaceae. I. *Mycologia*, **53**, 512–523.
Wickerham, L. J. & Burton, K. A. (1954). A classification of the relationship of *Candida guilliermondii* to other yeasts by a study of their mating types. *Journal of Bacteriology*, **68**, 594–597.
Wickerham, L. J. & Duprat, J. (1945). A remarkable fission yeast, *Schizosaccharomyces versatilis* nov.sp. *Journal of Bacteriology*, **50**, 597–607.
Wickerham, L. H., Lockwood, L. B., Pettijohn, O. G. & Ward, G. E. (1944). Starch hydrolysis and fermentation by the yeast *Endomycopsis fibuliger*. *Journal of Bacteriology*, **48**, 413–427.
Widra, A. & Delamater, E. D. (1955). The cytology of meiosis in *Schizosaccharomyces octosporus*. *American Journal of Botany*, **42**, 423–435.
Wieben, M. (1927). Die Infektion, die Myzelüberwinterung und die Kopulation bei Exoasceen. *Fortschritte auf dem Gebiete der Pflanzen-Krankheiten, Berlin*, **3**, 139–176.
Willetts, H. J. (1969). Structure of the outer surfaces of sclerotia of certain fungi. *Archiv für Mikrobiologie*, **69**, 48–53.
Willetts, H. J. (1971). The survival of fungal sclerotia under adverse environmental conditions. *Biological Reviews, Cambridge*, **46**, 387–407.
Willetts, H. J. & Wong, A. L. (1971). Ontogenetic diversity of sclerotia of *Sclerotinia sclerotiorum* and related species. *Transactions of the British Mycological Society*, **57**, 515–524.
Williams, P. G., Scott, K. J. & Kuhl, J. L. (1966). Vegetative growth of *Puccinia graminis* f.sp. *tritici in vitro*. *Phytopathology*, **56**, 1418–1419.
Williams, P. H. (1966). A cytochemical study of hypertrophy in clubroot of cabbage. *Phytopathology*, **56**, 521–524.
Williams, P. H. & McNabola, S. S. (1967). Fine structure of *Plasmodiophora brassicae* in sporogenesis. *Canadian Journal of Botany*, **45**, 1665–1669.
Williams, P. H. & Yukawa, Y. B. (1967). Ultrastructural studies on the host–parasite relations of *Plasmodiophora brassicae*. *Phytopathology*, **57**, 682–687.
Williams, S. T., Gray, T. R. G. & Hitchen, P. (1965). Heterothallic formation of zygospores in *Mortierella marburgensis*. *Transactions of the British Mycological Society*, **48**, 129–133.
Williams, W. T. & Webster, K. R. (1970). Electron microscopy of the sporangium of *Phytophthora capsici*. *Canadian Journal of Botany*, **48**, 221–227.
Willoughby, L. G. (1956). Studies on soil chytrids. I. *Rhizidium richmondense* sp.nov. and its parasites. *Transactions of the British Mycological Society*, **39**, 125–141.
Willoughby, L. G. (1957). Studies on soil chytrids. II. On *Karlingia dubia* Karling. *Transactions of the British Mycological Society*, **40**, 9–16.
Willoughby, L. G. (1958). Studies on soil chytrids. III. On *Karlingia rosea* Johanson and a multi-operculate chytrid parasitic on *Mucor*. *Transactions of the British Mycological Society*, **41**, 309–319.
Willoughby, L. G. (1962). The fruiting behaviour and nutrition of *Cladochytrium replicatum* Karling. *Annals of Botany, London*, N.S., **26**, 13–36.
Willoughby, L. G. (1968). Atlantic salmon disease fungus. *Nature, London*, **217**, 872–873.
Willoughby, L. G. (1969). Salmon disease in Windermere and the River Leven; the fungal aspect. *Salmon and Trout Magazine*, **186**, 124–130.
Willoughby, L. G. (1970). Mycological aspects of a disease of young perch in Windermere. *Journal of Fish Biology*, **2**, 113–116.
Willoughby, L. G. & Archer, J. F. (1973). The fungal spora of a freshwater stream and its colonization pattern on wood. *Freshwater Biology*, **3**, 219–239.

Willoughby, L. G. & Pickering, A. D. (1977). Viable Saprolegniaceae spores on the epidermis of the salmonid fish *Salmo trutta* and *Salvelinus alpinus*. *Transactions of the British Mycological Society*, **68**, 91–95.

Wilsenach, R. & Kessel, M. (1965). On the function and structure of the septal pore of *Polyporus rugulosus*. *Journal of General Microbiology*, **40**, 397–400.

Wilson, C. M. (1952a). Meiosis in *Allomyces*. *Bulletin of the Torrey Botanical Club*, **79**, 139–160.

Wilson, C. M. (1952b). Sexuality in the Acrasiales. *Proceedings of the National Academy of Sciences of the United States of America*, **38**, 659–662.

Wilson, C. M. (1953). Cytological study of the life cycle of *Dictyostelium*. *American Journal of Botany*, **40**, 714–718.

Wilson, C. M. & Flanagan, P. W. (1968). The life cycle and cytology of *Brachyallomyces*. *Canadian Journal of Botany*, **46**, 1361–1367.

Wilson, C. M. & Ross, I. K. (1957). Further cytological studies in the Acrasiales. *American Journal of Botany*, **44**, 345–350.

Wilson, I. M. (1952). The ascogenous hyphae of *Pyronema confluens*. *Annals of Botany, London*, N.S., **16**, 321–339.

Wilson, J. F., Garnjobst, L. & Tatum, E. L. (1961). Heterokaryon incompatibility in *Neurospora crassa* – micro-injection studies. *American Journal of Botany*, **48**, 299–305.

Wilson, J. G. M. (1976). Immunological aspects of fungal disease in fish. In *Recent Advances in Aquatic Mycology*, 573–601. Editor: E. B. Gareth Jones. London: Elek Science.

Wilson, M. & Henderson, D. M. (1966). *British Rust Fungi*. 384 pp. Cambridge University Press.

Wilson, R. W. & Beneke, E. S. (1966). Basidiospore germination in *Calvatia gigantea*. *Mycologia*, **58**, 328–332.

Winge, Ö. & Lausten, O. (1937). On two types of spore germination, and on genetic segregations in *Saccharomyces*, demonstrated through single spore cultures. *Comptes Rendus des Travaux du Laboratoire Carlsberg*, **22**, 99–116.

Winge, Ö. & Roberts, C. (1949). A gene for diploidization in yeasts. *Comptes Rendus des Travaux du Laboratoire Carlsberg*, **24**, 341–346.

Wogan, G. N. (ed.) (1965). *Mycotoxins in Foodstuffs*. 291 pp. Cambridge, Mass.: M.I.T. Press.

Wolf, F. T. (1974). The cultivation of rust fungi upon artificial media. *Canadian Journal of Botany*, **52**, 767–772.

Wolfe, M. S. (1972). The genetics of barley mildew. *Review of Plant Pathology*, **51**, 507–522.

Wong, A.-L. & Willetts, H. J. (1975). A taxonomic study of *Sclerotinia sclerotiorum* and related species: Mycelial interactions. *Journal of General Microbiology*, **88**, 329–334.

Wood, J. L. (1953). A cytological study of ascus development in *Ascobolus magnificus* Dodge. *Bulletin of the Torrey Botanical Club*, **80**, 1–15.

Woodham-Smith, C. (1962). *The Great Hunger. Ireland 1845–9*. London: Hamish Hamilton.

Woodward, R. C. (1927). Studies on *Podosphaera leucotricha* (Ell. & Ev.) Salm. I. The mode of perennation. *Transactions of the British Mycological Society*, **12**, 173–204.

Wormald, H. (1921). On the occurrence in Britain of the ascigerous stage of a 'brown rot' fungus. *Annals of Botany, London*, **35**, 125–135.

Wormald, H. (1954). The brown rot diseases of fruit trees. *Ministry of Agriculture Technical Bulletin No. 3*. 113 pp.

Woycicki, Z. (1927). Über die Zygotenbildung bei *Basidiobolus ranarum* Eidam. II. *Flora, oder Allgemeine Botanische Zeitung*, **122**, 159–166.

Yaegashi, H. & Udagawa, S. (1978). The taxonomical identity of the perfect state of *Pyricularia grisea* and its allies. *Canadian Journal of Botany*, **56**, 180–183.

Yanagishima, N. (1969). Sexual hormones in yeast. *Planta*, **87**, 110–118.

Yarwood, C. E. (1941). Diurnal cycle of ascus maturation of *Taphrina deformans*. *American Journal of Botany*, **28**, 355–357.

Yarwood, C. E. (1973). Pyrenomycetes: Erysiphales. In *The Fungi: An Advanced Treatise*, **IVA**, 71–86. Editors: G. C. Ainsworth, F. K. Sparrow & A. S. Sussman. New York and London: Academic Press.

Yarwood, C. E. (1978). History and taxonomy of powdery mildews. In *The Powdery Mildews*, 1–37. Editor: D. M. Spencer. London, New York, San Francisco: Academic Press.

Yendol, W. G. & Paschke, J. D. (1965). Pathology of an *Entomophthora* infection in the Eastern subterranean termite, *Reticulitermes flavipes* (Kollar). *Journal of Invertebrate Pathology*, **7**, 414–422.

Yerkes, W. D. & Shaw, G. C. (1959). Taxonomy of the *Peronospora* species on Cruciferae and Chenopodiaceae. *Phytopathology*, **49**, 499–507.

Yoo, B., Calleja, G. B. & Johnson, B. F. (1973). Ultrastructural changes of the fission yeast (*Schizosaccharomyces pombe*) during ascospore formation. *Archiv für Mikrobiologie*, **91**, 1–10.

Youatt, J. (1973a). Chemical nature of discharge papillae in *Allomyces*. *Transactions of the British Mycological Society*, **61**, 179–180.

Youatt, J. (1973b). Sporangium production by *Allomyces* in new chemically defined media. *Transactions of the British Mycological Society*, **61**, 257–263.

Youatt, J., Fleming, R. & Jobling, B. (1971). Differentiation in species of *Allomyces*: the production of sporangia. *Australian Journal of Biological Sciences*, **24**, 1163–1167.

Young, E. L. (1943). Studies on *Labyrinthula*. The etiologic agent of wasting disease of eel grass. *American Journal of Botany*, **30**, 586–593.

Yu, C. C. (1954). The culture and spore germination of *Ascobolus* with emphasis on *A. magnificus*. *American Journal of Botany*, **41**, 21–30.

Yuill, E. (1950). The numbers of nuclei in conidia of Aspergilli. *Transactions of the British Mycological Society*, **33**, 324–331.

Zak, B. (1976). Pure culture synthesis of *Pacific Madrone* ectendomycorrhizae. *Mycologia*, **68**, 362–369.

Zalokar, M. (1959). Growth and differentiation of *Neurospora* hyphae. *American Journal of Botany*, **46**, 602–610.

Zambettakis, C. (1970). Les formes imparfaites des Ustilaginées. *Revue de Mycologie*, **35**, 158–175.

Zambettakis, C. (1973). Recherches sur la germination des téliospores des Ustilaginales. I. Différents modes de germination selon l'espèce et le milieu utilisé. *Bulletin Trimestriel de la Société Mycologique de France*, **89**, 253–275.

Zoberi, M. H. (1961). Take-off of mould spores in relation to wind speed and humidity. *Annals of Botany, London*, N.S., **25**, 53–64.

Zycha, H., Siepmann, R. & Linnemann, G. (1969). Mucorales. Eine Beschreibung aller Gattungen und Arten dieser Pilzgruppe. 355 pp. Lehre: J. Cramer.

Index

Absidia, 202, 210, 213; *A. glauca*, 88, 89, 196, 207, 213; *A. orchidis*, 213; *A. spinosa*, 203, 207, 213
Acaulospora, 235
acervuli, 526, 566, 567
Achlya, 143, 148, 157; *A. ambisexualis*, 155, 157–60; *A. bisexualis*, 157, 159, 160; *A. colorata*, 149, 154, 155; *A. heterosexualis*, 157, 159; *A. klebsiana*, 148
Acrasiales, 10–11, 59
Acrasidae, 8, 10–11
acrasin (cyclic AMP), 13–14, 19
Acrasiomycetes, 7–19; classification, 8
Acrasis rosea, 10
Acrosporium, 290
Acytostelium, 19
adhesoria, 47
aecia, 500; diffuse (cacomata), 515; of *Puccinia*, 505, 506, 507; uredinoid, 513, 514
aecidiosori, 500
aecidiospore, 500
aeciospores, 401, 500, 501; of *Puccinia*, 505, 506, 507
Aessosporon, perfect state of *Sporobolomyces*, 499
aethalia, 33
aflatoxin, carcinogen produced by *Aspergillus flavus*, 309
Agaricaceae, 432–3
Agaricales, 91, 407, 415, 417–19, 430–8; classification, 429–30; fruit bodies, 420–9
Agaricus, 419, 420; *A. bisporus* (cultivated mushroom), 65, 397, 408, 409, 417, 432, (fruit body) 423–8; *A. campestris*, 404, 405, 409, 421, 432; *A. xanthodermus*, 433
agglutinating factor, of *Hansenula*, 278
aggregation, of myxamoebae of *Dictyostelium*, 11, 12, 13, 15
Agrocybe praecox, 408
Agropyron, leaf-stripe smut of, 492
air bubbles, trapped in conidia of aero-aquatic fungi, 539–40, 542
Alatospora acuminata, 527, 528, 529
Albuginaceae, 161, 162, 186–90
Albugo, 161; *A. bliti*, 189; *A. candida* (*Cystopus candidus*), 186, 187, 188; *A. tragopogi*, 186, 190
alder, leaf blister of, 280
Aleuria aurantia, 380, 384
aleuriospores (thalloconidia), 304–5, 333
algae, in lichens, 367
Allomyces, 64, 125, 126; *A. anomalus* (*Brachy-Allomyces*), 126, 132, 133, 134; *A. arbusculus*, *A. macrogynus* (*Eu-Allomyces*), 75, 78–9, 126–9, 133; *A. javanicus* (hybrid), 129, 130; *A. moniliformis*, *A. neo-moniliformis* (*Allomyces cystogenus*), 126, 130–2, 133
Alternaria, conidial state of *Pleospora*, 393, 394; *A. alternata* (*A. tenuis*), 555; *A. brassicae*, 555; *A. brassicicola*, 555, 556; *A. triticina*, 555
alternation of generations, 93, 129
Amanita, 418, 419; *A. excelsa*, 433; *A. fulva*, 432, 433; *A. muscaria* (fly agaric), 432; *A. phalloides* (death cap), 417, 432; *A. rubescens*, 418, 419, 432; *A. vaginata*, 421, 432; *A. verna*, 417, 432; *A. virosa*, 400, 432
Amanitaceae, 432
amanitin, toxic cyclopeptide, 432
Amanitopsis, = *Amanita*, 432
ambrosia beetle, *Endomycopsis* as food for larvae of, 279
amino acids: chemotactic response to, 147; requirement for (*Allomyces*), 126
p-aminobenzoic acid, mutant of *Neurospora* unable to synthesise, 251
ammonia: liberated by *Mucor plumbeus*, stimulates sporangial production in *Pilobolus*, 219; nematocidal, 550; as source of nitrogen, 100
amoebae, fungal parasites of, 236
cAMP (acrasin), 13–14, 19
Anabaena, *Blastocladiella* parasitic on, 134
Anemone: leaf smut of, 484, 492; *A. nemorosa*, *Plasmopara* disease of, 183, 184; *A. tuberosa*, black rot of, 359, 360
aneuploid nuclei, 18
angiocarpy, primary and secondary, 418, 419
Anguillospora, 533, 539; *A. crassa*, 527, 536; *A. furtiva*, 536; *A. longissima*, conidial state of *Massarina* sp., 533, 535, 536; *A.* sp., 527

anisokont flagella, 27, 43

annellides, annellophores, 506, 519–20, 521, 522, 524

annulus: apical ring of ascus in some Pyrenomycetes, 261; remains of partial veil in Agaricales, 419, 429, 432

antheridia: amphigynous and paragynous, 171–2; androgynous, diclinous and monoclinous, 154; of Ascomycetes, 257, 300, 304, 305, 351, 374–5, 378; epigynous, 139; of Oomycetes, 141, 142, 153, 173, 189, (*Pythium*) 167, 168; sex hormones and, in *Achlya*, 157–60

antheridiol, sex hormone of female *Achlya*, 159

antibiotics: intolerance of, 169; production of, 310, 333, 430

Aphanomyces, 143, 150, 152, 155; *A. astaci,* 150; *A. euteiches,* 142, 143, 150; *A. patersonii,* 150

aphids, *Entomophthora* in, 241

Aphyllophorales, 407, 417, 438–60

Apiocrea chrysosperma, perfect state of *Sepedonium,* parasite of *Boletus* and *Paxillus,* 437

Aplanes, 152

aplanetic forms, of Saprolegniaceae, 152

aplanogametes, 117

Aplanopsis, 152

aplanospores, 84, 152

Apodachlya, 140, 156

apogamous life cycle, 39–40

apophysis, 9, 10

Aporpium caryae (previously *Polyporus caryae*), 439, 464

apothecia, 71, 72, 263; of Discomycetes, 364, 368, 369, 371, 373, 374, 377, 379, (*Sclerotinia*) 358, 359, 362, 363

appendages: of ascospores, 324, 325, 334: of basidiospores, (*Nia*), 470; of cleistothecia of Erysiphales, 285, 291, 292, 294, 295, 296, 297, 298; of oidia, 412

apple: brown rot of, 360; canker of, 345; mildew of, 283, 294

appressoria, 176, 228, 229, 288, 538; of *Entomophthora,* 244, 245

arbuscules (haustoria of mycorrhizas), 234, 235

Arcyria, 33; *A. denudata,* 34, 37–8

Armillariella (*Armillaria*), 419; *A. mellea* (honey fungus), 67–8, 395, 409, 417, 430–1, 433

Arrhenatherum elatius (false-oat grass), loose smut of, 484, 485

Arthrobacter spp., induce fruiting in *Agaricus,* 426

Arthrobotrys, 551; *A. anchonia,* 547; *A. dactyloides,* 547, 550; *A. oligospora,* 543, 546, 550; *A. robusta,* 546

arthroconidia, 269, 270, 444; dry oidia developing as chains of, 412; *see also* arthrospores

Arthroderma, 299

arthropods, fungal parasites in guts of, 191; *see also* insects

arthrospores (arthroconidia): of Ascomycetes, 269, 270, 279, 290, 377–9, 385; of Basidiomycetes, 412, 444, 461, 462

Articulospora tetracladia, 529, 530

asci, 71, 84, 248; bituncate and unitunicate, 72, 261–2, 389; development of, 257–63, 334; explosive discharge of, 84, 248, 260, 262; operculate and inoperculate, 261, 358

Ascobolus, 259, 263, 322, 376–9, 381, 384; *A. carbonarius,* 376; *A. furfuraceus* (*A. stercorarius*), 261, 376, 377, 378, 379; *A. immersus,* 263, 377, 379, 381; *A. scatigenus* (*A. magnificus*), 378

ascocarps, 70, 71–2; of Eurotiales, 298, 299, 300, 307, 309; gymnohymenial and cleistohymenial development of, 379; sclerotioid, 312; *see also* apothecia, cleistothecia, perithecia

ascogenous hyphae, of Ascomycetes, 257, 258; of *Gymnoascus,* 300; of *Sordaria,* septal pores in, 322

ascogonia, 257; of Discomycetes, 362, 374–5, 378, 379; of Eurotiales, 300, 304, 305, 306, 307; of Pleosporales, 389; of Pyrenomycetes, 316, 320–1, 324, 329, 331–2, 334, 340, 351, 356

Ascomycetes (Ascomycotina), 97, 248–9; cell walls of, 59; classification of, 264–5; complementation in, 249–51; development of asci of, 257–63; fruit bodies of, 71, 72, 263–4; Fungi Imperfecti with perfect states in, 524, 533, 537; GC content of DNA of, 495; heterokaryosis in, 248, 251–2; life cycles of, 92; mating behaviour in, 255–7; nuclear migration in, 254–5; recombination in, 252–4; septate hyphae of, 57; spores of, *see* ascospores, conidia

ascospores (sexually produced), 71, 84–5, 248; development of, 257–63, 272, 275–7, 322, 324–5; with double flange, of *Emericella,* 308, 309, and *Eupenicillium,* 312; explosive discharge of, 84, 248, 260–1, 263; ribbed, of *Neurospora,* 327, 328, 329; viability of, 264

ascostroma, fruit body of Loculoascomycetes, 389

ash, *Daldinia* parasitic on, 334, 335

Aspergillus, form-genus, 298, 300; parasexual

Introduction to Fungi